EFFECT OF THE OCEAN ENVIRONMENT ON MICROBIAL ACTIVITIES

EFFECT OF THE
ON

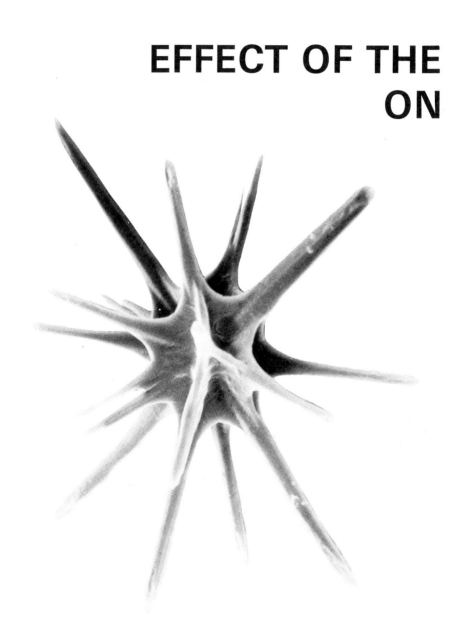

OCEAN ENVIRONMENT MICROBIAL ACTIVITIES

Edited by
R. R. Colwell
Department of Microbiology
University of Maryland
College Park, Maryland

and

R. Y. Morita
Department of Microbiology and
School of Oceanography
Oregon State University
Corvallis, Oregon

UNIVERSITY PARK PRESS
Baltimore · London · Tokyo

University Park Press
International Publishers in Science and Medicine
Chamber of Commerce Building
Baltimore, Maryland 21202

Copyright © 1974 by University Park Press

Printed in the United States of America by The Maple Press Company

Library of Congress Cataloging in Publication Data

United States-Japan Conference on Marine Microbiology,
  2d, University of Maryland, 1972.
  Effect of the ocean environment on microbial activities.

  1. Microbial ecology—Congresses. I. Colwell,
Rita R., 1934-   ed. II. Morita, Richard Y., ed.
III. Title. [DNLM: 1. Marine biology—Congresses.
2. Microbiology—Congresses. 3. Oceanography—
Congresses. QW80 U53e 1972] ·
QR100.U54 1972   576'.15   74-8639
ISBN 0-8391 0702.1

CONTENTS

ACKNOWLEDGMENTS

The Second United States-Japan Conference on Marine Microbiology was sponsored by the Office of International Programs, National Science Foundation, and the Japan Society for the Promotion of Science. The advice and excellent cooperation of the Japanese coordinators, Drs. H. Kadota and N. Taga, contributed significantly to the Conference's success. The editors are grateful to all the participants who made it a useful and productive interaction between the scientists of both countries.

Assistance in editing the manuscripts was provided by R. Berger, W. Corpe, E. Cota-Robles, H. Ehrlich, J. Landau, C. Litchfield, D. Pratt, C. Remsen, J. Sieburth, S. Stanley, and H. Stevenson. Special thanks are due Dr. J. Elbert O'Connell, Program Director, U.S.-Japan Cooperative Science Program.

Funds for social activities were provided by the following companies, to whom we express our thanks: the American Instrument Co., Inc.; Bellco Glass Co.; Haynie Products, Inc.; Intertechnique Instruments, Inc.; Ivan Sorvall, Inc.; New Brunswick Scientific Co.; New England Nuclear; Richard-Allen Medical Industries, Inc.; Rila Products, Inc.; and Van Waters and Rogers/Will Scientific.

Proceedings of the Second United States-Japan Conference on Marine Microbiology
University of Maryland, 25-30 August 1972

The Conference was dedicated to Professor Claude E. ZoBell for his invaluable scientific contributions in the field of marine microbiology over the past 40 years.

PREFACE

The first United States-Japan Conference was held in Tokyo, Japan, 15–19 August 1966, setting the stage for this Conference. Much new information has been accumulated by investigators of both countries in the intervening 6 years. The large-scale involvement of microbes in the biosphere has been neglected by microbiologists who tend to favor the test tube approach to microbiology. Although marine microbiology is becoming more important from an ecological viewpoint, the organisms primarily responsible for degradation of organic compounds in the sea are bacteria. Unfortunately, the number of people working in marine microbiology is not great. Thus, in comparison with developments in the field of molecular biology, progress in marine microbiology has been slow and somewhat uneven. The purpose of this Second United States-Japan Conference, then, was to identify areas of deficiency in the knowledge of marine microbiology and to indicate those problems requiring serious attention by research workers in the field. It is granted that synoptic marine microbiology is essential before the sophisticated techniques of molecular biology can be applied, but it is our hope that this as yet undeveloped field of science may reach the forefront of research endeavors within the next decade.

We have permitted the authors complete freedom in the preparation of their manuscripts. Editing purposely has been kept to a minimum in order to permit expression of ideas and new data. Formal journal papers, in our opinion, are often too restrictive, and the forum provided by a publication such as this volume can be an important mechanism for expression of new concepts and presentation of research in progress. We have attempted to provide an accurate transcription of the Conference proceedings, but errors, no doubt, will creep in despite our best efforts, and, for those errors or omissions that may occur, we offer our apologies to the participants.

PARTICIPANTS

United States

Berger, Leslie R., Department of Microbiology, University of Hawaii, Honolulu, Hawaii 96822

Carlucci, A. F., Scripps Institution of Oceanography, University of California, La Jolla, California 92038

Colwell, Rita R., Department of Microbiology, University of Maryland, College Park, Maryland 20742

Corpe, William A., Department of Biological Sciences, Barnard College of Columbia University, New York, New York 10027

Cota-Robles, Eugene H., Department of Microbiology, Pennsylvania State University, University Park, Pennsylvania 16802

Ehrlich, Henry L., Department of Biology, Rensselaer Polytechnic Institute, Troy, New York 12181

Fell, Jack, Institute of Marine Science, University of Miami, Miami, Florida 33149

Graikoski, John T., U. S. Department of Commerce, Biological Laboratory, Milford, Connecticut 06460

Kaneko, Tatsuo, Department of Biology, Georgetown University, Washington, D.C. 20007

Kim, Juhee, Department of Microbiology, California State University, Long Beach, California 90840

Landau, Joseph V., Department of Biology, Rensselaer Polytechnic Institute, Troy, New York 12181

Litchfield, Carol D., Department of Bacteriology, Rutgers College, New Brunswick, New Jersey 08903

Morita, Richard Y., Department of Microbiology, Oregon State University, Corvallis, Oregon 97331

Pratt, Darrell, Department of Bacteriology, University of Maine, Orono, Maine 04473

Remsen, Charles C., Department of Biology, Woods Hole Oceanographic Institution, Woods Hole, Massachusetts 02542

Sieburth, John M., Narragansett Marine Laboratory, Graduate School of Oceanography, University of Rhode Island, Kingston, Rhode Island 02881

Stevenson, L. Harold, Department of Biology, University of South Carolina, Columbia, South Carolina 29208

Wiebe, William, Department of Microbiology, University of Georgia, Athens, Georgia 30601

Japan

Ezura, Yoshio, Department of Fisheries, Hokkaido University, Hakodate, Japan

Ishida, Yuzaburo, Laboratory of Microbiology, Department of Fisheries, Kyoto University, Kyoto, Japan

Kadota, Hajime, Department of Fisheries, Faculty of Agriculture, Kyoto University, Kyoto, Japan

Kakimoto, Daiichi, Faculty of Fisheries, Kagoshima University, Kagoshima, Japan

Kimura, Takahisa, Laboratory of Microbiology, Department of Fisheries, Hokkaido University, Hakodate, Japan

Maruyama, Yoshiharu, Department of Agricultural Chemistry, University of Tokyo, Bunkyo-ku, Tokyo, Japan

Simidu, Usio, Institute of Food Microbiology, Chiba University, Narashino, Chiba, Japan

Sugahara, Isao, Department of Fisheries, Prefectural University of Mie, Tsu, Japan

Taga, Nobuo, Ocean Research Institute, University of Tokyo, Nakano-ku, Tokyo 164, Japan

Foreign

Albright, Lawrence J., Department of Biological Sciences, Simon Fraser University, Burnaby 2, Canada

Hamilton, Robert, Fisheries Research Board of Canada, 501 University Crescent, Winnipeg 19, Manitoba, Canada

Stanley, Simon O., Dunstaffnage Marine Research Laboratory, P.O. Box 3, Oban, PA 34 4AD, Scotland

Figure 1. Left to right: H. Kadota, J. Fell, W. Corpe, and N. Taga.

Figure 2. Left to right (facing camera): W. Wiebe, R. Hamilton, R. Y. Morita, R. Colwell, and A. F. Carlucci.

Figure 3. Front bench, left to right: W. Corpe and D. Pratt; second bench, left to right: J. Landau, T. Kimura, and U. Simidu; third bench, left to right: Y. Ezura and I. Sugahara; fourth bench, left to right: R. Wright and C. Litchfield; and fifth bench: H. Stevenson.

Figure 4. Front bench, left to right: E. Cota-Robles, Y. Ushida, and H. Kadota; second bench, left to right: J. Landau, T. Kimura, and D. Pratt; and third bench, left to right: Y. Ezura and I. Sugahara.

Figure 5. Left to right: J. Landau, T. Kimura, U. Simidu, Y. Ezura, and H. Stevenson.

Figure 6. First bench: A. F. Carlucci; second bench, left to right: C. Remsen, K. Kakimoto, and Y. Ushida; third bench, left to right: L. R. Berger, L. Albright, and J. Kim; fourth bench, left to right: J. Oliver and T. Kaneko; sixth bench, left to right: S. Stanley and J. Graikoski; and seventh bench, H. Ehrlich.

I

SALINITY EFFECTS AND INTERACTIONS

SALT REQUIREMENTS FOR GROWTH AND FUNCTION OF MARINE BACTERIA

DARRELL PRATT

Department of Microbiology
University of Maine

Seawater is for bacterial species an environment which is uniquely constant with respect to its ionic composition. Those organisms indigenous to the sea not only have the capacity to function and to grow in this habitat but many also show an obligate adaptation to its salinity. In samples of seawater from areas free of terrestrial contamination, often more than 90% of the bacteria found require nutrient media prepared with seawater. This growth requirement is usually large and readily detectable; it can usually be demonstrated in complex media containing as contamination at least sufficient amounts of the essential ions to allow the growth of comparable non-marine bacteria. Studies of the nature of this growth requirement for seawater are not numerous but some have been intensive and well executed. Two reviews having direct bearing on the ionic requirements of marine bacteria have been published (MacLeod, 1965; 1968), and several concerned with the general topic of halophilic bacteria are available (Ingram, 1957; Larsen, 1962).

The objective of this introductory chapter is not to make a thorough review but rather to outline the state of our present information, a process which must convince the reader that, in spite of some excellent and sophisticated studies, we have at best made a good beginning.

ION REQUIREMENTS FOR GROWTH

When those bacterial species which required seawater for growth on primary isolation were investigated, they were found to differ in two respects from most comparable non-marine species. They underwent rapid cytolysis when diluted in suitably hypotonic solutions or in distilled water and they required Na^+ for growth. The obvious connection here could easily be made and a simple hypothesis stated. Marine bacteria require Na^+ for growth because it is essential to maintain the osmotic environment for the protection of cellular integrity. Twenty-five years ago most biologists, with no further information, would have agreed that this was indeed a very likely explanation. Now, however, with more evidence, it is apparent that this hypothesis accounts at best for only part of the total explanation. The gist of the matter is that, where investigated, the Na^+ requirement of marine bacteria is not completely replaceable by other solutes, which it would be if the simple hypothesis were the explanation.

MacLeod and Onofrey (1957) reported that a marine bacterium desig-nated B-16 (*Pseudomonas* sp.) required 0.2–0.3 M NaCl for the optimum extent and rate of growth. In their work a carefully prepared, chemically defined basal medium was used. The requirement for Na^+ was reduced slightly for this and two other marine isolates in media containing sucrose as an osmotically active solute. These investigators concluded that the Na^+ require-ment was highly specific and that only a small component of the function was osmotic. Pratt and Austin (1963) studied the problem of replacement of Na^+ by other solutes using *Vibrio* MB 22. The approach was to determine if osmotic equivalence of solutes could be observed with suboptimal concentra-tions, an equivalence which might be masked by inhibitions at higher concen-trations. It was recognized that failure to replace an effective solute by a nonspecific solute had to be interpreted cautiously, since the substitute, in addition to being osmotically active, could not be inhibitory (or stimulatory) at relatively high concentrations. These workers found that this organism would allow a rather substantial partial replacement of Na^+ by K^+, Mg^{2+}, and sucrose, although both Mg^{2+} and sucrose were inhibitory in concentrations equivalent to the osmolarity of Na^+ in seawater. When these observations were extended to three additional organisms, only one gave completely similar results. In two, both Mg^{2+} and sucrose were inhibitory at 0.2 M, the test concentration; however, in all cultures the sparing action of K^+ was equivalent to Na^+ after the minimal Na^+ requirement was satisfied. Using a plating procedure and samples of seawater as inocula, Pratt (1963) observed that approximately half of the bacteria present would permit a substantial replacement of the Na^+ requirement with KCl or sucrose. Tedder (1966) studied 20 isolates using a chemically defined medium; the concentration of Na^+ required to give the optimum growth rate ranged from 0.2 to 0.5 M. Using 0.05 M Na^+ as a suboptimal concentration, she was able, by the addition of K^+ to give a total salt concentration equivalent to the optimum Na^+ concentration, to enhance the growth rate of each of the isolates. The increases ranged from only a few per cent to being equivalent to that produced by the optimum Na^+ concentration. Suitable controls were made to determine that the enhancement by K^+ was not the result of a specific K^+ deficiency in the medium. The observations suggested that in all these organisms Na^+ could serve two functions, one being specific and the other nonspecific. Variation in the concentration of Na^+ required to satisfy the specific requirement would account for the differences in the sparing effect of K^+. Thus the simple osmotic explanation of the Na^+ requirement for growth of marine bacteria does not suffice.

The requirement of Na^+ for growth shown by many marine bacteria has been found to be very stable. A variety of efforts to adapt marine bacteria to media lacking Na^+ have met with indifferent results. Pratt and Waddell (1959) reported the finding of what appeared to be Na^+-independent mutants by plating large numbers of a marine pseudomonad on a medium lacking Na^+. Since a complex medium was employed, it is not possible to be certain that these organisms were indeed Na^+ independent. MacLeod and Onofrey (1963) were able to adapt a marine pseudomonad to grow in a Trypticase medium without added Na^+ but the adapted strain still required the usual Na^+ concentration when grown in a chemically defined medium. The Trypticase medium was found by analysis to contain 0.028 M Na^+. The earlier observations of ZoBell and Mitchner (1938) that marine isolates on growing in laboratory media became adapted to a Na^+-free medium should not be overlooked, and the problem should be studied further. The requirement for Na^+ is found also in halophilic bacteria isolated from non-marine environments (Larsen, 1962) and has been reported in two instances among non-halophilic, non-marine bacteria (Sistrom, 1960; Bryant et al., 1959), the latter two organisms being *Rhodopseudomonas* sp. and *Bacteriodes succinogenes.*

Marine bacteria like other bacteria require K^+ and Mg^{2+} for growth (MacLeod and Onofrey, 1957); however, many differ in that their requirements for K^+ and Mg^{2+} may be greater than those of comparable gram-negative, non-marine bacteria. Using 0.1% Trypticase (BBL) as the basal medium, Tyler et al. (1960) were able to separate 96 marine isolates which required seawater on initial isolation into four classes based on their growth requirement for Na^+, K^+, and Mg^{2+} (Table 1). These results suffered from being highly empirical since they were obtained using a complex nutrient medium

Table 1. Separation
of marine bacterial isolates on the
basis of principal cation requirements[a]

Class	Cation requirement	Number of strains
I	Na^+	19
II	Na^+, K^+	25
III	Na^+, Mg^{2+}	17
IV	Na^+, K^+, Mg^{2+}	35

[a] Basal medium was Trypticase (BBL) in distilled water; indicated cations were added to give the concentration found in seawater. Incubated with shaking at 32°C for 20 hr. Compiled from Tyler et al. (1960).

which was undoubtedly contaminated with the cations under investigation. Such studies allow us to identify the growth requirements for cations which are present in insufficient quantities to meet the needs of the test organism. The requirement for Mg^{2+} has been shown to be sparsed by Ca^{2+} in the case of one marine bacterium and required for others (MacLeod and Onofrey, 1957). Recently, using salt-free casein hydrolysate as basal nutrient, we observed a Ca^{2+} requirement in one of the species studied by Tyler et al. (1960) using Trypticase as nutrient; this requirement had not been seen in the earlier work. The presence of unknown amounts of Ca^{2+} in the basal medium may mask Ca^{2+} requirements and modify Mg^{2+} requirements for growth.

A problem exists with respect to using complex or chemically defined media for determining the ionic requirements of marine bacteria. If only a few cultures are being studied, then a chemically defined medium which supports the growth of all test organisms will certainly be the method of choice. On the other hand, when natural populations are to be studied or large numbers of cultures are to be characterized, then a chemically defined medium may be too restricted in its nutritional versatility and a complex medium may serve better. Ideally the concentration of the ions of interest should be determined so that the basal amounts are known. Obviously an organism cannot be said to require Na^+ if the only evidence is that it grows in a medium containing unknown amounts of, but no added, Na^+. If cultures for study are picked on the basis that they require a complex medium prepared with seawater as compared with the same medium prepared with distilled water, then one can anticipate that these bacteria will have a relatively high requirement for Na^+ and perhaps other cations. Whether defined media or analyzed complex media are used, the investigator should be aware of the restrictions imposed. Ideally a chemically defined medium of established cation content having an infinitely wide nutritional versatility is desired.

PHYSIOLOGICAL FUNCTION OF Na^+

The Na^+ requirements for the metabolic activity of cells of marine bacteria have been found to be similar to those for their growth. The requirement is specific, although in some isolates it can be partially replaced by similar solutes. In cell-free extracts or with purified enzymes from marine bacteria no evidence for a specific requirement for Na^+ for activity has been observed. The respiration of specific substrates by washed suspensions of a marine bacterium (MacLeod et al., 1958) was found to require both Na^+ and K^+; however, the concentration of Na^+ required was observed to be a function of the substrate being oxidized. The maximum rate of oxidation of aliphatic monocarboxylic acids or an oxidizable sugar required 0.05 M Na^+; however, for dicarboxylic acids three to four times as much Na^+ was required. When the cell-free enzymes of the tricarboxylic acid cycle were studied, none were found to require Na^+ for activity, but aconitase and isocitric dehydrogenase required a medium having a suitable ionic strength (MacLeod et al., 1960). Using a marine vibrio, Pratt and Happold (1960) observed that, although Na^+ was required for the tryptophanase activity of whole cells, it was not essential to the activity of cell-free extracts. In the latter, K^+ and pyridoxal phosphate were required for indole production. The activity of intact cells also required the presence of K^+. The amount of Na^+ giving optimum activity with intact cells was inhibitory to indole formation by the cell-free preparations. From the information available thus far the ionic requirements of the enzymes of marine bacteria do not differ from those of similar non-marine forms. Whatever the function of Na^+ is in the growth of marine bacteria, it would seem to be related to the activities of the intact cell.

The possibility that Na^+ was essential to membrane function in marine bacteria followed from the observation that the requirement was for growth and metabolic activity of intact cells. Drapeau and MacLeod (1963) demonstrated that cells of the marine bacterium B-16 required Na^+ for the intracellular accumulation of the non-metabolizable analogues, α-aminoisobutyric acid (AIB) and D-fucose. This requirement for Na^+ could not be satisfied by other related ions. The concentration of Na^+ required for the optimum uptake of [^{14} C] AIB was similar to that required for the optimal oxidation of L-alanine;

furthermore, *L*-alanine prevented the uptake of [^{14}C] AIB. Parallel behavior was observed with *D*[^{14}C] fucose and the metabolizable sugar *D*-galactose. These results were interpreted to mean that two transport systems were involved, one of which functioned for *L*-alanine and AIB and a second which functioned for *D*-galactose and *D*-fucose. The quantitative requirement for Na$^+$ for growth of B-16, using *L*-alanine as the sole carbon source, was similar to that required for the oxidation of *L*-alanine and transport of AIB. The requirement for Na$^+$ for growth with *D*-galactose as the sole carbon source was similar to that for *L*-alanine and much higher than that required for the oxidation of *D*-galactose and the transport of *D*-fucose.

Payne (1958) observed that *Vibrio natriegens* was capable of producing induced enzymes for the oxidation of galacturonic acid. The ions essential to the induction of the enzyme were K$^+$ and Na$^+$. Cells preincubated with Na$^+$ and substrate were fully induced and the addition of K$^+$ to the incubation system resulted in the prompt oxidation of galacturonate. He concluded that Na$^+$ was essential to the formation of the transport system and that K$^+$ was essential to the oxidation system. A second interpretation has been offered by MacLeod (1968); he has suggested that induction of the permease system could depend on the transport of the substrate by basal level of permease in the un-induced cell, which would depend on the presence of Na$^+$. Thus induction would seem to depend on Na$^+$, but in reality only transport would be Na$^+$ dependent. Using the same organism, Rhodes and Payne (1962) observed that the induction of resting cells to the metabolism of mannitol and *L*-arabinose required Na$^+$, K$^+$, and Mg^{2+}. Increased concentrations of KCl (0.26 M) in the absence of Na$^+$ or Mg^{2+} allowed induced enzyme formation and oxidation of the substrate. They concluded that the increased K$^+$ made the cells more permeable and that substrate could enter without the specific transport system. Thus the induction of the transport system and its Na$^+$ dependence would not be needed. Studies using intact cells are of necessity complex and difficult to interpret. Certainly the observations of MacLeod and his co-workers using non-metabolizable substrates gave a clear picture of Na$^+$ dependence for transport of substrates.

THE LYTIC PHENOMENON

Most marine bacteria lyse when they are in distilled water or in solutions suitably hypotonic to seawater (Pratt and Riley, 1955; MacLeod and Matula, 1962; DeVoe and Oginsky, 1969*a*). The cytolysis is characterized by a decrease in the turbidity of the suspension, the release of UV adsorbing material, and a loss in viability. Variation among species is significant, with some showing only the slightest effects and others rather drastic changes

(Table 2) (Pratt *et al.*, 1959). The decrease in optical density occurs in two stages: the first occurs within a few seconds of mixing and accounts for the greatest part of the turbidity loss; the second phase is slower and requires an hour or more to complete. The simplest explanation is that cytolysis results from osmotic shock of bacteria having cell walls inherently weaker than those of non-marine bacteria. If this is the case, then solutes in the suspending medium, equiosmolar to the internal salt concentration, should prevent lysis by restraining the diffusion of water into the cell. Tyler *et al.* (1960), in a study of 100 strains of marine bacteria, found considerable strain variation, but, in general, observed that Mg^{2+} was the most effective in preventing lysis, with Na^+ next, and with K^+ much the least effective (Table 3). The turbidity of 60% of the strains was maintained by 0.05 M $MgCl_2$; however, Riley (1955), with a marine pseudomonad, found a considerable loss of viability and Thompson *et al.* (1970) have reported that cells of *Pseudomonas* sp. B-16 were plasmolysed by this treatment. The plasmolysis was accompanied by a loss of intracellular K^+ and could be prevented or reversed by the presence of this ion. When compared with Na^+ on an osmolar basis, Mg^{2+} has been found to be several times more effective in preventing lysis of marine bacteria (MacLeod and Matula, 1962; DeVoe and Oginsky, 1969a). DeVoe and Oginsky (1969a) have observed that pretreatment of cells in 1.0 M NaCl greatly increased their susceptibility to lysis in solutions of KCl. In contrast, pretreatment of the cells in 0.05 M $MgCl_2$ maintained the cells in 1 M KCl. More

Table 2. Cytolytic damage as reflected in turbidity, leakage of cytoplasmic components, and viability

Culture	Suspending medium	Residual[a] turbidity (%)	UV[b] adsorption	Survival (%)
Mb 21	Seawater	100	0.08	—
Mb 21	Distilled water	78	0.15	0.47
Mb 22	Seawater	100	0.08	—
Mb 22	Distilled water	47	0.35	5×10^{-4}
Mb 28	Seawater	100	0.05	—
Mb 28	Distilled water	35	0.21	5×10^{-6}

[a] Residual turbidity as compared with suspension in seawater.
[b] OD of supernatant fluid measured at 265 nm.

Table 3. Variation in the
lytic response of marine bacteria[a]

Test solution	Number of strains displaying the indicated residual turbidity (%)		
	100–90%	89–50%	49–0%
H_2O	9	39	52
0.05 M NaCl	10	36	54
0.05 M KCl	2	14	84
0.05 M $MgCl_2$	60	39	1
0.5 M NaCl	61	34	5
0.5 M KCl	14	55	31
0.5 M $MgCl_2$	89	9	2

[a] Cells were grown on the surface of agar containing 1% Trypticase (BBL) in seawater. After growth the cells were washed from the surface with artificial seawater. From the heavy suspension 0.1 ml was pipetted into the test solution. Residual turbidity was compared with that obtained in seawater taken as 100%.

extensive lysis has often been observed in dilute KCl solutions than in distilled water. The presence of K^+ in low concentrations and the absence of Mg^{2+} both seem to enhance the lytic process.

The cell envelope structure of *Pseudomonas* sp. B-16 has been studied (Forsberg *et al.*, 1970*a,b*). They reported it to consist of the following three layers: (1) a loose outer layer, (2) an outer double track membrane, and (3) an underlying layer. Chemical and electron microscopic evidence was presented in support of these views. Washing the cells in 0.5 M NaCl, followed by suspension in sucrose, removed the outer layers, leaving rod-shaped structures surrounded by the cytoplasmic membrane and a thin rigid layer of peptidoglycan. These structures, termed mureinoplasts, lost UV-absorbing material when suspended in 0.05 M $MgCl_2$ or 0.5 M NaCl. A balanced solution containing 0.3 M NaCl, 0.05 M $MgCl_2$ and 0.01 M KCl was required for their stability. D'Aoust and Kushner (1971) reported that Na^+ or Mg^{2+} would maintain the outer layers of the cell envelope of a red psychrophilic marine bacterium, but that Mg^{2+} was required in order to maintain the cytoplasmic membrane.

The various observations are difficult to fit into an uncomplicated osmotic shock explanation of the lytic susceptibility of marine bacteria. Another suggestion which has been advanced is that lysis of these bacteria results from a lack of cations. The subunits of the cell envelope are considered to have a large number of negatively charged groups; these are the free phosphates of the phospholipids and terminal carboxyl groups associated with amino acids. These negatively charged groups repel each other, weakening the structure of the cell envelope (Abram and Gibbons, 1960; Buckmire and MacLeod, 1965; DeVoe and Oginsky, 1969b). Cations in the suspending medium screen the free negative charges. The concept rests experimentally on the observations that isolated cell envelopes of various strains have been found to require cations to maintain their integrity. DeVoe and Oginsky (1969b) found that pretreatment of the envelopes with 0.05 M $MgCl_2$ prevented their subsequent disintegration in distilled water; on the other hand, pretreatment with 1 M NaCl greatly enhanced their disintegration in distilled water. The presence of Na^+ with Mg^{2+} in the pretreatment solution increased the effect of distilled water. They proposed that Mg^{2+} holds negatively charged subunits of the envelope together by divalent ionic bridges. Pretreatment with 1 M NaCl resulted in exchange of Na^+ for Mg^{2+} and in the disappearance of the Mg^{2+} bridges. A similar role for Mg^{2+} has been postulated in explanation of the EDTA induced cytolysis of some gram-negative non-marine bacteria (Asbell and Eagon, 1966). Buckmire and MacLeod (1965) felt that in envelopes of B-16, Na^+ stabilized the peptidoglycan layer by screening negative charges. Lysis of cells caused by ion deficit alone can be conceived; however, the combination of the effects of ion deficit on cell wall strength and osmotic shock provide perhaps a more attractive hypothesis.

The internal Na^+ concentration of *Pseudomonas* sp. B-16 has been found to be the same as that of the growth medium or suspending medium. This would indicate that Na^+ should not be osmotically effective since it can penetrate the cytoplasmic membrane (Takacs, Matula, and MacLeod, 1964). If cells are mixed in distilled water, the Na^+ rapidly diffuses from the cell, but as long as the internal concentration exceeds the external concentration, water will diffuse into the cell. Viewed in this way the possibility that Na^+ can contribute to the initial instantaneous osmotic shock must be considered. This is more apparent if one considers that the external volume in the usual test system is enormous as compared with the water space of the cell. Other internal solutes such as K^+, small molecules, and colloids also contribute to the internal osmotic pressure.

A unified picture of the lytic process in suspensions of marine bacteria in hypotonic solutions could be a rapid osmolysis of the cells whose cell walls were weakened by the ion deficit, which would account for the rapid phase of lysis; this could be followed by slower continued disintegration of cell envelopes because of the lack of cations, Mg^{2+} in particular. In this explanation the ineffectiveness of lower concentrations of K^+ would rest on this ion being rapidly penetrating and on its effectiveness in displacing Mg^{2+}.

SUMMARY

The obligate growth requirement of marine bacteria for NaCl cannot be ascribed to a simple osmotic need. The general finding of a specific requirement for Na^+ precludes the simple osmotic explanation. However, that some osmotic component may be involved is indicated by the partial replacement of Na^+ by K^+ and other solutes in maintaining the optimum growth rate. The specific requirement for Na^+ for substrate transport shown in *Pseudomonas* sp. B-16 may explain the reason for the specific growth requirement for Na^+, if this should be a general finding in other marine bacteria. The relationship of the lytic phenomenon to the salt requirement of marine bacteria is not clear, but obviously salts play an essential role in maintaining cellular integrity. Mg^{2+} apparently plays a most important role with respect to the maintenance of cell envelopes. It should be emphasized that the circumstances bringing on osmotic shock will rarely be found in the natural environment, since the internal and external salt concentrations, where investigated, are very similar. A possibility is that these bacteria have exaggerated versions of the cation requirements for cell envelope integrity found in non-marine, gram-negative bacteria; thus, we might anticipate the relatively high concentrations of Mg^{2+} required for growth by many marine bacteria. The difficulty is that at present too much generalization rests on too little observation; too few investigators have been involved in trying to understand the physiology of marine bacteria to have produced a reliable, coherent picture. Finally, almost nothing concrete can be said concerning the selective advantage of the obligate requirements for salinity shown by these marine bacteria. Because of their predominance in the environment one must assume that such an advantage exists.

LITERATURE CITED

Abram, D., and N. E. Gibbons. 1960. Turbidity of suspensions and morphology of red halophilic bacteria as influenced by sodium chloride concentration. Can. J. Microbiol. 6:535–543.

Asbel, M. A., and R. G. Eagon. 1966. Role of multivalent cations in the organization, structure, and assembly of the cell wall of *Pseudomonas aeruginosa*. J. Bacteriol. 92:380.

Brown, A. D. 1964. Aspects of bacterial response to the ionic environment. Bacteriol. Rev. 28:296–329.

Bryant, M. P., I. M. Robinson, and H. Chu. 1959. Observations on the nutrition of *Bacteroides succinogenes*—a ruminal cellulytic bacterium. J. Dairy Sci. 42:1831 −1847.

Buckmire, F. L. A., and R. A. MacLeod. 1965. Nutrition and metabolism of marine bacterium. XIV. On the mechanism of lysis of a marine bacterium. Can. J. Microbiol. 11:677−691.

D'Aoust, J. Y., and D. J. Kushner. 1971. Structural changes during lysis of a psychrophilic marine bacterium. J. Bacteriol. 108:916−927.

DeVoe, I. W., and E. L. Oginsky. 1969*a*. Antagonistic effect of monovalent cations in maintenance of cellular integrity of a marine bacterium. J. Bacteriol. 98:1355−1367.

DeVoe, I. W., and E. L. Oginsky. 1969*b*. Cation interactions and biochemical composition of the cell envelope of a marine bacterium. J. Bacteriol. 98:1368−1377.

DeVoe, I. W., J. Thompson, J. W. Costerton, and R. A. MacLeod. 1970. Stability and comparative transport capacity of cells, mureinoplasts, and true protoplasts of a gram negative bacterium. J. Bacteriol. 101:1014−1026.

Drapeau, G. P., and R. A. MacLeod. 1963. Na^+-dependent active transport of aminoisobutyric acid into the cells of a marine pseudomonad. Biochem. Biophys. Res. Commun. 12:111−115.

Forsberg, C. W., J. W. Costerton, and R. A. MacLeod. 1970*a*. Separation and localization of cell wall layers of a gram negative bacterium. J. Bacteriol. 104:1338−1353.

Forsberg, C. W., J. W. Costerton, and R. A. MacLeod. 1970*b*. Quantitation, chemical characteristics, and ultrastructure of the three outer cell wall layers of a gram-negative bacterium. J. Bacteriol. 104:1354−1368.

Ingram, M. 1957. Micro-organisms resisting high concentrations of sugars or salts. *In* R. E. O. Williams and C. C. Spier (eds), Microbial Ecology, pp. 90−133. Cambridge University Press, Cambridge, U.K.

Larsen, H. 1962. Halophilism. *In* I. C. Gunsalus and R. Y. Stanier (eds), The Bacteria: A Treatise on Structure and Function, Vol. 4, pp. 297−342. Academic Press, New York.

MacLeod, R. A. 1965. The question of the existence of specific marine bacteria. Bacteriol. Rev. 29:9−23.

MacLeod, R. A. 1968 On the role of inorganic ions in the physiology of marine bacteria. *In* M. R. Droop and E. J. Ferguson Wood (eds), Advances in Microbiology of the Sea, Vol. 1, pp. 95−126. Academic Press, New York.

MacLeod, R. A., C. A. Claridge, A. Hori, and J. F. Murray. 1958. Observations on the function of sodium in the metabolism of a marine bacterium. J. Biol. Chem. 232:829−834.

MacLeod, R. A., A. Hori, and S. M. Fox. 1960. Nutrition and metabolism of marine bacteria. IX. Ion requirements for obtaining and stabilizing isocitric dehydrogenase from a marine bacterium. Can. J. Biochem. Physiol. 38:693–701.

MacLeod, R. A., and T. I. Matula. 1962. Nutrition and metabolism of marine bacteria. XI. Some characteristics of the lytic phenomenon. Can. J. Microbiol. 8:883–896.

MacLeod, R. A., and E. Onofrey. 1957. Nutrition and metabolism of marine bacteria. III. The relation of sodium and potassium to growth. J. Cell. Comp. Physiol. 50:389–402.

MacLeod, R. A., and E. Onofrey. 1963. Studies on the stability of the Na^+ requirement of marine bacteria. In C. H. Oppenheimer (ed.), Symposium on Marine Microbiology, pp. 481–489. Charles C Thomas, Springfield, Illinois.

Payne, W. J. 1958. Studies on bacterial utilization of uronic acids. III. Induction of oxidative enzymes in a marine isolate. J. Bacteriol. 76: 301–307.

Payne, W. J. 1960. Effects of sodium and potassium ions on growth and substrate penetration of a marine pseudomonad. J. Bacteriol. 80:696–700.

Pratt, D. B. 1963. Specificity of the solute requirement by marine bacteria on primary isolation from seawater. Nature (London) 199:1309.

Pratt, D. B., and M. Austin. 1963. Osmotic regulation of the growth rate of four species of marine bacteria. In C. H. Oppenheimer (ed.), Symposium on Marine Microbiology, pp. 629–637. Charles C Thomas, Springfield, Illinois.

Pratt, D. B., M. Bielling, and M. E. Tyler. 1959. Variation in osmotic fragility of marine bacteria. Bact. Proc. 59:11.

Pratt, D. B., and F. C. Happold. 1960. Requirements for indole production by cells and extracts of a marine bacterium. J. Bacteriol. 80:232–336.

Pratt, D. B., and W. H. Riley. 1955. Lysis of a marine bacterium in salt solutions. Bact. Proc. 55:26.

Pratt, D. B., and G. Waddell. 1959. Adaptation of marine bacteria to media lacking sodium chloride. Nature (London) 183:1208–1209.

Rhodes, M. E., and W. J. Payne. 1962. Further observations on the effects of cations on enzyme induction in marine bacteria. Antonie van Leeuwenhoek. 28:121–128.

Riley, W. H., Jr. 1955. A study of factors causing lysis of a marine bacterium. M.S. Thesis. Univ. Florida, Gainesville.

Sistrom, W. R. 1960. A requirement for sodium in the growth of Rhodopseudomonas spheroides. J. Gen. Microbiol. 22:778–785.

Takacs, F. P., T. I. Matula, and R. A. MacLeod. 1964. Nutrition and metabolism of marine bacteria. XIII. Intracellular Na^+ and K^+ concentrations in a marine pseudomonad. J. Bacteriol. 87:510–518.

Tedder, S. 1966. Replacement of sodium ions by potassium ion in the growth of selected marine organisms. M.S. Thesis. Univ. Florida, Gainesville.

Thompson, J., J. W. Costerton, and R. A. MacLeod. 1970. K^+-Dependent deplasmolysis of a marine pseudomonad plasmolyzed in a hypotonic solution. J. Bacteriol. 102:843–854.

Tyler, M. E., M. C. Bielling, and D. B. Pratt. 1960. Mineral requirements and other characters of selected marine bacteria. J. Gen. Microbiol. 23:153–161.

ZoBell, C. E., and H. D. Mitchner. 1938. A paradox in the adaptation of marine bacteria to hypotonic solutions. Science 87:328–329.

MINERAL REQUIREMENTS OF MARINE BACTERIA

DAIICHI KAKIMOTO, KATUMI OKITA, and MICHIKO INAI

Faculty of Fisheries
Kagoshima University

None of the marine isolates studied grew in nutrient broth without the addition of NaCl. When the nutrient broth was supplemented with 3% NaCl *Vibrio* and *Photobacterium* spp. grew well. No growth of *Pseudomonas* was observed after 4 days incuation in nutrient broth supplemented with 3% NaCl. Species of *Bivrio* and *Photobacterium* grew in ZoBell's 2216E medium without addition of NaCl. Addition of NaCl to ZoBell's 2216E medium inhibited species of *Vibrio* at 6% NaCl whereas growth of *Pseudmonas, Alcaligenes,* and *Achromobacter* occurred in 7.5% NaCl; a few strains of these genera were able to grow in 12% NaCl after 48 hours incubation. In peptone water, without NaCl, only *Vibrio* and *Photobacterium* grew, but when peptone water was made up with one-sixth strength artificial seawater all the isolates showed the same growth pattern and no growth response differences between the genera were seen. The effect of substituting K^+ for Na^+ was examined in species of *Vibrio* and *Photobacterium*. A number of isolates were examined for lytic susceptibility to SLS (sodium lauryl sulfate) detergent. *Pseudomonas* species were most susceptible, the marine strains being more sensitive than the terrestrial species. *Vibrio* and *Photobacterium* were much more resistant than *Pseudomonas*. *Bacillus cereus* var. *mycoides* was resistant to SLS but showed reversible sensitivity to cationic detergent. The hexosamine content of the cell wall was inversely proportional to the SLS lytic susceptibility of marine bacteria.

In spite of several studies (ZoBell and Upham, 1944; Bukatsch, 1936; MacLeod *et al.,* 1954; MacLeod and Onofrey, 1956), there are many unresolved problems concerning the mineral requirements of marine bacteria. The authors studied the mineral requirements of marine isolates collected at two points off Cape Sata (about 58 and 100 miles respectively from Kuroshio) in summer 1971. A total of 67 isolates were obtained and from these 33 strains were selected for further examination. In the present studies, the authors used various media in which the growth of marine isolates was affected mainly by the inorganic constituents. As a result of these experiments, it is concluded that the mineral requirement of marine isolates depends not upon the individual species, but rather upon the specific mineral requirements of each genus. It was also found that the lytic susceptibility of marine isolates to an anionic detergent, sodium lauryl sulfate (SLS) (Gilby, 1957), varied from one bacterial genus to another, that the degree of lysis depended upon the salt concentrations in the culture suspension, and that suspensions of the marine isolates were usually more susceptible to SLS than corresponding terrestrial organisms. In these experiments the authors suggested that the SLS lytic susceptibility is inversely proportional to the hexosamine content in the cell wall.

MATERIALS AND METHODS

The authors studied the mineral requirements of marine isolates collected at two stations C and B: latitude 31°00′N, longitude 130°48′E, and latitude 31°00′N, longitude 132°38′E, respectively. The authors classified the isolates according to Shewan's (1963) method. Salinity effects were studied in strains isolated from station C and the media used were as follows.

1. Nutrient broth containing 3% NaCl, which had the following composition: peptone, 10 g; yeast extract, 3.5 g; NaCl, 30 g, dissolved in 1 liter of distilled water. The pH of the medium was adjusted to 7.6. NaCl-free nutrient broth had the same composition but without the addition of NaCl.
2. ZoBell 2216E modified medium I in which the original concentration of 3% NaCl was replaced by various concentrations from 0% to 12%.
3. 1% Peptone solution containing 3% NaCl for the most simple organic nitrogen medium.
4. ZoBell 2116E modified medium II (Hidaka, 1965). In this medium the original concentration of artificial seawater (ASW) was diluted to one-sixth strength ASW to examine the growth effect of various minerals other than NaCl.

Table 1. The characteristics of marine isolates and names of identified genera[a]

Strain no.	Form	Gram strain	Motility	Flagella	Hugh and Leifson	Spore	Pigment	Luminescence	0/129	Oxidase	Genus[b]
C-1	R	–	+	M	O	–	–	–	–	+	Pseudomonas
C-2	R	–	+	M	F	–	–	–	–	+	Achromobacter
C-3	R	–	+	M	O	–	–	–	–	+	Pseudomonas
C-4	R	–	+	M	O	–	–	–	–	+	Pseudomonas
C-5	R	–	+	M	O	–	–	–	–	+	Pseudomonas
C-6	R	–	+	M	F	–	–	+	–	+	Photobacterium
C-7	R	–	+	M	O	–	–	–	–	+	Pseudomonas
C-8	R	–	+	M	O	–	–	–	–	+	Pseudomonas
C-9	R	–	+	M	F	–	–	–	–	+	Achromobacter
C-10	R	–	+	M	O	–	–	–	–	+	Pseudomonas
C-11	R	–	+	M	O	–	–	–	–	+	Pseudomonas
C-12	R	–	+	M	F	–	–	–	+	+	Vibrio
C-13	R	–	+	M	F	–	–	–	+	+	Vibrio
C-14	R	–	+	M	O	–	–	–	–	+	Pseudomonas
C-15	R	–	+	M	O	–	–	–	–	+	Pseudomonas
C-16	R	–	+	M	O	–	–	–	–	+	Pseudomonas

Strain	Shape			Flagella	Metabolism						Organism
C-17	R	—	+	P	F	—	—	+	+	+	Photobacterium
C-18	R	—	+	M	F	—	—	—	+	+	Vibrio
C-19	R	—	+	M	N	—	—	—	—	+	Alcaligenes
C-20	R	—	+	M	F	—	—	—	—	+	Vibrio
C-21	R	—	+	M	N	—	—	—	—	+	Alcaligenes
C-22	R	—	+	M	O	—	—	—	+	+	Pseudomonas
C-23	R	—	+	M	O	—	—	—	—	+	—[b]
C-24	R	—	+	M	O	—	—	—	—	+	Pseudomonas
C-25	R	—	+	M	O	—	—	—	—	+	Pseudomonas
C-26	R	—	+	M	O	—	—	—	—	+	Pseudomonas
C-27	R	—	+	M	O	—	—	—	—	+	Pseudomonas
C-28	R	—	+	M	F	—	—	—	+	+	Vibrio
C-29	R	—	+	M	F	—	—	—	+	+	Vibrio
C-30	R	—	+	M	O	—	—	—	+	+	—[b]
C-31	R	—	+	P	F	—	—	+	+	+	Photobacterium
C-32	R	—	+	M	O	—	—	—	—	+	Pseudomonas
C-33	R	—	+	M	O	—	—	—	+	+	Pseudomonas

[a] Abbreviations used: R, rod; S, sphere; +, positive; —, negative; M, monotrichous; P, peritrichous; A, atrichous; O, oxidative; N, negative; F, fermentative; NG, no growth.

[b] —, Unidentified.

Table 2. The characteristics of marine isolates and names of identified genera[a]

Strain no.	Form	Gram strain	Motility	Flagella	Hugh and Leifson	Spore	Pigment	Luminescence	0/129	Oxidase	Genus[b]
B-1	R	–	+	M	O	–	–	–	–	+	Pseudomonas
B-2	R	–	+	M	N	–	–	–	+	+	–
B-3	R	–	+	M	N	–	–	–	–	+	Alcaligenes
B-4	R	–	+	M	F	–	–	–	+	+	Vibrio
B-5	R	–	+	M	F	–	–	–	–	+	Achromobacter
B-6	R	–	+	M	F	–	–	–	–	+	Achromobacter
B-7	R	–	+	M	O	–	–	–	–	+	Pseudomonas
B-8	R	–	+	M	F	–	–	–	+	+	Vibrio
B-9	R	–	+	M	NG	–	–	–	+	+	–
B-10	R	–	+	M	O	–	–	–	–	+	Pseudomonas
B-11	R	–	+	M	F	–	–	–	–	+	Achromobacter
B-12	R	–	+	M	F	–	–	–	+	+	Vibrio
B-13	R	–	+	M	F	–	–	–	–	+	Achromobacter
B-14	R	–	+	M	NG	–	–	–	–	+	Achromobacter
B-15	R	–	+	P	F	–	–	+	–	+	Photobacterium
B-16	R	–	+	M	F	–	–	–	+	+	Vibrio

B-17	R	–	+	M	F	–	–	+	+	*Vibrio*
B-18	R	–	+	M	N	–	–	–	+	*Alcaligenes*
B-19	R	–	+	M	F	–	–	–	+	*Achromobacter*
B-20	R	–	+	M	F	–	–	–	+	*Achromobacter*
B-21	R	–	+	M	F	–	–	–	+	*Achromobacter*
B-22	R	–	+	M	F	–	–	+	+	*Vibrio*
B-23	R	–	+	M	O	–	–	–	+	*Pseudomonas*
B-24	R	–	+	M	F	–	–	+	+	*Vibrio*
B-25	R	–	+	M	F	–	–	+	+	*Vibrio*
B-26	S	+	–	A	NG	–	–	–	–	—
B-27	R	–	–	M	NG	–	–	+	+	—
B-28	R	–	+	M	F	–	–	+	+	*Vibrio*
B-29	R	–	+	M	N	–	–	–	+	*Alcaligenes*
B-30	R	–	+	M	F	–	–	–	+	*Achromobacter*
B-31	R	–	+	M	NG	–	–	–	+	—
B-32	R	–	+	M	F	–	–	+	+	*Vibrio*
B-33	R	–	+	M	O	–	–	+	+	*Pseudomonas*
B-34	R	–	+	M	F	–	–	–	+	*Achromobacter*

[a] Abbreviations used: R, rod; S, sphere; +, positive; –, negative; M, monotrichous; P, peritrichous; A, atrichous; O, oxidative; N, negative; F, fermentative; NG, no growth.

[b] –, Unidentified.

Measurement of SLS Lysis

The concentration of SLS added to cell suspensions was affected by the fact that SLS precipitated at high salt concentrations. The actual amount of SLS used was determined by suspending *Pseudomonas* strain 1055-1 in ZoBell's 2216E medium containing 7% NaCl and determining the lowest level of SLS necessary to give lysis after 5 minutes at 25°C without causing precipitation of the SLS. It was found that the addition of 0.05 ml of 0.5 M SLS in aqueous solution to 10 ml of cell suspension in ZoBell's 2216E medium gave a satisfactory level of lysis.

The SLS lytic susceptibility of marine isolates was measured using suspensions of bacteria which had been grown at different concentrations of NaCl. The degree of lysis was calculated according to the following equation:

$$\%\text{SLS lysis} = \frac{A - B}{A} \times 100$$

In this equation A is optical density of growth suspension after incubation for 24 hours and B is residual turbidity after keeping the suspension in SLS for 30 minutes.

Since SLS lytic susceptibility is believed to be due to the lysis of bacterial surface components, mainly the hexosamine polymers and the proteins in the cell walls, it may be possible to determine the role of these surface components in marine bacteria if the relationship between lytic susceptibility and the content of these components in the bacterial surface is known. Accordingly, the hexosamine was determined using a colorimetric method (Boas, 1953). Isolation of the cell wall was carried out by the modified method of Yoshida *et al.* (1961), in which the cell wall is isolated by sonication and then subjected to glycerin density gradient centrifugation. The bacteria used were as follows: marine *Pseudomonas* C-1, C-2, C-4, C-5, and C-19; marine *Vibrio* C-13, C-18, C-20, and C-29; marine *Photobacterium* C-6 and C-31; *Pseudomonas fluorescens*, *Pseudomonas aeruginosa*, *Pseudomonas chlororaphis* (terrestrial *Pseudomonas*); *Bacillus cereus* var. *mycoides* (gram-positive).

The incubation of the test cultures was carried out at 25°C for the marine isolates and strain 1055-1, at 30°C for both *Bacillus cereus* var. *mycoides* and *Pseudomonas fluorescens*, and at 37°C for both *Pseudomonas aeruginosa* and *Pseudomonas chlororaphis*.

RESULTS AND DISCUSSION

According to Shewan's classification, as shown in Tables 1 and 2, of the 67 isolates, 22 strains were identified as *Pseudomonas,* 12 strains as *Achromo-*

bacter, 4 strains as *Photobacterium,* 16 strains as *Vibrio,* 5 strains as *Alcaligenes,* and 8 strains were unidentified.

In the study of salinity effects, 33 strains from station C were used, and results obtained are shown in Figs. 1–6. Most marine *Pseudomonas* grew little in the nutrient broth enriched with NaCl, but marine isolates of both *Vibrio* and *Photobacterium* grew well. The growth of *Achromobacter* and *Alcaligenes* was between the above two groups. None of the isolates grew in

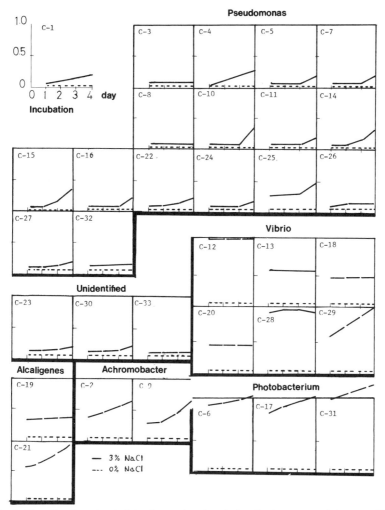

Fig. 1. Growth of marine isolates in nutrient broth. Incubation was carried out at 25°C for 4 days, and turbidity of growth was assessed at 540 mμ.

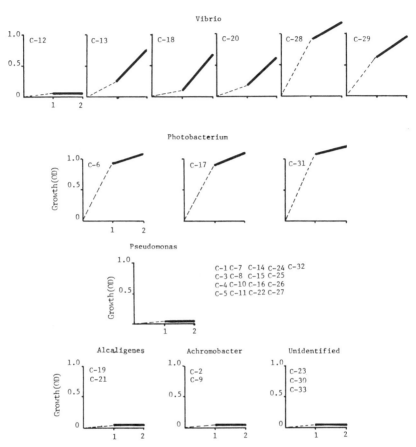

Fig. 2. Growth of marine isolates in ZoBell 2216E modified medium I (NaCl 0%). Incubation was carried out at 25°C for 2 days, and turbidity of growth was assessed at 540 mμ.

NaCl-free nutrient broth (Fig. 1). Most marine isolates grew well in the ZoBell 2216E. In contrast, in ZoBell 2216E without NaCl, most *Vibrio*, except C-12, and *Photobacterium* grew well in spite of the absence of NaCl, while only slight growth was observed in the other genera after 3–4 days incubation (Fig. 2).* In the ZoBell 2216E containing various concentrations of NaCl from 1 to 3%, there was no difference in the growth of strains of all genera; they all grew well after 1 day or more incubation.

* NB. The components of ZoBell's medium and nutrient broth may contain quite high levels of NaCl.

However, no growth was obtained with *Vibrio* and *Photobacterium* in the medium supplemented with 6% NaCl. The growth of *Pseudomonas, Alcaligenes,* and *Achromobacter* was not affected at this salt concentration. In 7.5% NaCl, the growth behavior of the latter three genera showed a lag period, and after 48 hours incubation growth was the same as that observed

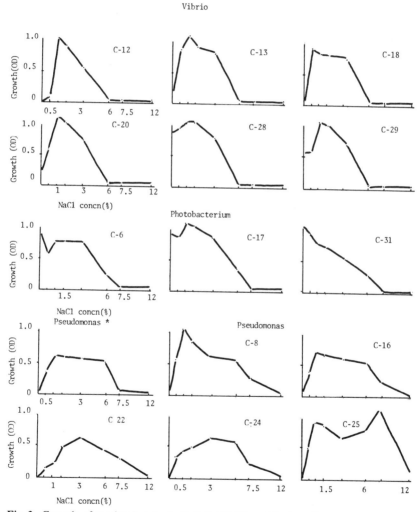

Fig. 3. Growth of marine isolates in ZoBell 2216E modified medium I (NaCl 0–12%). Incubation was carried out at 25°C for 2 days, and turbidity of growth was assessed at 540 mμ. All *Pseudomonas* except C-8, C-16, C-22, C-24, and C-25.

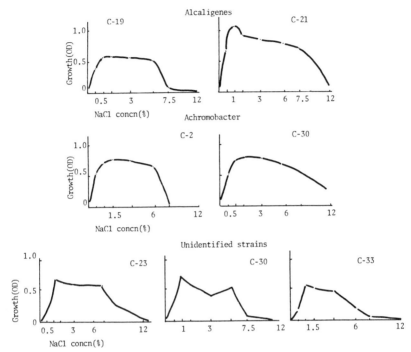

Fig. 4. Growth of marine isolates in ZoBell 2216E modified medium I (NaCl 0–12%). Incubation was carried out at 25°C for 2 days, and turbidity of growth was assessed at 540 mμ.

at 24 hours in 6% NaCl. Several strains were able to grow after lag periods in the medium containing 25% NaCl (Figs. 3 and 4). The growth in 1% peptone water supplemented with 3% NaCl was limited to the two genera *Vibrio* and *Photobacterium* (Fig. 5). In order to ascertain the effect of various minerals, except NaCl, Zobell's 2216E medium was used at one-sixth ASW concentration, except for NaCl, which was maintained at the full (3%) concentration. As shown in Fig. 6, no marked difference between each isolate was observed in such a medium except for two strains, C-12 and C-33. From this result, it is assumed that most marine isolates can grow in dilute ASW containing even a limited amount of minerals, if the medium contains enough minerals to make them grow. If Na^+ is replaced by K^+ in ZoBell 2216E medium there is no apparent effect on species of *Vibrio* and *Photobacterium* apart from *Vibrio* strain C-12, but K^+ cannot replace Na^+ in *Pseudomonas, Alcaligenes,* and *Achromobacter* (Fig. 7). Some caution must be observed in interpreting these data since contamination of the medium with Na^+ is unavoidable. The result of replacement of Na⁺ by Li⁺ was not the same as that of Na⁺ by K⁺ in the case of *Vibrio;* that is, four of the *Vibrio* strains did not grow when Na⁺ was replaced by Li⁺, whereas all but one strain grew when K⁺ was substituted for Na⁺ (Fig. 8).

From these experiments on the mineral requirements of these marine bacteria, it is concluded that most marine isolates require 1–3% NaCl for growth. It can also be shown that other minerals, especially divalent cations,

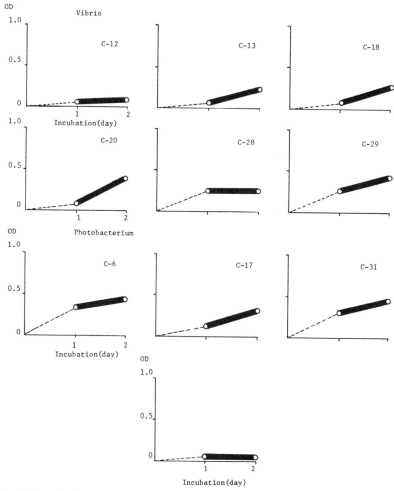

Fig. 5. Growth of marine isolates in peptone water containing 3% NaCl. Incubation was carried out at 25°C for 2 days, and turbidity of growth was assessed at 540 mμ.

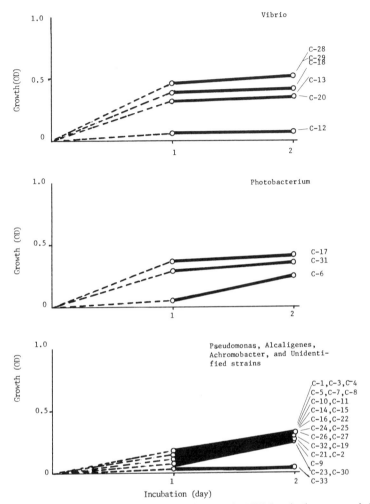

Fig. 6. Growth of marine isolates in one-sixth strength ASW. Incubation was carried out at 25°C for 2 days, and turbidity of growth was assessed at 540 mμ.

are indispensable for growth. It is presumed that NaCl requirement or tolerance varies according to the genus, but, in general, *Pseudomonas* tolerate wider concentrations of NaCl than *Vibrio* or *Photobacterium*. These results are in agreement with those obtained by other workers. The role of NaCl on the growth of marine bacteria may be concerned with the following functions: NaCl or Na$^+$ contributes to the oxidation of substrates (MacLeod *et al.*, 1960) as an activator; it may be the activator for ATPase (MacLeod, 1968).

As mentioned above, one of the important roles of Na⁺ is in relation to
the bacterial surface. For this reason an attempt has been made to see if there
is any relationship between the composition of some components of the cell
surface and the NaCl requirement for growth. Cells were lysed with SLS and
the results obtained are shown in Figs. 9 and 10. As shown in the figures,

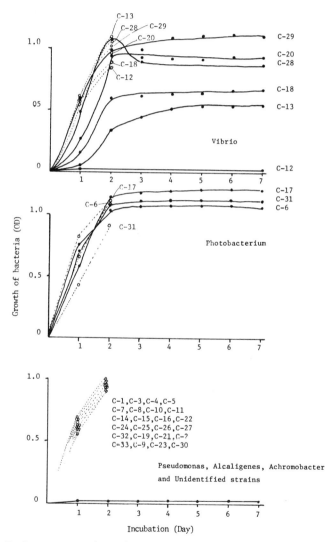

Fig. 7. Replacement of Na⁺ by K⁺. Incubation was carried out at 25°C for 7 days, and
turbidity of growth was assessed at 540 mμ. Dotted lines represent the controls.

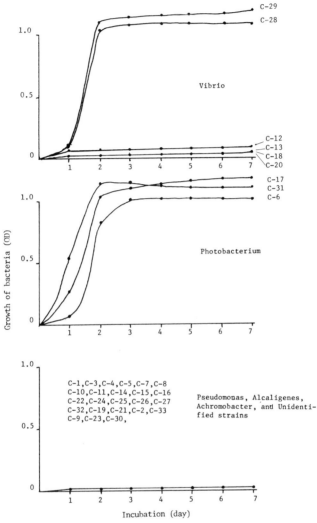

Fig. 8. Replacement of Na⁺ by Li⁺. Incubation was carried out at 25°C for 7 days, and turbidity of growth was assessed at 540 mμ. The controls used in this examination were the same as those used in the experiments shown in Fig. 7.

various lytic patterns were observed and, generally speaking, maximal lysis was observed in the cell suspension of maximal growth. The features of lysis were specific to each genus; for example, *Pseudomonas* had higher susceptibility to lysis at a wider range of salt concentrations than the other genera.

Some species of *Vibrio* were unaffected by SLS and only slight lysis was observed with *Photobacterium* species. On the other hand, terrestrial *Vibrio* has high susceptibility to SLS. The logarithmic average of the lysis ratio of each genus is shown in Fig. 11.

Finally, in order to determine the relationship between the lytic susceptibility and hexosamine content, the hexosamine content in the cell wall was

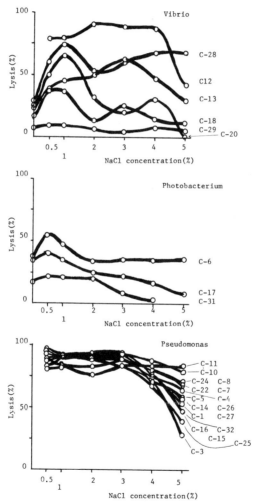

Fig. 9. Lytic susceptibility of marine isolates. Lytic susceptibility was measured by turbidity decrease at 540 mμ after the bacterial suspension was lysed with SLS for 30 minutes.

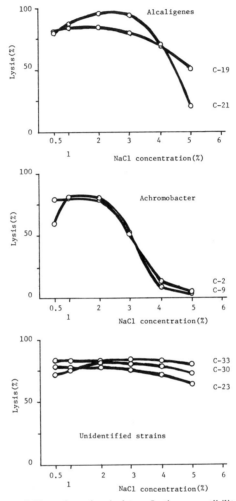

Fig. 10. Lytic susceptibility of marine isolates. Lytic susceptibility was measured by turbidity decrease at 540 mμ after the bacterial suspension was lysed with SLS for 30 minutes.

measured colorimetrically and the lytic susceptibility of the cell wall was observed by decrease in turbidity. Figure 12 shows the results of density gradient centrifugation in glycerin. The SLS lysis of the cell wall is shown in Table 3. In order to compare the lytic susceptibility in both SLS (an anionic detergent) and benzalkonium (a cationic detergent), lysis by the latter detergent was assessed by the same method as for SLS. The results are shown in Table 4. These results show that, for intact cells, many marine isolates are

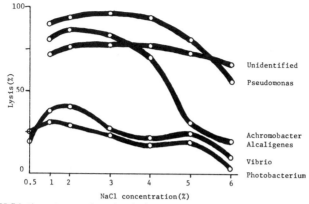

Fig. 11. SLS-lytic patterns of marine isolates. The figure shows the mean value of each genus.

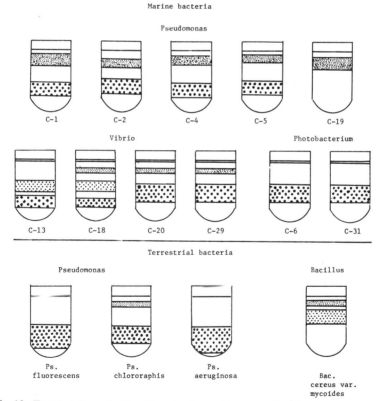

Fig. 12. Fractionation of cell wall by density gradient centrifugation. The cell wall was disrupted by sonication and subjected to glycerin gradient centrifugation in a 1 M KCl and 40% glycerin solution.

more easily lysed in the anionic detergent SLS than are terrestrial organisms. However, in the case of the cell wall prepared by density gradient the difference in lytic susceptibility between the cell walls of marine isolates and those of terrestrial organisms became uncertain. *Bacillus cereus* var. *mycoides* was the only organism examined which was susceptible to cationic detergent. The lytic susceptibility of various groups of test bacteria was in the following order: marine *Pseudomonas*, terrestrial *Pseudomonas*, marine *Vibrio*, marine *Photobacterium*, *Bacillus cereus* var. *mycoides*. The authors concluded that the cell walls of marine *Pseudomonas* isolates were lysed more easily than those of terrestrial ones, and that the upper layer of cell wall prepared by

Table 3. SLS lytic susceptibility of bacteria

| | Lysis (%)[a] | | | |
| | | Cell-wall density gradient | | |
Strain	Intact cell	Upper	Middle	Lower
Marine *Pseudomonas*				
C-1	89.3	95.5	–	85.0
C-2	82.1	92.2	87.5	–
C-4	85.3	90.3	–	83.3
C-5	82.9	91.7	–	87.5
C-19	84.3	90.0	–	–
Marine *Vibrio*				
C-13	43.4	–	–	–
C-18	30.4	–	–	–
C-20	28.0	77.5	74.2	–
C-29	11.3	–	77.8	–
Marine *Photobacterium*				
C-6	25.6	–	77.2	–
C-31	16.3	–	71.7	–
Pseudomonas aeruginosa	53.0	–	–	90.7
Pseudomonas chlororaphis	65.2	77.3	–	71.1
Pseudmonas fluorescens	56.8	–	–	93.1
Bacillus cereus var. *mycoides*	0	78.4	47.1	–

[a] Lysis (%) = A − B)/A × 100. Intact cell: A, optical density of growth suspension after incubation for 24 hr; B, residual turbidity after lysing the suspension with SLS for 30 min. Cell wall: A, optical density of cell wall suspension; B, residual turbidity after lysing the suspension with SLS for 30 min.

Table 4. Lytic susceptibility of bacteria by cationic detergent[a]

Strain	Intact cell	Cell wall density gradient		
		Upper	Middle	Lower
Marine *Pseudomonas*				
C-1	—	—		—
C-2	—	—	—	
C-4	—	—		—
C-5	—	—		—
C-19	—	—		
Marine *Vibrio*				
C-13	—		—	
C-18	—		—	
C-20	—		—	
C-29	—		—	
Marine *Photobacterium*				
C-6	—		—	
C-31	—		—	
Pseudomonas aeruginosa	—			—
Pseudomonas chlororaphis	—	77		—
Pseudomonas fluorescens	—			—
Bacillus cereus var. *mycoides*	42	—	87.1	

[a] 0.1 ml of 0.5 M benzalkonium chloride was added into 10 ml of culture suspension, and the turbidity decrease was assessed after 30 minutes. The lytic susceptibility induced by this detergent was calculated by the same equation as that for SLS.

density gradient was much more susceptible than the lower layer in the same *Pseudomonas*. The lysis of bacteria by chemical detergent is supposed to be mainly a reflection on the cell-wall structure of the living organisms. Since hexosamine and protein are important components of the cell-wall structure, the authors studied the hexosamine content of the marine isolates and the terrestrial organisms. The results are shown in Table 5. Strains rich in hexosamine were resistant to SLS lysis and this is particularly shown in

Table 5. Relation between hexosamine content and SLS lytic susceptibility

Strain	Hexosamine content (%)[a]			Lysis (%)			
	Upper	Middle	Lower	Intact Cell	Upper	Middle	Lower
Marine							
Pseudomonas							
C-1			0.76	89.3	95.5		85.0
C-2	0.46		0.34	82.1	92.2	87.5	
C-4			0.78	85.3	90.3		83.3
C-5	1.00		1.08	82.9	91.7		75.0
C-19	0.94			84.3	90.0		
Marine *Vibrio*							
C-13		2.00		43.3		72.3	
C-18		1.46		30.4		72.4	
C-20		1.44	—	28.0		74.2	
C-29		1.42		11.3		77.8	
Marine							
Photobacterium							
C-6		2.10		25.6		77.2	
C-31		1.46		16.3		71.7	
Pseudomonas			4.84	53.0			90.7
aeruginosa							
Pseudomonas	5.38		4.36	65.2			71.1
chlororaphis							
Pseudomonas			5.28	56.8	77.3		93.1
fluorescens							
Bacillus cereus	14.4		35.20	0	78.4	47.1	
var. *mycoides*							

[a] Hexosamine was counted by dry matter.

Table 6. Relation between hexosamine content and lytic susceptibility by synthetic detergents

Genus	Lysis (%)			Hexosamine content (%)[a]
	SLS		Cationic	
	Intact cell	Cell wall	cell wall	
Marine *Pseudomonas*	84.8	86.8	—	0.8
Marine *Vibrio*	28.3	74.2	—	1.58
Marine *Photobacterium*	21.0	74.0	—	1.98
Terrestrial *Pseudomonas*	54.9	91.9	—	5.06
Bacillus cereus var. *mycoides*	—	47.1	87.1	35.2

[a] Hexosamine was determined by measuring dry weight.

Bacillus cereus var. *mycoides.* A similar relation was observable in the other strains. Table 6 shows the summarized results. The authors concluded that lytic susceptibility to chemical detergents can be used as an indicator not only to classify bacteria in the available taxonomic groups, but also to show the characteristics of marine isolates in comparison with terrestrial bacteria.

LITERATURE CITED

Boas, N. F. 1953. Determination of hexosamine. J. Biol. Chem. 204:553–562.

Bukatsch, F. 1936. Requirement of K as well as Na. Sbr. Akad. Wiss. Wien Mathnaturw. K1. Abt. I, 141:259–276.

Gilby, A. R. 1957. The physical chemistry of bacterial protoplasts. Ph.D. Thesis. Univ. Cambridge, U.K.

Hidaka, T. 1965. Mineral requirement of marine bacteria. Mem. Fac. Fish. Kagoshima Univ. 14:127–180.

MacLeod, R. A. 1968. On the role of inorganic ions. *In* M. R. Droop and E. J. Ferguson Wood (eds), Advances in Microbiology of the Sea, Vol. I, pp. 95–123. Academic Press, New York.

MacLeod, R. A., C. A. Claridge, A. Hori, and J. F. Murray. 1959. The function of sodium in the metabolism of marine bacteria. J. Biol. Chem. 232:829–834.

MacLeod, R. A., A. Hori, and S. M. Fox. 1960. Nutrition and metabolism of marine bacteria. Can. J. Biochem. Microbiol. Technol. Eng. 3(2):151–159.

MacLeod, R. A., and E. Onofrey. 1956. Nutrition and metabolism of marine bacteria. Can. J. Microbiol. 3:753–759.

MacLeod, R. A., and E. Onofrey. 1957. Efforts to replace the Na by various elements. J. Cell. Comp. Physiol. 50:389–402.

MacLeod, R. A., E. Onofrey, and M. E. Norris. 1954. Nutrition and metabolism of marine bacteria. J. Bacteriol. 68:680–686.

Shewan, J. M. 1963. Classification of marine bacteria. *In* C. H. Oppenheimer (ed.), Symposium on Marine Microbiology, pp. 449–521. Charles C Thomas, Springfield, Illinois.

Yoshida, A., C. G. Heden, B. Cedergren, and L. Edebo. 1961. Preparation of undigested bacterial cell walls. J. Biochem. Microbiol. Technol. Eng. 3(2):151–159.

ZoBell, C. E., and H. C. Upham. 1944. Definition of marine bacteria. Bull. Scripps Instn. Oceanogr. Tech. Ser. 5:239–292.

VARIATIONS IN THE SALT REQUIREMENT FOR THE OPTIMUM GROWTH RATE OF MARINE BACTERIA

DARRELL PRATT and SUSAN TEDDER

Department of Microbiology
University of Maine
and
Department of Microbiology
University of Florida

The Na$^+$ requirement for the optimal growth rate of 20 cultures of marine bacteria was determined. Thirteen cultures were isolated by a procedure designed to select those having a high Na$^+$ requirement. Five were isolated from estuarine environments and had potentially a lower Na$^+$ requirement. The remaining two were previously described stock cultures. The results were consistent with the interpretation that marine halophiles have both a specific requirement for Na$^+$ and a non-specific requirement for solute which could be satisfied by KCl. The growth rate of all isolates was enhanced by the addition of KCl to suboptimal concentrations of NaCl. None of the isolates grew without the addition of Na$^+$.

Several years ago we reported the finding that a component of the Na$^+$ requirement of marine *Vibrio* MB 22 could be satisfied by other solutes and seemed in part a function of water activity (Pratt and Austin, 1963). Since this finding was not in agreement with reports dealing with other isolates, we speculated that this variation resulted from species differences (MacLeod, 1965; MacLeod and Onofrey, 1957; Payne, 1960). This interpretation was supported by the results of a simple experiment in which samples of seawater were inoculated onto agar media containing, in addition to nutrients, various concentrations of NaCl and diverse solutes. Approximately 50% of the organisms growing on the control seawater medium failed to grow on a medium containing 0.025 M NaCl and 0.4 M KCl. Nearly all failed to grow in the medium containing 0.4 M KCl alone. Thus, half the organisms present would permit a partial substitution of KCl for NaCl; the rest would not (Pratt, 1963). Our principal effort in the present study was directed to two groups of cultures. The members of one had a potentially high Na^+ requirement, and the other a potentially lower Na^+ requirement for optimum growth. These strains have been compared with respect to their Na^+ requirement and the ability of KCl to spare the requirement. The results were in agreement with the idea that Na^+ satisfies both a specific and a non-specific requirement. A considerable quantitative diversity exists among species with respect to these requirements.

MATERIALS AND METHODS

To isolate cultures of marine bacteria requiring a potentially higher concentration of Na^+ for growth, samples of seawater were spread onto a chemically defined medium containing 0.4 M NaCl. After incubation, the colonies were replicated by means of velveteen templates (Lederberg and Lederberg, 1952) to media containing 0.025 M NaCl, with and without the presence of 0.4 M KCl. Colonies which failed to grow were randomly selected. Thirteen cultures were isolated and characterized as *Pseudomonas* species. All were oxidase-positive (Kovacs, 1956), polarly flagellated, gram-negative rods. All attacked glucose oxidatively (Leifson, 1963) and were resistant to the vibriostatic agent 0/129 (Shewan *et al.*, 1954). These organisms have been designated as the SAT series. The second group of potentially low salt requiring organisms, the GW series consisting of five cultures, were isolated in Prof. M. E. Tyler's laboratory from samples of water taken from the estuary of the Waccasassa River in Florida. These were gram-negative, polarly flagellated rods. They were fermentative, aerogenic, and resistant to the vibriostatic agent. Four of the five were oxidase-negative; all produced a red water-insoluble pigment. As familiar organisms to serve as controls, *Vibrio* MB 22 and *Pseudomonas* MB 29 (Tyler *et al.*, 1960) were included in the study. All of the cultures were maintained on agar slants containing 1% Trypticase (BBL) in artificial seawater.

Growth in liquid media was estimated by measuring the increase in optical density at 625 nm. The cultures were grown in 250 ml Erlenmeyer flasks equipped with an attached tube for measuring the turbidity. The volume of the medium was 20 ml; the flasks were incubated at 25°C with constant shaking; the optical density was measured at intervals. The basal medium contained the following components: 0.005 M MgCl, 0.005 M K_2SO_4, 0.0003% $FeSO_4 \cdot 7H_2O$, 0.1% sucrose, 0.2% succinic acid, 0.1% glutamic acid, and 0.005 M $K^+PO_4^{3-}$ (pH 7.4). The glutamic and succinic acids were neutralized with ammonium hydroxide before their addition to the medium. Additions of NaCl and KCl were made to the basal medium as the experimentation indicated. Inocula were grown in a medium containing 0.1% Trypticase in artificial seawater plus 0.001% $FeSO_4 \cdot 7H_2O$ and 0.01 M Tris buffer (tris(hydroxymethyl)aminomethane), pH 7.4. After 15 hr growth at 25°C, the cells were harvested by centrifugation, washed once, and resuspended in artificial seawater. An inoculum of 0.2 ml was used for each flask and the initial optical density was measured immediately after inoculation. All experimental cultures contained at least 0.004 M Na^+ as contamination from the inoculum.

RESULTS

The typical response of a marine bacterium to Na^+ resembles in part that of an organism to a stimulatory growth factor. The total amount of growth is not influenced, but rather the rate of growth is a function of Na^+ concentration. However, there appears to be a threshold below which growth does not occur, a property which differs from the response of an organism to a stimulatory growth factor of the strepogenin type. Some typical results of growth in a liquid medium by *Vibrio* MB 22 are shown in Table 1. The total growth in this particular experiment was limited by the nutrient supplied by

Table 1. Control of growth rate by NaCl concentration[a]

Incubation time (hr)	Optical density		
	0 NaCl	0.1 M NaCl	0.4 M NaCl
0	0.02	0.02	0.02
3	0.02	0.06	0.13
7	0.02	0.25	0.50
24	0.04	0.64	0.60

[a] Basal medium: Trypticase, 0.1%; $MgCl_2$, 0.05 M; K_2SO_4, 0.005 M; $FeSo_4 \cdot 7H_2O$, 0.0003%. The inoculum was grown in the basal medium with 0.05 M NaCl and washed in an artificial seawater containing 0.05 M NaCl.

Table 2. Effect of NaCl on
the growth rate of isolate SAT 4[a]

NaCl	Optical density after indicated incubation time			
(M)	0 hr	7 hr	11 hr	24 hr
0	0.02	0.01	0.01	0.02
0.05	0.03	0.04	0.05	0.08
0.1	0.02	0.02	0.07	0.08
0.2	0.02	0.06	0.09	0.35
0.3	0.05	0.07	0.16	0.88
0.4	0.06	*0.17*	*0.60*	0.88
0.5	0.03	0.18	0.58	0.88
0.6	0.04	0.16	0.60	0.92

[a] Italicized values indicate point at which optimal growth rate occurs.

Table 3. Replacement by KCl of the optimal concentration of NaCl for isolate SAT 4

NaCl	KCl	Optical density at hours indicated			
(M)	(M)	0 hr	7 hr	11 hr	24 hr
0.40	0	0.04	0.16	0.52	0.77
0.35	0.05	0.04	0.16	0.53	0.78
0.30	0.10	0.05	0.16	0.52	0.74
0.25	0.15	0.03	0.16	0.51	0.78
0.20	0.20	0.04	0.12	0.38	0.65
0.15	0.25	0.04	0.11	0.39	0.67
0.10	0.30	0.05	0.10	0.32	0.68
0.05	0.35	0.05	0.08	0.34	0.78
0	0.40	0.04	0.03	0.07	0.01

0.1% Trypticase. In the absence of added Na⁺, growth failed; however, the discrepancies between 0.1 and 0.4 M, which were clearly shown during the first 7 hr of incubation, were eliminated after 24 hr. In the absence of added NaCl no growth occurred even after 24 hr despite the certainty of some Na⁺ contamination in the Trypticase and in the inoculum. These results have been presented to illustrate the typical nature of the response to Na⁺ shown by these bacteria.

For each of the test cultures and the control cultures, two determinations were made. In the first, the concentration of NaCl giving the optimal growth rate for the particular strain was estimated. The results shown in Table 2 give as an example the experimental data as observed for isolate SAT 4. The minimal concentration of NaCl allowing the optimal growth of this organism was 0.4 M. In the second, the concentration of NaCl required for optimal growth in the presence of KCl was determined. The amount of KCl in each flask was sufficient to maintain a total salt concentration equal to the

Table 4. Concentration of NaCl required for optimal growth of selected marine bacteria in the presence or absence of KCl

Culture No.	NaCl (M)		Culture No.	NaCl (M)	
	Without KCl	With KCl		Without KCl	With KCl
SAT6	0.3	0.25	SAT5	0.4	0.35
SAT9	0.3	0.25	SAT7	0.4	0.25
SAT10	0.3	0.20	SAT8	0.4	0.30
SAT11	0.3	0.25	GW11	0.3	0.05
SAT12	0.3	0.15	GW12	0.2	0.20
SAT13	0.3	0.25	GW13	0.2	0.15
SAT1	0.4	0.35	GW14	0.2	0.15
SAT2	0.4	0.25	GW15	0.3	0.10
SAT3	0.4	0.30	MB22	0.3	0.10
SAT4	0.4	0.25	MB29	0.4	0.15

previously determined optimal concentration for NaCl. The parameters for this experiment and the results for isolate SAT 4 are given in Table 3. The concentration of Na^+ giving optimal growth in this case was 0.25 M.

Considerable variation was observed among the organisms studied with respect to both determinations. The optimal concentrations for NaCl alone, without sparing amounts of KCl, ranged from 0.2 to 0.4 M. In the presence of compensating amounts of KCl, the NaCl optima ranged from 0.05 to 0.35 M. With both values, the higher concentrations were associated with the SAT strains which were isolated as high Na^+ requirers (Table 4). As an estimate of the specific requirement for NaCl, the response to 0.05 M NaCl with and without a compensating KCl were compared. The measurements were taken early in the growth period while differences in growth rate were still being

Table 5. Enhancement of the growth rate of marine bacteria by the addition of KCl to media containing suboptimal concentrations of NaCl

| Culture No. | Growth rate[a] | | | Culture No. | Growth rate[a] | | |
	Without KCl	With KCl	Control (KCl only)		Without KCl	With KCl	Control (KCl only)
SAT6	6	32	8	SAT5	3	72	5
SAT9	10	68	14	SAT7	11	57	5
SAT10	6	45	10	SAT8	17	5	5
SAT11	6	40	10	GW11	18	100	17
SAT12	4	37	10	GW12	14	56	32
SAT13	2	45	6	GW13	15	83	21
SAT1	21	53	14	GW14	22	81	18
SAT2	6	47	4	GW15	25	81	33
SAT3	6	12	4	MB22	2	97	25
SAT4	8	65	13	MB29	2	63	3

[a] Values in the table are percentages of the growth rate obtained with the optimal concentration of NaCl. The concentration of NaCl in each case was 0.05 M; the concentration of KCl was sufficient to give a total concentration of salt equivalent to optimal concentration of NaCl for the culture.

Table 6. Contrast of the specific
NaCl requirement of two marine isolates
as shown by the replacement capacity of KCl

NaCl (M)	SAT 3 (OD)[a]		GW 11 (OD)[b]	
	Without KCl	With KCl	Without KCl	With KCl
0	0.02	0.03	0.06	0.13
0.05	0.04	0.10	0.10	0.50
0.1	0.11	0.40	0.24	0.46
0.2	0.16	0.66	0.26	0.46
0.3	0.25	0.75	0.56	0.47
0.4	0.70	0.83	—	—

[a] SAT 3: KCl added to give a total concentration of 0.4 M. Optical density measurements made after 10 hr incubation.

[b] GW 11: KCl added to give a total concentration of 0.3 M. Optical density measurements made after 5 hr incubation.

expressed. The results are presented as the percentage of the rate obtained with the optimal amount of NaCl (Table 5). The enhancement effect of KCl varies from being very small with certain of the SAT strains to being complete with GW 11 and MB 22. All isolates show some enhancement; those in the GW series in general show the most, and the SAT series the least. The results from the studies using the two extreme strains are compared in Table 6. One, SAT 3, had a minimal specific requirement of about 0.3 M NaCl, while that of the other, GW 11, was only 0.05 M NaCl.

CONCLUSIONS

Bacterial cultures isolated from marine environments were shown to vary considerably with respect to the concentration of Na^+ required for the optimal growth rate. Cultures in the SAT series on isolation appeared to have a highly specific requirement for Na^+; however, even these strains grew more rapidly in the presence of suboptimal amounts of NaCl when the total salt concentration was maintained by KCl. Thus these marine halophiles have both a specific and non-specific component of the solute requirement. For a culture to grow at its optimal rate both requirements must be satisfied.

LITERATURE CITED

Kovacs, N. 1956. Identification of *Pseudomonas pyocyanea* by the oxidase reaction. Nature (London) 178:703.

Lederberg, J., and E. M. Lederberg. 1952. Replica plating and indirect selection of bacterial mutants. J. Bacteriol. 63:399.

Leifson, E. 1963. Determination of carbohydrate metabolism of marine bacteria. J. Bacteriol. 85:1183–1184.

MacLeod, R. A. 1965. The question of the existence of specific marine bacteria. Bacteriol. Rev. 29:9–23.

MacLeod, R. A., and E. Onofrey. 1957. Nutrition and metabolism of marine bacteria. III. The relation of sodium and potassium to growth. J. Cell. Comp. Physiol. 50: 389–402.

Payne, J. 1960. Effects of sodium and potassium ions on growth and substrate penetration of a marine pseudomonad. J. Bacteriol. 80:464–471.

Pratt, D. B., 1963. Specificity of the solute requirement by marine bacteria on primary isolation from seawater. Nature (London) 199:1309.

Pratt, D. B., and M. Austin. 1963. Osmotic regulation of the growth rate of four species of marine bacteria. *In* C. H. Oppenheimer (ed.), Symposium on Marine Microbiology, pp. 629–637. Charles C Thomas, Springfield, Illinois.

Shewan, J. M., W. Hodgkiss, and J. Liston. 1954. A method for the rapid differentiation of certain non-pathogenic, asporogenous bacilli. Nature (London) 173:208.

Tyler, M. E., M. C. Bielling, and D. B. Pratt. 1960. Mineral requirements and other characters of selected marine bacteria. J. Gen. Microbiol. 23:153–161.

LOCALIZATIONS AND SALT MODIFICATIONS OF PHOSPHOHYDROLASES IN SLIGHTLY HALOPHILIC *VIBRIO ALGINOLYTICUS*

TSUTOMU UNEMOTO, MAKI HAYASHI, YOSHIMICHI KOZUKA, and MAKOTO HAYASHI

Laboratory of Microbiological Chemistry
The Institute of Food Microbiology
Chiba University

The cells of slightly halophilic *Vibrio alginolyticus* were fractionated into periplasmic, envelope, and cytoplasmic fractions by means of osmotic shock and lysis procedures, and the distributions of phosphohydrolases were examined. The enzymes $2'3'$-cyclic phosphodiesterase ($3'$-nucleotidase) and alkaline phosphatase were located in the periplasm, while $5'$-nucleotidase was bound to the cell envelope. Inorganic pyrophosphatase and acid phosphatase were recovered in the cytoplasmic fraction, but the localization of the latter enzyme was not precisely determined.

In studying salt modifications of enzymes, it was found that $3'$-nucleotidase and $5'$-nucleotidase required salts for their activity. Acid phosphatase was very unstable in the absence of salts and was stabilized by their presence. Pyrophosphatase was inactivated and inhibited by salts. The enzymes associated with the periplasm or the envelope seemed to be more halophilic than those of the cytoplasm. The phosphohydrolases isolated from *Vibrio alginolyticus* showed a specificity for monovalent anions whose order of effectiveness was demonstrated to follow the lyotropic series, $Cl^- > Br^- > NO_3^- > ClO_4^- > SCN^-$.

The effects of salts on the enzyme kinetics were explained by the mechanism of the single substrate-single modifier case presented by Frieden (1964).

INTRODUCTION

Marine bacteria have a highly specific requirement for inorganic ions in their nutrition and metabolism (MacLeod, 1968). Recently, Thompson and MacLeod (1971) revealed the functions of Na^+ and K^+ in the active transport of a-amino-isobutyrate in a marine pseudomonad. However, the role of inorganic ions on the enzyme activities of marine bacteria has not been extensively studied. Using cell-free extracts of a marine pseudomonad, MacLeod et al. (1958) found that, among the enzymes of tricarboxylic acid cycle, succinic dehydrogenase (EC 1.3.99.1) and fumarase (EC 4.2.1.2) were more active in the absence of added salts than in their presence. Citrate synthetase (EC 4.1.3.7) and a-ketoglutarate dehydrogenase (EC 1.2.4.2) also were not stimulated by added salts, but aconitase (EC 4.2.1.3) and isocitrate dehydrogenase (EC 1.1.1.42) were stimulated non-specifically by any of a number of salts (MacLeod and Hori, 1960). Isolated cell envelopes of marine bacteria have been shown to possess adenosine triphosphatase activity which was also non-specifically stimulated by a number of different salts (Drapeau and MacLeod, 1963). This enzyme was characterized as cation-activated nucleotidase (EC 3.1.3.5) by Hayashi and Uchida (1965) and by Thompson et al. (1969) without consideration of the effect of anions. It is apparent that, unlike the enzymes from extreme halophiles, those from marine bacteria do not necessarily require salts for the manifestation of their activities. Enzymes which do require salts for their maximum activities may be non-specifically activated by inorganic ions as a result of their ionic strength. Unemoto and Hayashi (1969) purified $2',3'$-cyclic phosphodiesterase having $3'$-nucleotidase activity from slightly halophilic marine *Vibrio alginolyticus* and found that monovalent anions, especially Cl^-, inversely modified the two activities residing in the same protein molecule. Hayashi et al. (1970) reinvestigated $5'$-nucleotidase in the cell envelope of this organism and found it to be an anion-activated $5'$-nucleotidase. From this it is apparent that many of the enzymes isolated from marine bacteria seem to be influenced by inorganic ions. We therefore studied the mode of action of inorganic ions on the phosphohydrolases in a slightly halophilic *Vibrio alginolyticus* with special reference to their cellular localizations and to the effect of anions.

MATERIALS AND METHODS

Culture

Vibrio alginolyticus 138-2, used in these studies, was originally isolated from marine fish and was classified by Dr. K. Aiso. It is identical to *Beneckea alginolytica,* the name recently proposed by Baumann *et al.* (1971). This organism forms transparent and opaque colony variants. The former variant was used throughout these experiments.

Medium and Growth Conditions

The culture was maintained by monthly transfer on slants of a medium containing 0.5% polypeptone (Daigo-eiyo), 0.5% yeast extract (Daigo-eiyo), 0.4% K_2HPO_4, 3.0% NaCl, and 1.5% agar. The pH was adjusted to 7.4. The liquid medium employed contained the same constituents but with the agar omitted and an addition of 0.2% glucose. A 1/100 volume of the overnight liquid culture was inoculated into a fresh medium and incubated on a reciprocal shaker at 37°C. Growth was followed by measuring optical density at 550 nm using a Bausch-Lomb Spectronic 20. Any sample having an optical density above 0.3 was appropriately diluted with the growth medium.

For the induction of alkaline phosphatase (EC 3.1.3.1), a medium containing 0.2% polypeptone, 10 mM K_2SO_4, 5.0 mM $MgSO_4$, 20 mM Tris-HCl buffer (pH 7.4), 0.2% glucose, 0.2% glutamic acid, and 3.0% NaCl was used. This medium also contained 0.16 mM inorganic phosphate (Pi).

Procedure for the Cell Fractionations

The procedure of osmotic shock originally devised by Neu and Heppel (1965) was used to isolate the enzymes which may reside around or between the outer and the inner double-track layers of the cell envelope (periplasm). Since the cells of *Vibrio alginolyticus* are very fragile in a hypotonic environment, the procedure was modified to prevent significant lysis of the cells during osmotic shock. The cells were harvested in the logarithmic phase of growth by centrifugation at 10,000 X g for 10 min at 4°C and washed twice by centrifugation in a cold 1.0 M NaCl containing 50 mM Tris-HCl buffer (pH 7.4). The washed cells were suspended in a hypertonic medium containing 1.0 M NaCl, 0.5 M sucrose, and 10 mM Tris-acetate buffer (pH 7.2). The cell suspension was centrifuged and the sedimented cells were rapidly dispersed in a hypotonic medium containing 0.25 M NaCl, 10 mM $MgCl_2$ and 10 mM Tris-acetate (pH 7.2) and centrifuged. The supernatant was removed and used as a fraction of periplasm. The cells were lysed by suspending them in a medium containing 10 mM NaCl, 2 mM $MgCl_2$, and 10 mM Tris-acetate (pH 7.2) (lysis medium). After treatment with deoxyribonuclease, the lysate was centrifuged at 10,000 X g for 20 min and the supernatant was used as a cytoplasmic fraction. The pellets were washed twice with the lysis medium and used as an envelope fraction.

Electron Microscopy

The cells treated by the hypertonic, hypotonic, and lysis media as described above were fixed by the method of Ryter and Kellenberger (1958) in the presence of the same solutes at the same concentrations to which the cells had previously been exposed. They were dehydrated in a graded series of acetone and finally embedded in Vestopal W. Ultrathin sections were cut by a Porter Blum Model 2B Ultramicrotome with glass knives. The sections were post-stained in a saturated solution of uranyl acetate and lead hydroxide (Karnovsky, 1961). Electron micrographs were taken with a Hitachi HU-12 electron microscope at 75 kV accelerating voltage. Negatively stained cell envelopes were prepared by staining with neutralized 1% phosphotungstic acid on carbon coated collodion film.

Enzyme Purifications

$2',3'$-Cyclic phosphodiesterase ($3'$-nucleotidase) was purified as described previously (Unemoto and Hayashi, 1969). A 2000-fold purified preparation was obtained. It possessed the specific activity of 170 units/mg protein with $3'$-AMP as a substrate.

$5'$-Nucleotidase was prepared as previously described (Hayashi et al., 1970).

Alkaline phosphatase was purified from the shock medium by using DEAE-Cellulose, DEAE-Sephadex, and Sephadex G-100. A 500-fold purified preparation was obtained in these experiments starting with the total lysate.

Acid phosphatase (EC 3.1.3.2) was purified from the cytoplasmic fraction by using $(NH_4)_2 SO_4$ and pH fractionations and DEAE-Sephadex. Since the enzyme was very unstable in the absence of salts, further purifications were not carried out. A 20-fold purified preparation was obtained.

Inorganic pyrophosphatase (EC 3.6.1.1.) was purified from the cytoplasmic fraction by using $(NH_4)_2 SO_4$ and pH fractionations, Sephadex G-100, and by DEAE-Sephadex. A 100-fold purified preparation was obtained. Details of the purifications of alkaline phosphatase, acid phosphatase and pyrophosphatase will be published elsewhere.

Enzyme Assays

$2',3'$-Cyclic phosphodiesterase, $3'$-nucleotidase, and $5'$-nucleotidase were assayed as previously described (Unemoto and Hayashi, 1969; Hayashi et al., 1970).

Alkaline phosphatase and acid phosphatase were assayed spectrophotometrically at $37°C$ using p-nitrophenyl phosphate as a substrate. Para-nitrophenyl phosphate was recrystallized from 85% ethanol before use. For the assay of alkaline phosphatase, the reaction mixture contained 20 mM CAPS-NaOH buffer (pH 9.0–10.8), p-nitrophenyl phosphate, and enzyme in a final volume of 3.0 ml. The reaction was followed by measuring the absorbance at 405 nm using Hitachi two-wavelength double-beam spectrophotometer 356. For the assay of acid phosphatase, the standard reaction mixture contained 25 mM BES-Tris buffer (pH 5.5), 5.0 mM p-nitrophenyl phosphate, 50 mM $MgCl_2$ and enzyme in a final volume of 1.0 ml.. This was incubated at 37°C for 5 min and the reaction was stopped by adding 2.0 ml of a solution containing 0.1 M NaOH and 0.125 M Na_4EDTA. The absorbance was then measured at 405 nm. Alkaline-EDTA solution was used to prevent the precipitation of excess Mg^{2+} at alkaline pH.

Pyrophosphatase was assayed as follows. The reaction mixture (0.5 ml), containing 50 mM Tris-BICINE buffer (pH 8.7), 2.0 mM Na_4PP_i, 4.0 mM $MgCl_2$, and enzyme, was incubated at 37°C for 5 min. The reaction was stopped by adding 0.5 ml of 24% $HClO_4$ and the mixture was cooled in ice water to avoid the hydrolysis of PP_i. Liberated P_i was measured by the method of Allen (1940) in a final volume of 3.0 ml. When the PP_i and $MgCl_2$ concentrations were varied, the concentrations of Mg^{2+}, PP_i^{4-}, $MgPP_i^{2-}$, and Mg_2PP_i in the reaction mixture were calculated by using the stability constants for $MgPP_i^{2-}$ and Mg_2PP_i given by Lambert and Watters (1957).

One unit of activity was defined as that amount which hydrolyzes 1.0 μmole of substrate in 1 min.

RESULTS

Properties of the Phosphohydrolases Isolated from *Vibrio alginolyticus*

2′,3′-Cyclic phosphodiesterase hydrolyzed ribonucleoside 2′,3′-cyclic phosphates, ribonucleoside 3′-phosphates, and p-nitrophenyl phosphate. The activities for these substrates could not be separated. They were associated with the same protein. The other properties have been described by Unemoto and Hayashi (1969).

5′-Nucleotidase hydrolyzed all 5′-ribonucleotides to the corresponding ribonucleosides and P_i. The P_i released from 1 μmole each of AMP, ADP, and ATP amounted to 1, 2, and 3 μmoles, respectively. The other properties have been described by Hayashi *et al.* (1970).

Alkaline phosphatase was induced by limiting Pi in the growth medium. The enzyme catalyzed the hydrolysis of a wide range of phosphate esters. Its optimum pH was 10.3. In the presence of high concentrations of appropriate alcohols serving as phosphate acceptors, such as Tris and diethanolamine, it catalyzed transphosphorylation reactions, as in the case of *Escherichia coli*.

Acid phosphatase also catalyzed the hydrolysis of a wide range of phosphate esters. Its optimum pH was at 5.5, and its best substrate was *p*-nitrophenyl phosphate.

Pyrophosphatase catalyzed the hydrolysis of PP_i. The optimum pH of the enzyme activity was at 8.7 and was influenced by the concentrations of $MgCl_2$ and $Na_4 PP_i$. When the reciprocal of the initial velocity was plotted against the reciprocal of the concentration of $MgPP_i^{2-}$ in the reaction mixture, a straight line was obtained. Thus, the true substrate is $MgPP_i^{2-}$. Free Mg^{2+} acted as an activator and increased K_m and V_m for the substrate, saturating at about 3.0 mM (unpublished).

Localizations of the Phosphohydrolases in *Vibrio alginolyticus*

Figure 1 shows the distributions of the phosphohydrolases in the cellular fractions of *Vibrio alginolyticus*. 3′-Nucleotidase (cyclic phosphodiesterase) and alkaline phosphatase were recovered in the periplasmic fraction. Since

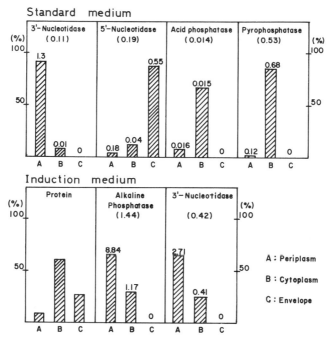

Fig. 1. Distribution of phosphohydrolases in cellular fractions of *Vibrio alginolyticus*. The cell fractionations were carried out as described in Materials and Methods. Cross-hatched bars indicate the percentage recovery of the enzyme activity relative to that present in the total lysate. Numbers in parentheses are specific activities (units/mg protein) in the total lysate and those on the bars are specific activities in each fraction.

there was no significant lysis of the cells, despite their exposure to osmotic shock, these enzymes may be located in the periplasmic space, as in the case of *Escherichia coli* (Neu and Heppel, 1965; Nossal and Heppel, 1966) and other Enterobacteriaceae (Neu and Chou, 1967). The low recovery of the alkaline phosphatase in the shock medium was caused by the fact that the cells grown in the induction medium are much more resistant to osmotic shock than those grown in a rich medium. This can be seen from the results obtained for 3'-nucleotidase.

The 5'-nucleotidase was mainly recovered in the envelope fraction, and no significant activity was removed by the washing of the envelope. Thus this enzyme seemed to be located in the envelope, bound to its structure.

Acid phosphatase was recovered in the cytoplasmic fraction. The total recovery of the activity was low owing to its unstable nature in the absence of salts, which will be mentioned later. Pyrophosphatase was mainly recovered in the cytoplasmic fraction.

When the distribution of these enzymes in *Vibrio alginolyticus* was compared to their distribution in *E. coli,* identical results were found in the cases of 3'-nucleotidase (cyclic phosphodiesterase), alkaline phosphatase, and pyrophosphatase. However, 5'-nucleotidase of *Vibrio alginolyticus* was located in the envelope and acid phosphatase in the cytoplasmic fractions. Thus the localizations of these two enzymes were apparently different from those in *Escherichia coli* in which these enzymes have been reported to be located in the periplasmic space (Neu and Heppel, 1965). With respect to acid phosphatase, it is possible that this enzyme is located at the inner side of the envelope and that it is released to the cytoplasmic fraction during lysis of the cells. Further details of the ultrastructural localization of this enzyme are under investigation.

Figure 2 shows electron micrographs of *Vibrio alginolyticus* treated by hypertonic, hypotonic, and lysis media. When compared with cells in the logarithmic phase of growth, no significant morphological changes were observed as a result of the treatment in the hypertonic medium. The envelope structure was typical of many gram-negative bacteria (Murray *et al.,* 1965), consisting of two double-tracked structures (Fig. 2*a*). When the cells were osmotically shocked by transferring them to the hypotonic medium, distinct separation of the outer double-track layer away from the inner double-track layer was observed without release of the cytoplasmic materials (Fig. 2*b*). Thus the enzymes located around or between the two layers may be released into the shock medium. When the cells were transferred to the lysis medium, large breaks could be seen in the two layers resulting in the formation of vesicular structures along the periphery of the cells (Fig. 2*c*). Complete loss of the intracellular materials was observed in most of the cells exposed to the

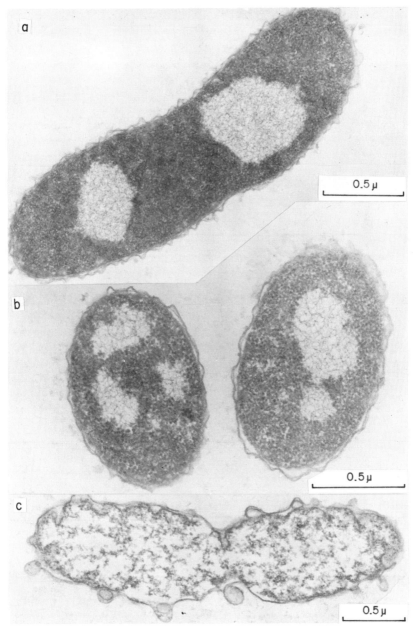

Fig. 2. Electron micrographs of *Vibrio alginolyticus* in (*a*) hypertonic, (*b*) hypotonic, and (*c*) lysis media.

lysis medium, but the cells retained their rod shape. The ultrastructural changes in the cell envelope of marine bacteria caused by various treatments have been extensively studied by Costerton *et al.* (1967), Forsberg *et al.* (1970*a,b*), and Thompson *et al.* (1970).

Effect of Salts on the Enzyme Activities

2′,3′-Cyclic Phosphodiesterase (3′-Nucleotidase). As shown in Fig. 3*a,* the 3′-nucleotidase was most effectively activated by the Cl^-, and it was saturated at about 200 mM. The SO_4^{2-} ion, however, showed no activation at all. The shapes of the activation curves conformed to Michaelis-Menten kinetics. On the other hand, as shown in Fig. 3*b,* the cyclic phosphodiesterase was inhibited by Cl^-. However, the SO_4^{2-} ion showed no inhibition. Thus it is apparent that the presence of Cl^- is essential for modifications of the two activities of the enzyme.

Table 1 shows the effect of salts on the 3′-nucleotidase activity. Mono-valent cations such as Li^+, Na^+, K^+, Rb^+, Cs^+, and $Tris^+$ could be substituted for one another, but NH_4^+ was a less effective substitute. Among monovalent anions, Cl^- was the most effective, the order of effectiveness being $Cl^- > Br^- = I^- > NO_3^- > HCOO^- > CH_3COO^-$. On a molar basis, $MgCl_2$ was much more effective than NaCl, but the maximum velocities obtained at the optimum concentrations of NaCl and $MgCl_2$ were identical. No further activation was observed with $MgCl_2$ in the presence of a saturating concentration of NaCl or *vice versa.* Since $MgSO_4$ showed no activating effect, Cl^- was essential for the activation even in the presence of Mg^{2+}

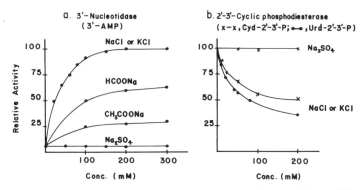

Fig. 3. Effect of salts on the cyclic phosphodiesterase. *a,* 3′-Nucleotidase activity. The reaction mixture (1.0 ml) contained 20 mM Tris-H_2SO_4 buffer (pH 8.3), 0.05 μg enzyme, 2.0 mM of 3′-AMP and various concentrations of salts as indicated. *b,* 2′,3′-Cyclic phosphodiesterase activity. The reaction mixture contained 20 mM Tris-H_2SO_4 (pH 8.4 for cytidine-2′,3′-P and pH 8.6 for uridine-2′,3′-P), 1.0 mM substrate, and various concentrations of salts as indicated.

Table 1. Effect of salts on the hydrolysis of
3'-AMP by the 2',3'-cyclic phosphodiesterase[a]

Salt added (200 mM)	Relative activity (%)	Salt added (50 mM)	Relative activity (%)
None	7	$MgCl_2$	100
LiCl	102	$CaCl_2$	58
NaCl	100	$SrCl_2$	100
KCl	103	$Mg(CH_3COO)_2$	39
RbCl	99	$MgSO_4$	6
CsCl	95		
Tris-HCl	98		
NH_4Cl	82		
KBr	93		
KI	92		
$NaNO_3$	65		
HCOONa	58		
CH_3COONa	28		

[a] The reaction mixture (1.0 ml) contained 20 mM Tris-H_2SO_4 (pH 8.3), 0.05 μg enzyme, 2 mM 3'-AMP and 50 or 200 mM of salts, as indicated. The activity in the presence of 200 mM NaCl was the same as that in the presence of 50 mM $MgCl_2$.

5'-Nucleotidase. As indicated in Fig. 4, 5'-nucleotidase activity was increased by pre-incubation of the freshly prepared envelope at alkaline pH in the presence of salts. The order of effectiveness for the pre-activation was $NH_4^+ \gg Li^+ > Na^+ > K^+ = Rb^+ = Cs^+$. Although it is not shown here, when the anions were added as ammonium salts at a concentration of 0.1 M and at a pH of 9.0, no significant differences in the effect of Cl^-, Br^-, I^-, and acetate were observed. Thus species of monovalent cations are more important than those of anions for the pre-activation of the 5'-nucleotidase.

Figure 5 shows negatively stained preparation of freshly prepared and pre-activated envelopes of *Vibrio alginolyticus*. The sheetlike structure of the envelope could be seen more clearly in the pre-activated preparation (Fig. 5b) than in the untreated one (Fig. 5a). This might be caused by removal of some materials from the envelope by the pre-activation treatment. However, the 5'-nucleotidase, detached from the envelope by Triton X-100, was also pre-activated by the same treatment. Thus the activation was not the result of solubility changes in the envelope. It seemed that the pre-activation was caused by the changes in the enzyme protein itself.

5'-Nucleotidase showed no activity in the absence of divalent cations and an appropriate anion. Figure 6 indicates the effect of Mg^{2+} and anions. As

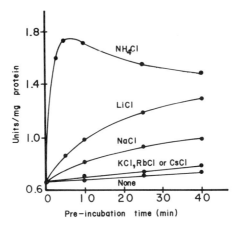

Fig. 4. Effect of salts on the pre-activation of 5'-nucleotidase in the cell envelopes of *Vibrio alginolyticus.* The mixture, containing 50 mM Tris-acetate (pH 9.0), 10 mM $MgCl_2$, freshly prepared envelope, and 0.5 M each of salts indicated above, was incubated at 37°C. At intervals, samples were rapidly cooled to 5°C and the activity for 5'-AMP was measured under standard conditions.

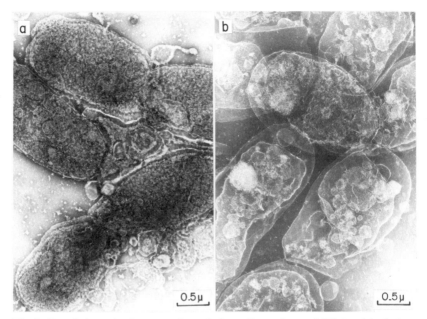

Fig. 5. Negatively stained preparation of the cell envelope of *Vibrio alginolyticus.* *a,* Freshly prepared envelope. *b,* The envelope suspended in 20 mM NH_4Cl and 50 mM Tris-acetate (pH 9.0) was frozen at −16°C and stored for 60 min, during which 5'-nucleotidase was completely pre-activated.

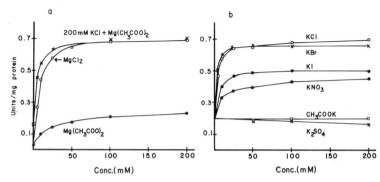

Fig. 6. Effect of Mg^{2+} and several anions on the hydrolysis of 5′-AMP. The reaction mixture (1.0 ml) contained 4.0 mM Tris-AMP, 50 mM Tris-acetate (pH 7.4), 80 μg of the unactivated enzyme and various concentrations of salt indicated. In Fig. 6*b*, all the reaction mixtures contained 100 mM magnesium acetate in addition to the salt indicated.

shown in Fig. 6*a*, maximum activation was obtained by the addition of 100 mM $MgCl_2$. The activation, however, was low with magnesium acetate. In the presence of 200 mM KCl, an activation curve similar to that for $MgCl_2$ was obtained. Thus, both Mg^{2+} and Cl^- are required for the maximum activity. Figure 6*b* indicates the effect of several anions in the presence of saturating concentrations of Mg^{2+}. Among monovalent anions, Cl^- was the most effective, the order of effectiveness being $Cl^- = Br^- > I^- > NO_3^-$. Acetate and SO_4^{2-} showed no effect on the activity. When 200 mM each of monovalent cations were added as chloride salts in the presence of 100 mM magnesium acetate, no significant differences in the activity were observed among Li^+, Na^+, K^+, Rb^+, Cs^+, and NH_4^+. Therefore, no particular species of monovalent cation was required for the 5′-nucleotidase. When the pre-activated enzyme was used, the same patterns of ionic requirements as described above were obtained.

Alkaline Phosphatase. Alkaline phosphatase of *Vibrio alginolyticus,* similar to the enzyme of *Escherichia coli,* did not require salt for its activity, but it was slightly modified by the presence of salt. The effects of salts were influenced by pH and substrate concentrations. As indicated in Table 2, the activation was higher at pH 8.0 than at the optimum pH (10.3) and it was unrelated to the species of monovalent cations. The activity, however, was inhibited by the presence of SCN^- or ClO_4^-. Thus, this enzyme seemed to be influenced by monovalent anions. The effect of substrate concentrations will be described later.

Acid Phosphatase. Acid phosphatase was very unstable in the absence of salts. As indicated in Fig. 7, one-half of the original activity was lost within 15 sec at 37°C in the absence of salts. In the presence of 400 mM each of

Table 2. Effect of salts
on the alkaline phosphatase

Salt added	pH 10.3[a]	pH 8.0[b]
None	100	100
NaCl	120	255
KCl	119	253
RbCl	120	269
KBr	123	238
KNO_3	121	208
KSCN	69	18
$NaClO_4$	59	0

[a] 20 mM CAPS-NaOH, 2 mM p-nitro-
phenyl phosphate, and 300 mM
salt.

[b] 25 mM Veronal buffer, 1 mM
p-nitrophenyl phosphate, and 600
mM salt.

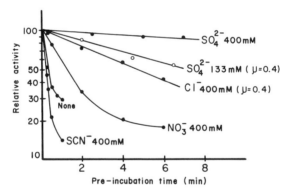

Fig. 7. Effect of salts on the stabilization of acid phosphatase. The mixture, containing 25 mM BES-Tris buffer (pH 5.5), enzyme, and sodium salts of the several anions indicated, was incubated at 37°C. At intervals, the activity was assayed under the standard conditions.

sodium salts of several anions, the most effective protection was obtained with SO_4^{2-}. When compared at the same ionic strength, SO_4^{2-} was slightly better than Cl^-. However, SCN^- accelerated the inactivation. The order of effectiveness for the stabilization was as follows: $SO_4^{2-} > Cl^- > Br^- > NO_3^- > SCN^-$. When 400 mM each of chloride salts of the monovalent cations were used, no significant differences in the stabilization were observed with Li^+,

Na^+, K^+, or Rb^+. The protective effect of SO_4^{2-} has been reported for isocitrate dehydrogenase of marine bacteria (MacLeod *et al.*, 1960).

Since the reaction rate of this enzyme decreased during the assay in the presence or absence of low concentrations of salts, the effect of salts on the initial velocity could not be determined.

Pyrophosphatase. In contrast to the acid phosphatase, pyrophosphatase was inactivated in the presence of salts. As indicated in Fig. 8a, the half-life of the enzyme was 3 min at 37°C in the presence of 200 mM NaCl. The remaining activities after pre-incubation at 37°C for 3 min in the presence of 200 mM each of sodium salts of Cl^-, Br^-, NO_3^-, ClO_4^-, and SCN^- were 48, 42, 31, 14, and 6% of the original activity, respectively. The chloride salts of Li^+, Na^+, K^+, and Rb^+ were nearly equally effective in inactivating the enzyme.

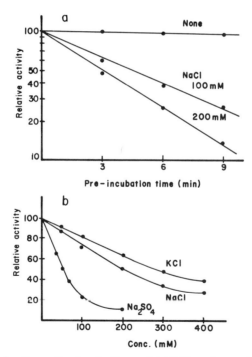

Fig. 8. Effect of salts on the inactivation and inhibition of pyrophosphatase. *a*, The mixture, containing 10 mM Tris-BICINE buffer (pH 8.7), enzyme, and NaCl as indicated was incubated at 37°C. At intervals, a portion of the reaction mixture was withdrawn and immediately assayed under standard conditions. *b*, The reaction mixture, containing 1.0 mM Na_4PP_i, 4.4 mM $MgCl_2$, 50 mM Tris-BICINE (pH 8.7), enzyme, and various concentrations of salts as indicated, was assayed. Results were expressed as relative activity.

No significant inactivation occurred after the start of the enzymatic reaction, indicating the protective effect of the substrate. Thus, the effect of salts on the initial velocity could be measured. As indicated in Fig. 8b, pyrophosphatase was inhibited by the presence of salts. Thus, 50% inhibition was observed by 200 mM NaCl. Na_2SO_4 effectively inhibited the activity and, at the same ionic strength, it was still more effective than NaCl. In the presence of 200 mM each of chloride salts of Li^+, Na^+, K^+, and Rb^+, the activity was inhibited 74, 52, 36 and 39%, respectively. On the other hand, inhibitory effects were nearly unrelated to the species of monovalent anions. Thus, the inhibition seemed to be affected by monovalent cations.

Effect of Salts on the Enzyme Kinetics

Figure 9 represents the activity-pH curves for 3'-nucleotidase in the presence and absence of NaCl. There was a small shift in optimum pH with increasing NaCl concentrations. When the activities at the different NaCl concentrations are compared at their optimum pH, it is seen that the activation by Cl^- was very marked. This Cl^- activation cannot be explained in terms of a shift in the dissociation constants of the active site of the enzyme controlling the pH curve.

Figure 10a represents the effect of NaCl on the 3'-nucleotidase activity. It shows that Cl^- has no effect on K_m for the substrate. The increase in V_m was greatest at about 200 mM NaCl. The activities of the other 3'-nucleotidase were also modified in a manner quite similar to the case with 3'-AMP.

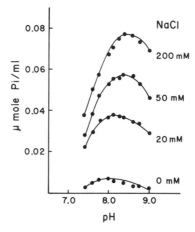

Fig. 9. Effect of the concentrations of NaCl on the optimum pH for 3'-nucleotidase. The reaction mixture (1.0 ml) contained 20 mM Tris-H_2SO_4, 2.0 mM 3'-AMP, 0.05 μg of enzyme, and NaCl as indicated.

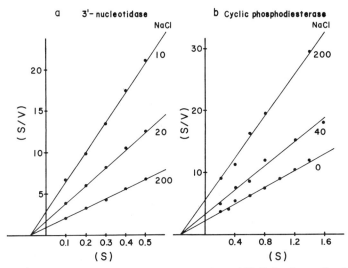

Fig. 10. *S/v versus S* plots at different concentrations of NaCl for the cyclic phospho-diesterase. *a*, 3'-AMP as substrate. *v* is expressed as μmoles P_i per 10 min per 0.05 μg of enzyme and *S* as the concentration of 3'-AMP in mM. *b*, Uridine-2',3'-P as substrate. *v* is expressed as μM of Uridine-2',3'-P hydrolyzed per min per 0.1 μg of enzyme and *S* as the concentration of Uridine-2',3'-P in mM.

Addition of $MgCl_2$ modified the activity similarly to NaCl except that the activation was greatest at about 20 mM. Figure 10*b* indicates the effect of NaCl on the hydrolysis of uridine-2',3'-P. As in the case of 3'-nucleotidase, Cl^- had no effect on K_m and the decrease in V_m was greatest at about 200 mM. Similar results were obtained in the hydrolysis of cytidine-2',3'-P. The maximum velocities in the presence and absence of 200 mM NaCl, and the concentration of NaCl to give half-maximum velocities are shown in Table 3.

Figure 11 indicates the effect of salts on the 5'-nucleotidase. The K_m's for the substrate and for Mg^{2+} were unaffected by the presence of salts, as in the case of 3'-nucleotidase. These results clearly indicate that Cl^- activation of 3'-nucleotidase and 5'-nucleotidase and Cl^- inhibition of the cyclic phospho-diesterase affect only V_m for the substrate.

Figure 12 shows the effect of NaCl on the alkaline phosphatase. All the reciprocal plots (1/*v* versus 1/*S*), where V is the initial velocity expressed in m units/3 ml (= x 10^{-3} units/3 ml) and S is the concentration of p-nitrophenyl phosphate in mM, were linear in the presence and absence of NaCl and they crossed at high substrate concentrations. Thus, NaCl inhibits the activity at low substrate concentrations and activates at saturating substrate concentrations. Figure 13 shows the effect of pH on K_m and V_m. The kinetic constants

Table 3. Summary of optimum pH, K_m, V_m, and $M_{1/2}$
for several substrates of the cyclic phosphodiesterase[a]

Substrate	Optimal pH	Without NaCl		With 200 mM NaCl		$M_{1/2}$ (mM)
		K_m (mM)	V_m	K_m (mM)	V_m	
3'-CMP	7.8	0.01	27	0.01	180	10
3'-UMP	7.6	0.05	6	0.05	170	18
3'-AMP	8.3	0.08	11	0.08	170	20
3'-GMP	8.1	0.10	9	0.10	260	50
Di-p-NPP	8.3	0.45	170	0.20	286	20
C-cyclic-p	8.4	0.10	380	0.10	200	30
U-cyclic-p	8.6	0.36	1560	0.36	600	40

[a] Reaction velocities at various concentrations of each substrate were measured at the optimum pH of the substrate using 20 mM Tris-SO$_4$ buffer with or without 200 mM NaCl. V_m was expressed as μmoles of substrate hydrolyzed per min per mg protein. $M_{1/2}$ represents the concentration of NaCl equivalent to the velocity lying halfway between the velocity in the absence of NaCl and the velocity in the presence of saturating amounts of NaCl.

were markedly affected by pH. The plots of K_m versus pH and V_m versus pH in the presence of NaCl are parallel to those in its absence. Thus, NaCl seemed to modify the enzyme activity by shifting these curves toward higher K_m and V_m within the pH range examined. These results were similar to those obtained with alkaline phosphatase of animal origin (Fernley and Walker, 1965).

DISCUSSION

Table 4 summarizes the localizations and the salt modifications of the phosphohydrolases isolated from slightly halophilic *Vibrio alginolyticus*. Among the enzymes catalyzing similar chemical reactions, various kinds of salt modifications of enzymes were observed. For example, 3'-nucleotidase and 5'-nucleotidase required salts for their activity while acid phosphatase was only stable in the presence of high concentrations of salts. Pyrophosphatase, however, was inactivated and also inhibited by salts. Such a wide variation in the effect of salts on enzymes would not have been seen with the enzymes isolated from non-halophilic bacteria. The enzymes from extremely halophilic bacteria have been shown to require high concentrations of salt for both activity and stability (Brown, 1964; Larsen, 1967; Kushner, 1968). On the other hand, the enzymes from slightly halophilic marine bacteria have been shown to vary in their requirement for salt for their activity (MacLeod, 1968). The most characteristic property of the slight halophiles seems to be the coexistence of enzymes having different kinds of salt responses in the same cell.

The salt dependence of the enzymes appeared to be correlated with their location in the cell. The enzymes associated with the periplasmic space or the envelope structure seemed to be more halophilic than those of the cytoplasm. Although the precise location of the acid phosphatase was not determined,

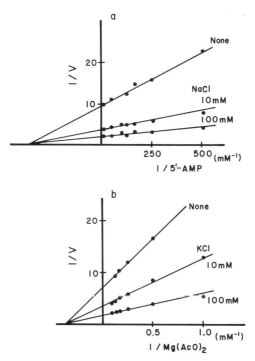

Fig. 11. Effect of salt on the kinetics of 5'-nucleotidase. *a,* The reaction mixture (1.0 ml) contained 50 mM Tris-acetate (pH 7.4), 50 mM magnesium acetate, enzyme, various concentrations of 5'-AMP, and NaCl as indicated above. The initial velocity (V) was expressed in m units/ml. *b,* The reaction mixture (1.0 ml) contained 50 mM Tris-acetate (pH 7.4), 0.1 mM 5'-AMP, enzyme, various concentrations of magnesium acetate, and KCl as indicated above. The initial velocity (v) was expressed in m units/ml.

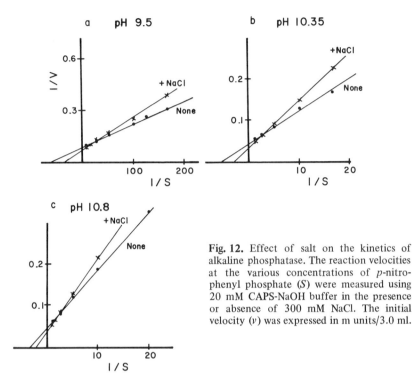

Fig. 12. Effect of salt on the kinetics of alkaline phosphatase. The reaction velocities at the various concentrations of p-nitrophenyl phosphate (S) were measured using 20 mM CAPS-NaOH buffer in the presence or absence of 300 mM NaCl. The initial velocity (v) was expressed in m units/3.0 ml.

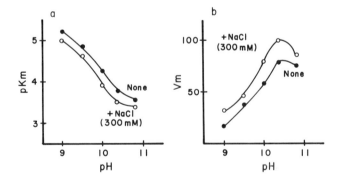

Fig. 13. Effect of pH on the K_m and V_m for the alkaline phosphatase in the presence and absence of NaCl. K_m was expressed in M and converted to $-\log K_m (pK_m)$. V_m values were expressed as relative activities.

Table 4. Summary of the localizations and salt modifications of phosphohydrolases isolated from *Vibrio alginolyticus*

Enzyme	Localization	Type of modification	Order of effectiveness
3'-Nucleotidase	Periplasm	Activation	$Cl^- > Br^- = I^- > NO_3^-$
5'-Nucleotidase	Envelope	Pre-activation	$NH_4^+ \gg Li^+ > Na^+ > K^+ = Rb^+ = Cs^+$
		Activation	$Cl^- = Br^- > I^- > NO_3^-$
Alkaline phosphatase	Periplasm	Activation	$Cl^- = Br^- = NO_3^- > none > SCN^- > ClO_4^-$
Acid phosphatase	Cytoplasm[a]	Stabilization	$Cl^- > Br^- > NO_3^- > none > SCN^-$
Pyrophosphatase	Cytoplasm	Inactivation	$Cl^- < Br^- < NO_3^- < ClO_4^- < SCN^-$
		Inhibition	$Li^+ > Na^+ > K^+ = Rb^+$

[a] Not precisely determined at present.

this enzyme is probably located at the inner side of the envelope. Alkaline phosphatase was not inhibited by high concentrations of salt (0.5–1.0 M). On the other hand, pyrophosphatase was inactivated and inhibited by salt. Most of the cytoplasmic enzymes of the tricarboxylic acid cycle have been reported to be active in the absence of added salt, but some of them were stimulated by the presence of salt (MacLeod, 1968). High concentrations of salt were inhibitory to these enzymes. Thus, the enzymes of cytoplasmic origin seemed to show no strict requirement of salt for activity. The evaluation of the relationship between the location and salt effect awaits further studies on other enzymes.

Lanyi and Stevenson (1970) pointed out the importance of hydrophobic interactions in the activity and stability of menadione reductase from extreme halophiles. Two mechanisms for the action of salts, namely charge-shielding effect at low concentrations (0.1–0.2 M) and salting-out effect at high concentrations (3–4 M) have been discussed in detail with respect to cytochrome oxidase (Lieberman and Lanyi, 1971) and threonine deaminase (Lieberman and Lanyi, 1972). As indicated in Table 4, the phosphohydrolases of the slightly halophilic *Vibrio alginolyticus* showed a specificity for monovalent anions. Thus, the charge-shielding effect is not sufficient to account for the effect of the salts. Although the salt concentrations required were not as high as in the case of extreme halophiles, the anion specificity was demonstrated to follow the lyotropic series, $Cl^- > Br^- > NO_3^- > ClO_4^- > SCN^-$, the order of effectiveness in activating and stabilizing the enzymes being the same as the order of salting-out power, and the order of effectiveness in inactivating the pyrophosphatase being the same as the order of salting-in power. These results suggest that the effect of salt in stabilizing hydrophobic interactions in the protein molecule should also be considered with respect to enzymes of slight halophiles.

The SO_4^{2-} ion acted differently, depending on the enzymes. For example, SO_4^{2-} showed no effect on the activities of cyclic phosphodiesterase (3'-nucleotidase) and 5'-nucleotidase, but it was more inhibitory than Cl^- for alkaline phosphatase and pyrophosphatase. In the stabilization of acid phosphatase, SO_4^{2-} was more effective than Cl^-. These results cannot be explained on its lyotropic nature alone.

Cation specificity was observed in the pre-activation of 5'-nucleotidase and in the inhibition of pyrophosphatase. It has been shown that Li^+ and Na^+ are much more effective than K^+ and Rb^+ in the prevention of cell lysis of halophilic bacteria (Christian and Ingram, 1959; Abram and Gibbons, 1961; MacLeod and Matula, 1962; Buckmire and MacLeod, 1965; De Voe and Oginsky, 1969). Thus there might be some structures in the enzymes sensitive to monovalent cations. In this case, the differences in the polarizing power of monovalent cations seems to be one of the important factors.

Frieden (1964) presented a useful method for the interpretation of kinetic data in the single substrate-single modifier case. Suppose an inorganic ion modifies the activity by binding to a site other than the active sites of the enzyme; the modifications of the activity may be explained by the following mechanism:

(1) $E + S$ = ES; $ES \xrightarrow{\ k_5\ } E + P$

(2) $E + M$ = EM;

(3) $ES + M$ = EMS; $EMS \xrightarrow{\ k_6\ } EM + P$

(4) $EM + S$ = EMS

S is the substrate and M is the modifier. The general form of the equation for the above mechanism derived by the rapid equilibrium treatment is as follows:

$$\frac{V_0}{(E)_0} = \frac{k_5(1 + k_6 M/k_5 K_3)/(1 + M/K_3)}{1 + \dfrac{K_1\,(1 + M/K_2)}{S\,(1 + M/K_3)}}$$

where K_1 to K_4 are the dissociation constants describing steps (1)–(4), and k_5 and k_6 are the rate constants for the breakdown of the ES and EMS complexes, respectively. When this mechanism was applied to the present results, the activation of 3'-nucleotidase by Cl⁻ (Fig. 10a) was found to correspond to the case where $K_2/K_3 > k_5/k_6$ and $K_2 = K_3$. From Fig. 10a and Table 3, the following kinetic constants for 3'-AMP hydrolysis can be obtained: $K_2 = K_3 = 20$ mM; $K_1 = K_4 = 0.88$ mM; $k_5 = 11$ units/mg protein and $k_6 = 170$ units/mg protein. The inhibition of the cyclic phosphodiesterase by Cl⁻ (Fig. 10b) was found to correspond to the case where $K_2/K_3 <$ k_5/k_6 and $K_2 = K_3$. The kinetic constants are as follows: $K_2 = K_3 = 40$ mM; $K_1 = K_4 = 0.36$ mM; $k_5 = 1560$ units/mg protein and $k_6 = 600$ units/mg protein. Thus, according to this mechanism, the binding of the substrate (or the modifier) to the enzyme is unaffected by the presence of modifier (or the substrate) and the activation in 3'-nucleotidase and the inhibition in the cyclic phosphodiesterase are caused by the differences in the rate constants for the breakdown of ES (k_5) and EMS (k_6), which represent the rate-determining steps in this mechanism. From this, it can be concluded that Cl^- inversely modifies the two activities in the same protein molecule by effecting configurational changes in the enzyme structure so as to activate the $3'$-nucleotidase on the one hand and to inhibit the cyclic phosphodiesterase on the other. The relationship between the two active sites has been reported by Unemoto *et al.* (1969).

Similarly, the results on the alkaline phosphatase (Fig. 12) correspond to the case where $K_2/K_3 < k_5/k_6$; $K_2 < K_3$; and $k_5 < k_6$. This means that the substrate is less tightly bound to the enzyme in the presence of the modifier than in its absence. This is also true for the binding of the modifier to the enzyme. Since $k_5 < k_6$, the modifier acts as an activator only in the presence of saturation concentrations of the substrate. At low substrate concentrations, the modifier acts as an inhibitor by increasing K_m for the substrate.

Actually, the true nature of the modified enzymes (EMS) cannot be inferred from the kinetic analyses, and further studies on the interactions between the enzyme proteins and inorganic ions, especially monovalent anions, are necessary to explain the mechanism of salt modifications in the enzymes from slightly halophilic marine bacteria.

More recent work on the mode of salt modification on alkaline phosphatase and pyrophosphatase is given in the papers by Hayashi *et al.* (1973) and Unemoto *et al.* (1973).

LITERATURE CITED

Abram, D., and N. E. Gibbons. 1961. The effect of chlorides of monovalent cations, urea, detergents, and heat on morphology and the turbidity of suspensions of red halophilic bacteria. Can. J. Microbiol. 7:741–750.

Allen, R. J. L. 1940. The estimation of phosphorus. Biochem. J. 34B:858.

Baumann, P., L. Baumann, and M. Mandel. 1971. Taxonomy of marine bacteria: The genus *Beneckea*. J. Bacteriol. 107:268–294.

Brown, A. D. 1964. Aspects of bacterial response to the ionic environment. Bacteriol. Rev. 28:296–329.

Buckmire, F. L. A., and R. A. MacLeod. 1965. Nutrition and metabolism of marine bacteria. XIV. On the mechanism of lysis of a marine bacterium. Can. J. Microbiol. 11:677–691.

Christian, J. H. B., and M. Ingram. 1959. Lysis of *Vibrio costicolus* by osmotic shock. J. Gen. Microbiol. 20:32–42.

Costerton, J. W., C. Forsberg, T. I. Matula, F. L. A. Buckmire, and R. A. MacLeod. 1967. Nutrition and metabolism of marine bacteria. XVI. Formation of protoplasts, spheroplasts, and related forms from a gram-negative marine bacterium. J. Bacteriol. 94:1764–1777.

De Voe, I. W., and E. L. Oginsky. 1969. Antagonistic effect of monovalent cations in maintenance of cellular integrity of a marine bacterium. J. Bacteriol. 98:1335–1367.

Drapeau, G. R., and R. A. MacLeod. 1963. Nutrition and metabolism of marine bacteria. XII. Ion activation of adenosine triphosphatase in membranes of marine bacterial cells. J. Bacteriol. 85:1413–1419.

Fernley, H. N., and P. G. Walker. 1965. Kinetic behaviour of calf-intestinal alkaline phosphatase with 4-methylumbelliferyl phosphate. Biochem. J. 97:95–103.

Forsberg, C. W., J. W. Costerton, and R. A. MacLeod. 1970a. Separation and localization of cell wall layers of a gram-negative bacterium. J. Bacteriol. 104:1338–1353.

Forsberg, C. W., J. W. Costerton, and R. A. MacLeod. 1970b. Quantitation, chemical characteristics, and ultrastructure of the three outer cell wall layers of a gram-negative bacterium. J. Bacteriol. 104:1354–1368.

Frieden, C. 1964. Treatment of enzyme kinetic data. I. The effect of modifiers on the kinetic parameters of single substrate enzymes. J. Biol. Chem. 239:3522–3531.

Hayashi, M., and R. Uchida. 1965. A cation activated adenosine triphosphatase in cell membranes of halophilic *Vibrio parahaemolyticus*. Biochim. Biophys. Acta 110:207–209.

Hayashi, M., T. Unemoto, and M. Hayashi. 1973. pH- and anion-dependent salt modifications of alkaline phosphatase from a slightly halophilic *Vibrio alginolyticus*. Biochim. Biophys. Acta 315:83–93.

Hayashi, M., T. Unemoto, Y. Kozuka, and M. Hayashi. 1970. Anion-activated 5′-nucleotidase in cell envelopes of a slightly halophilic *Vibrio alginolyticus*. Biochim. Biophys. Acta 220:244–255.

Karnovsky, M. J. 1961. Simple methods for "staining with lead" at high pH in electron microscopy. J. Biophys. Biochem. Cytol. 11:729–732.

Kushner, D. J. 1968. Halophilic bacteria. Advan. Appl. Microbiol. 10:73–99.

Lambert, S. M., and J. I. Watters. 1957. The complexes of magnesium ion with pyrophosphate and triphosphate ions. J. Amer. Chem. Soc. 79:5606–5608.

Lanyi, J. K., and J. Stevenson. 1970. Studies of the electron transport chain of extremely halophilic bacteria. IV. Role of hydrophobic forces in the structure of menadione reductase. J. Biol. Chem. 245:4074–4080.

Larsen, H. 1967. Biochemical aspects of extreme halophilism. *In* A. H. Rose and J. F. Wilkinson (eds), Advances in Microbial Physiology, Vol. 1, p. 97. Academic Press, New York.

Lieberman, M. M., and J. K. Lanyi. 1971. Studies of the electron transport chain of extremely halophilic bacteria. V. Mode of action of salts on cytochrome oxidase. Biochim. Biophys. Acta 245:21–33.

Lieberman, M. M., and J. K. Lanyi. 1972. Threonine deaminase from extremely halophilic bacteria. Cooperative substrate kinetics and salt dependence. Biochemistry 11:211–216.

MacLeod, R. A. 1968. On the role of inorganic ions in the physiology of marine bacteria. *In* M. R. Droop and E. J. Ferguson Wood (eds), Advances in Microbiology of the Sea, p. 95. Academic Press, New York.

MacLeod, R. A., C. A. Claridge, A. Hori, and J. F. Murray. 1958. Observations on the function of sodium in the metabolism of a marine bacterium. J. Biol. Chem. 232:829–834.

MacLeod, R. A., and A. Hori. 1960. Nutrition and metabolism of marine bacteria. VIII. Tricarboxylic acid cycle enzymes in a marine bacterium and their response to inorganic salts. J. Bacteriol. 80:464–471.

MacLeod, R. A., A. Hori, and S. M. Fox. 1960. Nutrition and metabolism of marine bacteria. IX. Ion requirements for obtaining and stabilizing isocitric dehydrogenase from a marine bacterium. Can. J. Biochem. Physiol. 38:693–701.

MacLeod, R. A., and T. I. Matula. 1962. Nutrition and metabolism of marine bacteria. XI. Some characteristics of the lytic phenomenon. Can. J. Microbiol. 8:883–896.

Murray, R. G. E., P. Steed and H. E. Elson. 1965. The location of the mucopeptide in sections of the cell wall of *Escherichia coli* and other gram-negative bacteria. Can. J. Microbiol. 11:547–560.

Neu, H. C., and J. Chou. 1967. Release of surface enzymes in Enterobacteriaceae by osmotic shock. J. Bacteriol. 94:1934–1945.

Neu, H. C., and L. A. Heppel. 1965. The release of enzymes from *Escherichia coli* by osmotic shock and during the formation of spheroplasts. J. Biol. Chem. 240:3685–3692.

Nossal, N. G., and L. A. Heppel. 1966. The release of enzymes by osmotic shock from *Escherichia coli* in exponential phase. J. Biol. Chem. 241:3055–3062.

Ryter, A., and E. Kellenberger. 1958. Etude au microscope electronique de plasma contenant de l'acide desoxyribonucleique. I. Les nucleotides des bacteries en croissance active. Z. Naturforsch. B 13:597–605.

Thompson, J., J. W. Costerton, and R. A. MacLeod. 1970. K^+-Dependent deplasmolysis of a marine pseudomonad plasmolyzed in a hypotonic solution. J. Bacteriol. 102:843–854.

Thompson, J., M. L. Green, and F. C. Happold. 1969. Cation-activated nucleotidase in cell envelopes of marine bacterium. J. Bacteriol. 99:834–841.

Thompson, J., and R. A. MacLeod. 1971. Function of Na^+ and K^+ in the active transport of α-aminoisobutyric acid in a marine pseudomonad. J. Biol. Chem. 246:4066–4074.

Unemoto, T., and M. Hayashi. 1969. Chloride ion as a modifier of $2',3'$-cyclic phosphodiesterase purified from halophilic *Vibrio alginolyticus*. Biochim. Biophys. Acta 171:89–102.

Unemoto, T., F. Takahashi, and M. Hayashi. 1969. Relationship between the active sites of $2',3'$-cyclic phosphodiesterase with $3'$-nucleotidase activity purified from *Vibrio alginolyticus.* Biochim. Biophys. Acta 185:134–142.

Unemoto, T., M. Tanaka, and M. Hayashi. 1973. Effect of free magnesium and salts on the inorganic pyrophosphatase purified from a slightly halophilic *Vibrio alginolyticus.* Biochim. Biophys. Acta 327: 490–500.

II

TEMPERATURE EFFECTS AND INTERACTIONS

TEMPERATURE EFFECTS ON MARINE MICROORGANISMS

RICHARD Y. MORITA

Department of Microbiology and School of Oceanography
Oregon State University

BACKGROUND INFORMATION

All organisms have a maximum, minimum, and optimum temperature for growth. The environmental temperature is directly related to the ability of the organisms to function. However, many microorganisms grow better at temperatures from 10 to 20°C above the environmental temperature from which they were isolated (ZoBell, 1961; Braarud, 1961). Furthermore, the effect of temperature on the physical and chemical environment in which microbes function should not be overlooked.

Water covers approximately 72% of the earth's surface and most of the water is marine. More than 90% (by volume) of the marine environment is colder than 5°C (ZoBell, 1962). The range in temperature found in this environment is from −3°C to about 42°C in the tropics (Sverdrup *et al.*, 1946). The water mass above the thermocline is characterized by variation in temperature, depending upon geographic location, whereas the area below the thermocline is uniformly cold (−1.5 to 4.5°C).

The general response of microorganisms to temperature follows the Q_{10} rule. The literature on the effect of temperature on microorganisms is well documented by Rose (1967), Ingraham (1962), and Johnson *et al.* (1954), and on marine psychrophilic bacteria by Morita (1966). However, other aspects of temperature should not be overlooked, such as its effect on water structure (Klotz, 1965) and the state of water within cells (Ling, 1967). A review of the temperature effect of marine microorganisms is presented by Oppenheimer (1970).

The maximum growth temperature for bacteria taken from tropical and temperate waters (except the thermophiles) is only a few degrees above their environmental temperature (ZoBell, 1962). Although spore-forming thermophilic bacteria (McBee and McBee, 1956; Bartholomew and Rittenberg, 1949; Bartholomew and Paik, 1966) and sulfate-reducing bacteria (Trüper, 1969) have been isolated from the marine environment, they are probably inactive except in a few isolated localities such as the Red Sea where geothermal activity takes place at the bottom.

The thermosensitivity of marine bacteria was first demonstrated by ZoBell and Conn (1940). Most marine microorganisms could not stand the plating temperature of agar (45°C). Heating of seawater and sediment samples to 30°C for 10 min was sufficient to kill approximately 25% of the bacteria, while a temperature of 40°C for 10 min killed 80% of the bacteria. An obligate psychrophile, *Vibrio marinus* MP-1, in nutrient medium was rendered non-viable when exposed to 28.9°C for 6.25 hr (Morita and Haight, 1964). Hagen *et al.* (1964) noted that their obligate psychrophile was rendered non-viable when exposed to temperatures of 20–30°C. Because of their thermosensitivity, most marine bacteria can be classified as facultative psychrophiles or obligate psychrophiles. Morita (1966) presents a discussion of the various terms employed to designate cold-loving organisms.

Much of the data obtained on marine bacteria have been the result of employing temperatures from close to room temperature to about 28°C (ZoBell, 1946; Wood, 1965; Oppenheimer, 1968). Such studies are appropriate with reference to the warmer surface waters as well as near-shore environments in temperate climates. However, in considering the cold environment of the ocean, studies indicate that cells should be grown at their environmental temperature if inferences are to be made concerning the deep sea environment (J. J. Haight and Morita, 1966; Albright and Morita, 1972). Growth of organisms at the optimum temperature for growth differs from that at 4 or 5°C, as evidenced by the ability of the cells to take up oxygen in the presence of glucose, viability retention under thermal stress, leakage of intracellular material, serine deamination under pressure, and substrate saturation.

Many concepts have been advanced as to why organisms of the various thermal groups cannot tolerate elevated temperatures which eventually lead to the death of the cell. These concepts are (1) the disruption of intracellular organization (Ingraham and Bailey, 1959), (2) accelerated use of the intracellular amino acid pool (Hagen and Rose, 1962), (3) changes in extent of cellular lipid saturation (Kates and Hagen, 1964), (4) enzyme inactivation by elevated temperatures (Edwards and Rettger, 1937; Langridge and Morita, 1966), (5) inactivation of enzyme-forming system(s) (Nashif and Nelson, 1953; Upadhay and Stokes, 1963), (6) loss of permeability control (Morita and Burton, 1963), and (7) thermally induced leakage of intracellular components (R. D. Haight and Morita, 1966; Kenis and Morita, 1968). The question is whether all these events occur at one time or whether there is a sequence of events in the inability of certain thermal groups to withstand thermal stress. Equally important is the answer to the question as to why obligately psychrophilic bacteria function so well at environmental temperatures which, at times, can be as great as 20°C below their optimum for growth.

Since the major portion of the marine biosphere is cold, the obligately psychrophilic marine bacteria are now receiving more attention. They have been found below the thermocline in many areas as well as at both polar regions (Morita and Haight, 1964; Sieburth, 1967; Morita and Burton, 1971; Morita et al., 1971). Low temperature is not inimical to their growth and large cell numbers can be obtained in a short time (Morita and Albright, 1965). Data being accumulated by various investigators definitely indicate that psychrophiles play a major role in the cycles of matter. If cold water is upwelled, there is also the possibility that thermal lysis may release into the water various ectocrine compounds necessary for other organisms.

In the intertidal region, especially where the seawater is diluted with freshwater run-off, the salinity-temperature interaction becomes very important. Stanley and Morita (1968) demonstrated that the maximum growth temperature varies with the salinity. All the salt-requiring organisms that have been tested in our laboratory display the salinity-temperature relationship with reference to the maximum growth temperature. The interaction has been further studied by Cooper and Morita (1972) in terms of the net protein synthesis. Within limits, the protein synthetic mechanism can tolerate a higher temperature when the salinity is increased, and the same can be said for the synthesis of RNA from uracil.

LITERATURE CITED

Albright, L. J., and R. Y. Morita. 1972. Effects of environmental parameters of low temperature and hydrostatic pressure on L-serine deamination by Vibrio marinus. J. Oceanogr. Soc. Jap. 28:63–70.

Bartholomew, J. W., and G. Paik. 1966. Isolation and identification of obligate thermophilic sporeforming bacilli from ocean basin cores. J. Bacteriol. 92:635–638.

Bartholomew, J. W., and S. C. Rittenberg. 1949. Thermophilic bacteria from deep ocean bottom cores. J. Bacteriol. 57:659.

Braarud, T. 1961. Cultivation of marine organisms as means of understanding environmental influences on populations. Oceanography, AAAS Publ. No. 67:271–298.

Cooper, M. F., and R. Y. Morita. 1972. Interaction of salinity and temperature on net protein synthesis and viability of *Vibrio marinus*. Limnol. Oceanogr. 17:556–565.

Edwards, O. F., and L. R. Rettger. 1937. The relationship of certain respiratory enzymes to the maximum growth temperatures of bacteria. J. Bacteriol. 34:489–515.

Hagen, P. O., D. J. Kushner, and N. E. Gibbons. 1964. Temperature-induced lysis in a psychrophilic bacterium. Can. J. Microbiol. 10:813–822.

Hagen, P. O., and A. H. Rose. 1962. Studies on the biochemical basis of low maximum temperature in a psychrophilic *Cryptococcus*. J. Gen. Microbiol. 27:89–99.

Haight, J. J., and R. Y. Morita. 1966. Some physiological differences in *Vibrio marinus* growth at environmental and optimal temperatures. Limnol. Oceanogr. 11:470–474.

Haight, R. D., and R. Y. Morita. 1966. Thermally induced leakage from *Vibrio marinus,* an obligately psychrophilic marine bacterium. J. Bacteriol. 92:1388–1393.

Ingraham, J. L. 1962. Temperature relationship. *In* I. C. Gunsalus and R. Y. Stanier (eds), The Bacteria: A Treatise on Structure and Function, Vol. 4, pp. 265–296. Academic Press, New York.

Ingraham, J. L., and G. R. Bailey. 1959. Comparative study of the effect of temperature on metabolism of psychrophilic and mesophilic bacteria. J. Bacteriol. 77:609–613.

Johnson, F. H., H. Eyring, and M. J. Polissar. 1954. The Kinetic Basis of Molecular Biology. John Wiley, New York.

Kates, M., and P. O. Hagen. 1964. Influence of temperature on fatty acid composition of psychrophilic and mesophilic *Serratia* species. Can. J. Biochem. 42:481–488.

Kenis, P. R., and R. Y. Morita. 1968. Thermally induced leakage of cellular material and viability in *Vibrio marinus,* a psychrophilic marine bacterium. Can. J. Microbiol. 14:1239–1244.

Klotz, I. M. 1965. Role of water structure in macromolecules. Fed. Proc. 24:S24–S33.

Langridge, P., and R. Y. Morita. 1966. Thermolability of malic dehydrogenase from the obligate psychrophile *Vibrio marinus*. J. Bacteriol. 92:418–423.

Ling, G. N. 1967. Effects of temperature on the state of water in the living cell. *In* A. H. Rose (ed.), Thermobiology, pp. 5–24. Academic Press, New York.

McBee, R. H., and V. H. McBee. 1956. The incidence of thermophilic bacteria in Arctic soils and waters. J. Bacteriol. 71:182–185.

Morita, R. Y. 1966. Marine psychrophilic bacteria. Oceanogr. Mar. Biol. Annu. Rev. 4:105–121.

Morita, R. Y., and S. D. Burton. 1963. Influence of moderate temperature on growth and malic dehydrogenase activity of a marine psychrophile. J. Bacteriol. 86:1025–1029.

Morita, R. Y., and S. D. Burton. 1971. Occurrence, possible significance and metabolism of obligate psychrophiles in marine waters. *In* D. W. Hood (ed.), Organic Matter in Natural Waters, pp. 275–285. Inst. Mar. Sci. Univ. Alaska Publ. No. 1.

Morita, R. Y., P. A. Gillespie, and L. P. Jones. Microbiology of Antarctic sea water. Antarct. J. U.S. 7:157.

Morita, R. Y., and R. D. Haight. 1964. Temperature effects on growth of an obligately psychrophilic bacterium. Limnol. Oceanogr. 9:103–106.

Nashif, S. A., and F. E. Nelson. 1953. The lipase of *Pseudomonas fragii*. II. Factors affecting lipase production. J. Dairy Sci. 36:471–480.

Oppenheimer, C. H. (ed.) 1968. Marine Biology, Vol. 4, Unresolved Problems in Marine Microbiology. New York Academy of Sciences, New York.

Oppenheimer, C. H. 1970. Temperature. *In* O. Kinne (ed.), Marine Ecology, Vol. 1, Pt. 1, pp. 347–361. John Wiley (Interscience), New York.

Rose, A. H. 1967. Thermobiology. Academic Press, New York.

Sieburth, J. McN. 1967. Seasonal selection of estuary bacteria by water temperature. J. Exp. Mar. Biol. Ecol. 1:98–121.

Stanley, S. O., and R. Y. Morita. 1968. Salinity effect on the maximal growth temperature of some bacteria isolated from marine environments. J. Bacteriol. 95:169–173.

Sverdrup, H. U., M. W. Johnson, and R. H. Fleming. 1946. The Oceans, p. 843. Prentice-Hall, New Jersey.

Trüper, H. G. 1969. Bacterial sulfate reduction in the Red Sea hot brines. *In* E. T. Degens and D. A. Ross (eds), Hot Brines and Recent Heavy Metal Deposits in the Red Sea, pp. 263–271. Springer-Verlag, Berlin.

Upadhay, J., and J. L. Stokes. 1963. Temperature-sensitive formic hydrogen-lyase in a psychrophilic bacterium. J. Bacteriol. 85:177–185.

Wood, E. J. F. 1965. Marine Microbial Ecology. Chapman & Hall-Reinhold, New York.

ZoBell, C. E. 1946. Marine Microbiology. Chronica Botanica Co., Waltham, Mass.

ZoBell, C. E. 1962. Importance of microorganisms in the sea. *In* Proc. Low Temp. Microbiol. Symp., Camden, New Jersey, pp. 107–132. Campbell Soup Co., Camden, New Jersey.

ZoBell, C. E., and J. E. Conn. 1940. Studies on the thermosensitivity of marine bacteria. J. Bacteriol. 40:223–238.

TEMPERATURE-SALINITY EFFECTS UPON THE GROWTH OF MARINE BACTERIA

YUZABURO ISHIDA, AKIHIKO NAKAYAMA, and HAJIME KADOTA

Laboratory of Microbiology
Department of Fisheries
Kyoto University

Increased incubation temperatures resulted in an increase in the minimum and optimum salinities for the growth of several bacterial strains isolated from seawater and from solar salts. However, a similar effect was not observed for several strains from marine mud and for *Vibrio parahaemolyticus*. The uptake of [^{14}C]glutamic acid by cells of a moderate halophile, *Flavobacterium* sp. 126, and cells of a marine bacterium, *Pseudomonas* sp. 1055-1, in various salinities was influenced by the growth temperature but not by the incubation temperature during substrate uptake. The minimum salinity for the uptake of substrates by resting cells increased as the growth temperature of the cells was increased. The salinities required for the uptake of [^{14}C]serine and [^{14}C]proline were not influenced by either the growth or test temperature, although these amino acids were equivalent to glutamic acid in supporting the growth of this organism. The concentration of NaCl below which cell suspensions of an extreme halophile, *Brevibacterium* sp. 185, lost viability was increased when the temperature of the suspensions was increased. This was in contrast to the behavior of a marine pseudomonad and of a moderate halophile. The salinities below which loss of turbidity and release of UV-absorbing substances occurred were not influenced by temperature in any of the test organisms. The results suggested that the observations concerning the influence of temperature on the minimum and optimum salinities for growth could be best explained in terms of alterations in transport activity.

INTRODUCTION

In the littoral zone, where freshwater streams enter the sea, the microorganisms are often subjected to drastic changes in salinity and temperature. Ritchie (1957) showed that the growth of marine fungi was inhibited by the combination of low temperature and high salinity or by the combination of high temperature and low salinity and that the optimum salinities for their growth were raised by increased temperatures. According to Stanley and Morita (1968), the salinity of the growth medium had a marked effect on the maximum growth temperatures of four strains of bacteria isolated from marine sources. Similar observations have been made on several bacterial species isolated from foods (Ingram, 1957; Goldman *et al.,* 1963). This paper deals mainly with a comparison of the effects of temperature on the minimum and optimum salinities for growth of a strain of marine bacterium, a moderate halophile, and an extreme halophile. The mechanism of the observed temperature-salinity interaction is discussed.

MATERIALS AND METHODS

Organisms

The principal organisms employed were a marine bacterium, *Pseudomonas* sp. 1055-1 (Hidaka and Sakai, 1968), an extreme halophile, *Brevibacterium* sp. 185, from solar salt, and a moderate halophile, *Flavobacterium* sp. 126, also from solar salt. Other organisms studied were *Flavobacterium* sp. 164 (solar salt), *Micrococcus* sp. 149 (solar salt) (Ishida and Fujii, 1970; Ishida, 1970*a,b*), *Achromobacter* sp. F2a42 (marine mud) and *Pseudomonas* sp. B3a16 (marine mud) (Kurata and Kimata, 1968). In addition *Vibrio parahaemolyticus* 0-5 (Hidaka and Sakai, 1968) and *Vibrio* sp. Fl which was isolated from the intestinal content of red sea bream (Sera *et al.,* 1972) were also studied.

Culture Media

Medium I for bacteria isolated from solar salt was composed as follows: polypeptone, 7.5 g; beef extract, 5 g; $MgSO_4 \cdot 7H_2O$, 3 g; KCl, 2 g; $FeSO_4 \cdot 7H_2O$, 0.01 g; $MnCl_2 \cdot 4H_2O$, 0.1 mg. NaCl was added as required and distilled water to a total volume of 1000 ml (pH of the medium after autoclaving was between 7.3 and 7.5). The stock cultures were maintained on agar medium containing 5% NaCl at 20°C for a halotolerant bacterium, and 15% NaCl at 20°C for halophilic bacteria.

Medium II for the other strains of bacteria was composed as follows: polypeptone, 5 g; yeast extract, 1 g; K_2HPO_4, 0.1 g; $FeSO_4 \cdot 7H_2O$, 0.1 g. Natural seawater was added to a total volume of 1000 ml (pH 7.5).

Medium III was composed as follows: vitamin-free casamino acids or an amino acid, 5 g; K_2HPO_4, 0.01 g. Artificial seawater was added to a total volume of 1000 ml (pH 7.5). The composition of artificial seawater (ASW) was as follows: NaCl, 24 g; KCl, 0.7 g; $MgCl_2 \cdot 6H_2O$, 10.8 g; Na_2SO_4, 4.0 g; $CaCl_2 \cdot 2H_2O$, 1.2 g; 250 mg/liter $FeCl_3 \cdot 6H_2O$ solution, 0.3 ml; trace metals, 1 ml; distilled water to a total volume of 1000 ml. The trace metals mixture consisted of the following: $ZnSO_4 \cdot 7H_2O$, 50 mg; $MnCl_2 \cdot 4H_2O$, 400 mg; $CoCl_2 \cdot 6H_2O$, 1 mg; $CuSO_4 \cdot 5H_2O$, 0.4 mg; H_3BO_3, 2000 mg; $Na_2MoO_4 \cdot 2H_2O$, 500 mg; distilled water to a total volume of 1000 ml.

Stock cultures were maintained on medium II at 20°C.

Estimation of Growth Rate

A 10-ml volume of culture medium I or II was kept for 1 hr on a rotary shaker at each temperature tested, before inoculating with 0.01 ml of the culture harvested at the logarithmic phase. Growth rate was estimated by following the change in optical density (OD) as measured with a Leitz colorimeter (Model M). Growth rates were expressed as the reciprocal of the doubling time.

Viability, Lysis, and Release of UV-Absorbing Material

Cells of *Flavobacterium* sp. 126 and *Brevibacterium* sp. 185 were harvested in the mid-logarithmic phase from cultures growing at 37, 30, or 20°C and in a medium containing 15% NaCl. The cells were washed three times with 10 mM Tris-HCl buffer (pH 7.5) containing 15% NaCl and were suspended in this solution in sufficient volume to give an OD_{600} of 10.00. Aliquots of 2.5 ml of the concentrated suspension were added to flasks containing 47.5 ml of 10 mM Tris-HCl buffer (pH 7.5) and various concentrations of NaCl which had been preincubated at 37, 30, or 20°C. The test suspensions were incubated at 37, 30, and 20°C for periods of 0, 20, 40, and 80 min; each of these suspensions was then divided into two equal portions, one for estimating lysis and the other for determining the release of UV-absorbing materials. Lysis was estimated as the percentage decrease in the OD of the test suspension as compared with that in 15% NaCl. The OD was measured at 600 nm using a Hitachi spectrophotometer, Model 101. The absorbance at 260 nm of the supernatant of the second portion was measured to estimate the release of UV-absorbing materials. The viability of these cell suspensions after 20 and 40 min incubation was estimated using a two-tube, decimal dilution method.

The marine bacterium *Pseudomonas* sp. 1055-1 was cultured in medium III containing casein hydrolysate at 35 or 20°C using a rotary shaker. The cells were harvested in the logarithmic phase of growth and were washed twice with 10 mM Tris-HCl buffer (pH 7.3) containing 75% ASW. The final suspensions in 75% ASW were adjusted to an OD $_{600}$ of 30.00. Volumes of 1 ml of the concentrated suspension were dispensed to flasks containing 100 ml of buffer and various strengths of ASW which had been preincubated at the test temperatures. The test suspensions were incubated at 35 and 25°C for 0, 10, 25, 50, 100, and 180 min. The turbidity, release of UV-absorbing substances, and viability were estimated by the same methods as used with the halophiles.

Uptake of Amino Acids

The halophiles *Brevibacterium* sp. 185 and *Flavobacterium* sp. 126 were grown at 37, 30, and 20°C and were harvested in the mid-logarithmic phase of growth. The concentrated cell suspensions contained 1.2 mg dry weight of cells per milliliter. Volumes of 1.0 ml of the suspensions were added to 19.0 ml of 10 mM Tris-HCl buffer (pH 7.5), adjusted to a selected concentration of NaCl which contained 1.0 mM L-[U-^{14}C]glutamic acid (0.1 μCi/μmole) for strain 126 and 0.5 mM L-[U-^{14}C]glutamic acid (0.25 μCi/μmole) for strain 185. The reaction mixtures were incubated at 0, 20, 30, and 37°C*; samples of 1.0 ml were withdrawn at intervals. The cells were collected on a Millipore filter (0.22 μ). The filters were washed three times with 2.0 ml portions of cold buffer and were dried at 60°C for 40 min. Substrate uptake was estimated by counting the filters in a liquid scintillation counter (Aloka LSC-502). The scintillation fluid was scintillator 1 (PPO, 4 g; POPOP, 100 mg; toluene, 1000 ml). The marine bacterium, *Pseudomonas* sp. 1055-1, was grown at 20 and 35°C and was harvested in the mid-logarithmic phase. The concentrated cell suspension had an OD $_{600}$ of 2.00–3.00. Volumes of 0.1 ml of the suspension were added to 1.75 ml of 10 mM Tris-HCl buffer (pH 7.5) containing various concentrations of ASW, 0.05 ml of an L-[U-^{14}C]amino acid (0.5 μCi/ml) and 0.1 ml of 1.0 mM L-amino acid in each tube. The reaction mixtures were incubated at 0, 20, and 35°C; after 10-min incubation the total volume was filtered through a membrane filter (Millipore 0.22 μ) and the collected cells were treated as described above. [U-^{14}C] protein hydrolysate (Radiochemical Centre, Amersham) was used without adding carrier hydrolysate.

---

* During periods of incubation, these two halophiles scarcely evolved carbon dioxide from glutamic acid.

RESULTS

Effects of Temperature on Growth

The minimum and optimum salinities for the growth of *Pseudomonas* sp. 1055-1 and *Vibrio* sp. F1 were increased when the growth temperature was increased. Similar observations were made with the halophilic bacteria *Flavobacterium* sp. 126 and 164 and *Brevibacterium* sp. 185 (Fig. 1). Such a temperature-salinity relationship was not observed with *Vibrio parahaemolyticus, Achromobacter* sp. F2a42, or *Pseudomonas* sp. B3a16 (Fig. 2). The minimum and optimum salinities for the growth of these organisms was not affected by the incubation temperature.

Viability, Lysis, Release of UV-Absorbing Substances

When the concentration of NaCl in suspensions of *Brevibacterium* sp. 185 was decreased below 5% at 20°C a loss in viability occurred; however, at 30°C the

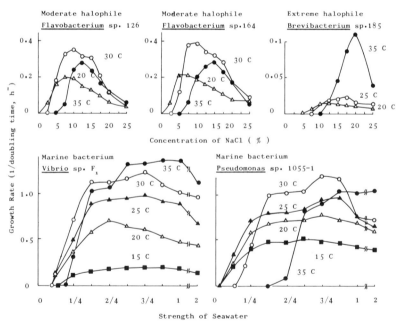

Fig. 1. Growths of *Pseudomonas* sp. 1055-1 and *Vibrio* sp. F1 (marine bacteria), *Flavobacterium* sp. 126 and *Flavobacterium* sp. 164 (moderately halophilic bacteria), and *Brevibacterium* sp. 185 (an extremely halophilic bacterium) at different temperatures on media containing various concentrations of NaCl or seawater.

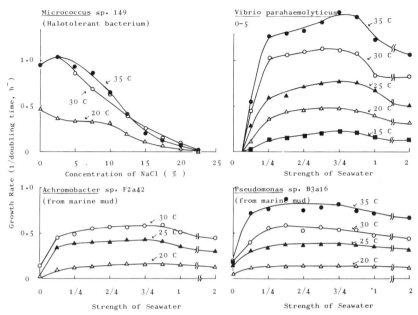

Fig. 2. Growths of *Achromobacter* sp. F2a42 and *Pseudomonas* sp. B3a16 (isolated from marine mud), *Micrococcus* sp. 149 (a halotolerant bacterium), and *Vibrio parahaemolyticus* 0-5 at different temperatures on media containing various concentrations of NaCl or seawater.

limiting concentration was 7.5% and at 37°C it was 15%. These results corresponded to the effect of temperature on optimum and minimum salinities for growth of this organism. No other temperature effects on viability were observed in either the moderate halophile or the marine pseudomonad. The limiting salinities below which the turbidity decreased or UV-absorbing substances were released were unaffected by the growth temperature or test temperature with the extreme halophile, the moderate halophile, or the marine bacterium.

Uptake of Amino Acids

Although the viability of *Brevibacterium* sp. 185 was largely lost at 37°C in concentrations of NaCl below 10%, the ability to take up [^{14}C] glutamic acid was retained. The optimum and minimum concentration of NaCl for glutamate uptake was increased in cells grown at higher temperatures (Table 1). Similarly in *Flavobacterium* sp. 126 the minimum concentration of NaCl required for the uptake of glutamic acid increased as the temperature at which the cells were grown was increased (Table 2). The temperature at

Table 1. Effect of growth and incubation temperatures on uptake of [14C]glutamic acid by *Brevibacterium* sp. 185, an extreme halophile

| Temperature (°C) | | Uptake rate ($10^{-3} \times$ μmole/60 μg dry wt./40 min)[a] | | | | | | | | |
| Growth | Incubation | Concentration of NaCl in Tris buffer (%) | | | | | | | | |
		0	2.5	5	7.5	10	12.5	15	20	25
37	30	0	0	0.25	—	0.97	—	2.02	1.33	0.85
30	37	0	0.13	0.42	0.68	0.68	1.25	0.92	0.80	0.34
	30	0	0.20	0.52	0.76	0.86	0.90	0.86	0.40	0.15
	20	0	0.08	0.26	0.82	0.52	0.52	0.18	0	0
20	30	0	0.10	0.35	—	0.88	—	1.20	0.58	0.45

[a] These values were subtracted from values at 0°C.

Table 2. Effect of growth and incubation temperatures on uptake of [^{14}C] glutamic acid by *Flavobacterium* sp. 126, a moderate halophile

Temperature (°C)		Uptake rate ($10^{-3} \times \mu$mole/60 μg dry wt./10 min)[a]						
		Concentration of NaCl in Tris buffer (%)						
Growth	Incubation	0	2.5	5	10	15	20	25
37	30	0	0	0	1.10	1.10	0.95	0.65
30	37	0	0	1.30	1.65	3.25	2.62	1.50
	30	0	0	2.60	3.33	2.95	2.20	1.80
	20	0	0	0.42	1.05	0.45	0.55	0.24
20	30	0	0.30	2.80	5.60	4.70	3.00	1.65

[a] These values were subtracted from values at 0°C.

which the uptake was observed did not affect the minimum and optimum salinities for uptake of the substrate. Uptake of [^{14}C] protein hydrolysate at various salinities into the cells of *Pseudomonas* sp. 1055-1, a marine bacterium, was markedly influenced by the temperature employed for the growth, but not by the salinity of the growth medium, and by the incubation temperature of the uptake experiment (Table 3). In this case, two minimum salinities for the uptake into the cells were observed when the cells were

Table 3. Effect of growth and incubation temperatures on uptake of [^{14}C] protein hydrolysate into cells of a marine bacterium *Pseudomonas* sp. 1055-1 suspended in Tris buffer containing various strengths of seawater

Salinity of growth (strength)	Temperature (°C)		Uptake rate (cpm $\times 10^{-3}$/2.5 min)					
			Strength of seawater					
	Growth	Incubation	0	1/8	2/8	3/8	4/8	1
1	20	35	1.2	8.8	5.3	6.6	8.5	4.6
		20	1.2	8.6	8.7	9.8	9.8	6.6
		0	1.2	6.8	–	–	–	2.8
1/2	35	35	1.5	9.6	8.6	3.0	1.8	9.1
		25	1.5	6.7	6.1	2.7	0.8	2.4
1		35	1.3	8.0	9.6	5.5	4.1	9.3
		25	1.3	11.8	11.6	9.1	9.1	10.6

grown at 35 and 20°C. The first minimum salinities for uptake by the cells grown at 35 and 20°C were at one-half and one-eighth strengths of seawater respectively; the minimal salinities for uptake corresponded reasonably with those for growth at 35 and 20°C. The second minimum salinities for uptake at 35 and 20°C were at the zero concentration.

Uptakes of single [^{14}C]amino acids at various salinities into the cell of *Pseudomonas* sp. 1055-1 grown at 35°C were examined. The [^{14}C]amino acids used were glutamic acid, serine, proline, phenylalanine, and valine. The former three amino acids supported the growth of this organism as a sole source of carbon and nitrogen in natural seawater media, but the others did not. The results of this experiment revealed that all the amino acids employed were incorporated into the cells at 35 and 20°C, even at a salinity of one-fourth strength of seawater at which this organism did not grow at 35°C. Among the amino acids employed only [^{14}C]glutamic acid showed a similar behavior to [^{14}C]protein hydrolyzate with respect to the uptake into the cells.

The effect of various inhibitors on the glutamic acid uptake was examined at salinities of full and one-fourth strength of seawater in all the

Table 4. Effect of some inhibitors on uptake of [^{14}C]glutamic acid by *Pseudomonas* sp. 1055-1, a marine bacterium[a]

	Inhibition rate (%)			
	35°C-grown cells		20°C-grown cells	
Inhibitors	Strength of seawater in Tris buffer solution			
	1	1/4	1	1/4
None	0	0	0	0
0.1 mM g-Stronphantin	9	7	−6	−4
3.3 μg/ml Antimycine	12	0	−17	−20
0.5 mM 2,4-DNP	46	40	47	43
20 mM NaN$_3$	78	63	85	83
3.0 mM KCN	86	75	88	93
Cold	38	—	—	59

[a] Reactions were held at 20°C for 10 min, except for the cold reaction which was done at 0°C.

suspensions. The cells used were grown at 35 and 20°C. As shown in Table 4 the uptake of [^{14}C]glutamic acid was inhibited by uncouplers such as sodium azide, 2,4-dinitrophenol, and potassium cyanide, but not by g-stronphantin and antimycin. There was no significant difference between the inhibition rates regardless of the salinity or growth temperature.

DISCUSSION

The minimum salinities for the growth of several strains of marine bacteria and halophilic bacteria were increased by raising the temperature of incubation. This effect was previously observed with halophilic and halotolerant bacteria by Ingram (1957), Gibbons and Payne (1961), and Goldman et al. (1963). It was observed with fungi by Ritchie (1957) and Ritchie and Jacobsohn (1963). Ingram (1957) suggested that this phenomenon was possibly common to many species. Ritchie and Jacobsohn (1963) found the temperature-salinity relationship in euryhaline or facultatively osmophilic fungi which were not necessarily either terrestrial or marine. In the present work several isolates from marine mud as well as *Vibrio parahaemolyticus* did not show a temperature-salinity interaction. Possibly one would have been observed if higher incubation temperatures had been used.

Possibly the mechanism of the observed increases in the minimum and optimum salinities resulting from increased incubation temperatures is related to changes in the properties of the cell membranes which influence the uptake of substrates. This seems particularly true of the moderate halophile *Flavobacterium* sp. 126. However, with the marine bacterium *Pseudomonas* sp. 1055-1 some conflicting behavior has been observed. Glutamic acid, serine, proline, and casein hydrolysate could each support the growth of this organism; however, the optimum salinity for the uptake of [^{14}C]serine and [^{14}C]proline was not influenced by either the temperature of growth or by the temperature at which the substrate uptake was observed.

It should be emphasized that the temperature at which the cells were grown controlled the minimum salinity for substrate uptake, rather than the temperature at which uptake of substrate was observed. The percentage of unsaturated fatty acids in membrane lipids has been observed to be a function of the growth temperature in other species (Okuyama, 1969) and to influence the rates of substrate transport in *Escherichia coli* (Shairer and Overath, 1969). Salt-dependent conformational changes in the proteins of the cell membrane of *Halobacterium salinarium* have been reported which may be related to the temperature-salinity effects (Hsia *et al.*, 1971). However, the implication of the alteration of membrane function in this phenomenon is still in the speculative stage, and, in fact, with the extreme halophile *Brevibac-*

terium sp. 185 the temperature-salinity effects on growth may be explained by parallel effects on heat resistance.

LITERATURE CITED

Gibbons, N. E., and J. I. Payne. 1961. Relation of temperature and sodium chloride concentration to growth and morphology of some halophilic bacteria. Can. J. Microbiol. 7:483–489.

Goldman, M., R. H. Deibel, and C. F. Niven, Jr. 1963. Interrelationship between temperature and sodium chloride on growth of lactic acid bacteria isolated from meat-curing brines. J. Bacteriol. 85:1017–1021.

Hidaka, T., and M. Sakai. 1968. Comparative observation of the inorganic salt requirements of the marine and terrestrial bacteria. Bull. Misaki Mar. Biol. Inst. Kyoto Univ. No. 12:125–149.

Hsia, J. C., P. T. S. Wong, and D. H. MacLennan. 1971. Salt-dependent conformational changes in the cell membrane of *Halobacterium salinarium.* Biochem. Biophys. Res. Commun. 43:88–93.

Ingram, M. 1957. Microorganisms resisting high concentrations of sugars or salts. Symp. Soc. Gen. Microbiol. 7:90–133.

Ishida, Y. 1970*a.* Growth behavior of halobacteria in relation to concentration of NaCl and temperature of environments. Bull. Japan. Soc. Sci. Fish. 36:397–401.

Ishida, Y. 1970*b.* Growth patterns of halobacteria in media containing salts, sugars or other compounds. Bull. Japan. Soc. Sci. Fish. 36:481–486.

Ishida, Y., and T. Fujii. 1970. Isolation of halophilic and halotolerant bacteria from solar salt. Bull. Japan. Soc. Sci. Fish. 36:391–396.

Kurata, A., and M. Kimata. 1968. Studies on marine bacteria producing vitamin B_{12} . I. On the distribution of marine bacteria producing vitamin B_{12} and the vitamin production of them. Bull. Res. Inst. Food Sci. Kyoto Univ. No. 31:26–34.

Okuyama, H. 1969. Phospholipid metabolism in *Escherichia coli* after a shift in temperature. Biochim. Biophys. Acta 176:125–134.

Ritchie, D. 1957. Salinity optima for marine fungi affected by temperature. Amer. J. Botany 44:870–874.

Ritchie, D., and M. K. Jacobsohn. 1963. The effects of osmotic and nutritional variation on growth of a salt-tolerant fungus. *In* C. H. Oppenheimer (ed.), Symposium on Marine Microbiology, pp. 286–299. Charles C Thomas, Springfield, Illinois.

Sera, H., Y. Ishida, and H. Kadota. 1972. Bacterial flora in the digestive tracts of marine fish. IV. Effect of H^+ concentration and gastric juices on survival of the indigenous bacteria. Bull. Japan. Soc. Sci. Fish. 38:859–863.

Shairer, H. U., and P. Overath. 1969. Lipids containing transunsaturated fatty acids change the temperature characteristic of thiomethylgalactoside accumulation in *Escherichia coli.* J. Molec. Biol. 44:209–214.

Stanley, S. O., and R. Y. Morita, 1968. Salinity effect on the maximal growth temperature of some bacteria isolated from marine environments. J. Bacteriol. 95:169–173.

INFLUENCE OF TEMPERATURE AND SALINITY ON THE AMINO-ACID POOLS OF SOME MARINE PSEUDOMONADS

SIMON O. STANLEY and C. M. BROWN*

Dunstaffnage Marine Research Laboratory

The amino-acid pools of marine bacteria are a function of the environment in which cultures are grown and vary both in size and composition with such parameters as growth rate, nitrogen source, growth limiting substrate, temperature, and salinity. Medium salinity (NaCl concentration) has the most marked effect on amino-acid pools and, in general, the pool content increases with an increase in NaCl concentration. In cultures in which the medium NaCl concentration was suddenly increased there was a rapid synthesis of glutamate (and then proline), these changes occurring within minutes of NaCl addition. Conversely, on decreasing the salinity, there was a rapid alteration in pool composition which under some conditions resulted in the excretion of substantial quantities of glutamate and proline into the environment. The possible significance of such excretion in natural waters is discussed.

INTRODUCTION

The internal amino-acid pools of bacteria contain only a limited number of components at concentrations greater than 1 mM and therefore do not reflect the composition of cellular proteins (Cowie, 1962; Tempest *et al.,* 1970; Brown and Stanley, 1972). Pool size and composition are controlled by the

* Present address: Department of Biological Sciences, University of Dundee.

environment and must be considered essential for cell function. The environmental parameter which most affects both pool size and pool composition is the medium salinity and this has led to the assumption that these amino acids serve some critical function in osmoregulatory processes (Tempest *et al.*, 1970; Brown and Stanley, 1972). Other parameters affecting the amino-acid pool include the nature and concentration of the medium nitrogen sources, the growth rate, and the growth temperature. This paper describes experiments in which the effects of temperature and medium salinity were determined on the amino-acid pools of some marine pseudomonads.

METHODS

Marine bacteria were isolated from near-shore environments by selective enrichment in an artificial seawater medium described previously (Brown and Stanley, 1972). The organisms used in this study were all psychrophyllic, gram-negative bacilli which we have termed "*Pseudomonas*-like organisms". Cultures were grown at a pH of 7.8 and a temperature of $10°C$ (unless otherwise stated) in a 1-liter capacity glass chemostat at dilution rates (equal to growth rates) of 0.05 or 0.1 h^{-1}. Harvesting of bacteria, extraction of pools, and quantitative assay of amino acids in both pools and spent media are described elsewhere (Stanley and Brown, 1972).

RESULTS AND DISCUSSION

Effect of Growth Temperature

The use of a chemostat enables the growth temperature to be varied while maintaining the growth rate of the organisms constant. With the bacterium MU$_3$ grown at a dilution rate of 0.05 hr^{-1} and a salt (NaCl) concentration of

Table 1. The effect of temperature on the amino-acid pool composition of the marine bacterium MU-3

| Temperature (°C) | Amino-acid pool composition (μg/ml culture) | | | | | |
	Asp	Glu	Pro	Ala	Leu	Gln
10	2.5	64.5	0.0	0.5	0.6	20.6
15	2.0	58.4	0.0	0.6	0.3	16.2
20	1.2	48.1	0.0	0.5	0.2	10.8

0.4 M, a decrease in the growth temperature resulted in an increase in the concentrations of most of the components of the amino-acid pool. The most marked increases, however, were shown in the concentrations of the major components glutamate and glutamine (Table 1). At a similar salt concentration the glutamate content of strain SW_4 remained constant when the temperature was decreased from 20 to 5°C while that of proline increased (Fig. 1). At a higher salt concentration (0.6 M) the proline content increased fourfold with a similar temperature decrease and in this case the glutamate content decreased almost twofold. While temperature, therefore, has a marked and specific effect on the composition of these pools, the physiological significance of these changes is not apparent.

Effect of Salinity

As demonstrated above, salinity has a very marked effect on pool composition and as shown in Fig. 1 the glutamate content of strain SW_4 grown at 5°C is sixfold greater at 0.6 M than at 0.4 M NaCl. A more detailed study with strains SW_4 and MU_3 (Figs. 2 and 3) revealed that the medium NaCl concentration had a continuous effect on the pool glutamate content of these

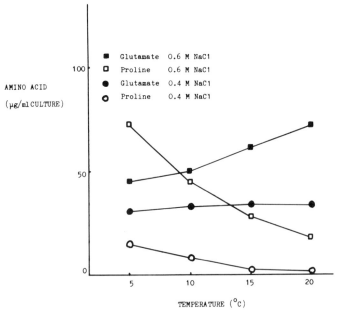

Fig. 1. Effect of temperature and salinity on the amino-acid pool of strain SW_4 grown under glucose limitation.

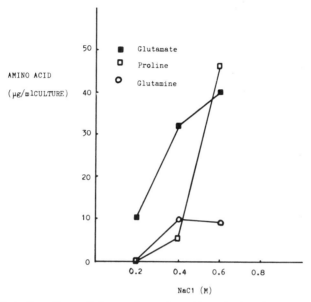

Fig. 2. Effect of medium salinity on the amino-acid pool of strain SW$_4$ grown under glucose limitation.

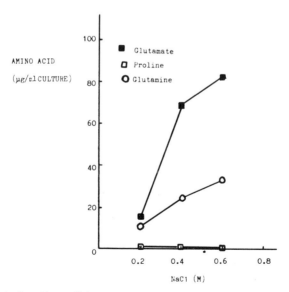

Fig. 3. Effect of medium salinity on the amino-acid pool of MU$_3$ grown under glucose limitation.

bacteria between 0.2 and 0.6 M. A marked difference between these two bacteria, however, was shown in the responses of their pool proline and glutamine contents. Thus while in MU_3 the glutamine content increased continuously with increased salinity and proline was not detected at any NaCl concentration, with SW_4 the glutamine content increased between 0.2 and 0.4 M NaCl but remained constant between 0.4 and 0.6 M. No proline was detected at 0.2 M NaCl, at 0.4 M only low levels were recorded, while at 0.6 M the proline content was higher than that of glutamate. Strain PL_1 was able to tolerate NaCl concentrations in the range 0.2–0.8 M (Fig. 4). The glutamate content increased between 0.2 and 0.4 M NaCl but decreased at higher values while the proline content (not detectable at 0.2 or 0.4 M) showed a marked increase between 0.4 and 0.8 M. It seems significant that in all three bacteria an increase in medium salinity was associated with an increased content of pool amino acids, and of the pool components only glutamine, glutamate, and, in some cases, proline were involved. Now glutamine and proline are precursors and/or products of the metabolism of glutamate and it is conceivable that the variations in the contents of these components reflect the control mechanisms exerted by the cell to maintain the glutamate concentration demanded by the imposed environmental conditions. A role for glutamate in osmoregulation in *Aerobacter aerogenes* and *Bacillus subtilis*

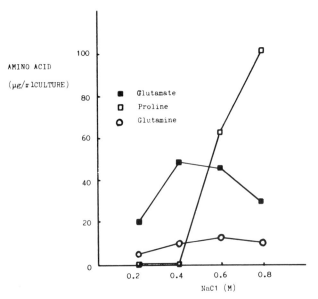

Fig. 4. Effect of medium salinity on the amino-acid pool of PL_1 grown under glucose limitation.

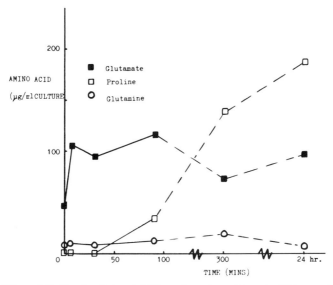

Fig. 5. Effect of NaCl pulse (0.4 to 0.8 M) on the amino-acid pool of PL$_1$ growing under carbon limitation.

var. *niger* has been postulated (Tempest *et al.,* 1970; Brown and Stanley, 1972), and it seems likely that the glutamate anion is synthesized in proportion to the K$^+$ cation which is accumulated within these organisms depending upon the medium salinity. Whether this same relationship holds for the marine bacteria is unknown.

Strain PL$_1$ was chosen for a more detailed study in order to determine the time response of pool composition to changes in medium salinity and the fate of the pool components in organisms subjected to either an increase or decrease in salinity as might occur in estuarine and near-shore environments owing to mixing of freshwater and seawater.

A NaCl "pulse", which increased the concentration to 0.8 M, was added at zero time to steady-state chemostat cultures growing under either carbon (glucose) or nitrogen (nitrate) limitation in the presence of 0.4 M NaCl. The incoming medium was adjusted from 0.4 to 0.8 M NaCl at the same time. Samples of culture were taken at time intervals for analysis of pool amino acids. In the case of the carbon-limited culture (Fig. 5), the immediate effect of the NaCl pulse was to double the glutamate pool content within 5 min. Thereafter the glutamate content remained constant for 90 min, before falling to approach the new steady state level at 5 and 24 hr. No proline was detected until after the increase in glutamate had occurred, and there was a steady increase in the content of this component between 30 min at 24 hr.

The glutamine content remained relatively constant throughout. The system was more complex, however, with nitrogen-limited cultures (Fig. 6) which at 0.4 M NaCl contained only trace amounts of glutamine and no detectable proline. The glutamine content rose to a maximum at 90 min but at 5 and 24 hr approached zero (only trace amounts of glutamine were present in pools from nitrogen-limited cultures at 0.8 M NaCl). The glutamate content similarly reached a maximum value at 90 min and then fell as the culture approached steady state at 0.8 M NaCl. It was noticeable that the increase in glutamate was much slower in nitrogen-limited than in carbon-limited cultures. As in the carbon-limited system, however, the proline content of nitrogen-limited cultures did not increase significantly until after the glutamate (and in this case glutamine) levels had reached their maximum values. The speed of response of the carbon-limited culture to increased NaCl suggests that there is either a rapid *de novo* synthesis of glutamate (from nitrate) or an equally rapid production of this amino acid from some intracellular component. In a similar experiment with *Aerobacter aerogenes* (Tempest *et al.,* 1970) it was found that the addition of NaCl directly increased the *de novo* synthesis of glutamate, since such synthesis was strongly dependent on the presence of a nitrogen source. A similar response is evident in the present experiments in which the rate of increase in the pool glutamate content was much slower in nitrogen-limited than in carbon-limited chemostat cultures. It is also evident that the pool proline content is controlled by and probably derived from glutamate, since only when the latter reaches its highest level does proline begin to accumulate.

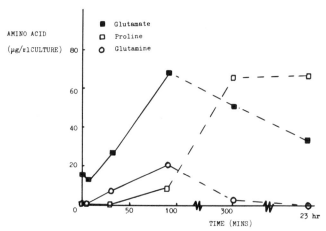

Fig. 6. Effect of NaCl pulse (0.4 to 0.8 M) on the amino-acid pool of PL_1 growing under nitrogen limitation.

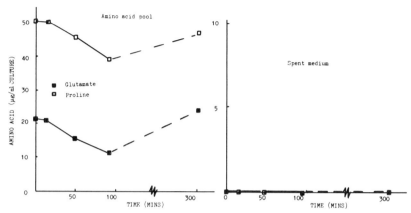

Fig. 7. Effect of decreased salinity (control, 0.8 to 0.8 M) on the glucose-limited cells of PL$_1$ grown at 0.8 M NaCl.

The short-term appearance of glutamine in the pools of nitrogen-limited organisms during the transition between 0.4 and 0.8 M NaCl but not in steady-state samples at either salinity is consistent with the above discussion, since glutamine is a known intermediate in the *de novo* synthesis of glutamate (and hence proline) in this organism (Brown *et al.*, 1973).

In order to study the effect of a decrease of salinity, 125 ml samples of culture of carbon-limited bacteria growing at 0.8 M NaCl were harvested by centrifugation, resuspended in 250 ml amounts of similar medium containing alternatively 0.8, 0.4, or 0.2 M NaCl, and incubated batchwise with stirring at 10°C. Samples were taken at time intervals between commencing the experiment and 5 hr, and the amino-acid contents of both the pools and the spent media assayed. All cultures showed a growth lag over the first 90 min after harvesting, but then grew at similar rates, independent of the medium salinity, reaching similar cell yields after 24 hr. In the control culture (0.8 M to 0.8 M) as shown in Fig. 7 there was a fall in the contents of both glutamate and proline over the time period when the culture was in lag phase. The initial pool concentrations were regained, however, after 5 hr when logarithmic growth had commenced. At no time were amino acids excreted into the spent medium even in trace amounts. As demonstrated above, cultures of this organism do not contain proline in detectable amounts at 0.4 M NaCl, while the content at 0.8 M is high. Results obtained with cells subjected to a decrease in salinity from 0.8 to 0.4 M are shown in Fig. 8. The pool proline content decreased rapidly with time and fell to zero after 90 min. Over this time period, however, the glutamate content remained constant showing the selectivity of these changes. Between 90 min and 5 hr the glutamate content

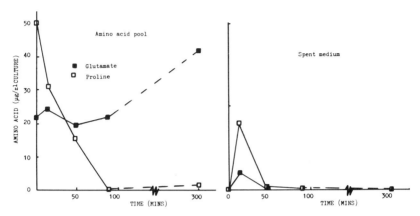

Fig. 8. Effect of decreased salinity (0.8 to 0.4 M) on carbon-limited cells of PL₁ grown at 0.8 M NaCl.

increased to approach the level required for batch growth at 0.4 M NaCl. The fall in the proline content over the first 50 min was accompanied by the excretion of small but significant quantities of both proline and glutamate into the spent medium. These were presumably reabsorbed, as by 50 min only trace amounts remained in the spent medium.

Finally, the most marked changes were apparent with a decrease in salinity from 0.8 to 0.2 M (Fig. 9). The proline content fell from over 50 to less than 5 μg/ml culture within 30 min and to zero within 90 min. There was also a marked decrease in the glutamate content within 13 min, although

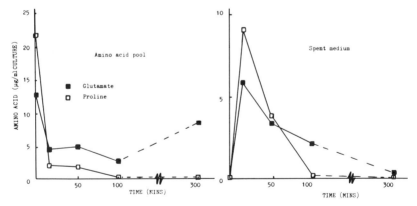

Fig. 9. Effect of decreased salinity (0.8 to 0.2 M) on carbon-limited cells of PL₁ grown at 0.8 M NaCl.

this then increased toward the growth requirement of 0.2 M NaCl. At 13 min there were appreciable quantities of both proline and glutamate excreted into the spent medium and, although these were steadily reabsorbed, they were still present in small amounts at 48 and 90 min. A similar experiment was then performed with cells grown under nitrogen limitation at a salinity of 0.8 M NaCl and the results obtained (Figs. 10–12) were very similar to those with cells derived from carbon-limited cultures. Thus there was significant excre-

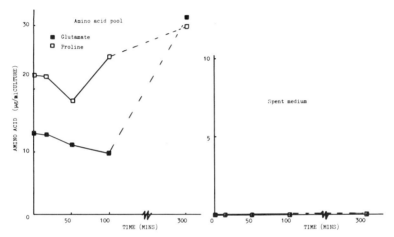

Fig. 10. Effect of decreased salinity (control, 0.8 to 0.8 M) on nitrogen-limited cells of PL_1 grown at 0.8 M NaCl.

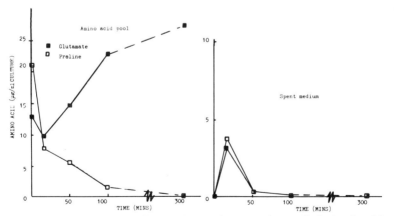

Fig. 11. Effect of decreased salinity (0.8 to 0.4 M) on nitrogen-limited cells of PL_1 grown at 0.8 M NaCl.

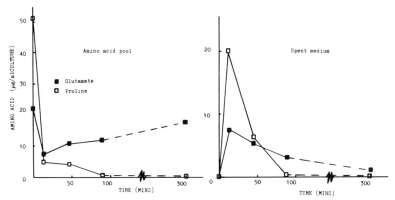

Fig. 12. Effect of decreased salinity (0.8 to 0.2 M) on nitrogen-limited cells of PL_1 grown at 0.8 M NaCl.

tion of pool glutamate and proline from cells that had been subjected to a decrease to 0.4 and 0.2 M NaCl during the lag phase of growth (up to 100 min) and these components were again reabsorbed so that only trace amounts remained 5 hr after commencing the experiment. The lack of excretion of pool components in the control culture (0.8 to 0.8 M) and more generally in normal chemostat cultures at any salinity (see Brown and Stanley, 1972) demonstrates the tight control exerted on these pools and their absolute requirement for growth in a particular environment. Excretion (when this occurred) was always specific and did not reflect shock excretion which would be typified by a non-specific loss of components and possible loss of viability. Indeed no loss of viability was evident in any culture used in the present experiments and, moreover, after the lag phase growth rates were equivalent at 0.2 and 0.8 M NaCl. Quantitatively, excretion did not account for the loss of pool materials; it is assumed that while very rapid mechanisms exist for increasing the pool content with an increased salinity, then an equally rapid mechanism exists to decrease the pool content as required with a minimal loss by excretion. It may be, for example, that in the shift from 0.8 to 0.4 M NaCl the increase in pool glutamate is directly at the expense of pool proline. Despite these control mechanisms, however, significant excretion did occur and, while in the present experiments the excreted amino acids were soon reabsorbed and presumably assimilated into non-pool components, in the natural environment these amino acids would be available for reabsorption by the whole microbial, etc., population. Indeed if this process is widespread in marine bacteria (and, as is tempting to speculate, also in phytoplankton and macroalgae) then under stable salinity conditions, e.g., in estuaries and sea lochs subject to freshwater inflow, these organisms could conceivably contribute to the organic nitrogen content of the environment.

LITERATURE CITED

Brown, C. M., D. S. Macdonald-Brown, and S. O. Stanley. 1973. The mechanisms of nitrogen assimilation in pseudomonads. Antonie van Leeuwenhoek. 39:89.

Brown, C. M., and S. O. Stanley. 1972. Environment-mediated changes in the cellular content of the 'pool' constituents and their associated changes in cell physiology. *In* A. C. R. Dean, S. J. Pirt, and D. W. Tempest (eds), Environmental Control of Cell Synthesis and Structure. Academic Press, London.

Cowie, D. B. 1962. *In* J. T. Holden (ed.), Amino Acid Pools. Elsevier, Amsterdam.

Stanley, S. O., and C. M. Brown. 1972. Nitrogen metabolism in marine bacteria: Intracellular amino acid pools. (In preparation.)

Tempest, D. W., J. L. Meers, and C. M. Brown. 1970. Influence of environment on the content and composition of microbial free amino acid pools. J. Gen. Microbiol. 64:171.

PROCEDURE FOR THE ISOLATION OF PSYCHROPHILIC MARINE BACTERIA

KENICHI TAJIMA, KATSUNOBU DAIKU, YOSHIO EZURA,
TAKAHISA KIMURA, and MINORU SAKAI

Laboratory of Microbiology
Faculty of Fisheries
Hokkaido University

INTRODUCTION

Psychrophilic bacteria are of great interest in the fields of taxonomy, physiology, and biochemistry, and they also play an important role in the cycling of various substances, especially in the ocean and landmass during wintertime. In recent years psychrophilic bacteria have attracted keen interest from the seafood-processing industry, particularly with respect to seafood hygiene.

Since most of the psychrophilic bacteria are highly proteolytic and lipolytic in nature (Ogawa, 1966*a,b,c,d,e*; Shaw and Shewan, 1968), they can exert a substantial impact upon the durable quality and shelf life of chilled seafoods (ZenYoji, 1969). However, since knowledge of psychorphilic bacteria is mainly based on studies of dairy microbiology (Thomas, 1969), there is some doubt as to the direct application of the information to the seafood-processing industry. Thus it has become greatly urgent in Japan that basic studies on psychrophilic marine bacteria be promoted because Japan relies heavily on marine products as a source of animal protein.

In this paper, we report basic studies on psychrophilic bacteria in coastal seawaters (which are deemed to be almost completely non-polluted). These studies have been done to develop a method for the isolation and culture of psychrophilic bacteria occurring in seawater. The following describes the results of our studies.

MATERIALS AND METHODS

Seawater Samples

The seawater samples employed in this study were collected at the tip of a breakwater protruding about 100 m offshore from the beach of Nanaehama in a suburb of Hakodate City. We took 24 samples (approximately 50 ml each) of surface seawater during a period of 1 year, from early June 1968 to mid-June 1969. The seawater samples were immediately chilled and brought as quickly as possible to our laboratory for tests.

Each seawater sample was properly diluted with sterilized artificial seawater (Nakamura, 1957) and 0.2 ml of each was spread onto a ZoBell 2216E agar plate (Morita and ZoBell, 1955). The plates were divided into four groups labeled (1) 4 days at 40°C, (2) 4 days at 35°C, (3) 6 days at 25°C, and (4) 10 days at 15°C, and were properly incubated. The average number of viable bacteria per milliliter was determined in a prescribed manner (Welfare Ministry of Japan, 1959). A plate from each of the groups above was selected and all of the colonies (30–40) within a marked sector of each plate were individually transferred to a master plate. These were incubated at the isolation temperatures. Each of the resulting plates was then replicated (Hirota, 1963) to seven daughter plates, each daughter plate being subcultured at one of seven temperatures: 40, 35, 25, 15, 8, 5, and 0°C. Meanwhile, plates were checked for colonies at 2, 4, 6, 7, 10, 14, and 21 days after inoculation, to evaluate the range of temperature for growth of bacteria thus isolated. The procedures above mentioned are illustrated in Fig. 1.

In carrying out this series of tests, the following method was adopted to calculate the number of viable bacteria which had been expected to occur in 1 ml of seawater.

As seen in Table 1, portions of the same test sample were plated at 15, 25, 35, and 40°C, followed by individual transfer of all colonies in a designated segment of individual plates. The range of temperature for growth of each colony group was evaluated by the replication method mentioned above. Colonies so obtained were grouped according to their range of temperature for growth. The proportion of the total number of isolates in each group was calculated. From this ratio and the total number of viable bacteria enumerated by standard methods, the probable number of bacteria having each of the defined ranges of temperature for growth was determined. The

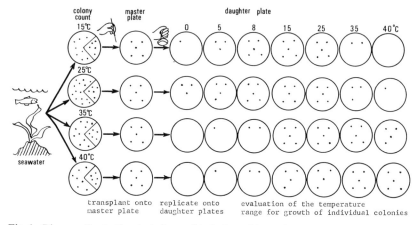

Fig. 1. Diagram illustrating technique of isolation of bacteria and evaluation of growth temperature range.

sum of these values for any of the four isolation temperatures represents the maximum probable number of viable bacteria in the seawater sample. For example, the total probable number of viable bacteria estimated to exist per milliliter of seawater listed in Table 1 amounted to 38,000.

RESULTS AND DISCUSSION

Figure 2 demonstrates the temperature of seawater at sampling time and the estimates made of the number of viable bacteria in the seawater samples tested at the four culture temperatures. The maximum temperature of these test samples was 22°C, from late August to mid-September, and the minimum was about 5°C from late February to early March. The average temperature between July and early December was above 12°C, and that of the subsequent months until May was a little below 12°C, which gradually rose after May until summer.

No significant correlation was observed during the year between seasons and the total probable number of viable bacteria evaluated in accordance with the number explained in the testing method. However, a very specific difference could be observed between the warm-water period (over 12°C) and the cold-water period (below 12°C) in terms of the number of viable cell counts evaluated at various culture temperatures. The viable count obtained by incubation at lower temperatures was consistently larger than that at higher temperatures, especially during the cold-water period.

Table 1. Procedure for the calculation of total probable number of viable bacteria in seawater sample[a]

Incubation temp. and period	15°C, 10 days		25°C, 6 days		35°C, 4 days		40°C, 4 days	
Viable count per ml	22,000		25,000		5800		2400	
No. of picked colonies	33		40		30		25	
Growth temp. range	B/A	C	B/A	C	B/A	C	B/A	C
40–0°C	1/33 = 3.0%	660	8/40 = 20.0%	5000				
–5					7/30 = 23.3%	1400	3/25 = 12.0%	290
–8					5/30 = 16.7	970	3/25 = 12.0	290
–15					13/30 = 43.3	2500	2/25 = 8.0	200
–25					1/30 = 3.3	190	13/25 = 52.0	1200
–35					1/30 = 3.3	190	2/25 = 8.0	200
							2/25 = 8.0	200
35–0°C	8/33 = 24.2	5300	12/40 = 30.0	7500				
–5	5/33 = 15.2	3300	7/40 = 17.5	4400	2/30 = 6.6	380		
–8					1/30 = 3.3	190		
–15								
25–0°C	1/33 = 3.0	660	1/40 = 2.5	630				
–5	10/33 = 30.0	6700	1/40 = 2.5	630				
–8	7/33 = 21.2	4700	11/40 = 27.5	6900				
–15	1/33 = 3.0	660						

[a]

$$\frac{\text{No. of colonies at each temperature range (B)}}{\text{No. of picked colonies (A)}} \times \text{viable count} = \text{Probable number of viable bacteria having each temp. range for growth (C)}$$

$$\text{Sum of maximum value of C (italic figures) in each temp. range} = \text{Total probable number of viable bacteria per ml seawater tested}$$

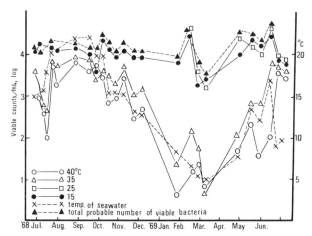

Fig. 2. Seasonal changes of seawater temperature, viable count in seawater evaluated at various culture temperatures, and total probable numbers of viable bacteria calculated.

The number of viable bacteria obtained by incubation at 15 or 25°C roughly coincided with the total probable number of viable bacteria regardless of the season. This result supports use of the method we proposed for calculating the total probable number of viable bacteria.

It was also suggested that high-temperature incubation ($\geq 35°C$) gave substantial errors in the enumeration of viable bacteria, especially during the cold-water period.

Figure 3 shows the ratio of the number of viable bacteria at each of the four culture temperatures *versus* the calculated total probable number of viable bacteria.

Under conditions of incubation at higher temperatures such as 40 or 35°C, slightly higher ratios were obtained during the warm-water period, but they dropped sharply as the temperature of the water dropped. Few values could be obtained during the cold-water period, but the values of the ratios rose sharply when the incubation was conducted at 15°C. During the period from February to May 1969, the particular ratios which were low in the 15°C incubation were comparably high, 90–100%, at the 25°C incubation. On the basis of the above results, it was concluded that either 15 or 25°C incubation should be adopted throughout the year to enumerate viable bacteria in seawater, but during the cold-water period, both 15 and 25°C incubation should be used.

The upper temperature for growth was evaluated for psychrophilic isolates obtained at 15°C which could grow below 8°C.

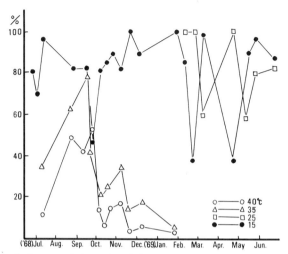

Fig. 3. Seasonal changes of the ratio of the number of viable bacteria counted at each culture temperature to the total probable number of viable bacteria calculated.

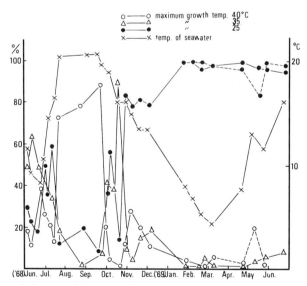

Fig. 4. Seasonal changes of the ratio of psychrophilic bacteria which were classified by upper limit of growth temperature to the viable counts evaluated at 15 and 25°C. Dashed line: isolates at 25°C.

Figure 4 shows the seasonal change in the proportion of organisms tested at their respective maximum temperatures for growth. Approximately 80% of the psychrophilic bacteria had a higher upper limit for growth temperature (40°C) in the warm-water period from mid-July to mid-September. When the water temperature began to fall, from mid-September to late October, the number of bacteria having a slightly lower upper limit for growth temperature (35°C) increased. The number of bacteria with a lower upper limit for growth temperature (25°C) was found to be very small during the warm-water period, but this increased sharply as the water temperature fell after late October. Hardly any bacteria having a higher upper limit for growth temperature could be detected at this time. The latter bacteria amounted to 90—100% of material during the cold-water period. This trend was similar to that for bacteria isolated at 25°C.

An exact definition of psychrophilic bacteria has yet to be made. If bacteria whose lower limit for growth is below 8°C are regarded as psychrophiles, then the above results indicate that for seawater the facultative psychrophile is in the majority during the warm-water period and the obligate psychrophile in the majority during the cold-water period.

The probable number of viable bacteria was estimated in the above group and their proportion of the total probable number of viable bacteria was then evaluated. These results are compared with the total number of psychrophilic bacteria, and are plotted against the month of their isolation in Fig. 5.

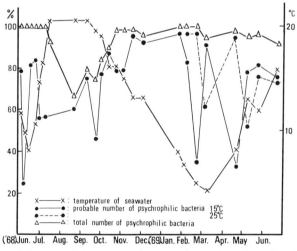

Fig. 5. Seasonal changes of the ratio of the probable number of psychrophilic bacteria evaluated at 15 and 25°C, and four culture temperatures, to the calculated total probable numbers of viable bacteria.

Regardless of the results for the upper growth temperature limits of the bacteria during the warm-water period, the above data roughly coincided with the proportion of the psychrophilic bacteria in the total probable number of viable bacteria. In contrast, the two did not always coincide during the cold-water period. However, as pointed out in Fig. 3, it was postulated that the isolation of psychrophilic bacteria from seawater samples and the determination of viable bacteria in them could be successfully made during the cold-water period by incubating the cultures at both 15 and 25°C and whichever showed a larger number of viable bacteria.

LITERATURE CITED

Hirota, Y. 1963. The experimental methods of genetics. Protein Nucl. Acid Enzyme (Tokyo) 8(9):532.

Morita, R. Y., and C. E. ZoBell. 1955. Occurrence of bacteria in pelagic sediments collected during the Mid-Pacific Expedition. Deep Sea Res. 3:66.

Nakamura, H. 1957. Marine microorganisms. *In* T. Taniya *et al.* (eds), Microbiology Handbook, pp. 610−623. Gihodo, Tokyo.

Ogawa, M. 1966a. Psychrophilic bacteria. Jap. Food Sci. 5(1):64.

Ogawa, M. 1966b. Psychrophilic bacteria. Jap. Food Sci. 5(2):62.

Ogawa, M. 1966c. Psychrophilic bacteria. Jap. Food Sci. 5(3):71.

Ogawa, M. 1966d. Psychrophilic bacteria. Jap. Food Sci. 5(4):66.

Ogawa, M. 1966e. Psychrophilic bacteria. Jap. Food Sci. 5(5):67.

Shaw, B. G., and J. M. Shewan. 1968. Psychrophilic spoilage bacteria of fish. J. Appl. Bacteriol. 31:89.

Thomas, S. B. 1969. Methods of assessing the psychrophilic bacterial count of milk. J. Appl. Bacteriol. 32:269.

Welfare Ministry of Japan. 1959. The standard plate count method. *In* The Guidebook of Hygienic Inspection, III-1, pp. 129−136.

ZenYoji, H. 1969. Factors affecting the growth of psychrophilic microorganisms in food. Media Circle 14:187.

SEASONAL DIFFERENCES IN BACTERIAL COUNTS AND HETEROTROPHIC BACTERIAL FLORA IN AKKESHI BAY

YOSHIO EZURA, KATSUNOBU DAIKU, KENICHI TAJIMA,
TAKAHISA KIMURA, and MINORU SAKAI

Laboratory of Microbiology
Faculty of Fisheries
Hokkaido University

Many studies have been reported on the bacterial population in inshore waters. However, other than the work by Sieburth (1968) who studied the ecology of planktonic bacteria in Narragansett Bay, few papers have been published confirming the influence of environmental factors on bacterial populations.

As a JIBP-PM project, the present authors cooperated to determine bacterial counts and investigate microbial flora in joint studies of marine productivity in Akkeshi Bay. We attempted to determine the relationship between bacterial populations and environmental factors.

MATERIALS AND METHODS

Water and Mud Samples

Samples were collected in Akkeshi Bay in northeastern Hokkaido throughout 1967, 1969, and 1971 (Fig. 1; Table 1). The sampling was made with a J-Z bacteriological water sampler (ZoBell, 1941) from depths of 10 and 20 m. The surface water was collected with sterilized bottles. In 1971, mud samples were collected with a grab sampler. The samples collected were placed in a chilled box for 40–50 min, during transfer to the Akkeshi Marine Station, Faculty of Science, Hokkaido University, for the bacteriological analyses.

Enumeration of Bacterial Number

The appropriately diluted samples were inoculated into ZoBell 2216E (Morita and ZoBell, 1955) liquid medium, ZoBell medium with nitrate or with starch, and media for nitrite formers and nitrate formers (Kimata, Kawai, and Yoshida, 1961). The inoculated media were incubated at 25°C for 3 weeks, with the exception of those in 1971, which were incubated at 20°C for 3 weeks. The total viable bacterial counts and the number of ammonia formers, nitrate reducers, starch hydrolyzers, and nitrite and nitrate formers were determined by the most probable number (MPN) method.

Isolation of Strains

The appropriately diluted samples studied in 1969 were plated onto ZoBell 2116E solid medium. Incubation was done at 25°C for 1 week. The colonies

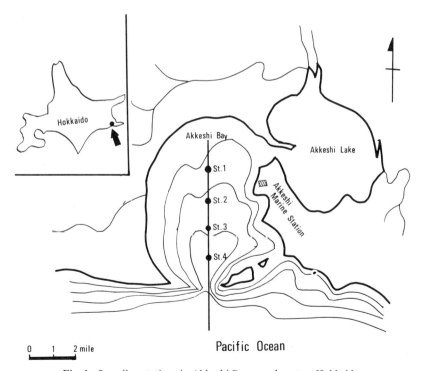

Fig. 1. Sampling stations in Akkeshi Bay, northeastern Hokkaido.

Table 1. Date of the sampling

Date	Hour	Depth (m)	Station
12 June 1967	11:10	0	1–4
	11:10	10	1–4
	15:10	20	1–4
16 Aug. 1967	14:00	0	3
	14:00	10	3
	14:00	20	3
	17:05	0	3
	17:05	10	3
	17:05	20	3
17 Oct. 1967	14:40	0	3
	14:40	10	3
	14:40	20	3
16 Nov. 1967	10:40	0	3
	10:40	10	3
	10:40	20	3

Date	Hour	Depth (m)	Station
12 May 1969	13:45	0	3
	13:45	10	3
	13:45	20	3
13 May 1969	10:45	0	3
	10:45	10	3
	10:45	20	3
28 Aug. 1969	14:00	0	3
	14:00	10	3
	14:00	20	3
	14:15	0	3
	14:15	10	3
	14:15	20	3
28 Oct. 1969	14:15	0	3
	14:15	10	3
	14:15	20	3
29 Oct. 1969	10:35	0	3
	10:35	10	3
	10:35	20	3
2 Dec. 1969	9:55	0	3
	9:55	10	3
	9:55	20	3
	10:05	0	3
	10:05	10	3
	10:05	20	3

Date	Hour	Depth (m)	Station
27 May 1971	10:00	0	3
	10:00	10	3
	10:00	20	3
25 Aug. 1971	10:55	0	3
	10:55	10	3
	10:55	20	3
	10:55	Mud	3
26 Oct. 1971	9:45	0	3
	9:45	10	3
	9:45	20	3
	9:45	Mud	3
7 Dec. 1971	10:45	0	3
	10:45	10	3
	10:45	20	3
	10:45	Mud	3
24 Feb. 1972	10:30	0	3
	10:30	10	3
	10:30	20	3
	10:30	Mud	3

were picked, purified by streaking three times, and eventually 718 isolates were obtained.

Classification of Isolates

A scheme devised by Shewan *et al.* (1960) was applied to characterize the isolates with a life cycle recognized as belonging to the genus *Arthrobacter* (Sieburth, 1968) (an exception was the identification of pleomorphic organisms). Subsequently the isolates were grouped into three types by the Hidaka salt-requirement test (Hidaka, 1964): terrestrial type (T), halophilic type (H), and marine type (M).

    Among some parameters of the environmental conditions in Akkeshi Bay investigated by other members of Akkeshi Bay Research Group (1968, 1970), water temperature, salinity, dissolved oxygen, nutrient salts, phytoplankton pigment, and particulate materials were considered in the bacteriological analyses.

RESULTS AND DISCUSSION

Figure 2 shows the water temperature and salinity variations at a station in the central part of Akkeshi Bay in 1967. Water temperature increased gradually from June to early September at all depths with a maximum temperature of 19°C in surface water. Through this period, a stable thermocline was observed at a depth of 5–7 m. When seawater was cool in mid-September, convection currents developed which altered the temperature gradient. Salinity variation was not significant, a minimum appearing in slight salinity gradient was detected when the thermocline appeared.

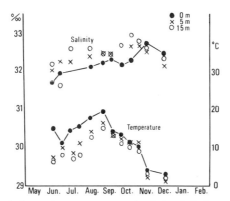

Fig. 2. Seasonal variations of temperature and salinity (o/oo) at central station 3 of Akkeshi Bay, 1967.

Viable bacterial counts were determined in the first samples from the four stations on 12 June 1967. The results are shown in Fig. 3. As it was found that the water samples collected from the four stations indicated almost identical bacterial counts, subsequent samplings were made only from the central station (3) of Akkeshi Bay. Viable bacterial counts in water were within the range 3.3×10^2 to 6.3×10^4 cells/ml throughout the 3 years (Fig. 4). In 1969, the counts attained a maximum in summer and decreased in the cold-water season. However, both in 1967 and 1971 the maximum counts were obtained in the cold-water seasons. The minimum count was obtained in the sample of 27 February 1972, when the water temperature was $0°C$. The incubation temperature at $20°C$ seemed to be too high for the growth of organisms in these samples. In the period during which a thermocline developed bacterial counts were much higher in both the bottom and surface waters than in middle-layer water. After the thermocline broke down, no difference in bacterial counts could be found between the three layers.

On the basis of the data obtained in 1967 regarding water temperature, salinity, dissolved oxygen, phosphate, nitrate, nitrite, ammonia, seston-carbon, chlorophyll a, and phaeophytin, an attempt was made to account for

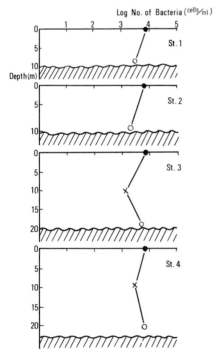

Fig. 3. Vertical distribution of total viable bacteria in water at the four stations on 12 June 1967.

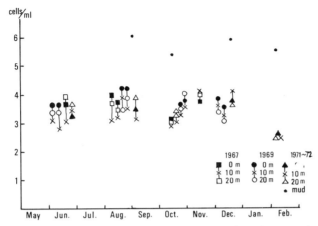

Fig. 4. Total number of viable bacteria counts in water and mud at central station 3 of Akkeshi Bay.

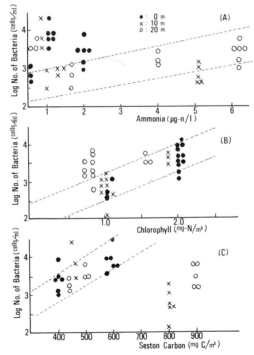

Fig. 5. Relationship between the number of bacteria and levels of (A) ammonia, (B) chlorophyll, (C) seston-carbon in Akkeshi Bay, 1967.

the relationships between the change of heterotrophic bacterial counts and the environmental factors. The change in bacterial counts correlated to some extent with levels of ammonia, seston-carbon, and chlorophyll a. As shown in Fig. 5, it was recognized that counts of bottom water were correlated with ammonia content while counts of surface and middle-layer waters correlated with chlorophyll a content. It is also apparent that bacterial counts correlated with seston-carbon content below 700 $\mu g/m^3$.

The counts of ammonia formers, nitrate reducers, and starch hydrolyzers are illustrated in Fig. 6. The counts of ammonia formers and nitrate reducers changed generally in accordance with that of heterotrophic bacteria. Starch hydrolyzer gradually increased from May to December both in 1967 and 1971; however, in 1969 it did not show such a tendency. The counts in mud were very high in every sample. In 1969, an inverse relationship was observed between ammonia-forming and nitrifying bacteria; that is, ammonia formers were high in number while nitrifying bacteria were low in number in May and October. However, the nitrifying bacteria increased and the ammonia formers decreased in August and December (Table 2). There was no apparent relation

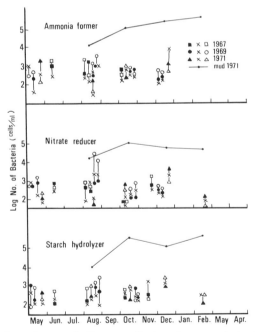

Fig. 6. The number of ammonia formers, nitrate reducers, and starch hydrolyzers in water and mud at central station 3 of Akkeshi Bay, 1967, 1969, 1971.

Table 2. The number of ammonia formers, nitrite formers, and nitrate formers in water samples collected at the central station, 1969

Date	Hour	Depth (m)	NH_3 former (cells/ml)	NO_2 former (cells/100 ml)	NO_3 former (cells/100 ml)
12 May	13:45	0	1100	0	1.8
		10	1100	0	0
		20	330	0	0
28 Aug.	14:15	0	330	6.8	6.1
		10	490	17	8.2
		20	1300	17	6.8
28 Oct.	14:15	0	1100	6.8	0
		10	790	2.0	0
		20	5400	4.0	0
2 Dec.	10:05	0	460	18	18
		10	330	0	18
		20	700	36	4

between the number of each of these types of bacteria and the environmental factors.

From 12 water samples collected in 1969, 718 strains of heterotrophic bacteria were isolated and were classified according to the determinative scheme of gram-negative rods recommended by Shewan *et al.* (1960). The frequency of the appearance of bacterial genera in microflora in seawater is shown in Table 3. In all samples, the prevailing genera were *Pseudomonas, Achromobacter,* and *Flavobacterium,* accounting for more than 80% of the 718 strains. The rate of occurrence of *Pseudomonas* was high in the samples, especially in the surface water in October and December and in the middle layer and bottom water in August. The microbial flora in water on 12 May were dominated by *Flavobacterium*; then the genus decreased on 28 August, while *Achromobacter* and *Vibrio* increased. On 28 October *Flavobacterium* and *Pseudomonas* were dominant isolates like those in May, but they were different from other samples because the proportion of gram-positive organisms was above 10% and *Aeromonas* and Enterobacteriaceae were found. *Pseudomonas* was prevalent in the surface water and *Vibrio* accounted for 20% of the organisms in bottom water on 2 December.

Through all samples, the following facts were observed conclusively. First, there was an inverse relationship between *Flavobacterium* and *Vibrio.* A

Table 3. The percentage composition of bacterial flora in water at central station 3 of Akkeshi Bay, 1969

Sample No.	Date	Depth (m)	No. isolates	Pseudo-monas	Aero-monas	Vibrio	Achromo-bacter	Flavo-bacterium	Entero-bacteriaceae	Micro-coccus	Cory-neform	Arthrobacter-like organisms
								Isolates (%)				
1	12 May	0	52	34.6	0	3.8	15.4	40.4	0	3.8	1.9	0
2		10	53	35.8	0	3.8	11.3	41.5	0	3.8	3.8	0
3		20	77	37.7	0	5.2	27.3	22.1	0	2.6	5.2	0
4	28 Aug.	0	46	28.3	0	10.9	37.0	8.7	0	2.2	6.5	6.5
5		10	35	49.2	0	8.6	25.7	5.7	0	5.7	5.7	5.7
6		20	96	40.6	0	7.3	19.8	14.6	0	2.1	0	15.6
7	28 Oct.	0	57	47.4	0	3.5	22.8	12.3	0	3.5	0	10.5
8		10	56	30.4	0	3.6	8.9	35.7	1.8	12.5	7.1	0
9		20	75	33.3	1.3	0	14.7	32.0	0	9.3	1.3	8.0
10	2 Dec.	0	62	61.3	0	3.2	17.7	11.3	0	0	3.2	3.2
11		10	49	40.8	0	6.1	14.3	30.6	0	4.1	4.1	0
12		20	60	31.7	0	20.0	26.7	11.7	0	1.7	6.7	1.7
	Total		718	38.8	0.1	6.1	19.9	22.3	0.1	4.2	3.5	5.0

similar phenomenon had been reported by Sieburth (1968). Second, the pattern of occurrence of micrococci was similar to that of *Flavobacterium*. It was suspected that carotinoid pigments were closely connected with the relationship between both genera, because most of the isolated micrococci produced the pigments.

On the basis of the Hidaka salt-requirement test, 53% of the isolates can be grouped into marine type, 30% as halophilic type, and 17% as terrestrial type. As shown in Fig. 7, marine-type isolates gradually increased from May to December, and it seems that the occurrence rate of the three types was closely related to the varying salinity of water. However, an increase was noted in terrestrial type in October. As mentioned above microbial flora differed in this period from other seasons. This might be due to some influence of the run-off from land on the microbial flora. It may be that the main constituents of bacterial population in the water of Akkeshi Bay are halophilic- and marine-type organisms which require NaCl or other salts to grow and so may be estuary or marine habitants.

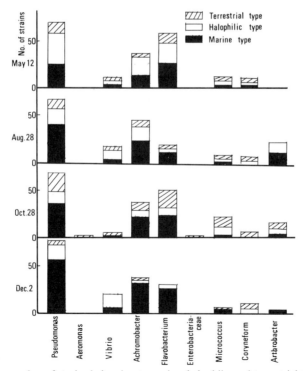

Fig. 7. The number of strains belonging to marine, halophile, and terrestrial type in each genera isolated from central station 3, 1967.

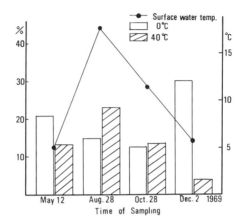

Fig. 8. Occurrence rate of psychrophilic bacteria (grown for 1 week at 0°C) and mesophilic bacteria (grown for 5 days at 40°C) observed at central station 3, 1969.

Occurrence rates of the psychrophilic bacteria which can grow after incubation for 1 week at 0°C and organisms capable of growth at 40°C for 5 days are shown in Fig. 8. Their occurrence is closely related to the change of water temperature. In December, 50% of *Pseudomonas* were psychrophilic organisms. Many organisms capable of growth at 40°C were *Flavobacterium* and *Achromobacter*. However, the strains used in this study were isolated at 25°C and a number of psychrophilic bacteria would not be isolated at this incubation temperature (Sieburth, 1967; Tajima *et al.*, 1971).

LITERATURE CITED

Akkeshi Bay Research Group. 1968. Studies on the productivities of bio-coenoses in northern cold waters. Progr. Rep. 1967–1968. 49 p.

Akkeshi Bay Research Group. 1970. Studies on the productivities of bio-coenoses in northern cold waters. Progr. Rep. 1969–1970. 67 p.

Hidaka, T., 1964. Studies on the marine bacteria. I. Mem. Fac. Fish. Kagoshima Univ. 12:135–152.

Kimata, M., A. Kawai, and Y. Yoshida. 1961. Studies on marine nitrifying bacteria (nitrite formers and nitrate formers). I. Bull. Jap. Soc. Sci. Fish. 27:593–597.

Morita, R. Y., and C. E. ZoBell, 1955. Occurrence of bacteria in pelagic sediments collected during the Mid-Pacific Expedition. Deep-Sea Res. 3:66–73.

Shewan, J. M., G. Hobbs, and W. Hodgkiss. 1960. A determinative scheme for the identification of certain genera of gram-negative bacteria, with special reference to the Pseudomonadaceae. J. Appl. Bacteriol. 23:379–390.

Sieburth, J. McN. 1967. Seasonal selection of estuarine bacteria by water temperature. J. Exp. Biol. Ecol. 1:98–121.

Sieburth, J., McN. 1968. Observations on planktonic bacteria in Narragansett Bay, Rhode Island. Proc. U.S. Jap. Sem. Mar. Microbiol., Bull. Misaki Mar. Biol. Inst. Kyoto Univ. 12:49–64.

Tajima, K., Y. Ezwia, and M. Sakai. 1971. Procedure for the isolation of psychrophilic marine bacteria. Bull. Fac. Fish. Hokkaido Univ. 22:73–79.

ZoBell, C. E. 1941. Apparatus for collecting water samples from different depths for bacteriological analysis. J. Mar. Res. 4:173–188.

LOW-TEMPERATURE INHIBITION OF SUBSTRATE UPTAKE

RICHARD Y. MORITA and GEORGE E. BUCK

Department of Microbiology and School of Oceanography
Oregon State University

Paul and Morita (1971) demonstrated that both low temperature and increased hydrostatic pressure inhibited the uptake of amino acids. The uptake of uracil was also inhibited by low temperature and increased pressure (see Baross *et al.*, 1974). The low-temperature portion of these investigations demonstrated the possibility that effect on uptake of substrates by bacterial cells might be a reason for the minimal temperature for growth of microorganisms.

The precise reasons for the minimal temperature for growth of microorganisms is not known. The effect of low temperature on the uptake of substrates by *Escherichia coli* has been studied by several investigators. Shaw and Ingraham (1967) reported an initial decrease in the uptake of a-methyl-[U-^{14}C]glucoside when *Escherichia coli* ML 30 cells were shifted from 37 to 10°C. However, full permeability to a-methylglucoside returned after 1 hr at 10°C. Piperno and Oxender (1968) concluded that the minimal temperature for leucine uptake in *Escherichia coli* K12 was near 4°C, while Kawasaki *et al.* (1969) demonstrated that uptake of [^{14}C]thiamine is inhibited at 0°C in cultures of *Escherichia coli* K12 and KG33.

The uptake of nutrients by the cell is governed by the membrane. The chemical make-up of membranes includes lipoproteins which contain hydrophobic groups. Hydrophobic groups are known to be affected by temperature (Kettman *et al.*, 1966) and therefore aggregation or disassociation of hydrophobic groups should produce an alteration in the membrane. This alteration in the membrane could then affect the uptake of essential materials into the cell. This investigation was undertaken to determine whether or not low temperature could affect the pattern of amino-acid uptake.

Escherichia coli was grown in nutrient broth (Difco) (8 g/liter) at 37°C without shaking. After 10 hr the cells (in log phase) were harvested by centrifugation and suspended in dilute nutrient broth (0.5 g/liter) so that the cell suspension gave an OD of 0.05 at 600 nm on a Bausch and Lomb Spectronic 20. Parallel cell suspensions (50 ml in 250 ml Erlenmeyer flasks) were incubated at 5 and 37°C for each amino acid employed before the addition of [^{14}C]-amino acid in order to allow the suspensions to equilibrate to the proper temperature. In all experiments, the ^{14}C-labeled compounds were added so that the final concentration was 0.01 μCi/ml. A 5-ml sample was taken immediately to correct for any amino acid that might adhere to the surface of the cells or the membrane filter. Five-milliliter samples (in duplicate) were taken after 1 hr incubation.

The samples were treated in an apparatus described by Hobbie and Crawford (1969) that allows for the determination of respired CO_2. This apparatus consists of a serum bottle with a cap which is fitted with a small plastic cup and rod assembly (Kontes Glass Co., Vineland, N.J., Cat. No. K-882320). The cup contained a piece of fluted filter paper. Sulfuric acid (0.4 ml, 0.1 N) was added to each serum bottle before the introduction of the samples. Immediately after a 5-ml sample was introduced, the cap with the plastic cup and rod assembly was secured. Ten minutes later 0.2 ml β-phenylethylamine was added to the cup with a syringe through the top of the serum bottle cap and allowed to remain for 2 hr in order to absorb the carbon dioxide.

The contents of each bottle were then filtered (Millipore, HA, 0.45 μm, 23 mm diameter), and the filter dried and placed in a scintillation vial. Scintillation fluor (0.3 g 1,4-bis-2-(4-methyl-5-phenyloxazolyl)-benzene, and 5.0 g 2,5-diphenyloxazole in 1 liter toluene) was added to the vials and the radioactivity was measured. The contents of the plastic cup were also placed in scintillation vials and fluor was added in order to determine the amount of $^{14}CO_2$ respired. Quenching was corrected with the channel-ratio method.

Table 1 shows the uptake of the various [^{14}C] amino acids and glucose, the amount of $^{14}CO_2$ respired and the total amount of ^{14}C uptake and respiration at 5 and 37°C. It can be seen that uptake of many of the amino acids ceases at 5°C. The minimal temperature for growth of *Escherichia coli* in nutrient broth was determined to be between 6 and 7°C. A small amount of radioactivity was detected in cells incubated with threonine and serine at 5°C. On the other hand, uptake was exceedingly good with all compounds

Table 1. Uptake and respiration of [^{14}C] glucose
and [^{14}C] amino acids at 5 and 37°C by *Escherichia coli*[a]

	Disintegrations/min					
	5°C			37°C		
Substrate	Cells	CO$_2$	Total	Cells	CO$_2$	Total
Glucose	1174	1096	2270	14,341	1772	16,113
Glutamic acid	163	3228	3391	3682	14,424	18,106
Aspartic acid	222	1901	2123	15,021	4783	19,804
Serine	564	402	966	32,952	2423	35,375
Threonine	375	94	470	31,342	1648	32,990
Proline	30	0	30	28,712	4339	33,111
Alanine	0	0	0	11,489	3999	15,488
Glycine	0	0	0	21,986	4524	26,510
Asparagine	3	6	9	2721	665	3384
Phenylalanine	5	0	5	9295	25	9320
Methionine	29	12	41	36,946	2	36,948
Tyrosine	88	0	88	12,453	0	12,453
Histidine	0	0	0	11,331	0	11,287
Lysine	0	1	1	10,934	69	11,003
Arginine	171	32	203	5390	608	5998

[a] Incubation period was 1 hr. The values obtained for the samples taken at zero time were subtracted from the values obtained for subsequent samples in order to correct for any radioactivity adhering to the surface of the cell or the membrane filters.

tested at 37°C. The uptake of glucose, glutamic acid, and aspartic acid remained high compared with the other substrates. Over 48% of the glucose that entered the cells was respired at 5°C, whereas 10.9% of the glucose was respired by the cells at 37°C. This situation was also true when glutamic and aspartic acids were added to the system at 5 and 37°C, since the amount respired was 95.7 and 29.1% at 5°C and 79.6 and 24.1% at 37°C, respectively. Thus, it appears that these substrates are mainly used by the cells at 5°C for energy of maintenance and not to synthesize new cells, since no growth occurred at 5°C.

In all temperature profile studies, the cell suspension (5 ml) as previously described was pipetted into test tubes. These test tubes were placed in a polythermostat and allowed to equilibrate to the selected temperature (15 min). The [^{14}C] amino acids were then added and the mixture allowed to incubate for an appropriate time. These samples were treated as described in the preceding experiment.

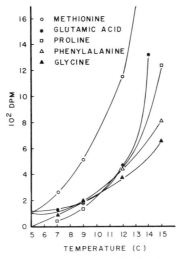

Fig. 1. Temperature profile for the uptake of [^{14}C]amino acids by *Escherichia coli.* Incubation period was 1 hr at various temperatures.

Figure 1 illustrates the uptake of proline, phenylalanine, glutamic acid, glycine, and methionine as a function of temperature. In all cases it can be seen that kinetics of amino-acid uptake was non-linear. Each amino acid had its own temperature profile in relation to the amount of glutamic acid taken up and the amount of $^{14}CO_2$ respired. It can be readily seen that, at lower temperatures, the amount of CO_2 respired is significantly greater than the amount incorporated into the cells, thereby indicating that most of the

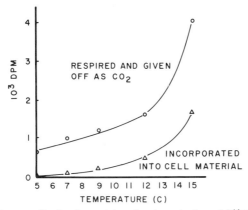

Fig. 2. Temperature profile for the uptake of and respiration of [^{14}C]glutamic acid by *Escherichia coli.* Incubation period was 1 hr at various temperatures.

glutamic acid is utilized as an energy source (Fig. 2). There was no difference in the amount of uptake of $[^{14}C]$proline at $5°C$ in cells incubated for 1 and 4 hr (Fig. 3). Above $5°C$, there is naturally an increase in the amount of proline taken up with time. However, the amount of CO_2 respired does not increase until the temperature is above $11°C$.

Our results are consistent with those of Shaw and Ingraham (1967) who demonstrated that permeability was unaffected by low temperature and that respiration continues (in our case, glucose, glutamic acid, and aspartic acid). Their results were based on the uptake of a-methylglucoside at $10°C$ and our present studies show that glucose continues to be accumulated and respired, but certain amino acids are not taken up by the cells. The data presented are also consistent with those of Piperno and Oxender (1968) who reported that the minimal temperature for uptake activity of leucine is near $4°C$.

It is recognized that *Escherichia coli* can grow on glucose and mineral salts containing ammonia salt. Since the size of the amino acid does not appear to govern its uptake at low temperature, the question of whether or not the ammonium salt would be taken up by the cells becomes interesting. However, when *Escherichia coli* is incubated in nutrient broth at $37°C$, the energy for the cells comes from the utilization of the amino acids. The amount of CO_2 respired from the addition of glutamic acid, aspartic acid, serine, threonine, proline, alanine, and glycine is quite large at $37°C$ (Table 1), which indicates that they are being used as an energy source. At $5°C$ only glutamic and aspartic acids appear to be used as energy sources. In nutrient broth at $5°C$, *Escherichia coli* probably does not obtain sufficient energy to permit growth. Since low temperature acts on *Escherichia coli* in limiting the amino acids that it can take up, then the low-temperature inhibition of

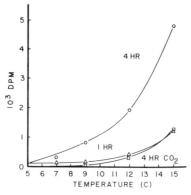

Fig. 3. Temperature profile for the uptake and respiration of $[^{14}C]$proline by *Escherichia coli*. Incubation periods were 1 and 4 hr at various temperatures.

amino-acid uptake or possibly other essential substances is probably a limiting factor at the minimal growth temperature in certain organisms.

In order to make certain that one essential amino acid can bring about the minimum temperature for growth in *Escherichia coli*, we are now employing an auxotroph (for leucine, arginine, histidine, and methionine). The preliminary data indicate that the same trend can be observed with the auxotroph grown at 37°C and then shifting the temperature to 5 or 6°C.

ACKNOWLEDGMENT

This investigation was supported by National Science Foundation research grant GA-28521.

LITERATURE CITED

Baross, J. A., F. J. Hanus, and R. Y. Morita. 1974. This conference.

Hobbie, J. E., and C. C. Crawford. 1969. Respiration corrections for bacterial uptake of dissolved organic compounds in natural waters. Limnol. Oceanogr. 14:528–532.

Kawasaki, T., I. Miyata, K. Esaki, and Y. Nose. 1969. Thiamine uptake in *Escherichia coli*. I. General properties of thiamine uptake system in *Escherichia coli*. Arch. Biochem. Biophys. 131:223–230.

Kettman, M. S., A. H. Nishikawa, R. Y. Morita, and R. R. Becker. 1966. Effects of hydrostatic pressure on the aggregation reaction of poly-L-valylribonuclease. Biochem. Biophys. Res. Commun. 22:262–267.

Paul, K. L., and R. Y. Morita. 1971. Effects of hydrostatic pressure and temperature on the uptake and respiration of amino acids by a facultatively psychrophilic marine bacterium. J. Bacteriol. 108:835–843.

Piperno, J. R., and D. L. Oxender. 1968. Amino acid transport systems in *Escherichia coli* K12. J. Biol. Chem. 243:5914–5920.

Shaw, M. K., and J. L. Ingraham. 1967. Synthesis of macromolecules by *Escherichia coli* near the minimal temperature of growth. J. Bacteriol. 94:157–164.

III

PRESSURE EFFECTS

HYDROSTATIC PRESSURE EFFECTS ON MICROORGANISMS

RICHARD Y. MORITA

Department of Microbiology and School of Oceanography
Oregon State University

BACKGROUND INFORMATION

The marine biosphere is characterized by hydrostatic pressures of 1 atm to approximately 1200 atm. The average depth of the ocean is estimated to be 3800 m (380 atm). All organisms living below the surface of the ocean are subjected to varying degrees of hydrostatic pressure, and hydrostatic pressure should be considered of equal importance with temperature and salinity as major factors affecting biological systems. Microbial life has been demonstrated in various regions of the ocean where the pressure reaches approximately 1150 atm (ZoBell and Morita, 1957; Seki and ZoBell, 1967).

The ideal gas law $(PV = nRT)$ and Le Chatelier's principle, when applicable to biological systems, should not be neglected. In the deep sea, pressure is the main variable (since temperature is nearly constant at ca. $5^{\circ}C$) influencing the chemical and physical environment in which the organisms live. Many of the pressure effects on biological systems can be explained in terms of molecular volume (ΔV) changes. Intracellular materials are subjected to a molecular volume change, depending on the pressure, temperature, ionic environment, and pH. Increases in molecular volume result from an increase in temperature, and increased pressure decreases the molecular volume. This situation is well illustrated for enzymes (Johnson et al., 1954; Haight and Morita, 1962; Morita and Haight, 1962; Morita and Mathemeier, 1964). The pressure-temperature relationship on molecular-volume change can also be used to interpret the growth data obtained by ZoBell and Johnson (1949). An increase in salinity was found to increase the ability of marine bacteria to divide under pressure (Albright and Hanigman, 1971). The various review articles dealing with hydrostatic pressure effects on biological systems are Johnson *et al.* (1954), Morita (1965, 1967, 1972), and ZoBell (1964). A review on various methods for pressurization was presented by Morita (1970). For information dealing with pressure effects on cellular systems of various organisms, the book edited by Zimmerman (1970) should be consulted.

The pressure that exists in the marine biosphere should be considered as "moderate" pressure (1 to ca. 1000 atm), and many reactions at moderate pressures are reversible when the pressure is released. At high pressures (above the kilobar range) many reactions are not reversible but become inactive (Johnson *et al.,* 1954). As a general rule of thumb, the pressure increases 1 atm for every 10 m in depth of the ocean. Most investigators employ "atm" as the unit of pressure measurement. Some of the older literature employs "psi" (pounds per square inch). The metric unit of pressure measurement is "newtons per square meter" (N/m^2). The pressure of dry air at $0°C$ at sea level at a latitude of $45°$ is 1.0 atm. One atm equals 1.01325×10^5 N/m^2, 14.696 psi, 760 mm Hg, or 1.0133 bars. For further details on other units of pressure measurement, Kinne (1972) should be consulted.

Many complicating factors arise when a system is subjected to pressure. Some of these complex factors are solubilities, pH changes, water structure, etc. (see Morita and Becker (1970) for more details). However, in most instances pressure or temperatures are the factors that are changed during the various experimental procedures and interpretation of the data is performed mainly in terms of these two factors. It should be kept in mind that there are complicating factors arising from the use of pressure in experiments.

ZoBell and Johnson (1949) coined the term "barophile" to designate those organisms that grew well at elevated pressures. Most surface forms which do not function at elevated pressure are termed "barophobic", while organisms capable of growth between 1 and 600 atm are considered "eury-baric" (ZoBell, 1968).

Pressure research was initiated in the latter part of the 1800's by Regnard (1884) and Certes (1884a,b). At that time researchers did not know that there was a temperature-pressure relationship, and consequently the data are difficult to interpret. Growth and viability studies on surface forms were conducted by ZoBell and Johnson (1949) who demonstrated that most of the microbes tested could not survive at 400 to 500 atm. The degree to which pressure and temperature affect the organisms appears to be dependent on the species in question. The barophilic forms grow at pressures from 500 to

1200 atm (ZoBell and Morita, 1957). A temperature-pressure relationship on growth, viability and tolerance was noted during the studies by ZoBell and Johnson (1949). Under pressure, morphological changes (filament formation, pleomorphism, cell size, etc.) as well as spore formation take place (ZoBell and Oppenheimer, 1950; Oppenheimer and ZoBell, 1952; ZoBell and Cobet, 1962, 1964; Boatman, 1967).

The tolerance of microbes to pressure depends to a large degree on the previous environmental temperature at which they are grown. It is important that cells are not grown at their optimal temperature (generally 10–20°C above the temperature from which they were isolated) and then subjected to isothermic and isobaric conditions of the deep sea (Johnson et al., 1954; Albright and Morita, 1972). If cells are grown at their optimal temperature for growth and then subjected to isothermic and isobaric conditions of the deep sea, both lower temperature and higher pressure become additive in terms of the molecular volume change. The tolerance to pressure is also dependent upon the medium employed, substrate, and salinity, as well as temperature. Within limits, bacteria (and enzymes) can tolerate higher temperatures under pressure than at 1 atm.

It is difficult to state why death results from the application of pressure to organisms which possess a complex and highly organized structure. Some of the possibilities that have been investigated are (1) pressure-induced lesion(s) in the TCA cycle so that the energy supply is shut off (Hill and Morita, 1964), (2) retardation of enzyme reactions, (3) retardation of macromolecular synthesis, and (4) inability to take up substrate.

Various microbial enzymes have been subjected to pressure and have been found to become more inactivate with increased pressure. A review of the various enzyme reactions under pressure was presented by ZoBell and Kim (1972). However, it should be noted that most of the enzyme reactions were carried out at temperatures between 20 and 35°C and not at deep-sea temperatures. Chitinase was found by Kim and ZoBell (1972) to be quite active at 4°C under pressure. However, Penniston (1971) postulated that monomeric enzymes are stimulated by pressure and multimeric enzymes are diminished. Evidence produced by Becker and Evans (1969) on β-galactosidase (multimeric enzyme) shows a reversal of the effect of pressure depending on whether the monovalent cation present is K^+ or Na^+. Whether the data obtained by Penniston (1971) on his enzymes can be extrapolated to other enzymes remains to be investigated.

Many studies have been conducted on the microbe's ability to synthesize RNA, DNA, and protein (Landau, 1966, 1970; Yayanos and Pollard, 1969; Albright and Morita, 1968; Albright, 1969) and all the data indicate that pressure inhibits the synthetic mechanism. Current investigations are focused on induction, transcription, and translation processes in cells (Arnold and Albright, 1971; Landau, 1967; Schwarz and Landau, 1972). Further information on this subject is presented in this book.

Paul and Morita (1971) presented evidence that amino-acid uptake was inhibited by pressure and low temperature, thereby implicating a membrane function. With increased pressures the ability of the cell to take up amino acids decreases but the ability to respire the amount of amino acid that enters the cell is not inhibited. Further information on this aspect is also included in this book.

LITERATURE CITED

Albright, L. J. 1969. Alternate pressurization-depressurization effects on growth and net protein synthesis by *Escherichia coli* and *Vibrio marinus*. Can. J. Microbiol. 15:1237–1240.

Albright, L. J., and J. F. Hanigman. 1971. Seawater salts–hydrostatic pressure effects upon cell division of several bacteria. Can. J. Microbiol. 17:1246–1248.

Albright, L. J., and R. Y. Morita. 1968. Effect of hydrostatic pressure on synthesis of protein, ribonucleic acid, and deoxyribonucleic acid by the psychrophilic marine bacterium, *Vibrio marinus*. Limnol. Oceanogr. 13:637–643.

Albright, L. J., and R. Y. Morita. 1972. Effect of environmental parameters of low temperature and hydrostatic pressure on L-serine deamination by *Vibrio marinus*. J. Oceanogr. Soc. Jap. 28:63–70.

Arnold, R. M., and L. J. Albright. 1971. Hydrostatic pressure effects on the translation stages of protein synthesis in a cell-free system from *Escherichia coli*. Biochim. Biophys. Acta 238:347–354.

Becker, V. E., and H. J. Evans. 1969. The influence of monovalent cations and hydrostatic pressure on galactosidase activity. Biochim. Biophys. Acta 191:95–104.

Boatman, E. S. 1967. Electron microscopy of two marine *Bacillus* spp. subjected to hydrostatic pressure during growth. Bacteriol. Proc. p. 25.

Certes, A. 1884a. Sur la culture, a l'abri des germes atmospheriques, des eaux et des sediments rapportes par les expeditions du Travailleur et du Talisman; 1882–1883. C. R. Acad. Sci. Paris 98:690–693.

Certes, A. 1884b. Note relative a l'action des hautes pressions sur la vitalite des micro-organisms d'eau douce et d'eau de mer. C. R. Soc. Biol. 36:220–222.

Haight, R. D., and R. Y. Morita. 1962. Interaction between the parameters of hydrostatic pressure and temperature on aspartase of *Escherichia coli*. J. Bacteriol. 83:112–120.

Hill, E. P., and R. Y. Morita. 1964. Dehydrogenase activity under hydrostatic pressure by isolated mitochondria obtained from *Allomyces macrogynus*. Limnol. Oceanogr. 9:243−248.

Johnson, F. H., H. Eyring, and M. J. Polissar. 1954. The Kinetic Basis of Molecular Biology. John Wiley, New York.

Kim, J., and ZoBell, C. E. 1972. Agarase, amylase, cellulase, and chitinase activity at deep-sea pressures and temperatures. J. Oceanogr. Soc. Jap.

Kinne, O. (ed.). 1972. Marine Ecology, Vol. 1, Part 3. Interscience, London.

Landau, J. V. 1966. Protein and nucleic acid synthesis in *Escherichia coli:* pressure and temperature effects. Science 153:1273−1274.

Landau, J. V. 1967. Induction, transcription and translation in *Escherichia coli:* a hydrostatic pressure study. Biochim. Biophys. Acta 149:506−512.

Landau, J. V. 1970. Hydrostatic pressure on the biosynthesis of macromolecules. *In* A. M. Zimmerman (ed.), High Pressure Effects on Cellular Processes, pp. 45−70. Academic Press, New York.

Morita, R. Y. 1965. Effect of hydrostatic pressure. *In* G. C. Ainsworth and A. S. Sussman (eds), The Fungi, pp. 551−558. Academic Press, New York.

Morita, R. Y. 1967. Effect of hydrostatic pressure on marine microorganisms. Oceanogr. Mar. Biol. Annu. Rev. 5:187−203.

Morita, R. Y. 1970. Application of hydrostatic pressure to microbial cultures. *In* J. R. Norris and D. W. Ribbons (eds), Methods in Microbiology, Vol. 2, pp. 243−258. Academic Press, New York.

Morita, R. Y. 1972. Pressure−Bacteria, fungi and blue green algae. *In* O. Kinne (ed.), Marine Ecology, Vol. 1, Part 3, pp. 1361−1388. Interscience, London.

Morita, R. Y., and R. R. Becker, 1970. Hydrostatic pressure effects on selected biological systems. *In* A. M. Zimmerman (ed.), High Pressure Effects on Cellular Systems, pp. 71−83. Academic Press, New York.

Morita, R. Y., and R. D. Haight. 1962. Malic dehydrogenase activity at 101°C under hydrostatic pressure. J. Bacteriol. 83:1341−1346.

Morita, R. Y., and P. F. Mathemeier. 1964. Temperature−hydrostatic pressure studies on partially purified inorganic pyrophosphatase activity. J. Bacteriol. 88:1667−1671.

Oppenheimer, C. H., and C. E. ZoBell. 1952. The growth and viability of sixty-three species of marine bacteria as influenced by hydrostatic pressure. J. Mar. Res. 11:10−18.

Paul, K. L., and R. Y. Morita. 1971. Effects of hydrostatic pressure and temperature on the uptake and respiration of amino acids by facultatively psychrophilic marine bacterium. J. Bacteriol. 108:835−843.

Penniston, J. T. 1971. High hydrostatic pressure and enzymic activity: inhibition of multimeric enzymes by dissociation. Arch. Biochem. Biophys. 142:322−332.

Regnard, P. 1884. Recherches experimentales sur l'influence des tres hautes pressions sur les organismes vivants. C.R. Acad. Sci. Paris 98:745−747. 98:745−747.

Schwarz, J. R., and J. V. Landau. 1972. Hydrostatic pressure effects on *Escherichia coli:* site of inhibition of protein synthesis. J. Bacteriol. 109:945–948.

Seki, H., and C. E. ZoBell. 1967. Microbial assimilation of carbon dioxide in the Japan Trench. J. Oceanogr. Soc. Jap. 23:182–188.

Yayanos, A. A., and E. C. Pollard. 1969. A study of the effects of hydrostatic pressure on macromolecular synthesis in *Escherichia coli.* Biophys. J. 9:1464–1484.

Zimmerman, A. M. (ed.). 1970. High Pressure Effects on Cellular Processes. Academic Press, New York.

ZoBell C. E. 1964. Hydrostatic pressure as a factor affecting the activities of marine microbes. *In* Y. Miyake and T. Koyama (eds), Recent Researches in the Fields of Hydrosphere, Atmosphere and Nuclear Chemistry, pp. 83–116. Muruzen, Tokyo.

ZoBell, C. E. 1968. Bacterial life in the deep sea. Bull. Misaki Mar. Biol. Inst. Kyoto Univ. 12:77–96.

ZoBell, C. E., and A. B. Cobet. 1962. Growth, reproduction and death rates of *Escherichia coli* at increased hydrostatic pressure. J. Bacteriol. 84:1228–1236.

ZoBell, C. E., and A. B. Cobet. 1964. Filament formation by *Escherichia coli* at increased hydrostatic pressure. J. Bacteriol. 87:710–719.

ZoBell, C. E., and F. H. Johnson. 1949. The influence of hydrostatic pressure on the growth and viability of terrestrial and marine bacteria. J. Bacteriol. 57:179–189.

ZoBell, C. E., and J. Kim. 1972. Effects of deep-sea pressure on microbial enzyme systems. Symp. Soc. Exp. Biol., The Effects of Pressure on Organisms 26:1–21.

ZoBell, C. E., and R. Y. Morita. 1957. Barophilic bacteria in some deep sea sediments. J. Bacteriol. 73:563–568.

ZoBell, C. E., and C. H. Oppenheimer. 1950. Some effects of hydrostatic pressure on the multiplication and the morphology of marine bacteria. J. Bacteriol. 60:771–781.

KINETIC REGULATION OF ENZYMIC ACTIVITY AT INCREASED HYDROSTATIC PRESSURE: STUDIES WITH A MALIC DEHYDROGENASE FROM A MARINE BACTERIUM

K. C. MOHANKUMAR* and LESLIE R. BERGER

Department of Microbiology
University of Hawaii

We recently described an optical pressure cylinder which can be used to follow reaction kinetics at hydrostatic pressures from 1 to 1000 atm approximately 10 sec following initiation of the reaction (Mohankumar and Berger, 1972). This apparatus makes it possible to measure enzyme kinetics under conditions of limited substrate. One would expect limited substrate in the intracellular and extracellular environments of most marine bacteria, most of the time. The effects of hydrostatic pressure on the activity of many enzymes have been reported. Among these, a number have dealt with dehydrogenases (for example, Hill and Morita, 1964; Carter *et al.*, 1971).

The recent work of Hochachka (1971) and his collaborators, using enzymes obtained from freshly caught fish, clearly indicates that the rates of enzymic reactions at increased hydrostatic pressure are often controlled by kinetic parameters, such as substrate concentration, rather than by changes in equilibria brought about by molecular-volume changes in activated complexes. This added dimension of physiological control may or may not be of

* This paper is taken in part from a thesis to be submitted in partial fulfillment of the requirements for the degree of Doctor of Philosophy in the Department of Microbiology at the University of Hawaii.

great importance in limiting or extending the range of environments in which a given species can exist. In this report some kinetics of an NAD-linked malic dehydrogenase (EC 1.1.1.37) obtained from a marine bacterium are presented.

MATERIALS AND METHODS

Enzyme was obtained as follows. Cells of a marine *Vibrio* (our strain 108) grown in basal mannitol-seawater medium were harvested during exponential growth, washed, and sonicated. One percent of an 11% streptomycin sulfate solution was added to the supernatant fluid of the cell extract to precipitate nucleic acids. Further purification was done by ammonium sulfate fractionation. Such preparations were free from NADH oxidase and malic enzyme (EC 1.1.1.39).

Enzyme reactions were assayed at 30°C in 0.05 M Tris-HCl buffer, pH 8.4 NAD or NADH was used in excess. Velocities of the reactions are expressed in OD per min at 340 nm.

RESULTS AND DISCUSSION

The rates of the reaction catalyzed by malic dehydrogenase were inhibited by excess malate or oxaloacetate (OAA; Fig. 1); excess NAD or NADH did not inhibit. Such kinetics are well known. In this instance, the data obtained with malate, but not with OAA, fit the model proposed by Walter (1965) shown below:

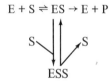

where E is free enzyme, S is free substrate, ES is active enzyme-substrate complex, ESS is inactive enzyme-substrate complex, and P is the product. An inactive complex of enzyme with two molecules of substrate is in equilibrium with the active enzyme-substrate complex. K_m is the Michaelis-Menten constant. K_s is the analogous constant for the formation of the inactive complex. Thus, at low substrate, ESS is negligible and may be neglected. At high substrate concentration much of the enzyme is tied up as inactive complex. The value of K_s may be determined graphically (see Fig. 2); the intercept of the abscissa is equal to K_s.

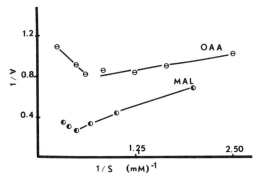

Fig. 1. Effect of concentration of malate and oxaloacetate on the rates of malic dehydrogenase activity. Assays were done at 1 atm and 30°C. Activity is expressed as change in OD per min measured at 340 nm. The ordinate scale for malate is reduced 10-fold.

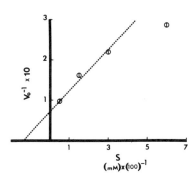

Fig. 2. Effect of high concentration of malate on the activity of malic dehydrogenase. Assay conditions as in Fig. 1. See text for explanation.

The effects of hydrostatic pressure at both low and high malate concentrations were measured on the rate of enzymic reaction. At low malate concentration the rate of reaction is inhibited by increased pressure at all temperatures tested. However, K_m decreases with increased pressure, partially offsetting the inhibition (Table 1). At high malate concentration the rate of reaction is enhanced by pressure, and K_s increases with increasing pressure also offsetting the decreased activity at the higher pressures (Fig. 3).

It is evident from Fig. 4 that the concentration of malate at which an optimum rate of enzymic activity occurs increases with increased hydrostatic pressure.

Table 1. Effects of hydrostatic pressure and temperature on the maximum velocity (V_m) and the Michaelis-Menten constant (K_m) of a malic dehydrogenase using malate as the substrate[a]

Temp. (°C)	1 atm		333		666		1000	
	K_m	V_m	K_m	V_m	K_m	V_m	K_m	V_m
18	9.0	0.8	7.2	0.6	5.9	0.5	5.88	0.16
23	10.5	0.9	10.0	0.7	7.4	0.7	6.25	0.7
30	11.0	2.2	10.5	2.0	8.0	1.7	7.8	1.4

[a] K_m is in mM; V_m is change in OD at 340/min.

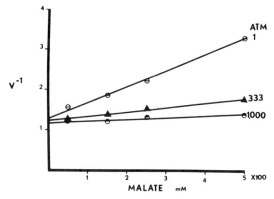

Fig. 3. Effect of malate concentration on the activity of malic dehydrogenase at various hydrostatic pressures. Other conditions as in Fig. 1.

A number of explanations for the above data exist. Pressure clearly enhances the formation of enzyme-substrate complex. At low substrate concentrations, the decreased rate exhibited at increased pressures could be due to denaturation of the enzyme, increase in the rate of reverse reaction (i.e., oxaloacetate to malate), or an increase in the energy of activation of the forward reaction. Experiments have shown the enzyme to be stable at increased pressure without substrate, and in the presence of either malate or NAD. Increased pressure also slows the conversion of oxaloacetate to malate (Fig. 5). In this case K_m increases with increased pressure. Thus the inhibition of the reaction going from malate to oxaloacetate is not due to the slow release of products from an enzyme-product complex.

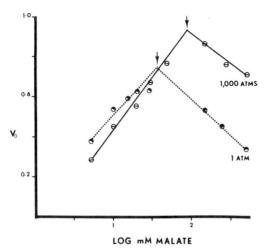

Fig. 4. Activity of malic dehydrogenase at various malate concentrations. The concentration of malate at which optimum activity occurs is indicated by the arrow. Conditions as in Fig. 1.

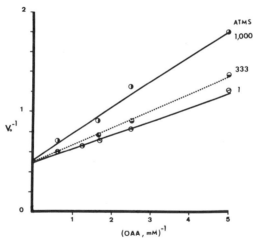

Fig. 5. Effect of oxaloacetate concentration on the activity of malic dehydrogenase at various hydrostatic pressures. Conditions as in Fig. 3.

In the data presented, increased hydrostatic pressure favors formation of an enzyme-malate complex but results in decreased enzymic activity. In a general way, organisms possessing such enzymes would be favored for their increased scavenging capacity for substrates, while at the same time maintaining their rates of reaction in balance with other metabolic activities since the rates of many enzymic reactions are reduced by increased pressures.

LITERATURE CITED

Carter, J. E., E. F. Graham, R. C. Lillehel, and P. L. Blackshear. 1971. The effect of high hydrostatic pressure and low temperature on lactic dehydrogenase and glutamic oxaloacetic transaminase. Cryobiology 8: 524–534.

Hill, E. P., and R. Y. Morita. 1964. Dehydrogenase activity under hydrostatic pressure by isolated mitochondria obtained from *Allomyces macrogynus*. Limnol. Oceanogr. 9:243–248.

Hochachka, P. W. 1971. Enzyme mechanisms in temperature and pressure adaptation of off-shore benthic organisms: the basic problem. Amer. Zoologist 11:425–435.

Mohankumar, K. C., and L. R. Berger. 1972. A method for rapid enzyme kinetic assay at increased hydrostatic pressure. Anal. Biochem. 49:336–342.

Walter, C. 1965. Steady-State Applications in Enzyme Kinetics, pp. 72–81. Ronald Press, New York.

COMPARATIVE EFFECTS OF PRESSURE ON
PROTEIN AND RNA SYNTHESIS IN
BACTERIA ISOLATED FROM MARINE SEDIMENTS*

RANDALL W. SWARTZ, JOHN R. SCHWARZ, and JOSEPH V. LANDAU

Department of Biology
Rensselaer Polytechnic Institute

Experiments reveal that protein and RNA synthesis in *Escherichia coli* are inhibited by pressure in a predictable manner. The response of protein synthesis to pressure suggests that the translation phase involves a rate-limiting reaction which occurs with a positive volume change of activation (ΔV^*) of approximately 100 cm^3/mole. The overall ΔV^* for the transcription phase is approximately 55 cm^3/mole. For translation in *Escherichia coli*, the rate-limiting process has been shown to involve the binding of aminoacyl-tRNA to the ribosome-mRNA complex. Similar studies of the responses of *Bacillus* 29 (Blake Plateau, 1700 m) and of *Pseudomonas bathycetes* (Marianas Trench, 10,000 m) show both to be more tolerant of increased pressure and decreased temperature than *Escherichia coli*. The quantitative results obtained with *Bacillus* 29 show an overall similarity to those of *Escherichia coli* for both transcription and translation. *Pseudomonas bathycetes* shows a much greater barotolerance than either *Bacillus* 29 or *Escherichia coli*. Furthermore, while

* The work presented is partially supported by a National Science Foundation Grant No. GA-28398. This paper is taken in part from a thesis to be submitted in partial fulfillment of the requirements for the degree of Doctor of Philosophy in the Department of Biology at Rensselaer Polytechnic Institute.

the response of transcription to pressure inhibition is quite similar to that of the other organisms studied, the response of translation in *Pseudomonas bathycetes* mimics that of transcription, suggesting that translation per se in this highly barotolerant strain is not inhibited directly by pressure.

It is suggested that the utilization of "response to pressure" of hybridized synthesizing systems could allow for a delineation of the barotolerant factor of marine bacteria indigenous to the deep-sea environment.

Marine microorganisms make up a substantial portion of the biomass of the oceans. Among these organisms the marine bacteria are important both as a food source and to complete the nitrogen and sulfur cycles (Wood, 1967; Droop and Wood, 1968). There is insufficient data about the ecology of the oceans to say much more about the specific involvement of bacteria. Heightened interest in the oceans as a source of natural resources (minerals, petroleum, animal protein) and as a waste-disposal site has accelerated interest in the aquatic environment: its flora, fauna, and ecology. The marine bacteria would seem to be a key element in this ecological system.

Many bacteria isolated from the oceans are able to grow at temperatures below 10°C (Quigley and Colwell, 1968) and hydrostatic pressures above 500 atm (Oppenheimer and ZoBell, 1952) which are lethal to terrestrial organisms. In spite of these environmental differences, marine bacteria have been shown to be morphologically and metabolically similar to their terrestrial cousins. ZoBell and Upham (1944) listed 60 new species of marine bacteria. Typically the temperature for their maximum growth rate was 20–25°C. For other marine organisms, such as *Vibrio marinus* strain MP-1, this temperature is 15–16°C (Morita and Haight, 1964). Many of these organisms, however, exhibit growth at temperatures as low as 3–5°C. For some time it was believed that the metabolic processes of organisms living at high hydrostatic pressure and low temperature must proceed more slowly than those of terrestrial forms. This prediction appears correct for some organisms, such as ZoBell and Morita's (1957) obligate barophilic sulfate reducers from the Weber Deep, which required as long as 10 months in a hyperbaric environment (700 atm) before growth at 4°C was demonstrable. Others, such as *Vibrio marinus,* are more similar in their metabolic rate to analogous terrestrial forms (Morita and Albright, 1965). Growth at moderate rates has been demonstrated for this organism at 3°C and at up to 300 atm. It should be noted that growth rates in the laboratory show only the limitations due to physical parameters. It is clear that the availability of metabolites plays an important role in determining the growth rate in the ocean. Growth at high hydrostatic pressure and low temperature has been demonstrated with many organisms. ZoBell and Johnson (1949) made a study of the effects of temperature and pressure on microbes from different environments. Several marine organisms were clearly distinguished by their marked tolerance for these conditions.

In spite of their apparent similarity to terrestrial forms, the deep-sea bacteria must exhibit at least a subtle difference at the molecular level, since their enzyme systems are operable at low temperature and high hydrostatic pressure. Under such conditions many terrestrial organisms are unable to maintain a stable metabolism. Morita (1957) has shown that several *Escherichia coli* enzymes are inhibited to varying degrees as pressure is increased and that increased pressure also has a substantial effect on enzymes from a marine *Vibrio*. Several amino-acid transport systems appear to be inhibited (Paul and Morita, 1971).

In the light of the marked inhibitory effects of increased hydrostatic pressure and reduced temperature on terrestrial organisms, studies were undertaken to determine how bacteria from the marine sediments were able to function. *Escherichia coli* was the terrestrial organism of choice. Good progress had been made in determining the inhibitory effect of pressure on the biosynthesis of protein and, as will be shown, experiments were underway which could delineate fairly well the site of such inhibition. Two marine organisms were selected, *Bacillus* 29 and *Pseudomonas bathycetes*.

Bacillus 29 is an unclassified strain obtained from Dr. Henry Ehrlich. It is indigenous to a moderate pressure environment. It was isolated from manganese nodules found on the floor of the Blake Plateau, approximately 200 miles east of the Georgia-Florida coast at a depth of 1700 m. This corresponds to a pressure of 170 atm. The temperature is presumably between 4 and 8°C judging from the depth, although there are subsurface currents in that region which could alter this somewhat.

Pseudomonas bathycetes, obtained from Dr. R. Colwell, was isolated from the sediment of both the Philippine and the Marianas Trenches. The depth of isolation was about 10,000 m which corresponds to 1000 atm pressure. The temperature at that depth is between 3 and 5°C. Our studies were performed on the type strain $C_2 M_2$ (ATCC #23597).

THE UNDERLYING THEORY

The basic outline of both protein and ribonucleic acid synthesis in bacterial systems is well known. Each of these processes requires the formation of specific molecular complexes, the kinetics of which may be interpreted in terms of activated-complex theory.

The effects of temperature and hydrostatic pressure on chemical reactions may be explained by the activated-complex theory of chemical reaction rates (Laidler, 1969). This theory is based on the concept of an energy barrier to the occurrence of a reaction and on the existence of a hypothetical activated complex. The concentration of activated complex determines the rate of the reaction.

The energy change of activation (ΔE^*) is the energy required for the formation of activated complex from reactants. It is at this point that the effect of thermal energy on reaction rates is felt. By analogous reasoning, hydrostatic pressure may be viewed as affecting the concentrations of activated complex through an effect on the ease of achieving the *volume change of activation* (ΔV^*).

If such a molecular interaction involves a volume increase, the effect of hydrostatic pressure application will be to retard the reaction. A reaction, the formation of whose activated complex involves a volume decrease, would be accelerated. If activated-complex formation involves no volume change, the rate should not be affected by pressure.

The rates of specific biological processes may be dependent on numerous chemical-reaction mechanisms, thus obscuring a definitive analysis of pressure effects. If, however, a single reaction step may be considered as rate limiting, and if its mechanism is not too complex, then the equation $\Delta V^* = [2.3\,RT \times \log(k_{p1}/k_{p2})]/(P_2 - P_1)$ can be applied (Johnson *et al.,* 1954), and one might expect a semilog plot of rate *versus* pressure to be linear. The bacterial apparatus utilized for the synthesis of proteins is certainly one of the most carefully delineated and best understood cellular systems. Both hydrostatic pressure and temperature are known to influence its rate (Landau, 1966; Yayanos and Pollard, 1969; Albright, 1969). Moreover, this influence is, at least in part, a direct effect upon the translation phase of the system as shown by *in vivo* and *in vitro* work with *Escherichia coli* (Landau, 1967; Schwarz and Landau, 1972*a,b;* Arnold and Albright, 1971).

PROTEIN AND NUCLEIC ACID SYNTHESIS: WHOLE-CELL STUDIES

In our laboratory the effect of pressure on protein and nucleic acid synthesis in *Escherichia coli* has been studied (Landau, 1966, 1970). In general, it was found that the application of a specific pressure may result in either stimulation or inhibition of incorporation of ^{14}C-labeled amino acid into protein, depending on temperature; e.g., a pressure of 265 atm stimulates at 37°C, has no measurable effect at 27°C, and inhibits at 22°C, whereas a pressure of 400 atm has no effect at 37°C and inhibits at lower temperatures. These results have led to the postulation of (1) a primary rate-controlling reaction which involves an activated enzyme and is directly involved with protein synthesis, and (2) a reversible thermal inactivation of the enzyme, inhibited by pressure application, with the inactivation displaying a greater sensitivity to pressure.

On this basis, application of 265 atm at 37°C would have little repressive effect on the primary reaction but would produce a greater amount of activated enzyme which could be measured in terms of a stimulation of amino-acid incorporation. On the other hand, application of 440 atm would result in a greater enzyme supply, but it would, in turn, inhibit the primary reaction to a greater extent, resulting in no net change in the rate of reaction. From this it follows that at temperatures below 25°C the thermal inactivation step should be eliminated and the pressure effect should be directly upon the primary reaction. The data indicated the possibility of an effect on a rate-limiting reaction with a ΔV^* of approximately 100 cm^3/mole.

The stimulation or lack of inhibition by pressure at certain temperatures might be the result of a controlling function associated with translation. The translational machinery would seem to be capable of a higher rate of synthesis than is normally needed at atmospheric pressure in log phase above 25°C. Thus, the system must be partially repressed in order to allow positive control. As this repression is removed, either by increased pressure or decreased temperature, the pressure response is linear on the plot of log incorporation rate *versus* pressure.

The effect of pressure at 37°C on the incorporation of radioactively labeled uridine into RNA is superficially similar to that on protein synthesis, with at least one notable difference. At 680 atm protein synthesis is totally inhibited while RNA synthesis, though markedly decreased, continues at a distinctly measurable rate (Landau, 1966).

More detailed studies have now been performed on this process and the results are shown in Fig. 1. Analysis of the linear portions of this graph reveals that the rate-limiting step for transcription in *Escherichia coli* has a volume change of activation of 55 cm^3/mole at both 25 and 30°C. Thus the separate effects of pressure on transcription and translation would seem to be easily distinguishable in *Escherichia coli*.

In order to delineate further which phase of protein synthesis pressure may be primarily affecting, the induced synthesis of a specific and measurable protein, β-galactosidase, was studied (Landau, 1967). This study also showed that a specific, functional protein could be synthesized under pressure. The data again indicated that at pressures above 265 atm, where the inhibitory effect is measurable, the effect may be considered to be on a rate-limiting reaction with a ΔV^* of 100 cm^3/mole. Under the specific conditions of these experiments this volume increase of activation was found to be associated with the translation phase of protein biosynthesis. During the course of these experiments it was also found that the formation of the inducer-repressor complex (Gilbert and Muller-Hill, 1966) was readily susceptible to pressure and that the pressure effect could be measured at all levels above 1 atm. Thus,

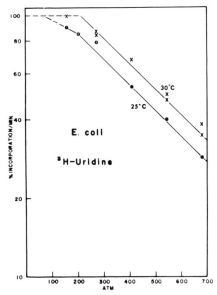

Fig. 1. [³H]Uridine incorporation into the cold TCA-insoluble cell fraction of *Escherichia coli.*

there was direct evidence that some important biochemical processes may be measurably affected by a relatively small increase in hydrostatic pressure.

Among the possible steps of translation at which inhibition by pressure could occur are amino-acid permeability, amino-acid activation and amino-acyl-tRNA formation, initiation, polysomal integrity, aminoacyl-tRNA binding to polysomes, peptide bond formation, translocation, and termination. Experiments have been performed to test certain of these possibilities.

An inhibition of amino-acid permeability is not the primary effect of pressure upon *Escherichia coli* protein synthesis. This has been shown in two ways. Using whole-cell preparations and labeled amino acids, permeability was demonstrated during a 5-min period of incubation for pressure and control cultures. The results showed that at 680 atm L-[¹⁴C]leucine, L-[¹⁴C]phenylalanine, and L-[¹⁴C]amino acid mixture entered and combined with the acid-soluble pool material of the cell. Also the amount of label entering the cell was at least the equivalent of that at atmospheric pressure. Thus, permeability to these amino acids did not seem to be the limiting factor in protein synthesis (Schwarz and Landau, 1972a). The effect of pressure on translation in a cell-free system is reported below. It will be seen that the pressure responses are quantitatively similar to those of whole-cell preparations (Schwarz and Landau, 1972*b*). Since there is no permeability barrier in the cell-free system, permeability cannot be a factor in pressure inhibition of translation in *Escherichia coli.*

The amino acylation of specific tRNA was shown to occur in cells kept at 680 atm pressure. The production of aminoacyl-tRNA was identical to that of control samples maintained at atmospheric pressure. Since 680 atm totally inhibits protein synthesis in *Escherichia coli,* it was concluded that amino-acyl-tRNA formation was not the site of inhibition by pressure (Schwarz and Landau, 1972*a*).

The effect of 680 atm upon polysomal integrity was also examined. Polysomes were extracted from pressurized and control cultures and run on linear 15–30% sucrose gradients. Polysome profiles as shown by optical densities were similar for pressure and control samples. In some experiments pressure even seemed to enhance polysome preservation. In any case, poly-somes were found to be stable under 680 atm pressure (Schwarz and Landau, 1972*a*).

Experiments on nascent peptide formation were performed in a similar system. Cells were pulse labeled with [^{14}C]amino acids immediately before and after application of 680 atm pressure. Results showed that if cells were allowed to form nascent peptides at atmospheric pressure, subsequent appli-cation of 680 atm did not remove the peptide already formed. Once pressure is applied, however, elongation of the growing peptide chain is inhibited (Schwarz and Landau, 1972*a*).

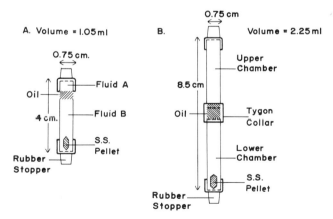

Fig. 2. Miniature reaction vessels for the addition of material to a cell-free system under pressure. (After Schwarz and Landau, 1972*b*.)

CELL-FREE EXPERIMENTS WITH *ESCHERICHIA COLI* K-12

In order to determine the particular step or steps in the translation process which was inhibited by pressure, it was necessary to examine the responses of cell-free protein-synthesizing systems (Nirenberg, 1963). Experimentation with these systems required the development of pressure vessels suitable for experimentation on sample volumes about one-tenth that of the vessels used for whole-cell experiments. The ability to mix components under pressure was also required. A system was developed which met these requirements, was easily produced, and quite inexpensive. The vessels are diagrammed in Fig. 2.

Initially, a cell-free system directed by synthetic messenger (poly U) was studied; then natural messenger (MS_2 viral) RNA was used (Schwarz and Landau, 1972b). As shown in Fig. 3, the results with both systems were identical. A log plot of incorporation per 5 min *versus* pressure shows no effect up to 136 atm. Between 204 and 544 atm there is a straight line indicating progressively greater inhibition. Above 544 atm, the system is completely shut off. The volume change of activation calculated from the straight line portion of the curve between 204 and 544 atm is about 100 cm^3/mole for both the poly U and MS_2 systems. The inhibition was completely and immediately reversible at all pressure levels tested.

The quantitative similarity between the results for *Escherichia coli in vivo* and *in vitro* protein-synthesis experiments is evidence that the cell-free system is a valid model for the whole-cell response and that the primary site of the pressure effect is one of or a combination of the steps of the translation process. Further use of the cell-free system also made possible experiments designed to pinpoint more precisely the rate-limiting step or steps. The complex initiation process of *in vivo* translation is eliminated in a poly U directed *in vitro* system, but retained in an MS_2 directed system. Since the response of both these systems to pressure is identical, it seems likely that the initiation process is not the primary site of pressure inhibition (Schwarz and Landau, 1972b).

Puromycin has been shown to act by competing with the incoming aminoacyl-tRNA for the carboxy-terminal end of the bound peptidyl-tRNA (Smith *et al.*, 1965). In this process a peptide bond is formed between the antibiotic and the released peptide chain (Gottesman, 1967). When labeled puromycin was added to a poly U directed protein-synthesizing system under 680 atm pressure the peptidyl [³H] puromycin extractable from the pressure sample was 84% of that from the atmospheric control (Schwarz and Landau, 1972b). There is no doubt, therefore, that peptide-bond formation, per se,

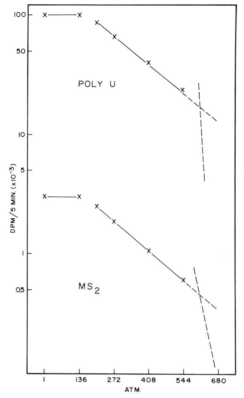

Fig. 3. The effect of pressure on cell-free protein synthesis. (After Schwarz and Landau, 1972*b*.)

can occur at a pressure which totally inhibits protein synthesis, and to a degree comparable with that found in controls maintained at atmospheric pressure.

The binding of phenylalanyl-tRNA to the ribosomes of a poly U directed cell-free synthesizing system was measured. The rate of this binding at atmospheric pressure and 37°C is linear within the first 8 min. This reaction is reversibly inhibited by pressure (Schwarz and Landau, 1972*b*). Arnold and Albright (1971) had shown that aminoacyl-tRNA binding was inhibited by pressure, but their experiments could not differentiate between an inability to bind at an unoccupied ribosomal site and an inability to bind at an already occupied site. Thus the rate-limiting step could have involved either binding per se, or some other step such as translocation, which must occur in order to create an open tRNA receptor site on the ribosome. In the experiment performed by Schwarz and Landau (1972*b*), it was shown that the pressure inhibition was caused by an inability to bind at a free ribosomal site. A log plot of binding rate *versus* pressure level for phenylalanyl-tRNA binding (Fig. 4) was identical to those of [^{14}C] phenylalanine incorporation into polypeptide in both the poly U directed and the MS$_2$ directed cell-free protein-synthesizing systems. Thus the binding reaction involving aminoacyl-tRNA with mRNA and ribosomes appears to be the rate-limiting reaction in translation which is inhibited by pressure.

EFFECTS OF HYDROSTATIC PRESSURE AND
VARYING TEMPERATURE ON SOME MARINE BACTERIA

With the *Escherichia coli* system as a basis for comparison and as a model for experimental procedure, *in vivo* experiments were performed on the two marine bacteria previously designated, i.e., *Bacillus* 29 and *Pseudomonas bathycetes*.

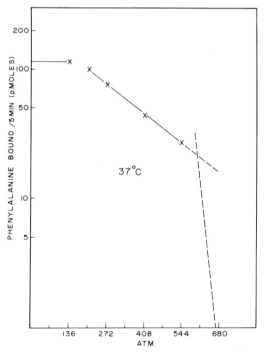

Fig. 4. The effect of pressure on the binding of [^{14} C] phenylalanyl-tRNA to the mRNA-ribosome complex. (After Schwarz and Landau, 1972*b*.)

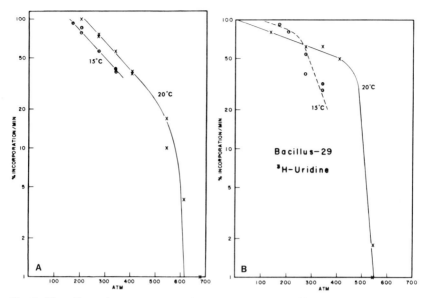

Fig. 5. The effect of pressure on the incorporation of (A) [¹⁴C] amino acids and (B) [³H] uridine into the cold TCA-insoluble cell fraction of *Bacillus* 29.

With each strain it was possible to obtain linear incorporation of ^{14}C mixed amino acids and [^{3}H] uridine into the cold TCA-insoluble cell fraction, which were taken to represent newly synthesized protein and RNA, respectively. Experiments were then performed to determine the effects of pressure upon these linear rates of transcription and translation. It was determined that hydrostatic pressure is inhibitory above a threshold pressure, as is the case with organisms indigenous to atmospheric pressure, but the pressure effects on *Bacillus* 29 and *Pseudomonas bathycetes* are different in several aspects. The results are shown in Figs. 5 and 6. Each semilog plot shows the rate of incorporation of [^{14}C] amino-acid mixture or [^{3}H] uridine into the cold acid insoluble cell fraction *versus* pressure. The rates are shown as a percentage of the rate at atmospheric pressure for the particular temperature under consideration. In these graphs the shift to the left as temperature is decreased reflects the fact that at reduced temperature the systems under study operate at a greater percentage of their maximum rate. Thus there is a reduced capacity for positive control.

With *Bacillus* 29 the overall nature of the semilog plot of [^{14}C] amino-acid incorporation at 20°C is similar to that of *Escherichia coli* at 37°C, although the rate of inhibition by pressure is greater (Fig. 5A). There is no inhibition below about 200 atm. Above this pressure there is an extensive

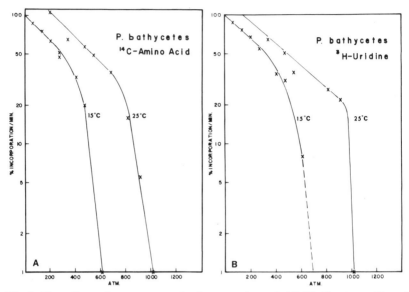

Fig. 6. The effect of pressure on the incorporation of (A) [^{14}C]amino acids and (B) [^{3}H]uridine into the cold TCA-insoluble cell fraction of *Pseudomonas bathycetes*.

linear portion of the curve which is closely paralleled by experiments at 15°C. The ΔV^* is about 120 cm^3/mole. An interesting aspect is that synthesis continues at a reduced rate until a pressure above 500 atm at 20°C, much higher than the critical pressure for *Escherichia coli* at this temperature. Thus comparison of the effects of pressure and temperature on *Escherichia coli* and *Bacillus* 29 show quantitative differences both in the kinetic effects and in the limiting effects of pressure.

Parallel studies have been performed to monitor transcription using [^{3}H]uridine as a label (Fig. 5B). As with translation, the Arrhenius-type plot for ΔV^* is linear and shows that pressure is inhibitory above a threshold pressure. Because the log plot is linear, a single reaction may be involved. A ΔV^* of approximately 40 cm^3/mole has been calculated for [^{3}H]uridine incorporation at 20°C.

Pseudomonas bathycetes protein synthesis responds similarly to the other organisms studied qualitatively, but certain quantitative aspects are quite different.

A semilog plot of [^{14}C]amino-acid incorporation indicates a ΔV^* for translation of about 55 cm^3/mole at both 25 and 15°C (Fig. 6A). At 25°C this portion of the curve, which begins at about 200 atm, extends out to 750 atm, and then drops off sharply as it reaches 1% of the atmospheric rate at about 1000 atm. The 15°C experiments have a linear slope, corresponding to 55 cm^3/mole up to a pressure of about 350 atm, which then drops off, reaching 1% of the atmospheric pressure rate at about 600 atm.

The response of transcription, as measured by incorporation of [^{3}H] uridine is quite similar to that of translation in *Pseudomonas bathycetes* at both temperatures and at all pressures examined (Fig. 6B). In both *Escherichia coli* and *Bacillus* 29, it appeared that direct pressure effects on transcription and translation could be distinguished from one another. This does not seem possible in *Pseudomonas bathycetes*. It is important to note that the ΔV^* for both transcription and translation in *Pseudomonas bathycetes* is identical to that for transcription in *Escherichia coli*. In all other systems studied, there is a substantial, direct effect of pressure on translation. This has been most clearly shown for *Escherichia coli* in the cell-free synthesizing system where no transcription is involved, but also appears to be true of the other systems since in all of them the ΔV^* for translation is substantially larger than that for transcription. It is possible that pressure may not directly affect the translation phase of protein synthesis in *Pseudomonas bathycetes*. Definitive information concerning the effect of pressure on translation might be obtained using a *Pseudomonas bathycetes* cell-free system. Here transcription is not a factor. Experiments with such a system have begun.

Since the responses of *Pseudomonas bathycetes* to pressure are quantitatively different from those of *Escherichia coli*, these differences might represent either adaptation to the marine environment or genus specificity and selection. Whole-cell studies of the pressure responses of a terrestrial pseudomonad, *Pseudomonas fluorescens*, are underway. If the response of this strain emulates that of *Escherichia coli*, then the response of *Pseudomonas bathycetes* would seem to suggest an adaptation to the abyssal marine environment. If *Pseudomonas fluorescens* behaves more like *Pseudomonas bathycetes*, this would suggest that pseudomonads are selected for by the high hydrostatic pressure of the marine environment.

Whether the results indicate genus specificity or adaptation, we believe that it now will be possible to identify the specific component of the synthesizing system which allows a pressure tolerance. This will be done by studying the response to pressure of several hybrid cell-free systems, each containing constituents from both barotolerant and non-barotolerant organisms.

LITERATURE CITED

Albright, L. J. 1969. Alternate pressurization-depressurization effects on growth and net protein synthesis by *Escherichia coli* and *Vibrio marinus*. Can. J. Microbiol. 15:1237–1240.

Arnold, R. M., and L. J. Albright. 1971. Hydrostatic pressure effects on the translation stages of protein synthesis in a cell-free system from *Escherichia coli.* Biochim. Biophys. Acta 238:347−354.

Droop, M. R., and E. J. F. Wood (eds). 1968. Advances in Microbiology of the Sea, Vol. I. Academic Press, New York.

Gilbert, W., and B. Muller-Hill. 1966. Isolation of the Lac Repressor. Proc. Nat. Acad. Sci. U.S. 56:1891−1898.

Gottesman, M. 1967. Reaction of ribosome-bound peptidyl transfer ribonucleic acid with aminoacyl transfer ribonucleic acid or puromycin. J. Biol. Chem. 242:5564−5571.

Johnson, F. H., H. Eyring, and M. J. Polissar. 1954. The Kinetic Basis of Molecular Biology, Chap. 9. John Wiley, New York.

Laidler, K. J. 1969. Theories of Chemical Reaction Rates, Chap. 3. McGraw-Hill, New York.

Landau, J. V. 1966. Protein and nucleic acid synthesis in *Escherichia coli:* pressure and temperature effects. Science 153:1273−1274.

Landau, J. V. 1967. Induction, transcription and translation in *Escherichia coli:* a hydrostatic pressure study. Biochim. Biophys. Acta 149:506−512.

Landau, J. V. 1970. Hydrostatic pressure on the biosynthesis of macromolecules. *In* A. M. Zimmerman (ed.), High Pressure Effects on Cellular Processses, pp. 45−70. Academic Press, New York.

Morita, R. Y. 1957. Effect of hydrostatic pressure on succinic, formic, and malic dehydrogenases in *Escherichia coli.* J. Bacteriol. 74:251−255.

Morita, R. Y., and L. J. Albright. 1965. Cell yields of *Vibrio marinus,* an obligate psychrophile at low temperature. Can. J. Microbiol. 11:221−227.

Morita, R. Y., and R. D. Haight. 1964. Temperature effects on the growth of an obligate psychrophilic marine bacterium. Limnol. Oceanogr. 9:103−106.

Nirenberg, M. W. 1963. Cell free protein synthesis directed by messenger RNA. *In* S. P. Colowick and N. O. Kaplan (eds), Methods in Enzymology, Vol. 6, pp. 17−23. Academic Press, New York.

Oppenheimer, C. H., and C. E. ZoBell. 1952. The growth and viability of sixty-three species of marine bacteria as influenced by hydrostatic pressure. J. Mar. Res. 11:10−18.

Paul, K. L., and R. Y. Morita. 1971. Effects of hydrostatic pressure and temperature on the uptake and respiration of amino acids by a facultatively psychrophilic marine bacterium. J. Bacteriol. 108:835−843.

Quigley, M. M., and R. R. Colwell. 1968. Properties of bacteria isolated from deep-sea sediments. J. Bacteriol. 95:211−220.

Schwarz, J. R., and J. V. Landau. 1972*a.* Hydrostatic pressure effects on *Escherichia coli:* site of inhibition of protein synthesis. J. Bacteriol. 109:945−948.

Schwarz, J. R., and J. V. Landau. 1972*b.* Inhibition of cell-free protein synthesis by hydrostatic pressure. J. Bacteriol. 112:1222−1227.

Smith, J. D., R. R. Traut, G. M. Blackburn, and R. E. Munro. 1965. Action of puromycin in polyadenylic acid-directed polylysine synthesis. J. Molec. Biol. 13:617—624.

Wood, E. J. F. 1967. Microbiology of Oceans and Estuaries, Chap. 10. Elsevier, New York.

Yayanos, A. A., and E. C. Pollard. 1969. A study of the effects of hydrostatic pressure on macromolecular synthesis in *Escherichia coli*. Biophys. J. 9:1464—1482.

ZoBell, C. E., and F. H. Johnson. 1949. The influence of hydrostatic pressure on the growth and viability of terrestrial and marine bacteria. J. Bacteriol. 57:179—189.

ZoBell, C. E., and R. Y. Morita. 1957. Barophilic bacteria in some deep sea sediments. J. Bacteriol. 73:563—568.

ZoBell, C. E., and H. C. Upham. 1944. A list of marine bacteria including descriptions of sixty new species. Bull. Scripps Inst. Oceanogr. 5:239—292.

HYDROSTATIC PRESSURE EFFECTS UPON
PROTEIN SYNTHESIS IN TWO BAROPHOBIC BACTERIA

LAWRENCE J. ALBRIGHT and MICHAEL J. HARDON

Department of Biological Sciences
Simon Fraser University

The effects of hydrostatic pressure on protein synthesis in *Escherichia coli* B/r and *Vibrio marinus* MP-1 were examined. Pressure inhibits protein synthesis in whole cells at least in part by reducing active transport of exogenous amino acids. Polypeptide synthesis in cell-free systems under pressure is less easily inhibited when amino-acid precursors are presented already attached to tRNA. Phenylalanyl-tRNA synthetase activity is resistant to inhibition by pressure; however, binding of aminoacyl-tRNA to ribosomes is suppressed, formation of the peptide bond is inhibited, and the apparent stability of the mRNA-ribosome complex is reduced. Moderate pressures appear to effect an increase in fidelity of translation of the genetic code.

The world ocean is an environment characterized by hydrostatic pressures ranging from 1 atm at the surface to approximately 1150 atm in the greatest known depths. The organisms which comprise the indigenous microbial flora of this environment have been classified according to their responses to and tolerances of these pressures (ZoBell, 1968; ZoBell and Johnson, 1949). First, there are the barophobes, organisms unable to grow at pressures greater than 400–600 atm, originating mostly in surface and inshore waters. Second, there are the barophiles, organisms which grow only at pressures of 500–1100 atm, originating in the deeper portions of the ocean. Last, a very few species of baroduric microbes have been reported. These organisms grow over the entire pressure range of 1–1100 atm.

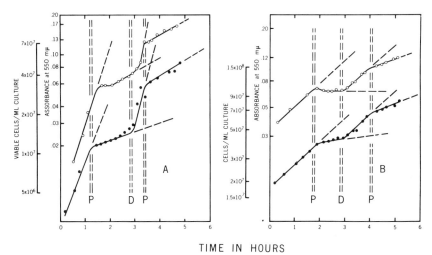

TIME IN HOURS

Fig. 1. The effect of the alternate application of hydrostatic pressures of 1 and 544 atm upon cell division (•) and culture absorbance (○) of (A) *Escherichia coli* B/r, and (B) *Vibrio marinus* MP-1. P and D indicate 544 and 1 atm pressure application respectively. (From Albright, 1969.)

Hydrostatic pressure is believed to inhibit the growth of barophobic bacteria in a number of ways, including the following: (1) by reducing the active transport of nutrients and diminishing cell permeability (Paul and Morita, 1971), (2) by altering the activities of anabolic and catabolic enzymes (Johnson *et al.,* 1954; Morita, 1967), and (3) by inhibiting macromolecular synthesis (Albright, 1969; Landau, 1970; Pollard and Weller, 1966).

Studies in this laboratory have involved two barophobic bacteria: the marine psychrophile *Vibrio marinus* MP-1, which grows from 1–420 atm*, and *Escherichia coli* B/r, which grows at pressures of 1–500 atm.

Albright (1969) showed that 544 atm pressure causes an immediate and almost complete inhibition of growth and cell division in both *Escherichia coli* and *Vibrio marinus,* followed by an almost immediate resumption of these processes upon pressure release to 1 atm (Fig. 1). It was also demonstrated that net protein synthesis is inhibited almost immediately by 544 atm, although RNA and DNA syntheses are more resistant to inhibition by this pressure (Fig. 2). Pressure release within 1.0–1.5 hr allow synthesis to resume at the rates observed at 1 atm before pressurization. Protein synthesis is more pressure sensitive than RNA synthesis; DNA synthesis is less sensitive than

---

* The maximum pressure to which *Vibrio marinus* grows is dependent upon the salinity of the medium. As the salinity is lowered, the maximum pressure for growth is decreased, e.g., no growth takes place above ca. 50 atm at salinity 2.5⁰/₀₀ (Palmer and Albright, 1970).

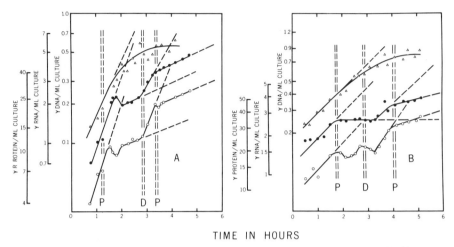

TIME IN HOURS

Fig. 2. The effect of the alternate application of hydrostatic pressures of 1 and 544 atm upon net protein (○), RNA (●) and DNA (△) synthesis by (A) *Escherichia coli* B/r, and (B) *Vibrio marinus* MP-1. P and D indicate 544 and 1 atm pressure application respectively. (From Albright, 1969.)

RNA synthesis. Accordingly, we have investigated the effects of pressure on protein synthesis more thoroughly.

Protein synthesis in whole cells requires an adequate supply of amino-acid residues. Exogenous sources of such residues may not be readily available to cells under pressure, since we have shown that active transport of phe[†] into *Escherichia coli* is sharply decreased as pressure is increased from 1 to 500 atm, with almost no uptake whatsoever at 600 atm or greater (Fig. 3). Paul and Morita (1971) obtained similar results in a study of the uptake of four amino acids by *Vibrio marinus* MP-38, demonstrating that uptake activity at 500 atm was less than 10% of the value observed at 1 atm. Endogenous sources of amino acids are also likely to be restricted in cells under elevated pressures, since pressure inhibits the catabolic processes which ultimately yield the energy and precursors required for *de novo* biosynthesis of amino acids (Johnson *et al.*, 1954; Morita, 1957; Haight and Morita, 1962). Since it is probable, however, that pressure also affects the further stages of protein synthesis, we have investigated several of these using cell-free systems from *Escherichia coli*.

In the first set of experiments, the effects of pressure on incorporation of phe into polypeptide in a poly U directed system were examined (Fig. 4).

† Abbreviations are as follows: phe, L-phenylalanine; leu, L-leucine; phe-tRNA, L-phenylalanyl-transfer RNA; leu-tRNA, L-leucyl-transfer RNA; poly U, polyuridylic acid.

Pressures of 100–500 atm cause an immediate, marked reduction in incorporation, with no polypeptide synthesis occurring at 600 atm or greater. In the second set of experiments, the effects of pressure on incorporation of phe presented to the system as phe-tRNA were examined (Fig. 5). Incorporation is reduced by only a small amount at pressures up to 300 atm, although a sharp reduction occurs at 400 atm or greater, with no incorporation taking

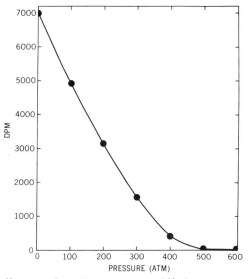

Fig. 3. Pressure effects on the active transport of [^{14} C] phe by whole cells of *Escherichia coli* B/r. (From Hardon and Albright, 1974.)

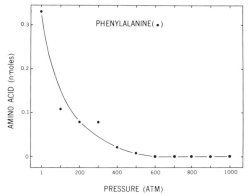

Fig. 4. Pressure effects on the incorporation of [^{14} C] phe into polypeptide. (From Hardon and Albright, 1974.)

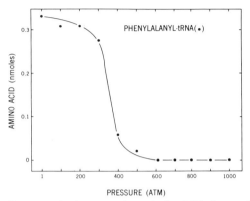

Fig. 5. Pressure effects on the incorporation of phe-tRNA into polypeptide. (From Hardon and Albright, 1974.)

place at pressures greater than 600 atm. These results are in contrast to the data obtained when phe is presented unattached to tRNA and the differences are summarized in Table 1. The amino-acid system responds immediately to pressure through a sharp reduction in incorporation. The aminoacyl-tRNA system appears to be more pressure resistant in the same range of pressures, although both systems eventually display minimum values at 400 and 500 atm which are quite similar. These data suggested that the final steps in protein synthesis, namely the binding of aminoacyl-tRNA to the ribosome and subsequent formation of the peptide bond, are, in fact, more pressure

Table 1. Comparison of the effects of pressure on phe and phe-tRNA incorporation into polypeptide[a]

Hydrostatic pressure (atm)	Phe (% incorporation at 1 atm)	Phe-tRNA (% incorporation at 1 atm)
100	46.5	97.2
200	25.5	93.0
300	13.5	82.6
400	6.0	16.5
500	2.1	3.0

[a] Tabulation of data of Figs. 4 and 5.

resistant than is the first step, namely the activation of amino acid to form aminoacyl-tRNA. Accordingly, the effects of pressure on these stages in protein synthesis were examined.

The effect of pressure on the poly U directed binding of the phe-tRNA to isolated *Escherichia coli* ribosomes is shown in Fig. 6. Pressures up to 600 atm cause a marked inhibition of binding of aminoacyl-tRNA to the poly U-ribosome complex. Pressures greater than 600 atm may cause a dissociation of the phe-tRNA which binds to the ribosomes before pressure is applied, for the following reason. These data are uncorrected for the binding of 8.5 pM of phe which occurs during the 1.5 min interval before pressurization, and for the 3.5 pM of phe which binds during the 0.5 min interval after pressurization. The total value, 11.9 pM, exceeds the binding observed at pressures greater than 600 atm.

A study of the kinetics of phe-tRNA binding indicated that the rate of binding is reversibly reduced at 400 atm, and almost completely inhibited at 600 atm (Fig. 7). One thousand atm appears to cause a complete and rapid dissociation of the phe-tRNA which binds in the 90 sec period before pressure is applied.

Subsequent experiments demonstrated a difference between the effects of pressure on the binding of phe-tRNA to ribosomes and the effects of

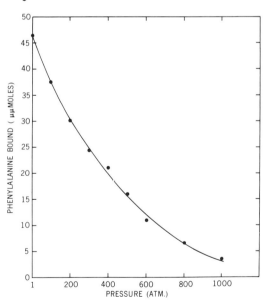

Fig. 6. Pressure effects on the poly U directed binding of [¹⁴C]phe-tRNA to *Escherichia coli* B/r ribosomes. (From Arnold and Albright, 1971.)

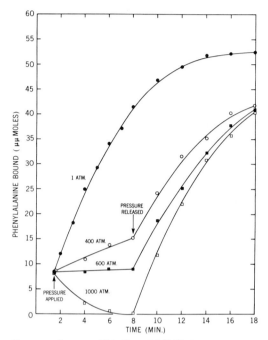

Fig. 7. Pressure effects on the rate of binding of [^{14}C] phe-tRNA to *Escherichia coli* B/r ribosomes. The mixed preparations were pressurized for 6.5 min after remaining at 1 atm for 1.5 min. The data are corrected for binding which occurs in the 35-sec interval between pressure release to 1 atm and reaction termination. (From Arnold and Albright, 1971.)

pressure on peptide-bond formation between adjacent phe residues (Fig. 8). Although both processes are inhibited by pressure, peptide-bond formation is more sensitive than aminoacyl-tRNA binding, as shown in Table 2. Here the ratio of phe bound to phe polymerized increases by a factor of almost 2, as pressure is increased from 1 to 600 atm.

The effects of pressure on amino-acid binding and peptide-bond formation may be a reflection of pressure-induced dissociation of messenger RNA (in this case, poly U) from the ribosome. However, the rate of binding of poly U to ribosomes is such that the reaction is virtually complete before pressure can be applied. This problem was approached indirectly, therefore, using the following procedure. Binding sites on ribosomes were first saturated with [^{14}C] poly U, then unlabeled poly U was added to the reaction mixture, and finally pressure was applied. Control experiments at 1 atm showed that no exchange occurred between the labeled and the unlabeled poly U. It was assumed that, at depressurization, competition of labeled and unlabeled poly

U for ribosomal binding sites would result in dilution of the label in proportion to the amount of dissociation occurring under pressure. The data presented in Fig. 9 show that an increase in the concentration of unlabeled poly U results in a greater dilution of bound label, and therefore support the

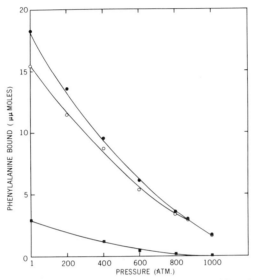

Fig. 8. The influence of pressure upon binding of phe-tRNA (○) and phe-tRNA (■) to *Escherichia coli* B/r ribosomes. Total phe binding (●) is the sum of the other curves. (From Arnold and Albright, 1971.)

Table 2. The effect of pressure on the ratio of phe bound to phe polymerized by the cell-free *Escherichia coli* B/r ribosomal preparation[a]

Hydrostatic pressure (atm)	Phe bound : phe polymerized
1	5.3
200	6.2
400	7.7
600	10.2

[a] Tabulation of data of Fig. 8.

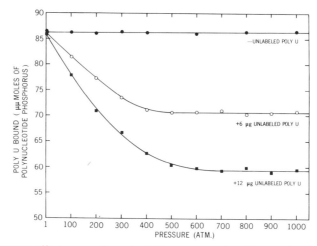

Fig. 9. Pressure effects upon the poly U-ribosome complex. The reaction mixture was treated with unlabeled poly U in the amounts indicated. (From Arnold and Albright, 1970.)

hypothesis that dissociation of the messenger from the ribosome occurs under pressure.

The aminoacyl-tRNA-synthetase reaction in *Escherichia coli* appears to be more pressure resistant than was expected. Palmer (1970) has shown that, although pressure does inhibit the activity of this enzyme, approximately 60% of the activity observed at 1 atm still remains at 600 atm, and that low levels of activity can be detected at pressures up to 1000 atm. These data suggest that aminoacylation is not a critically sensitive step in protein synthesis under pressure.

Further experiments have been undertaken to determine the effects of pressure on fidelity of translation of the genetic code, by examining the incorporation of phe and leu in separate poly U directed systems. In theory, poly U should direct only the incorporation of phe into polypeptide, but, in practice, it has been shown that leu is also incorporated. Thus, the cell-free system exhibits some amount of inherent ambiguity. The results of these experiments (Fig. 10) are in agreement with this, since both leu and phe were incorporated at 1 atm. Application of moderate hydrostatic pressures, from 100 to 300 atm, effects an immediate reduction in both cases, although leu incorporation appears to be inhibited more extensively than phe incorporation. This response is analogous to that observed previously when the incubation temperature was increased from 20 to 37°C in experiments performed at 1 atm (Fig. 11). An increase in temperature results in a marked stimulation of phe incorporation, but only a moderate stimulation of leu incorporation.

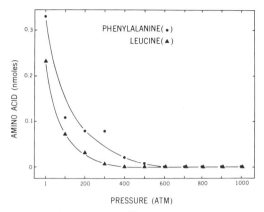

Fig. 10. Incorporation of [^{14}C]phe and [^{14}C]leu into polypeptide in poly U directed systems at 1–1000 atm. (From Hardon and Albright, 1974.)

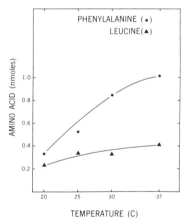

Fig. 11. Temperature-dependent incorporation of [^{14}C]phe and [^{14}C]leu into polypeptide. (From Hardon and Albright, 1974.)

Conversely, it might be said that a reduction in temperature increases the ratio of leu to phe, and thereby appears to increase the frequency of coding errors. Pressure and temperature data are summarized in Fig. 12, where the graph of the ratio of leu to phe incorporated under increasing temperature can be compared with the graph of the same ratio for increasing pressure. The similarity in response seems to indicate that, although hydrostatic pressure progressively inhibits net amino-acid incorporation, it may also function to reduce the frequency of coding errors in much the same way that an increase in temperature does.

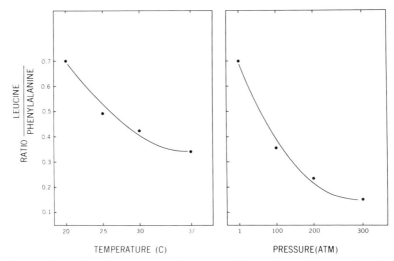

Fig. 12. The ratios of leu to phe incorporated into polypeptide as a function of temperature and pressure (derived from the data of Figs. 10 and 11). (From Hardon and Albright, 1974.)

Recent studies (Infante and Krauss, 1971; Infante and Baierlein, 1971; Waterson *et al.,* 1972) indicate that hydrostatic pressures up to 1200 atm encountered by sea-urchin ribosomes during ultracentrifugation induce dissociation of free ribosomes (75 S) to subunits (56 and 35 S), although dissociation was not observed when ribosomes were complexed with tRNA and/or mRNA. Hence pressure may affect the complex coupling of subunits into active ribosomes, as well as suppressing the synthesis of the proteins and nucleotides which comprise these subunits.

The data and observations discussed in this communication are summarized in Fig. 13. The following aspects of protein synthesis in *Escherichia coli* are known to be inhibited by increasing hydrostatic pressure from 1 to 1000 atm: (1) permeability and active transport of amino acids; (2) aminoacylation of tRNA; (3) stability of the mRNA-ribosome complex; (4) formation of the phe-tRNA mRNA-ribosome complex; (5) synthesis of the peptide bond. The biosynthesis of protein and ribonucleic acid precursors of ribosomes may also be inhibited. Other key processes in protein biosynthesis include (1) synthesis of tRNA, (2) synthesis of mRNA, and (3) synthesis of amino acids *de novo,* but the extent to which pressure affects these processes is unknown at the present time. In addition, moderate pressures appear to effect an increase in the fidelity of translation of the genetic code.

In summary, the influence of hydrostatic pressure on protein synthesis in bacterial systems is only partially understood. Indeed, a total understanding

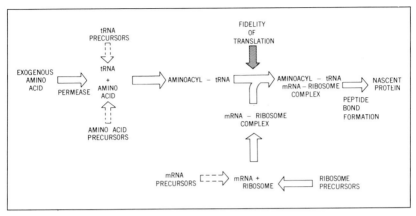

Fig. 13. Model of protein synthesis in *Escherichia coli,* indicating points of pressure effects. Reactions known to be inhibited by pressure are symbolized by open arrows. The size of the arrow is proportional to the degree to which the reaction proceeds under pressure. Pressure-induced increase in translational fidelity is denoted by a hatched arrow. Where pressure effects are unknown, reactions are symbolized by broken arrows.

of protein synthesis per se is yet to be achieved. Given that the translation stage alone involves the orderly interaction of more than 100 different macromolecules (Kurland, 1970), it is not surprising that this is so. Hopefully our understanding of the effects of hydrostatic pressure on protein synthesis will progress at a rate commensurate with the elucidation of the overall mechanism of protein synthesis.

ACKNOWLEDGMENTS

The authors gratefully acknowledge the technical assistance of Elaine Wilson and the financial support of the National Research Council of Canada.

LITERATURE CITED

Albright, L. J. 1969. Alternate pressurization–depressurization effects on growth and net protein, RNA and DNA synthesis by *Escherichia coli* and *Vibrio marinus.* Can. J. Microbiol. 15:1237–1240.

Arnold, R. M., and L. J. Albright. 1971. Hydrostatic pressure effects on the translation stages of protein synthesis in a cell free system from *Escherichia coli.* Biochim. Biophys. Acta 238:347–354.

Haight, R. D., and R. Y. Morita. 1962. Interaction between the parameters of hydrostatic pressure and temperature on aspartase of *Escherichia coli.* J. Bacteriol. 83:112–120.

Hardon, M. J., and L. J. Albright. 1974. Effects of hydrostatic pressure on protein synthesis and coding ambiguity in cell-free systems from *E. coli* B/r. (In press.)

Infante, A. A., and R. Baierlein. 1971. Pressure-induced dissociation of sedimenting ribosomes: effect on sedimentation patterns. Proc. Nat. Acad. Sci. 68:1780–1785.

Infante, A. A., and M. Krauss. 1971. Dissociation of ribosomes induced by centrifugation: evidence for doubting conformational changes in ribosomes. Biochim. Biophys. Acta 246:81–99.

Johnson, F. H., H. Eyring, and M. J. Polissar. 1954. The Kinetic Basis of Molecular Biology, 874 p. John Wiley, New York.

Kurland, G. C. 1970. Ribosome structure and function emergent. Science 169:1171.

Landau, J. V. 1970. Hydrostatic pressure on the biosynthesis of macromolecules. *In* A. Zimmerman (ed.), High Pressure Effects on Cellular Processes, pp. 45–70. Academic Press, New York.

Morita, R. Y. 1957. Effects of hydrostatic pressure on succinic, formic, and malic dehydrogenases in *Escherichia coli*. J. Bacteriol. 74:251–255.

Morita, R. Y. 1967. Effects of hydrostatic pressure on marine microorganisms. Oceanogr. Mar. Biol. Annu. Rev. 5:187–203.

Palmer, D. S. 1970. The effects of hydrostatic pressure and salt on cell growth and on phenylalanyl-tRNA synthetase activity of *Vibrio marinus* and *Escherichia coli*. M.Sc. Thesis. Simon Fraser Univ., Burnaby, British Columbia, Canada.

Palmer, D. S., and L. J. Albright. 1970. Salinity effects on the maximum hydrostatic pressure for growth of the marine psychrophilic bacterium, *Vibrio marinus*. Limnol. Oceanogr. 15:343–347.

Paul, K. L., and R. Y. Morita, 1971. Effects of hydrostatic pressure and temperature on the uptake and respiration of amino acids by a facultatively psychrophilic marine bacterium. J. Bacteriol. 108:835–843.

Pollard, E. C., and P. K. Weller. 1966. The effect of hydrostatic pressure on the synthetic processes in bacteria. Biochim. Biophys. Acta 112:573–580.

Waterson, J., M. L. Sopori, S. L. Gupta, and P. Lengyel. 1972. Apparent changes in ribosome conformation during protein synthesis. Centrifugation at high speed distorts initiation, pretranslocation and posttranslocation complexes to a different extent. Biochemistry 11:1377–1382.

ZoBell, C. E. 1968. Bacterial life in the deep sea. Bull. Misaki Mar. Biol. Inst. Kyoto Univ. Jap. 12:77–96.

ZoBell, C. E., and F. H. Johnson. 1949. The influence of hydrostatic pressure on the growth and viability of terrestrial and marine bacteria. J. Bacteriol. 57:170–189.

MEASUREMENT OF ACTIVE TRANSPORT BY BACTERIA AT INCREASED HYDROSTATIC PRESSURE

JOHN C. SHEN* and LESLIE R. BERGER

Department of Microbiology
University of Hawaii

In the seas, as in all environments, survival may depend to a large extent on the ability of a species to meet its nutritional and energetic requirements. For organisms which depend solely upon dissolved substrates, the utilization of nutrients in low concentrations requires special transferring or accumulating mechanisms. Solutes must be transferred from the outside to the inside of the cell. Cytoplasmic membranes are, universally, impermeable to most solutes by passive diffusion. Thus, special mechanisms have evolved which permit the transport of nutrients. Similar mechanisms function to control elimination of specific substances.

The best understood systems of active transport are the so-called "Roseman systems" for the transport of sugars (Kundig *et al.*, 1964). These systems have been found to transport a variety of sugars in a large number of bacterial species. Whether marine bacteria (as a group), and particularly those which are found at high hydrostatic pressures, utilize this method for accumulating substrates is not known.

* This work is taken in part from a thesis to be submitted in partial fulfillment of the requirements for the degree of Doctor of Philosophy in the Department of Microbiology, University of Hawaii.

The kinetics of these transport mechanisms in terrestrial bacteria are such that substrates saturate the systems usually in less than 1 min. This posed serious technical difficulties for performing studies at increased hydrostatic pressure.

This report describes a simple apparatus which permits introduction and removal of materials from a pressure cylinder in a few seconds. Preliminary data are presented to show the use of the apparatus for the measurement of active transport at increased hydrostatic pressure.

MATERIALS AND METHODS

Pressure Apparatus and Sampling Device

The basic pressure cylinder described by ZoBell and Oppenheimer (1950) is used with a modified top or cap (C in Fig. 1). The top is made thicker to permit the machining of a female high-pressure fitting (H). A short piece of high-pressure tubing is attached to the cap and a shut-off valve is mounted on its opposite end. This keeps the shut-off valve (not shown) out of the way of the sampling attachment. The sampling attachment is a capillary needle T-valve (cnv) designed for gas-chromatographic applications (Precision Sampling Corp., Baton Rouge, La.). The T-valve is brazed to a fitting (F) which is screwed into a threaded hole in the cylinder top. An O-ring makes a pressure-

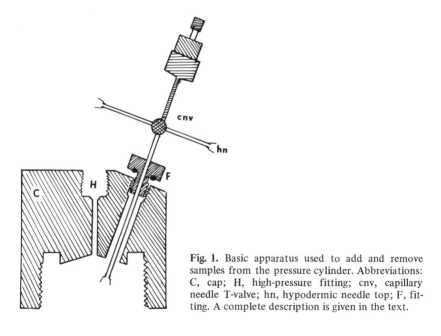

Fig. 1. Basic apparatus used to add and remove samples from the pressure cylinder. Abbreviations: C, cap; H, high-pressure fitting; cnv, capillary needle T-valve; hn, hypodermic needle top; F, fitting. A complete description is given in the text.

tight seal. The entire unit is removable should the valve seat need regrinding at the factory. A hypodermic needle top (hn) is brazed at each of the three ends of the capillary tubes on the valve. This permits attachment of syringes to and from which materials may be added or withdrawn. The syringe within the pressure cylinder serves as the reaction vessel. In this study 0.5-, 1.0-, and 2.0-ml syringes were used. It is possible to use 30-ml units. To introduce materials into the syringe within the cylinder, the needle valve is opened and the material is injected from the outside syringes. The valve is closed and hydrostatic pressure may be instantly applied from a prepressurized cylinder connected to the same hydraulic line. The entire procedure may take from 1 to 5 sec depending upon the volume introduced through the capillary tubing. To remove a sample which has been incubated at increased hydrostatic pressure, the pressure is released, the needle valve is opened and the fluid is withdrawn into the exterior syringe. The syringe may contain a substance to stop a reaction. Alternatively, the reaction may be stopped by addition of an agent to the reaction vessel (syringe) within the cylinder as described above.

Hydrostatic pressure must be released before cellular samples are removed in order to avoid lysis of the cells owing to rapid flow through the capillary openings. Release of pressure, removal of the cell suspension from the cylinder, and the deposition of the cells on a membrane filter can be accomplished in less than 10 sec.

a-Methylglucoside (aMG) is phosphorylated by the glucose transport system of *Escherichia coli* (ML308/225). It is accumulated, but not metabolized, by this organism. This system was used as a model to test the apparatus. The procedures followed were basically those of Winkler (1971). Efflux of the accumulated sugar from the cells was in some experiments reduced by preloading the cells with unlabeled aMG before addition of the ^{14}C-labeled aMG. After incubation, cells were placed on membrane filters, dried, and their ^{14}C activity determined in a scintillation counter.

In most experiments, 0.1 ml of cell suspension was introduced through the capillary valve and mixed with 0.1 ml substrate. Generally, 0.1-ml samples were placed on the membrane filters.

RESULTS AND DISCUSSION

The rate of uptake of aMG is influenced by hydrostatic pressure. Increased pressure decreased both the apparent rate of uptake and the total amount retained by the cells (Fig. 2). Figure 3 shows the rate of uptake of labeled aMG by the cells over the first 60 sec (solid lines) and the rate averaged over the first 2.5 min (dotted lines). Within the first 60 sec, the rates of uptake are nearly identical, but by 2.5 min reduced rates are evident at the increased

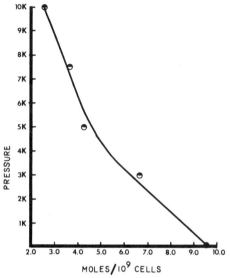

Fig. 2. Amount of α-methylglucoside taken up by cells of *Escherichia coli* (ML308/225) at various hydrostatic pressures over a 20-min period. (Pressures given in pounds per square inch; K indicates 1000.)

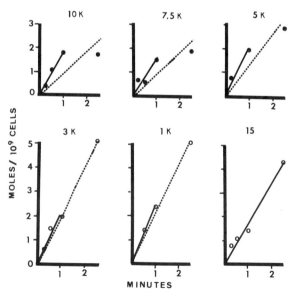

Fig. 3. Uptake of α-methylglucoside by cells of *Escherichia coli* (ML308/225). Solid line drawn over first 60 sec; dotted line averaging the first 2.5 min. At atmospheric pressure (15 psi) a single line is drawn.

pressures. The amount of aMG taken up by the cells after 20 min decreases with increased hydrostatic pressure (Fig. 4).

The effects of hydrostatic pressure were also measured on the rate of efflux of aMG from the cells. The procedure requires preloading with labeled aMG for 30 min. Loaded cells (0.1-ml samples) were then rapidly introduced into the pressure vessel and pressurized for 30 or 60 sec. Samples were then removed into unlabeled aMG, filtered, and washed. Figure 5 shows the results of such an experiment. Increased hydrostatic pressure increased the rate of efflux at about the same rate regardless of the specific pressure. This was not due to cell lysis within the apparatus. The data in Table 1 confirm qualitatively the data of Fig. 5. The ratio of aMG in the cells after 20-min incubation at various pressures is compared with that before incubation. Efflux is faster at increased hydrostatic pressure.

We do not wish to draw major conclusions from these few experiments. It is, however, likely that the smaller amount of aMG accumulated by the cells at increased pressure results in part from the increased rate of leakage evident from the efflux experiments. Though premature, it is tempting to draw other conclusions from these data alone. However, a method has been developed for rapidly adding or removing small volumes from vessels inside bulky pressure apparatus.

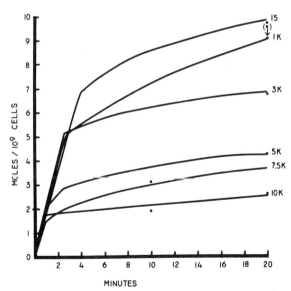

Fig. 4. Uptake of a-methylglucoside by cells of *Escherichia coli* (ML308/225) plotted against hydrostatic pressure. (Pressures given in pounds per square inch; K indicates 1000.)

Table 1. Efflux of a-methylglucoside from cells of *Escherichia coli*
(ML308/225) after 20 min preloading[a]

Pressure (psi)	Efflux ratio (final : initial)
10,000	0.17
7500	0.13
5000	0.14
3000	0.37
1000	0.24
15,000	0.73

[a] Data is given as the ratio of intracellular aMG
after a 20 min incubation at the indicated
pressure to that measured before pressurization.

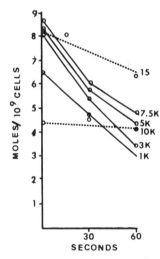

Fig. 5. Efflux of a-methylglucoside from cells of *Escherichia coli* as a function of
incubation time at various hydrostatic pressures. (Pressures given in pounds per square
inch; K indicates 1000.)

LITERATURE CITED

Kundig, W., S. Ghosh, and S. Roseman. 1964. Phosphate bound to histidine in a novel phosphotransferase system. Proc. Nat. Acad. Sci. 52:1067–1074.

Winkler, H. H. 1971. Efflux and steady state in α-methylglucoside transport in *Escherichia coli*. J. Bacteriol. 106:362–368.

ZoBell, C. E., and C. H. Oppenheimer. 1950. Some effects of hydrostatic pressure on the multiplication and morphology of marine bacteria. J. Bacteriol. 60:771–781.

EFFECTS OF HYDROSTATIC PRESSURE ON URACIL UPTAKE, RIBONUCLEIC ACID SYNTHESIS, AND GROWTH OF THREE OBLIGATELY PSYCHROPHILIC MARINE VIBRIOS, *VIBRIO ALGINOLYTICUS,* AND *ESCHERICHIA COLI*

JOHN A. BAROSS, F. JOE HANUS, and RICHARD Y. MORITA

Departments of Microbiology and Oceanography
Oregon State University

Two obligately psychrophilic antarctic vibrios, *Vibrio marinus* MP-1, *Vibrio alginolyticus* V-374, and *Escherichia coli* C-600 were subjected to hydrostatic pressures ranging from 1 to 800 atm in the presence of [^{14}C] uracil. At 15°C it was found that at moderate pressures (100–300 atm) the uptake of uracil was immediately reduced by approximately 70% in *Escherichia coli* and *Vibrio alginolyticus* and at 500 atm these mesophiles were incapable of taking up uracil. Similar results were found with *Vibrio marinus* at 5°C. The antarctic *vibrios* at 5°C were significantly more barotolerant than MP-1 since these psychrophiles showed considerable uptake activities at 500 atm. Antarctic isolate 18-500 was found to have an inducible uracil-uptake system which could be altered at very low pressures (100–300 atm) without any concomitant damage to the cells's ability to grow. In all cases, except *Escherichia coli,* the ratios of the amount of uracil taken up by the cells to the amount of uracil incorporated into RNA approximated 1, thus indicating that the RNA synthetic mechanism was less sensitive to pressure than the uptake system. Data are also presented which indicate that the inhibition of uracil uptake under pressure was not reversible upon pressure release to 1 atm. The specific pressures which caused irreversible impairment of uracil uptake varied with each bacterial strain. In general, the mesophiles were less sensitive to sustained pressure damage after pressure release and it is postulated that membrane damage under pressure is more severe at temperatures approximating the optimum growth temperature and that irreversible pressure damage to membranes is probably related to the ability of the organisms to synthesize membrane and/or other structural components involved in cellular division. A close correlation was also shown between the ability of the cells to divide and the rate of uracil uptake at each pressure for all bacterial test strains. The data provide strong evidence that the initial site of pressure damage to bacterial cells is the membrane in so far as this structure affects the uptake of organic substrates.

INTRODUCTION

The marine environment is characterized by extremes in hydrostatic pressure, temperature, and salinity. In the open ocean, however, at depths near the average (3800 m), or in hadal regions (10, 915), the variations encountered in the water temperature and the salinity are minor, whereas the pressures increase with depth to 1100 atm. In these regions, therefore, hydrostatic pressure is the primary physical parameter affecting the activities of organisms.

In general, many of the most representative bacteria isolated from oceanic waters are both psychrophilic and salt requiring. Barophilic bacteria, however, have been demonstrated but never isolated (ZoBell, 1961). As a consequence, most of the investigations of pressure effects on microorganisms have involved terrestrial species or bacteria isolated from estuarine or marine surface waters. In general, these observations showed that most species of bacteria tested had a decreased growth rate at 200 atm and that pressures above 500 atm either prevented growth or killed the organism (Oppenheimer and ZoBell, 1952; ZoBell and Johnson, 1949). In addition to bacterial growth responses, increased hydrostatic pressure has been shown to affect enzyme reaction rates (Johnson, 1970), and the synthesis of macromolecules including protein, ribonucleic acid, and deoxyribonucleic acid (Albright and Morita, 1968; Arnold and Albright, 1971; Landau, 1970; Yayanos and Pollard, 1969). Recently, it has been demonstrated that pressure also caused a marked decrease in the uptake of amino acids in the marine bacterium MP-38 (Paul and Morita, 1971).

It is axiomatic that the effect of both low temperature and increasing pressures are additive as a result of volume changes (ΔV) incurred by macromolecules such as enzymes (Morita, 1968). It is quite probable, therefore, that mesophilic and facultative psychrophilic bacteria would be completely inactive at temperatures below the thermocline (20–30°C below the optimum growth temperature) and at pressures exceeding 300 atm. Moreover, it should not be overlooked that the optimal growth temperatures for most of the bacteria, dinoflagellates, and diatoms taken from marine waters are 10–20°C higher than *in situ* temperatures (Braarud, 1961). This would indicate, therefore, that marine organisms residing in pressure-affected waters would be subjected to the additive effects of below optimum growth temperatures and high hydrostatic pressure. Many investigations on pressure effects on microorganisms, nevertheless, have utilized incubation temperatures which were optimum for the test organism, and therefore considerably higher than what could ever be encountered in marine environments (e.g., 37°C for *Escherichia coli*). Although many interesting and informative data have resulted from these studies, little can be inferred about the activities of organisms residing in pressure-affected marine environments.

In this investigation, the effect of hydrostatic pressure on the uptake of uracil and its incorporation into RNA was determined at *in situ* temperatures using three psychrophilic and two mesophilic bacterial strains. The data presented provide further evidence for the hypothesis that the initial site of pressure damage to cells is the membrane in so far as this structure affects the uptake of organic substrates. The synthetic systems for RNA synthesis were not significantly affected. Evidence is also presented which strongly suggests a difference in the kinetics of uracil uptake as a function of increasing pressures for each bacterial species.

MATERIALS AND METHODS

Bacterial Strains

Three psychrophilic marine *Vibrios,* designated *Vibrio marinus* MP-1, 18-500, and 18-300, one mesophilic *Vibrio, Vibrio alginolyticus* V-374, and *Escherichia coli* C-600 were employed in this investigation. The source and temperature range of growth of these isolates are described in Table 1.

Growth of Bacteria

The psychrophilic vibrios were grown at 5°C, whereas the mesophilic strains were cultivated at 15°C. The vibrio cultures were grown in Lib-X broth (Bacto-yeast extract, 1.2 g; Trypticase (BBL), 2.3 g; sodium citrate, 0.3 g; L-glutamic acid, 0.3 g; sodium nitrate, 0.05 g; ferrous sulfate, 0.005 g; Rila marine salts, 33 g; distilled water to 1 liter and adjusted to pH 7.5) with shaking in a Psychrotherm (New Brunswick). *Escherichia coli* C-600 was grown in Nutrient Broth (Difco) containing 1.0% Rila salts at a pH of 7.0 at 15°C.

Growth curves for vibrio strains under pressure were measured by spread plate counts (triplicate) using Lib-X agar (12 g agar per liter); *Escherichia coli* was grown in Nutrient Broth with 1% Rila salts and enumerated using

Table 1. Species designation, source, and temperature range of growth of the isolates employed

Strain	Genus-species	Temperature range of growth (°C)[a]	Source
MP-1	*Vibrio marinus* (ATCC 15381)	<0 to 20	Pacific Ocean
18-500	*Vibrio* sp.	<0 to 16	Antarctic waters[b]
18-300	*Vibrio* sp.	<0 to 13	Antarctic waters[b]
V-374	*Vibrio alginolyticus*	8 to 45	H. ZenYoji[c]
C-600	*Escherichia coli*	8 to 45	O. S. U. strain

[a] Determined in a polythermostat.

[b] *Eltanin* cruise number 46, 1970–1971.

[c] Department of Bacteriology, Tokyo-to Laboratories for Medical Science, Bureau of Health, Tokyo, Japan.

Nutrient Agar (Difco) with 1% Rila salts. Washed cell suspensions were loaded into 5-ml sterile disposal plastic syringes (Becton, Dickerson and Co.) which were capped with sterile sealed hypodermic needles. The syringes were placed in pressure cylinders which were immediately pressurized and incubated in a water bath set at the appropriate temperature. Samples were removed from separate pressure vessels at various time intervals up to 150 hr. Cell suspensions were serially diluted for plate counting using the appropriate growth medium as the diluent. All media and pipettes were prechilled to 0°C in an ice bath before use. Photomicrographs were made using a Leitz-Orthomat microscope and camera system.

Uptake of Uracil

All cultures were harvested for experimentation at the mid-log phase as measured by OD. The vibrio strains were washed twice in artificial seawater (Rila salts, 35 g/liter) containing 0.1% Bacto-yeast extract (Difco). *Escherichia coli* C-600 was similarly harvested using a 1% Rila salts solution supplemented with 0.1% yeast extract. The washed cells were resuspended in sufficient Rila salt-yeast extract solution to provide an OD of 0.25 at 600 nm using a Bausch and Lomb Spectronic 20 colorimeter. The cell suspensions were diluted 1 : 10 as the final concentration of cells ($1-5 \times 10^6$/ml). Throughout this whole procedure the cells and reagents were maintained at 0°C and at a pH of 7.5. Uniformly labeled [^{14}C] uracil (New England Nuclear) was added to the dilute cell suspension to a final concentration of 0.1 μCi/ml (0.4 μg/ml). The thoroughly mixed suspension of cells and uracil was loaded into 5 ml syringes as described and pressurized. Extreme care was taken to insure that all of the air bubbles were removed from the syringe. The uptake period under pressure for all strains was 2 hr at 5°C for the psychrophiles and 15°C for the mesophiles.

The uptake of uracil was terminated by ejecting the cell suspension from the syringe into sufficient sulfuric acid (2 N) to reduce the pH to 2. The acid-treated cells were incubated for 30 min at 5°C; however, we found maximum release of bound uracil within 15 sec after acid treatment. This procedure was found to be sufficient for releasing bound uracil from the cell surface. No evidence of cell lysis after acid treatment was found if the pH was not lower than 2 and the acid cell suspension was maintained at temperatures of 5°C or below. Fixed cells were captured on 0.45 μm Millipore filters; the filters were washed three times with approximately 10-ml portions of 3.5% Rila salts solution for the *Vibrio* sp. and 1.0% Rila salts for *Escherichia coli*.

The extent of reversible damage to the uracil uptake system was measured by pressurizing two sets of syringes at each pressure. After 2 hr the pressure was released; one set of syringes was fixed to determine the uptake while the cells were under pressure. The second set of syringes was maintained at atmospheric pressure for an additional 2 hr to determine the rate of uptake after pressure release.

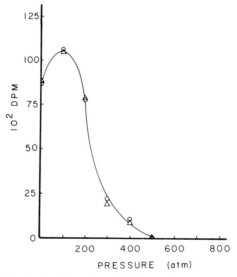

Fig. 1. Uracil uptake by *Vibrio marinus* MP-1 at various pressures at 5°C. ○ is DPM uracil taken up by cells; △ is DPM uracil incorporated into RNA.

Incorporation of Uracil into RNA

The proportion of [^{14}C]uracil incorporated into RNA was determined by exposing cell suspensions to uracil at various pressures as described. The cell suspensions were then added to an equal volume of cold 20% (w/v) trichloroacetic acid (TCA). This mixture was then incubated for 60 min at 0°C in an ice bath. The precipitated RNA was captured on Millipore glass prefilters (Kennell, 1967). The prefilters were then washed with three successive 10-ml portions of cold 10% TCA and dried. The amount of uracil incorporated into RNA was always compared with the total [^{14}C]uracil incorporated into the cells for each pressure.

All counting of ^{14}C-labeled material was performed with a Nuclear Chicago Mark I liquid scintillation counter in toluene fluor. All counts were corrected for counting efficiency by the channel-ratio method.

RESULTS

Uracil-Uptake Patterns

The differences in the kinetics of uracil uptake at increasing pressures by the five test organisms are shown in Figs. 1 through 5. Clearly, the patterns for uptake of uracil as a function of increasing pressures varied with each strain. In general, *Vibrio marinus* MP-1 (Fig. 1), *Escherichia coli* C-600 (Fig. 2), and

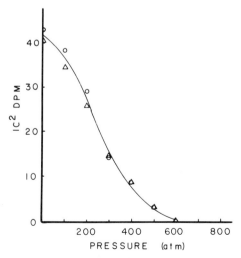

Fig. 2. Uracil uptake by *Escherichia coli* C-600 at various pressures at 15°C. ○ is DPM uracil taken up by cells; △ is DPM incorporated into RNA.

Vibrio alginolyticus V-374 (Fig. 3) were quite sensitive to pressures exceeding 100 atm at the temperatures employed. Thus the total uptake of uracil at 300 atm with respect to 1 atm (100%) was reduced to 27% for MP-1, 32% for V-374, and 33% for C-600, and at 500 atm there was no significant uptake of uracil by these three bacterial strains. In contrast, the two antarctic isolates (Figs. 4 and 5) behaved quite differently in that the uptake of uracil was stimulated at moderate pressures. The stimulation pattern for strain 18-500, however, was indicative of an inducible uracil-uptake system, since the results from a time-course experiment showed that this vibrio would proceed to take up uracil at 1 atm only after exposure to this substrate for 90 to 145 min, thus accounting for the low uptake at 1 atm. The ability to adapt was inactivated gradually at pressures ranging from 100 to 500 atm. This system will be more thoroughly discussed in conjunction with the pressure release studies. The antarctic isolate 18-300 (Fig. 5) showed increased uptake of uracil at 100 and 200 atm followed by a linear decrease in uptake from 300 to 800 atm. In contrast to MP-1 and the mesophilic strains, 18-300 showed only a 50% decrease in uptake activity at 500 atm.

Uracil-Uptake Reversibility

In the following set of experiments it is presumed that the degree of reversibility of uracil uptake at 1 atm after pressure release would be directly

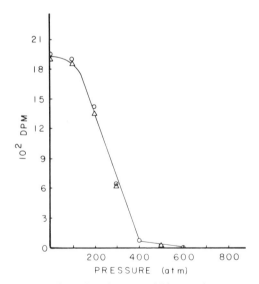

Fig. 3. Uracil uptake by *Vibrio alginolyticus* V-374 at various pressures at 15°C. ○ is DPM uracil taken up by cells; △ is DPM uracil incorporated into RNA.

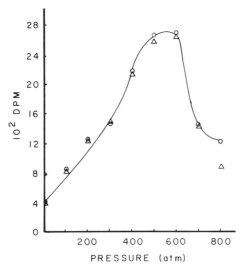

Fig. 4. Uracil uptake by antarctic psychrophile 18-500 at various pressures at 5°C. ○ is DPM uracil taken up by cells; △ is DPM uracil incorporated into RNA.

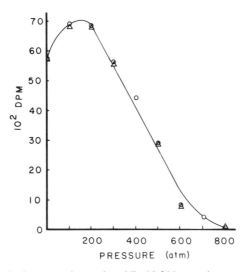

Fig. 5. Uracil uptake by antarctic psychrophile 18-300 at various pressures at 5°C. ○ is DPM uracil taken up by cells; △ is DPM uracil incorporated into RNA.

related to the "short-term" irreversible damage to the cell transport mechanism. In each experiment the cells were pressurized for 2 hr followed by depressurization to 1 atm for an additional 2 hr. The rate of uracil uptake following depressurization was compared with the uptake of unpressurized cells. Figures 6 through 10 show these results.

The reversibility of uracil uptake by MP-1 after pressure release is shown in Fig. 6. At all pressures the resumed rate of uptake at 1 atm was approximately the same as the uptake rate under each pressure. Thus MP-1 showed a 74% decreased uptake rate at 300 atm with respect to 1 atm and upon release of pressure the resumed uptake rate was approximately 40% of the 1 atm uptake rate. At 400 atm the uptake system in MP-1 was definitely injured since this vibrio could only recover 15% of the 1 atm uptake activity upon pressure release. Similar results are shown in Fig. 7 for the antarctic isolate, 18-300. This psychrophile retained the stimulated uptake rate at 100 atm but gradually decreased to 45% of the 1 atm uptake at 500 atm. At 600 atm or greater, 18-300 rapidly lost its ability to take up uracil.

The antarctic psychrophile, 18-500, showed very unexpected uptake behavior after pressure release (Fig. 8). At pressures of 1, 100, 500, and 600 atm, the rates of uracil uptake were approximately the same. At 200, 300, and 400 atm the rates of uracil uptake upon pressure release were markedly lower than the resumed uptake rate at 500 and 600 atm, for example. It is presumed that the mechanism for uracil uptake is irreversibly damaged at 500 and 600 atm whereas at 400 atm, for example, the uptake mechanism was in

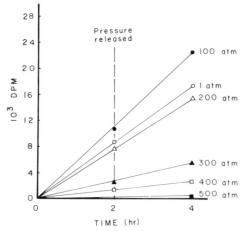

Fig. 6. Extent of reversibility of uracil uptake after pressure release to 1 atm by *Vibrio marinus* MP-1 at 5°C. Cells exposed to pressure for 2 hr and for an additional 2 hr at 1 atm after pressure release.

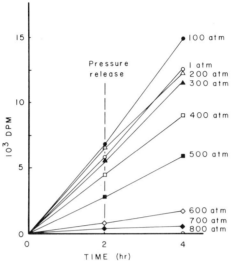

Fig. 7. Extent of reversibility of uracil uptake after pressure release to 1 atm by the antarctic psychrophile 18-300 at 5°C. Cells exposed to pressure for 2 hr and for an additional 2 hr at 1 atm after pressure release.

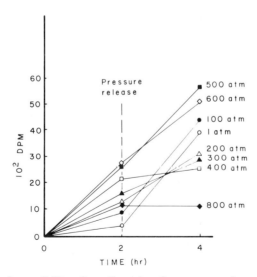

Fig. 8. Extent of reversibility of uracil uptake after pressure release to 1 atm by the antarctic psychrophile 18-500 at 5°C. Cells exposed to pressure for 2 hr and for an additional 2 hr at 1 atm after pressure release.

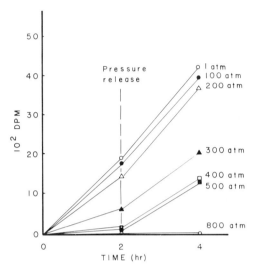

Fig. 9. Extent of reversibility of uracil uptake after pressure release to 1 atm by *Vibrio alginolyticus* V-374 at 15°C. Cells exposed to pressure for 2 hr and for an additional 2 hr at 1 atm after pressure release.

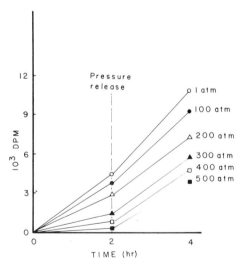

Fig. 10. Extent of reversibility of uracil uptake after pressure release to 1 atm by *Escherichia coli* C-600 at 15°C. Cells exposed to pressure for 2 hr and for an additional 2 hr at 1 atm after pressure release.

a sense "frozen" so that upon pressure release, the cells could adapt in a manner analogous to the 1 atm cells. At 200 and 300 atm varying degrees of this latter phenomenon took place. At 100 atm, there was no effect one way or the other and the cells behaved like their counterparts at 1 atm. At 800 atm, the uptake mechanism in 18-500 was definitely damaged.

The effects of pressure on the activities of microorganisms are more prominent at temperatures below their optimum growth temperature. ZoBell and Cobet (1964), for example, showed that *Escherichia coli* could divide at pressures up to 525 atm at 30°C, whereas at 20°C no growth occurred at 400 atm. In the present investigation, *Escherichia coli* and *Vibrio alginolyticus* were exposed to varying pressures at 15°C which is approximately 25°C below their optimum growth temperature (Figs. 9 and 10). This is reflected in the marked decreases in uracil uptake at moderate pressures (Figs. 4 and 5). The apparent damage to these mesophiles, as indicated by the resumed uptake rate after pressure release, was only minor in comparison with MP-1 grown near its optimum temperature. Thus at 500 atm, both V-374 and C-600 recovered greater than 50% of their uracil uptake capacity as compared with the 1 atm activity. This is more striking if it is considered that at 500 atm the uptake of uracil by these mesophiles was essentially zero.

Table 2 summarizes all the data pertaining to the percentage uptake of uracil at each pressure with respect to the uptake at 1 atm (100%), and the percentage uptake at 1 atm after pressure release for all five strains. With the exception of the antarctic psychrophile, 18-500, there is clear evidence that the rate of uracil uptake decreases with increasing pressures and that at pressures exceeding 500 atm the uptake of uracil is effectively shut off in all of the test bacterial strains.

Incorporation of Uracil into RNA

In the following experiments, the amount of uracil taken up by the cells was compared with the amount of uracil incorporated into RNA. The results from these experiments should indicate whether the system for uracil uptake was more susceptible to hydrostatic pressure than the RNA-synthetic system. If the RNA-synthetic mechanism was damaged at a specific pressure to a greater extent than the uracil-uptake mechanism, then it is possible that uracil could accumulate intracellularly as a uracil pool. There is, however, no evidence that a uracil pool exists in the test organisms, but, if pools did exist, it is quite likely that the acid-fixing treatment used to remove extracellular bound uracil would also release intracellular pools. Furthermore, if uracil pools did exist, it is unlikely that the amount of uracil in these pools would represent a significant proportion of the available substrate. Another possibility was that one or all of the test bacterial strains could metabolize uracil. Controls were run for each strain to determine if any CO_2 was respired from uracil. The procedure employed has been described elsewhere (Hobbie and Crawford, 1969). In all cases, there was no $^{14}CO_2$ evolved as a result of repiration. It is therefore assumed that the principal cellular role for uracil utilization is in the synthesis of RNA.

Table 2. Percentage uptake of [^{14}C] uracil under pressure and after pressure release relative to the total uptake at 1 atm (100%)

| Pressure (atm) | Variable[a] | Percentage uptake of uracil relative to uptake at 1 atm | | | | |
| | | Bacterial strains | | | | |
		MP-1	18-500	18-300	V-374	C-600
1	Pressurized	100	100	100	100	100
	Released	100	100	100	100	100
100	Pressurized	124	200	120	97	86
	Released	140	104	118	95	81
200	Pressurized	89	310	120	72	67
	Released	102	51	79	101	65
300	Pressurized	27	350	98	32	33
	Released	41	45	86	62	65
400	Pressurized	15	530	77	2.5	18
	Released	15	12	67	60	60
500	Pressurized	0	650	51	1	8
	Released	1	84	45	60	52
600	Pressurized	0	650	13	0	0
	Released	ND[b]	68	13	ND	ND
700	Pressurized	0	350	8	0	0
	Released	0	0	2	ND	ND
800	Pressurized	0	300	1.5	0	0
	Released	0	0	2	0	0

[a] Pressurized is the uptake of [^{14}C] uracil for 2 hr at various pressures; released is the uptake of [^{14}C] uracil for 2 hr at 1 atm after pressure release (total at 4 hr minus the total at 2 hr).

[b] ND is not determined.

Other investigators, working with either cell-free extracts (Arnold and Albright, 1971) or with whole cells (Schwarz and Landau, 1972), have demonstrated that the RNA-synthetic mechanisms were not inhibited by the moderate pressures shown in this study to decrease the amount of uracil taken in by the cells. Evidence presented by these investigators indicate that the RNA-synthetic mechanisms are not "shut off" gradually with increasing pressures but "abruptly" as a result of specific pressures such as 670 atm for *Escherichia coli* (Schwarz and Landau, 1972). In contrast, the amount of uracil taken in by these test bacterial strains gradually decreased with increasing pressures. Moderately low pressures such as 200 and 300 atm caused a decrease in the uptake of uracil in *Vibrio marinus* and in the antarctic vibrio 18-300; however, these organisms were capable of dividing at these pressures. It is reasonable to conclude, therefore, that moderate pressures do not irreversibly damage any of the synthetic system necessary for cellular division and that low pressures only affect the amount of substrate taken in by the cells.

Table 3 summarizes all the results pertaining to the total number of DPM from uracil taken up by the cells and the total DPM incorporated into RNA. The ratios of the total uptake in cells *versus* the total incorporation of uracil into RNA are included in Table 3. These data clearly show that all of the test organisms, except *Escherichia coli* C-600, incorporated uracil into RNA as soon as uracil entered the cell. The ratios, which ranged from 0.99 to 1.04, are within the range of experimental error. At 800 atm the antarctic psychrophile 18-500 apparently can transport more uracil into the cell than can be incorporated into RNA, thus accounting for the ratio of 1.40. It is assumed that with this psychrophile there is both membrane damage and damage to the RNA-synthesizing mechanisms at 800 atm.

The *Escherichia coli* strain behaved quite differently than the other test organisms in that there was a significant difference in the ratios of cellular uptake of uracil and the uracil as RNA at 1–200 atm. The possibility exists, therefore, that, at temperatures close to the minimum growth temperature for *Escherichia coli*, RNA synthesis may be at least as sensitive to pressure as the uracil-uptake systems. However, we have some preliminary information which would indicate that *Escherichia coli* C-600 irreversibly binds organic substrates, including uracil, to the cell surface. These substrates could not be dislocated by acid treatment.

Growth under Pressure

In order to determine the exclusiveness of membrane damage resulting from short-term exposure to hydrostatic pressure, it is necessary to assess the extent of cell viability and growth at the temperatures and pressures em-

Table 3. Total [14C] uracil incorporated into the cell compared with the total [14C] uracil incorporated into the RNA using 2-hr uptake periods

Bacterial strains	DPM incorporated into	Temperature (°C)	Amount of [14C] uracil (DPM) incorporated in 2 hr								
			Pressure (atm)								
			1	100	200	300	400	500	600	700	800
MP-1	Cells	5	8572	10,626	7708	2309	1338	No significant uptake			
	RNA		8654	10,624	7800	2292	1268				
	Ratio[a]		0.99	1.00	0.99	1.01	1.06				
18-500	Cells	5	410	826	1284	1433	2190	2665	2680	1441	1220
	RNA		402	824	1258	1441	2112	2580	2630	1428	864
	Ratio		1.02	1.00	1.02	0.99	1.04	1.03	1.02	1.01	1.40
18-300	Cells	5	5794	6960	6953	5675	4460	2944	755	460	87
	RNA		5570	6939	6846	5560	ND[b]	2963	771	ND	93
	Ratio		1.04	1.00	1.02	1.02	—	0.99	0.98	—	0.94
V-374	Cells	15	1975	1920	1430	645	49	20	0	—	—
	RNA		1910	1897	1406	643	ND	0	0	—	—
	Ratio		1.03	1.01	1.02	1.00	—	—	—	—	—
C-600	Cells	15	4380	3790	2900	1450	780	340	0	—	—
	RNA		4100	3370	2550	1460	780	340	0	—	—
	Ratio		1.07	1.13	1.14	0.99	1.00	1.00	—	—	—

[a] Ratio is total uracil in cells divided by total uracil as RNA.
[b] ND is not determined.

ployed in the uptake studies. If, for example, there was a loss in viability or lysis of cells after 2 hr exposure to pressure, it would be impossible to implicate membrane damage, or, for that matter, damage to the macromolecular synthesizing mechanisms as a result of pressure. In the following experiments, each of the test organisms was exposed to various pressures at the temperatures employed for the uptake studies in an effort to determine the maximum pressure allowing cellular division, and to determine the rate of death at pressures above the maximum growth pressure. The results from these experiments are summarized in Fig. 11 (A–E).

Within the 2-hr period used in the uptake studies there was no significant change in cell numbers of any of the test strains incubated under different pressures except 800 atm. At 800 atm there was approximately a 2 log reduction in the numbers of the psychrophiles MP-1, 18-500, and 18-300, whereas there was less than a 1 log reduction in the numbers of *Escherichia coli* C-600 or *Vibrio alginolyticus* V-374 at 800 atm.

The growth patterns differed markedly between each of the test strains. In general the marine psychrophiles were more sensitive to pressures which

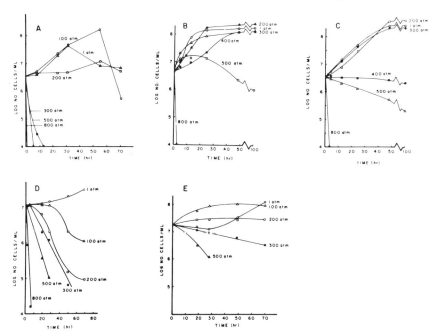

Fig. 11. Effect of pressure on cellular division of all bacterial test strains in a nutrient medium as indicated by colony count. A, *Vibrio marinus* MP-1 at 5°C; B, antarctic psychrophile 18-500 at 5°C; C, antarctic psychrophile 18-300 at 5°C; D, *Escherichia coli* C-600 at 15°C; E, *Vibrio alginolyticus* V-374 at 15°C.

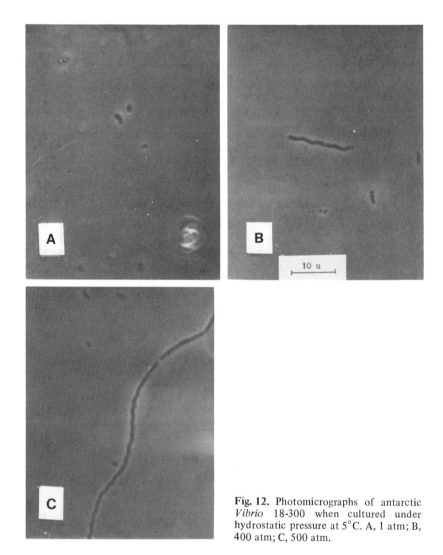

Fig. 12. Photomicrographs of antarctic *Vibrio* 18-300 when cultured under hydrostatic pressure at 5°C. A, 1 atm; B, 400 atm; C, 500 atm.

were higher than their maximum growth pressure. Thus MP-1 was only capable of dividing at 100 atm and only after a 30-hr lag at 200 atm. At pressures of 300 atm or greater, MP-1 cells rapidly lost their viability. The antarctic psychrophiles, on the other hand, were much more barotolerant than MP-1 at 5°C. These psychrophiles could tolerate 500 atm, and, in fact, 18-500 was capable of dividing at 500 atm. Antarctic isolate 18-300 did not divide at pressures above 300 atm even though there was little loss of cell viability of this strain at 400 and 500 atm.

In contrast to the psychrophilic isolates, the ability of the mesophiles *Escherichia coli* C-600 and *Vibrio alginolyticus* V-374 to divide at moderate pressures (100 and 200 atm) and at 15°C was significantly affected. Some evidence of division occurred at 100 atm in *Escherichia coli* but not in *Vibrio alginolyticus* cells. Furthermore, at pressures up to and including 300 atm, *Escherichia coli* cells were quite stable, whereas *Vibrio alginolyticus* cells gradually lost their viability.

Morphological changes were observed in all of the psychrophilic cultures at pressures which were limiting to their growth. Figure 12 shows some of the long forms observed with the antarctic vibrio 18-300 at 400 and 500 atm. Similar elongated cells were noted at 400 and 500 atm with vibrio 18-500, whereas these forms were noted in MP-1 at 200 and 300 atm. *Vibrio marinus* MP-1 formed spheroplast-like structures within 2 hr at pressures exceeding 300 atm. The antarctic isolates formed spheroplast-like structures within 8–16 hr exposure to pressures greater than 500 atm. No abnormal structural changes were observed in the mesophilic cultures at any pressure at 15°C.

The cumulative effect of both above-maximum growth temperatures and moderate pressure on the growth of psychrophilic marine bacteria is clearly illustrated in Fig. 13 using the antarctic isolate 18-300. Originally, 18-300 displayed a maximum growth temperature between 8 and 9°C thus accounting for its lack of growth at 10°C as shown in Fig. 13. It can be seen that these psychrophiles were capable of dividing at 10°C when cultured under 100 atm pressure. At 15°C and at 100 atm there was a significant increase in the number of viable cells when compared with the 1 atm culture. The cells cultured at both 10 and 15°C at 100 atm showed an increase in volume that was greater than 50 times the 1 atm grown cells (Fig. 14); furthermore, these cells were highly vacuolated. It is interesting to speculate that pressures

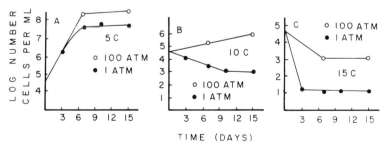

Fig. 13. Growth of antarctic *Vibrio* 18-300 at 5, 10, and 15°C both at 1 atm and at 100 atm as determined by plate counts.

exceeding 100 atm could compensate for the cellular-volume increases which result from exposure to above-maximum growth temperatures. This could be predictable if the ideal gas law $(PV = nRT)$ could be applied to biological systems.

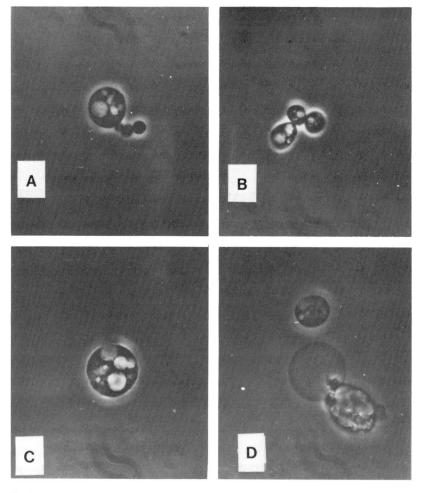

Fig. 14. Photomicrographs showing the morphological changes in antarctic *Vibrio* 18-300 when cultured at 10 and 15°C and 100 atm. A, 100 atm at 10°C; B, 100 atm at 10°C; C, 100 atm at 15°C; D, 100 atm at 15°C followed by 10 min exposure to 20°C at 1 atm. Magnification 665 ×.

DISCUSSION

There is little doubt that pressure exerts injurious effects on bacterial structural components and synthetic systems (Albright and Morita, 1972; Albright
and Morita, 1968; Arnold and Albright, 1971; Landau, 1970; Schwarz and
Landau, 1972; Yayanos and Pollard, 1969). All of these different pressure-
affected alterations, however, do not occur simultaneously at any given
pressure-temperature, nor can it be assumed that one alteration is contiguous
with any other alteration. What is important to a particular microorganism in
pressure-affected waters is the most pressure-sensitive cellular structure, synthetic system, or metabolic function. Paul and Morita (1971) have shown that
at moderate pressures (300–500 atm) the ability of marine bacteria to
transport amino acids is severely affected, whereas protein synthesis and
catabolism were not affected. The present study substantiates these findings
and further shows that RNA synthesis was not impaired at pressures which
definitely caused a decrease in the uptake of uracil and "short-term" irreversible membrane damage. Furthermore, these data strongly emphasize the
additive affects of below-optimum growth temperatures, and increased pressure on uracil uptake, membrane damage, and cellular division.

Previous investigations have shown that macromolecular synthesis, especially translational proteins involved in cellular division cycles, were the most
pressure-sensitive sites in *Escherichia coli* (Arnold and Albright, 1971;
Schwarz and Landau, 1972) and *Vibrio marinus* (Albright and Morita, 1968).
This is consistent with early work on the growth rates of various bacterial
species at different pressures (ZoBell and Johnson, 1949; ZoBell and Oppenheimer, 1950). It was assumed that, because various gram-negative bacteria
such as *Flavobacterium okeanokoites, Escherichia coli,* and *Serratia marinorubra* produced long filaments without cross-wall septa when grown at pressures near their pressure maximum, cellular-division mechanisms were the
most pressure sensitive, and thus limiting factors for growth under pressure.
There is little question that one or more structures or synthetic mechanisms
were affected at the relatively high pressures employed by these investigators;
however, concomitant with this limitation is the possibility that much less
substrate was being transported into the cells. The length of time the cells
were exposed to pressure and the incubation temperature would also be
influencing factors. Furthermore, Marquis *et al.* (1971) showed that with
Streptococcus faecalis different catabolic reactions displayed various levels of
barotolerance. Thus glucose-grown cells were less severely affected by pressure than cells grown in a ribose medium. These authors showed that

glucose-grown cells displayed a significantly greater activation volume (ΔV^*) for growth than for glycolysis at 408 atm. In *Streptococcus faecalis*, then, division was not a limiting factor under pressure and no evidence of bizarre morphological forms was observed at pressures which limited catabolic reactions.

It is apparent that different bacterial species display various levels of barotolerance (Oppenheimer and ZoBell, 1952; ZoBell and Johnson, 1949; ZoBell and Oppenheimer, 1950). The data presented in this report definitely substantiate these findings and further indicate that these differences are probably attributable to subtle variations between each species' membrane, thus reflecting a difference in the kinetics of substrate transport. Concordant with these differences may be the pressure susceptibility of either pre-formed membrane structures such as binding sites and transport proteins or their synthetic mechanisms. This would be consistent with previous investigations showing that hydrophobic groups are "altered" by moderate pressures (Kettman *et al.,* 1966). It is possible that the "short-term" irreversible membrane damage noted in this study could be attributed to changes in the hydrophobic portion of binding proteins. This could also be a partial explanation for the protective effect of organic substrates against pressure damage noted by other investigators (Paul and Morita, 1971). It would, therefore, seem to be a reasonable assumption that the breakdown of the inducible system in the antarctic psychrophile 18-500 involved an alteration of some structure or mechanism which was intimately associated with the outer portion of the cell's membrane. Pressures as low as 200 atm affected this breakdown even though this psychrophile was capable of dividing at 500 atm.

The temperature-pressure relationship has been discussed in many previous papers (Albright and Morita, 1972; Haight and Morita, 1962; Morita and Mathemeier, 1964; Paul and Morita, 1971; ZoBell and Oppenheimer, 1950). *Escherichia coli* has been shown to divide at 600 atm at 40°C but only at 300 atm at 20°C (ZoBell and Johnson, 1949). In the present investigation, *Escherichia coli* C-600 was capable of dividing only at 100 atm at 15°C. Additionally, there was no evidence of cellular division by *Vibrio alginolyticus* at pressures of 100 atm or greater at 15°C. The additive effects of pressure and temperature are obvious. What is interesting from these data is the fact that, when bacteria were grown near their optimum (7°C for 18-300; 10°C for 18-500; 15°C for MP-1) and exposed to pressures just above the maximum pressure allowing growth, the extent of "short-term" irreversible membrane damage was much more pronounced than was observed with the mesophilic organisms exposed to the same pressure at 15°C. This would strongly indicate that some aspect of growth and/or cellular division was damaged. This could possibly involve any macromolecule at the transcriptional or translational levels or DNA replication. It seems safe to assume, therefore, that the primary structure affected by pressure is the membrane and, depending on the temperature, the levels of pressure employed, and the particular bacterial strain, this effect could be irreversible due probably to damage in the macromolecular synthetic mechanisms. One interesting possibility from these data is that in the open ocean, where there are pressures and temperatures below the minimum for transport of substrates (3°–5°C) by terrestrial mesophiles such as human enteric bacteria, these microorganisms could persist for long periods of time, resulting possibly in the contamination of nekton and other indigenous fauna.

ACKNOWLEDGMENT

This investigation was supported by NSF grant GA-28521.

LITERATURE CITED

Albright, L. J., and R. Y. Morita. 1968. Effect of hydrostatic pressure on synthesis of protein, ribonucleic and deoxyribonucleic acid by the psychrophilic marine bacterium, *Vibrio marinus*. Limnol. Oceanogr. 13:637–643.

Albright, L. J., and R. Y. Morita. 1972. Effects of environmental parameters of low temperature and hydrostatic pressure on L-serine deamination by *Vibrio marinus*. J. Oceanogr. Soc. Jap. 28:63–70.

Arnold, R. M., and L. J. Albright. 1971. Hydrostatic pressure effects on the translation stages of protein synthesis in a cell-free system in *Escherichia coli*. Biochim. Biophys. Acta 238:347–354.

Braarud, T. 1961. Cultivation of marine organisms as a means of understanding environmental influences on populations. *In* M. Sears (ed.), Oceanography, pp. 271–298. Amer. Ass. Advan. Sci. Publ. No. 67, Washington, D.C.

Haight, R. D., and R. Y. Morita. 1962. Interaction between the parameters of hydrostatic pressure and temperature on aspartase of *Escherichia coli*. J. Bacteriol. 83:112–120.

Hobbie, J. E., and C. C. Crawford. 1969. Respiration correction for bacterial uptake of dissolved organic matter in natural waters. Limnol. Oceanogr. 14:528–532.

Johnson, F. H. 1970. The kinetic basis of pressure effects in biology and chemistry. *In* A. M. Zimmerman (ed.), High Pressure Effects on Cellular Processes, pp. 1–44. Academic Press, New York.

Kennell, D. 1967. Use of filters to separate radioactivity in RNA, DNA, and protein. *In* C. Grossman and K. Moldave (eds), Methods of Enzymology, Vol. 12, pp. 686–693. Academic Press, New York.

Kettman, M. S., A. H. Nishikawa, R. Y. Morita, and P. R. Becker. 1966. Effect of hydrostatic pressure on the aggregation reaction of poly-L-valyl ribonuclease. Biochem. Biophys. Res. Commun. 22:262–267.

Landau, J. V. 1970. Hydrostatic pressure on the biosynthesis of macromolecules. *In* A. Zimmerman (ed.), High Pressure Effects on Cellular Processes, pp. 45–70. Academic Press, New York.

Marquis, R. E., W. P. Brown, and W. O. Fenn. 1971. Pressure sensitivity of streptococcal growth in relation to catabolism. J. Bacteriol. 105:504–511.

Morita, R. Y. 1968. Distribution effects of temperature, pressure, currents, organic matter, living organisms, etc. *In* C. H. Oppenheimer (ed.), Symposium on Marine Biology, Vol. 4, p. 195. New York Academy of Sciences, New York.

Morita, R. Y., and P. F. Mathemeier. 1964. Temperature-hydrostatic pressures studies on partially purified inorganic pyrophosphatase. J. Bacteriol. 88:1667–1671.

Oppenheimer, C. H., and C. E. ZoBell. 1952. The growth and viability of sixty-three species of marine bacteria as influenced by hydrostatic pressure. J. Mar. Res. 11:10–18.

Paul, K. L., and R. Y. Morita. 1971. Effects of hydrostatic pressure and temperature on the uptake and respiration of amino acids by a facultatively psychrophilic marine bacterium. J. Bacteriol. 108:835–843.

Schwarz, J. R., and J. V. Landau. 1972. Hydrostatic pressure effects on *Escherichia coli:* Site of inhibition of protein synthesis. J. Bacteriol. 109:945–948.

Yayanos, A. A., and E. C. Pollard. 1969. A study of the effects of hydrostatic pressure on macromolecular synthesis in *Escherichia coli.* Biophys. J. 9:1464–1482.

ZoBell, C. E. 1961. Importance of microorganisms in the sea. *In* Proceedings of Low Temperature Microbiology Symposium, Camden, New Jersey, pp. 107–132. Campbell Soup Company, Camden, New Jersey.

ZoBell, C. E., and A. B. Cobet. 1964. Filament formation by *Escherichia coli* at increased hydrostatic pressures. J. Bacteriol. 87:710–719.

ZoBell, C. E., and F. H. Johnson. 1949. The influence of hydrostatic pressure on the growth and viability of terrestrial and marine bacteria. J. Bacteriol. 57:179–189.

ZoBell, C. E., and C. H. Oppenheimer. 1950. Some effects of hydrostatic pressure on the multiplication and morphology of marine bacteria. J. Bacteriol. 60:771–781.

ALGAL PHOTOSYNTHESIS AT CONSTANT pO_2 AND INCREASED HYDROSTATIC PRESSURE*

D. H. POPE† and LESLIE R. BERGER

Department of Microbiology
University of Hawaii

We recently reported an apparatus which controls the concentration of dissolved oxygen in closed containers at any desired value (Pope and Berger, 1973). At the same time the rate of oxygen production by cells in the container can be measured. The apparatus is designed to operate at increased hydrostatic pressures up to 1000 atm. It consists basically of two polarographic oxygen electrodes, one which measures, in effect, the concentration of dissolved oxygen and the other, a larger unit, which electrochemically consumes oxygen at a rapid rate. The latter electrode is servo-regulated such that, when the concentration of oxygen in solution exceeds the preset value, it begins to consume oxygen. When the concentration of dissolved oxygen is reduced to the preset value, the electrode shuts off. This device can measure the rates of oxygen production as a function of light intensity, CO_2 concentration, O_2 concentration, hydrostatic pressure, and/or other parameters, while maintaining the concentration of oxygen in solution constant, since the concentration of oxygen is known to affect the photosynthetic capacity of

* This paper was supported by Office of Naval Research Contract No. N00014-67-A-0387-0008 and intramural research grants from the University of Hawaii.
† This work is taken in part from a thesis to be submitted in partial fulfillment of the requirements for the degree of Doctor of Philosophy in the Department of Microbiology, University of Hawaii.

many organisms. We report, here, determination of the rates of oxygen evolution, carbon fixation, and growth by a number of algal types in axenic culture at various concentrations of dissolved oxygen at both ambient pressure and at increased hydrostatic pressures.

MATERIALS AND METHODS

All procedures were done by techniques described by Pope and Berger (1973). Axenic cultures of *Anacystis nidulans,* and species of *Phaeodactylum, Chlorella,* and *Cricosphera* were maintained routinely under fluorescent light in the laboratory. Exponentially growing cells were used in all experiments reported. Uptake of ^{14}C-labeled bicarbonate was done by conventional procedures. Activity was measured in a scintillation counter with the cells on the membrane filters totally immersed in scintillation fluid. Growth at increased hydrostatic pressure was determined turbidometrically.

RESULTS AND DISCUSSION

Figure 1 demonstrates the ability of the apparatus to control the concentration of dissolved oxygen at various hydrostatic pressures. As the figure demonstrates, control is achieved at each pressure tested. Figure 2 shows the effect of hydrostatic pressure on the rate of light-induced oxygen evolution in three algal species. In each case rates of oxygen determination were determined sequentially from the lowest pressure to the highest value. Cells were maintained for 1 hr at each pressure. Many determinations with *Anacystis* demonstrated that the shape of the pressure inhibition curve was essentially the same whether the experiment was conducted as described above or

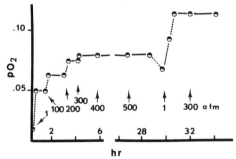

Fig. 1. Regulation of the concentration of dissolved oxygen at increased hydrostatic pressure. The initial density of *Anacystis nidulans* was 0.6×10^8 cells per ml. Pressure was applied at the times indicated by the arrows and maintained until change. ——, Light on, pO_2 regulated; - - - light on, pO_2 unregulated; · · · · light off.

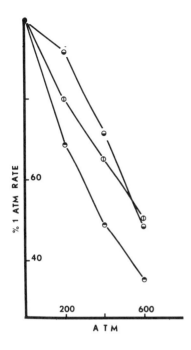

Fig. 2. Effect of hydrostatic pressure on the rate of oxygen evolution of three algal cultures. Exponentially growing cells were suspended in fresh medium and equilibrated with gas mixtures containing the indicated concentration of oxygen. The suspension was pressurized to the indicated value for 60 min during which time the rate of oxygen evolution was determined. ⊖, Species of *Phaeodactylum;* ⊙, Species of *Cricosphera;* ⊕, *Anacystis nidulans.* All determinations were at 25°C, 50% light saturation.

whether a separate sample of cells was used at each pressure. Furthermore, inhibition by hydrostatic pressure was not changed significantly whether the experiment was done in solutions containing 0.8, 2.0, or 8.4 mg/liter dissolved oxygen. The three organisms tested (Fig. 2) showed relatively small differences in their response to hydrostatic pressure. Figure 3 shows the results of hydrostatic pressure on $^{14}CO_2$ uptake. These experiments were done with that concentration of dissolved oxygen obtained by bubbling the culture with air. The concentration of oxygen was subsequently not controlled. Uptake of ^{14}C in all cases, however, was linear over a 2-hr period as determined by periodic sampling of the culture. As with oxygen evolution, the inhibition of CO_2 uptake by increased hydrostatic pressure is approximately linear over the range tested. However, the effect of hydrostatic pressure is more pronounced. Figure 4 summarizes results with *Anacystis nidulans.* The data for the rates of oxygen release at 5 and 60 min were obtained with an electrode similar to that used by Chandler and Vidaver (1971). The data for the rates of oxygen production at constant oxygen concentration and of CO_2 fixation are the same as shown in previous figures. Growth at increased hydrostatic pressure was determined in media which were initially air saturated. The concentration of oxygen was subsequently not controlled. Examination of the data in Fig. 4 leads to the following conclusions. The oxygen-evolving system appears to be less affected by and less sensitive to increased hydrostatic pressure than are the mechanisms of CO_2 fixation. This is in agreement with the conclusions of Vidaver (1969). The fact that CO_2 fixation and growth data are nearly superimposable is not surprising: without CO_2 fixation there can be no growth unless large endogenous food reserves are present and mobilizable. The role of such food reserves in the dark respiration of algae in the ocean, possibly at considerable depths, is the subject of current studies by this laboratory. Whether or not the oxygen-producing mechanism is stable and/or functional at high pressures is not relevant to the marine environment except, perhaps, in terms of the

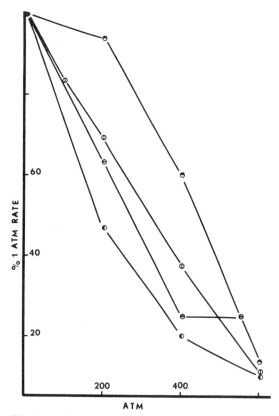

Fig. 3. Effect of hydrostatic pressure on the rate of $^{14}CO_2$ by four algal cultures. Exponentially growing cells were transferred to fresh medium and equilibrated with air. ^{14}C-labeled bicarbonate was added and the suspension pressurized. At intervals samples were removed from the cylinders, and the cells were placed on membrane filters and dried. ^{14}C activity was determined by scintillation counting. ◐, Species of *Phaeodactylum;* ◑, Species of *Chlorella;* ⊖, Species of *Cricosphera;* Φ *Anacystis nidulans.* Determination at 25°C, 50% light saturation.

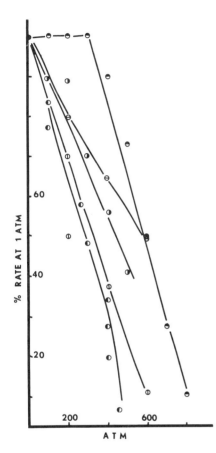

Fig. 4. Effect of hydrostatic pressure on growth, CO$_2$ uptake, and rates of oxygen evolution by *Anacystis nidulans.* ⊖, Rate of oxygen evolved after 5 min pressurization, initial pO$_2$ near zero, 750 foot-candles; ◐, same as above but measured after 60 min pressurization; ⊖, oxygen evolution after 3 hr pressurization with O$_2$ concentration maintained at 8.4 mg/liter; ◑, rate of CO$_2$ uptake, oxygen held at 8.4 mg/liter, 1150 foot-candles; ◐, growth, measured by turbidity.

survival of organisms which are transported within water masses from the surface to considerable depths only to reappear on the surface.

LITERATURE CITED

Chandler, M. T., and W. Vidaver. 1971. Stationary platinum electrode for measurement of oxygen exchange by biological systems under hydrostatic pressure. Rev. Sci. Instrum. 42:143–146.

Pope, D. H., and L. R. Berger. 1973. An apparatus to measure the rate of oxygen evolution while maintaining pO$_2$ constant during photosynthetic growth in closed culture vessels capable of operation at increased hydrostatic pressures. Biotech. Bioeng. 15:518.

Vidaver, W. 1969. Hydrostatic pressure effects on photosynthesis. Int. Rev. Gesamten Hydrobiol. 54:697–747.

RESPONSE OF SOME ACTIVITIES OF FERROMANGANESE NODULE BACTERIA TO HYDROSTATIC PRESSURE

HENRY L. EHRLICH

Department of Biology
Rensselaer Polytechnic Institute

The Mn(II) oxidizing *Arthrobacter* 37, which came from a ferromanganese nodule in shallow water (730 m), was found to grow in 0.5% peptone-seawater broth to varying extents during 54 hr of incubation at 25°C at hydrostatic pressures from 1 to 408 atm. This is a narrower range of pressures than that in which this organism oxidizes Mn(II) (1–476 atm). The Mn(II) oxidizing culture BIII 39, which came from a ferromanganese nodule in the deep sea (about 5000 m), was found to grow in 0.5% peptone-seawater broth to varying extents during 54 hr of incubation at 15°C at hydrostatic pressures from 1 to 408 atm. Changes in composition of the medium did not increase the pressure range which permitted growth by this organism. Generation times were determined for growth at 1 and 340 atm in 0.1% peptone-seawater and 0.1% gelatin-seawater broths. Both cell division and cell-wall synthesis were upset at 340 atm but not at 204 atm. The culture was able to oxidize Mn(II) at 4°C over a wider pressure range (1–476 atm) than it was able to grow at 15°C. The MnO_2 reducing culture BIII 32, which came from the same ferromanganese nodule as culture BIII 39, was found to grow in 0.5% peptone-seawater broth over a pressure range of 1–408 atm during 54 hr of incubation at 15°C. Cell division and cell-wall synthesis were upset at 340 but not at 204 atm. When its MnO_2 reductase was induced, culture BIII 39 was able to reduce MnO_2 at 4°C in a pressure range from 1 to 408 atm.

INTRODUCTION

My chief interest in marine microbiology at the present time is directed toward the region of the sediment-water interface of the deep sea. In particular, I am interested in the microbes, principally bacteria, which are associated with ferromanganese nodules. Such nodules may occur in great numbers in this environment in the world's oceans. As work from my laboratory has shown, some bacteria associated with the nodules can catalyze manganese accretion while some others can catalyze Mn(IV) oxide reduction if in the presence of a substrate that can serve as an appropriate electron donor (Ehrlich, 1970). Our previous studies have emphasized the distribution of bacteria in nodules and associated sediments (Ehrlich *et al.*, 1972) and an examination of some of the biochemical processes involved in Mn(II) oxidation and Mn(IV) reduction (Ehrlich, 1963, 1966, 1968; Ehrlich *et al.*, 1973; Trimble and Ehrlich, 1968, 1970). One of our current projects is an investigation of the influence of hydrostatic pressure on growth of the bacterial flora of manganese nodules and on the biochemical processes of Mn(II) oxidation and Mn(IV) reduction of which some of its members are capable. The initial results from these studies are the subject of this paper.

MATERIALS AND METHODS

Cultures

The organisms used in these studies included the Mn(II) oxidizing *Arthrobacter* 37 (NCIB 10328, National Collection of Industrial Bacteria, Torry Research Station, Aberdeen, Scotland) and the Mn(II) oxidizing culture BIII 39. The latter is a gram-negative, motile rod which, unlike *Arthrobacter* 37, does not grow in freshwater media. It does not ferment glucose, lactose, or sucrose, nor does it reduce nitrate or hydrolyze starch. It does liquefy gelatin. Also used was the MnO_2 reducing culture BIII 32. It is a gram-negative, motile, ovoid rod which does not ferment glucose, lactose, or sucrose, nor does it hydrolyze starch or reduce nitrate. It does liquefy gelatin. It possesses an inducible MnO_2 reductase system. Both cultures, BIII 39 and BIII 32, were isolated from the same Pacific deep-sea nodule.

Media

The following media were employed in studying the effect of pressure on growth: 0.5% peptone broth; 0.1% peptone broth; 1% gelatin broth; 0.5% gelatin broth; 0.1% gelatin broth; GA broth containing 1% glucose, 0.09%

$(NH_4)_2SO_4$, and 0.005% yeast extract. All broths were made up in natural seawater.

Plate counts were made on nutrient agar (Difco). One-milliliter inocula were introduced onto the plates in 3 ml of capping agar containing 4.0 g peptone, 2.4 g beef extract, and 1.2 g agar per liter. The basal and capping agars were prepared in natural seawater.

Growth Experiments

The method for studying growth at different hydrostatic pressures was adapted from the one described by ZoBell and Oppenheimer (1950). It consisted of incubating Neoprene-stoppered test tubes (82 × 14 mm) in water-filled pressure cylinders at various hydrostatic pressures at 15°C. The tubes contained an appropriate broth medium inoculated with about 100 cells per ml of a 24-hr broth culture. They were removed from their respective cylinders at different times, and the viable population that had developed in them was estimated by plate counting. Each determination at a given time and pressure required analysis of the content of a separately pressurized growth tube.

Induction of MnO_2 Reductase Activity in Culture BIII 32

To induce MnO_2 reductase activity, culture BIII 32 was grown at 15°C for 24 hr on seawater-nutrient agar slants containing 5 ml of 0.1 M $MnSO_4 \cdot H_2O$ per 100 ml medium in Roux flasks. The cells were harvested in 10-fold diluted seawater and washed by centrifugation at 12,000 × g and 4°C until the supernatant showed no further presence of manganese by the persulfate method (Ehrlich, 1966). The final cell crop was resuspended in 5 ml of 10-fold diluted seawater at 4°C.

Mn(II) Oxidation and MnO_2 Reduction under Hydrostatic Pressure

The procedures used for these studies were those described by Ehrlich (1971). However, the composition of the solution to study MnO_2 reduction was modified by omitting $NaHCO_3$. As a result of a modification worked out by Ghiorse (1972) in this laboratory, the reaction mixture of MnO_2 reducing experiments was sampled 10 min after addition of 0.05 ml of 10 N H_2SO_4 instead of 30 min as in previous studies with *Bacillus* 29. Both Mn(II) oxidation and MnO_2 reduction reactions were incubated for 17 hr at 4°C.

RESULTS

Growth of *Arthrobacter* 37 at 25°C and Different Hydrostatic Pressures

We had previously reported that *Arthrobacter* 37, which I isolated from an Atlantic ferromanganese nodule in 1961, oxidizes Mn(II) at 25°C in a

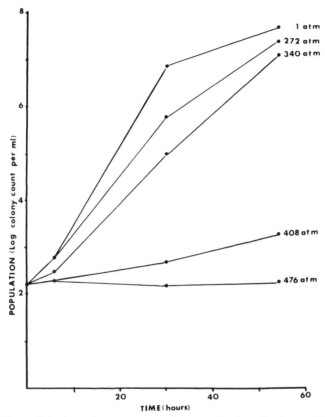

Fig. 1. Effect of hydrostatic pressure on the growth of *Arthrobacter* 37 in 0.5% peptone-seawater broth at 25°C.

hydrostatic pressure range from 1 to 476 atm (Ehrlich, 1971).* This raised the obvious question of whether this organism can also grow in this range of pressures at 25°C. This was tested by growing the culture in 0.5% peptone-seawater broth in pressure chambers as described in Materials and Methods. The results are summarized in Fig. 1. They show a sharp drop in growth rate between 340 and 408 atm, and a lack of growth, at least during 54 hr of incubation, at 476 atm. The organism thus failed to grow during 54 hr at a pressure that was lower than that at which it failed to oxidize Mn(II).

* The conversions of pressure units from pounds per square inch (psi) to atmospheres (atm) in Tables 1 through 4 in the paper by Ehrlich (1971) were made by setting 1 atm equal to 15 psi. For the purpose of the present paper, the measured psi units in the tables were recalculated as atm units by setting 1 atm equal to 14.7 psi, which is more exact. The latter conversion factor was also used for all pressure measurements in the present paper which were made in terms of psi units.

Growth of Culture BIII 39 at 15°C at Different Hydrostatic Pressures

Arthrobacter 37 came from a nodule from a relatively shallow environment (730 m) in the Atlantic Ocean. Yet, most nodules occur at much greater depths. Thus it became desirable to study a Mn(II) oxidizing organism which came from a nodule from a deep-sea environment (about 5000 m). I selected for this purpose culture BIII 39 which came from such an environment in the tropical Central Pacific Ocean. When this organism was grown at different hydrostatic pressures at 15°C in 0.5% peptone-seawater broth, it showed a marked difference in growth rate between 204 and 340 atm (Fig. 2). It did not grow significantly at 408, 476, or 612 atm. Indeed, the curves for these pressures suggest a slow die-off.

Effect of Different Culture Media on Growth of Culture BIII 39 at Different Hydrostatic Pressures and 15°C

Since the nodule from which culture BIII 39 was isolated came from a depth range of 4250–5200 m, growth should have occurred at 408 atm. One reason why no significant growth was observed at that pressure may be that growth rates are too slow for a significant population increase to have been observed in 54 hr. This possibility has not been tested. Another reason for the absence of significant growth at 408 atm and 15°C may be that the growth medium was not favorable at that pressure. Kriss and Mitskevich (1967) have reported on a change in barotolerance of *Pseudomonas* sp. strain 8113 with a change in culture-medium composition. Therefore, we tested growth in six different media as listed in Table 1 and described in "Materials and Methods". More luxuriant media were not tried because they would be very unrepresentative of the nutritional environment of the deep sea. ZoBell estimated the organic carbon content of deep-sea mud to range from 1 to 5 mg/m^2 (ZoBell, 1954, as cited by ZoBell and Morita, 1957).

The results in Table 1 show that at 15°C and 1 atm the extent of 54 hr of growth of culture BIII 39 was greatest in 0.5% peptone-seawater broth and smallest in 0.1% gelatin-seawater broth. Growth at 1 atm in GA broth was roughly comparable with that in 1% gelatin broth. A pressure increase to 204 atm caused a slight decrease in 54 hr of growth in the three media on which this effect was tested. A pressure increase to 272 atm also caused only a slight drop in the extent of 54 hr of growth in 0.1% peptone broth but caused an approximately 50-fold reduction in 0.1% gelatin broth. A pressure increase to 340 atm caused a 2- to 4-log decrease in growth depending on the medium. Growth at 408 atm appeared negligible, and at 476 atm it was absent.

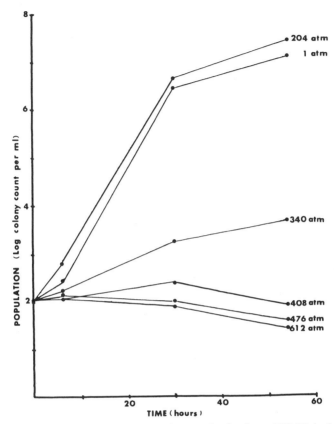

Fig. 2. Effect of hydrostatic pressure on the growth of culture BIII 39 in 0.5% peptone-seawater broth at 15°C.

It will be noted that the reproducibility of the plate counts for all media at 340 atm and for GA broth and 0.5% peptone broth at 408 atm was poor. For the 0.5% peptone broth at 340 atm, for instance, the standard deviation was ±0.84, which means that in individual experiments most plate counts fell in the range of 2.6×10^4 to 1.0×10^6 colonies per ml. This compares with a standard deviation for the same medium at 1 atm of ±0.10, which means that in individual experiments most plate counts fell in the range of 6.0×10^7 to 9.8×10^7 colonies per ml. The reason for this lack of reproducibility may be a weakened cell structure of the organisms at the higher pressures as noted below. The small deviations in the remaining media at the higher pressures is attributed to the absence of significant growth.

Table 1. Effect of pressure on 54-hr growth of culture BIII 39 in different media at 15°C[a]

Pressure (atm)	54-hr Growth					
	GA broth	0.5% peptone	0.1% peptone	1.0% gelatin	0.5% gelatin	0.1% gelatin
1	7.53 ± 0.07 (3)[b]	7.89 ± 0.10 (13)	7.67 ± 0.09 (9)	7.58 ± 0.08 (9)	7.45 ± 0.06 (4)	7.30 ± 0.19 (8)
204	7.31 ± 0.02 (2)	7.79 ± 0.13 (3)			7.13 ± 0.05 (2)	
272			7.11 ± 0.16 (2)			5.66 ± 0.04 (2)
340	4.82 ± 0.44 (2)	5.25 ± 0.84 (6)	3.75 ± 0.78 (3)	4.50 ± 0.67 (4)	4.07 ± 0.69 (2)	4.07 ± 0.80 (4)
408	2.76 ± 0.36 (2)	2.95 ± 0.51 (8)	2.62 ± 0.09 (2)	2.97 ± 0.20 (2)	2.32 ± 0.04 (3)	2.79 ± 0.11 (2)
476	2.51 ± 0.02 (2)	2.21 ± 0.08 (4)		2.37 ± 0.07 (2)	2.19 ± 0.04 (2)	

[a] The results are reported as the arithmetic mean of the log of the colony counts as determined by plating. The logs of all counts were normalized on the basis of an initial count of 4.3×10^2 per ml, whose log is 2.6335. Counts in excess of 1×10^7 per ml were not normalized because they fell into the range of those expected in the maximum stationary phase.

[b] The numbers in parentheses correspond to the number of independent determinations from which the mean was calculated. Where this number was 4 or greater, the range is reported as the standard deviation. In all other instances, the range reported represents the one into which the data at a given set of conditions fell.

Generation Times of Culture BIII 39 at 1 and 340 atm at 15°C

Generation times were determined for culture BIII 39 at 1 atm and 340 atm at 15°C in 0.1% peptone broth and 0.1% gelatin broth. Population changes were measured at 0, 16, 24, 40, and 48 hr. The generation times calculated from the results are shown in Table 2. As expected, at 1 atm reproducibility was good while at 340 atm it was poor. Nevertheless, I conclude that growth in 0.1% peptone broth was faster than in 0.1% gelatin broth, and by almost the same factor at 1 atm as at 340 atm. The generation time in GA broth was 5.7 ± 0.8 hr at 340 atm, thus comparable with growth in 0.1% peptone broth.

Morphological Changes of Culture BIII 39 at 340 atm

The morphologies of the cells of culture BIII 39 grown for 48 and 54 hr at 15°C from an initial population density of about 1000 cells per ml of GA broth were compared at pressures of 204 and 340 atm. It was found that the cells at 204 atm had a normal rod shape and retained motility while the cells at 340 atm tended to be non-motile and bloated, some appearing as spheroplasts, others being elongate and sausage shaped. These observations suggest that at 340 atm the culture encountered some difficulty in cell division and, more importantly, in cell-wall synthesis. The latter difficulty may have left the cells fragile and may account for the poor reproducibility in the growth experiments. A defective cell-wall structure may, however, be of a distinct advantage to the organism in oxidizing Mn(II) because the appropriate enzyme system appears to be associated with the cell surface (Ehrlich, 1968).

Table 2. Generation times of culture BIII 39 growing in peptone-seawater and gelatin-seawater broths at 15°C at atmospheric pressure and at 340 atm

Pressure (atm)	Generation time (hr)		g_{pep}/g_{gel} [a]
	0.1% peptone	0.1% gelatin	
1	1.6; 1.6	2.1; 2.2	0.74
340	5.4 ± 0.48	7.8 ± 1.71	0.69

[a] g_{pep}/g_{gel} is the ratio of the generation time in peptone broth to the generation time in gelatin broth at a given pressure.

Effect of Hydrostatic Pressure on Mn(II) Oxidation by Culture BIII 39 at 4°C

Mn(II) oxidation of culture BIII 39 was tested at various pressures by incubating at a temperature of 4°C for 17 hr. Table 3 shows that oxidation became unmeasurable between pressures of 476 and 544 atm. Since we did not incubate for longer than 17 hr, we cannot be certain whether the lack of significant oxidation at 544 atm means that the enzyme was totally inhibited. In any case, the highest pressure at which oxidation was observed was above the pressure at which growth was observed. This difference is the more remarkable when it is remembered that the growth studies were made at 15°C while the enzyme studies were made at 4°C. These findings raise the question whether this culture grows in the deep-sea environment at pressures above 340 atm by using Mn(II) oxidation as an energy source while reserving organic carbon for assimilation. We plan to examine this question in the future.

Table 3. Mn(II) oxidation by culture BIII 39 during 17 hr of incubation at 4°C at various hydrostatic pressures

Pressure (atm)	Cell no. on pad ($\times 10^9$)	ΔMn(II)[a] (nmole/ml)
1	6.1	50
	4.8	40
340	4.8	20
	4.3	40
408	3.8	50
	4.4	50
476	4.4	30
	4.1	30
544	4.1	Tr[b]
	4.3	Tr

[a] The difference in residual Mn(II) concentration between reaction vessels with and without bacterial cells.

[b] Tr, trace.

Table 4. Effect of pressure on
54-hr growth of culture BIII 32 in
0.5% peptone-seawater broth at 15°C

Pressure (atm)	54-hr growth log (colonies per ml)
1	7.93 ± 0.06 (4)[a]
204	7.75 ± 0.02 (3)
340	6.61 ± 0.68 (4)
408	3.21 ± 0.42 (3)
476	2.49 ± 0.13 (3)

[a] The numbers in parentheses denote the number of individual experiments from which the mean was calculated. For an explanation of how the ranges of the average logarithms of the counts were calculated see Table 1. The logarithm of the initial count was 2.6335.

Growth of the MnO$_2$-Reducing Culture BIII 32 at 15°C and Different Hydrostatic Pressures

Our previous studies on MnO$_2$ reduction were performed mainly with *Bacillus* 29, which came from a nodule from a shallow environment in the Atlantic Ocean (1700 m). It thus became desirable to study an MnO$_2$ reducer from a nodule from a deep-sea environment in the Pacific Ocean (about 5000 m). The organism that was selected was designated culture BIII 32. It came from the same nodule as culture BIII 39. When grown for 54 hr at different hydrostatic pressures in 0.5% peptone-seawater broth, it behaved more or less like culture BIII 39 (Table 4). The growth yield was considerably less at 408 atm than at 1, 204, or 340 atm. Although significant growth was noted at 408 atm, none was seen at 476 atm. As with culture BIII 39, growth yields in repeated experiments spread over a much wider range than at 1 atm.

Morphological Changes of Culture BIII 32 at 340 atm

Like culture BIII 39, culture BIII 32 showed a tendency for cells to appear bloated or as spheroplasts when the pressure at 15°C was raised to 340 atm for 48 or 54 hr. No such forms were seen after similar incubation at 204 atm. The frequency of occurrence of spheroplasts and bloated cells seemed somewhat less than with BIII 39. Whether this is attributable to a difference in medium (BIII 32 was studied morphologically in 0.5% peptone broth while BIII 39 was studied in GA broth), or whether BIII 32 is a little more resistant to 340 atm than BIII 39 has not yet been determined. As in the case of BIII 39, spheroplast formation by BIII 32 may indicate difficulty in cell-wall synthesis. A defective cell wall may have an advantage for the cell in facilitating enzymic MnO_2 reduction.

Effect of Hydrostatic Pressure on MnO_2-Reductase Activity of Culture BIII 32 at 4°C

When induced cells of culture BIII 32 were incubated for 17 hr at various hydrostatic pressures and 4°C, the results in Table 5 were obtained. It is seen that MnO_2-reductase activity became unmeasurable between pressures of 408 and 476 atm. Since we did not incubate for longer than 17 hr at 476 atm, we cannot say whether inhibition at this pressure was absolute. It is clear, however, that in these experiments the culture was actively reducing MnO_2 in the same pressure range as that in which growth took place, except that

Table 5. MnO_2 reduction by induced culture BIII 32 during 17 hr of incubation at 4°C at various hydrostatic pressures

Pressure (atm)	Cell no. on pad ($\times 10^9$)	Mn released (nmole/ml)
1	3.1	14
	3.2	18
204	3.0	22
	3.0	20
272	2.9	16
	3.2	18
340	3.0	16
	3.0	16
408	3.0	14
	3.0	8
476	3.1	0
	3.2	0

reduction was studied at 4°C while growth was studied at 15°C. These findings raise the question whether growth of this culture in the deep-sea environment is possible at 4°C under conditions where it uses MnO_2 as a terminal electron acceptor instead of O_2. We plan to test this problem in the future.

DISCUSSION

These studies show that Mn(II)-oxidizing and MnO_2-reducing activities can occur at hydrostatic pressures and temperatures which exist at the sediment-water interface of the deep sea. The range of pressures and temperatures at which MnO_2-reductase activity can be induced remains to be determined, however. The studies leave unresolved the question of growth of the test cultures at the deep-sea pressures and temperatures which they must have encountered in their natural habitat. Lack of significant growth at 15°C of BIII 39 at 408 atm and above and of BIII 32 at 476 atm and above may be accounted for in terms of (a) too slow a rate of growth at these pressures for a cell increases to be noted in 54 hr, (b) a loss of viability on depressurization of cells grown at these critical pressures, (c) unfavorable nutritional conditions, or (d) unfavorable environmental conditions other than temperature and pressure. The last-named possibility includes growth in seawater suspension instead of on the surface or in pores of ferromanganese nodules. Limitations in salinity are a less likely factor since natural seawater was used. However, Palmer and Albright (1970) and Albright and Henigman (1971) have observed salinity-pressure effects on cell division of a few marine bacteria. Both cultures BIII 39 and BIII 32 are able to grow at 4°C at 1 atm in 0.5% and 0.1% peptone-seawater broth and in 0.5% and 0.1% gelatin-seawater broth. Yet, at 10°C and 340 atm, culture BIII 39 does not grow significantly in 48 hr in 0.1% peptone broth or 0.1% gelatin broth.

An upper absolute hydrostatic pressure limit of growth that cannot be raised by chemical or physical manipulation must exist for any culture at a given temperature in its physiological range. This limit must be governed by inhibition of a key process which is vital to the cell. The work of Morita and co-workers suggests that an alteration in the cell membrane which prevents uptake of a vital metabolite, like an amino acid, may be such a key process. Thus, Paul and Morita (1971) reported that *Vibrio* MP-38, a facultative psychrophile, which grows at 1 atm in a temperature range from less than 0° to greater than 30°C, exhibits greatly reduced glutamate, phenylalanine, and glycine uptake at pressures of 500 atm in the range of 5–25°C, but shows much less affected respiratory rates on these amino acids, except for glycine at 5°C. On the other hand, the work of Landau and co-workers suggests that the inhibitory locus may be associated with protein synthesis. Thus, Landau (1966, 1970) found amino-acid incorporation by whole cells of *Escherichia coli* K-12 to be inhibited around 670 atm at 37°C, and Schwarz and Landau (1972) found that amino acid incorporation by cell-free preparations of *Escherichia coli* S-30 was similarly inhibited at 670 atm. Interestingly, Landau (1966, 1970) reported that nucleic-acid synthesis still proceeded at a measurable rate under these conditions in whole cells of *Escherichia coli* K-12. Where these limits lie for cultures BIII 39 and BIII 32 remains to be determined.

The observations on morphological changes under high hydrostatic pressure are not new. Such changes were first reported by ZoBell and Oppenheimer (1950). They noted cell elongation and attributed it to inhibition of cell division. Such morphological changes have also been reported by Oppenheimer and ZoBell (1952) and by Kriss and Mitskevich (1967). The latter found in addition that the composition of the medium affected the degree of cell elongation. None of these studies indicated spheroplast formation or bloating of cells suggesting inhibition of cell-wall formation. This aspect needs further examination.

ACKNOWLEDGMENTS

The very expert technical assistance of Alice R. Ellett in the experimental work of this paper is gratefully acknowledged. This research was aided by a grant, GA-28398, from the National Science Foundation.

LITERATURE CITED

Albright, L. J., and J. E. Henigman. 1971. Seawater salts-hydrostatic pressure effects upon cell division of several bacteria. Can. J. Microbiol. 17:1246−1248.

Ehrlich, H. L. 1963. Bacteriology of manganese nodules. I. Bacterial action on manganese in nodule enrichments. Appl. Microbiol. 11:15−19.

Ehrlich, H. L. 1966. Reactions with manganese by bacteria from ferromanganese nodules. Develop. Ind. Microbiol. 7:279−286.

Ehrlich, H. L. 1968. Bacteriology of manganese nodules. II. Manganese oxidation by cell-free extract from a manganese nodule bacterium. Appl. Microbiol. 16:197−202.

Ehrlich, H. L. 1970. The microbiology of manganese nodules. U.S. Clearing House. Fed. Sci. Tech. Inform. AD 1970, No. 716,508.

Ehrlich, H. L. 1971. Bacteriology of manganese nodules. V. Effect of hydrostatic pressure on bacterial oxidation of Mn(II) and reduction of MnO_2. Appl. Microbiol. 21:306−310.

Ehrlich, H. L., W. C. Ghiorse, and G. L. Johnson II. 1972. Distribution of microbes in manganese nodules from the Atlantic and Pacific Oceans. Develop. Ind. Microbiol. 13:57–65.

Ehrlich, H. L., S. H. Yang, and J. D. Mainwaring, Jr. 1973. Bacteriology of manganese nodules. VI. Fate of copper, nickel, cobalt, and iron during bacterial and chemical reduction of the manganese (IV) oxide matrix of nodules. Z. Allg. Mikrobiol. 13:39–48.

Ghiorse, W. C. 1972. Characterization of the MnO_2-reductase system in a marine Bacillus. Ph.D. Diss., Rensselaer Polytechnic Inst., Troy, N.Y.

Kriss, A. E., and I. N. Mitskevich. 1967. Effect of the nutrient medium on the tolerance of barotolerant bacteria to high hydrostatic pressures. Microbiology 36:203–206.

Landau, J. V. 1966. Protein and nucleic acid synthesis in *Escherichia coli:* pressure and temperature effects. Science 153:1273–1274.

Landau, J. V. 1970. Hydrostatic pressure on the biosynthesis of macromolecules. *In* A. M. Zimmerman (ed.), High Pressure Effects on Cellular Processes, pp. 45–70. Academic Press, New York.

Oppenheimer, C. H., and C. E. ZoBell. 1952. The growth and viability of sixty-three species of marine bacteria as influenced by hydrostatic pressure. J. Mar. Res. 11:10–18.

Palmer, D. S., and L. J. Albright. 1970. Salinity effects on the maximum hydrostatic pressure for growth of the marine psychrophilic bacterium *Vibrio marinus.* Limnol. Oceanogr. 15:343–347.

Paul, K. L., and R. Y. Morita. 1971. Effects of hydrostatic pressure and temperature on the uptake and respiration of amino acids by a facultative psychrophilic marine bacterium. J. Bacteriol. 108:835–843.

Schwarz, J. R., and J. V. Landau. 1972. Hydrostatic pressure effects on a cell-free synthesizing system of *Escherichia coli.* Abstr. Annu. Meeting, Amer. Soc. Microbiol., p. 30.

Trimble, R. B., and H. L. Ehrlich. 1968. Bacteriology of manganese nodules. III. Reduction of MnO_2 by two strains of nodule bacteria. Appl. Microbiol. 16:695–702.

Trimble, R. B., and H. L. Ehrlich. 1970. Bacteriology of manganese nodules. IV. Induction of an MnO_2 reductase system in a marine bacillus. Appl. Microbiol. 19:966–972.

ZoBell, C. E., and R. Y. Morita, 1957. Barophilic bacteria in some deep sea sediments. J. Bacteriol. 73:563–568.

ZoBell, C. E., and C. H. Oppenheimer. 1950. Some effects of hydrostatic pressure on the multiplication and morphology of marine bacteria. J. Bacteriol. 60:771–781.

SOME THOUGHTS ON THE NATURE OF BAROPHILIC BACTERIA

LESLIE R. BERGER

Department of Microbiology
University of Hawaii

This paper presents a series of thoughts on the nature of barophilic bacteria; it is not intended as a scientific report. No doubt many other investigators who are interested in this topic have shared many of the same ideas. Thus, I do not claim that this is entirely original thinking. At the same time, because I have no way of knowing who thought what and when, I will credit no individual with any original statement which may appear and I take full credit for mis-statements.

A barophilic bacterium, or barophile, as defined in this paper is any bacterium capable of growth at high hydrostatic pressure. Arbitrarily, I will define high pressure to be in excess of 600 atm. My definition of barophile includes the "baroduric" organisms as used by other investigators. At least four general approaches are employed to seek information about barophilic organisms.

1. The direct study. Unfortunately, until recently this most logical approach has been fraught with almost insurmountable technical difficulties. Indeed, there exists but a handful of reports directly dealing with these bacteria. Most of them have unfortunately been severely limited by the primitive techniques which had to be employed.

2. Pressurization of non-barophiles. This approach is largely deductive. Terrestrial and ordinary marine bacteria are subjected to quasi-hadal conditions and observed. Not surprisingly most organisms (perhaps all of them) exhibit some sort of anomalous metabolic behavior at the increased hydrostatic pressures whose cause may or may not be determinable. Once a pressure-sensitive lesion is characterized, one attempts to deduce from the information something of how genuine barophiles have evolved. Let us assume the following hypothetical example. It is found that in a large number of bacterial species, reaction X in the biosynthesis of substance Y is extremely sensitive to increased hydrostatic pressure. It is further shown that enzyme X' in the reaction is irreversibly inactivated owing to some conformational change in the protein. We deduce, then, that in the evolution of barophiles, some change in the structure of enzyme X' must have occurred or some alternate pathway which obviates its use must be operative. From the results of studies such as these, we shall probably conclude that we know something new and finite about barophiles.

3. Extrapolation. This method is also largely deductive. Barophilic organisms are difficult to work with because, in part, they are hard to obtain. For the study of some particular problem, organisms are obtained from surface waters and from waters at increasing depths to the limit of the sampling capability. When the results of the study are analyzed, any parameter which shows a real correlation with hydrostatic pressure is subjected to extrapolation to the desired barophilic conditions. Conclusions are then cautiously, but inevitably, drawn.

4. The last approach questions the premise that barophiles are organisms with particularly distinct or unusual properties. Rather than looking for differences in metabolic pathways, or in protein structure, etc., the approach is directed to the study of environmental differences (besides temperature and hydrostatic pressure). The approach also requires the investigator to question conventional experimental methods. The fact that he has always worked with non-barophiles must not bias the experimental approach. Much of what follows amplifies these statements.

Perhaps most neglected among those factors which all organisms share is (the passage of) time. I believe that microbiologists especially must start thinking of barophiles in terms of *an expanded time scale.* We gauge the growth of bacteria in terms of doubling times ranging from a few minutes to, at most, a few weeks. For the slower growing species, experience has taught us that, if proper growth conditions were known, generation times would generally be substantially reduced. Proper growth conditions usually mean those conditions which yield maximum growth rates or maximum cell crops.

These criteria are applied to the study of psychrophiles, mesophiles, and thermophiles. I believe we have applied them in our thinking about barophiles. Perhaps we have even sought to grow barophiles by manipulating conditions which in our past experience has resulted in increased growth rates and cell yields. I do not know of any major successes in the culture of barophiles. There is evidence to suggest that an expanded time scale is valid. In lieu of minutes or weeks, the generation time of a barophile may be years or decades. Recently, the sunken research submarine *Alvin* was raised after 8 months on the ocean floor at relatively low temperature and moderate hydrostatic pressure. Unsterile food which had remained in contact with natural seawater on board was recovered unspoiled. After a few days at atmospheric pressure at the same temperature, these foods were completely decomposed by the same microbial flora. This, of course, is a dramatic demonstration of lowered biological activity in deep oceanic environments. There are other kinds of evidence from which we may deduce that bacterial activity in hadal waters is slow. In old deep-water masses the concentrations of nutrients which are present are insufficient to permit rapid turnover of biological systems. Many deep-water masses are essentially limited to the nutrients originally present since only trivial amounts of additional organic matter are added as the result of transient animal life.

By definition the barophiles enjoy hydrostatic pressures above 600 atm. In the oceans this condition is accompanied by temperatures below 5°C. Bacteria living in such regions may be of two ecological types. Organisms of the first type are those which are associated with such animals as epiphytes, gastrointestinal parasites, or symbionts, and those which live associated with particulate detritus. The second ecological type includes those bacteria which are capable of free-living existence in open waters, surviving on the extremely low concentrations of dissolved nutrients. Among these are found, I believe, both heterotrophs and autotrophs, for example, nitrifiers. The low concentrations of dissolved nutrients would permit only extremely low rates of growth and they would support very low population densities. I doubt that there is much direct interaction between these free-living bacteria and other living organisms, animal or microbe. There are few filter feeders in deep, open, barren waters. Thus, these free-living bacteria may be the final scavengers of available nutrients.

Unless an organism is in a state of complete suspended animation, it will require energy and nutrients. With these it may grow and it will maintain itself. The slower a bacterium grows, the greater the relative amount of nutrient and energy which it will expend per cell generation for maintenance. Maintenance requirements are generally independent of growth rate. We must consider the very low nutrient concentration of open ocean water as a limiting factor for the rate of growth of free-living bacteria. Unless hydro-static pressure and low temperature alter the maintenance requirements of bacteria, all free-living forms in barren waters would starve to death; mainte-nance requirements would exceed the rate of nutrient uptake.

There is no need to present here a biochemical or biophysical treatment of the effects of hydrostatic pressure, temperature, and nutrient (substrate) concentration on the kinetics of cell growth (or enzyme activity) in micro-organisms. It suffices to draw some *gross generalizations.*

1. With non-barophilic organisms, the effect of increased hydrostatic pressure on most enzyme-catalyzed reactions is one of decreased activity. Assuming that all reactions in a cell function concertedly and in a like manner and extent, increased hydrostatic pressure would lower rates of growth and reproduction. Within limited bounds, this is the observed result.
2. The above argument holds also for the effect of decreasing temperature.
3. There is evidence that at high hydrostatic pressure and low temperature the enzymes of many non-barophilic bacteria are reversibly inactivated. This may be thought of as partial (or temporary) suspension of biological function. If maintenance reactions were more susceptible to such inactiva-tions than growth reactions, then such organisms would function more efficiently in environments of low nutrient concentration at increased hydrostatic pressure than they would at atmospheric pressure. If the reverse reasoning held true, barophiles would starve to death in their native nutrient environment at lowered hydrostatic pressure. This is an oversimplification, of course.
4. There is unequivocal evidence that the affinity of many enzymes for their specific substrates increases significantly as hydrostatic pressure is in-creased. If this phenomenon also provides true for active transport sys-tems, these organisms are more efficient as scavengers of residual nutrients at increased hydrostatic pressures.

Whether any of the above thoughts are correct in their entirety is problematic. They have led me, nevertheless, to a somewhat heretical conclu-sion: barophilic bacteria must be basically marine psychrophiles with much of the same diversity of form and function as their surface-dwelling counter-parts. No doubt some species have adapted exclusively to hadal existence as obligate animal parasites, as symbionts, possibly as attached forms, etc. There must exist a continuum of pressure ranges over which barophilic species can grow. This is, of course, analogous to the continuum of temperature ranges exhibited by terrestrial bacteria.

Hadal water masses which were once surface waters must contain the progeny of the indigenous surface flora. Looked at the other way, barophilic bacteria are part of the ordinary marine psychrophilic flora.

ACKNOWLEDGMENT

I wish to acknowledge many valuable and interesting conversations over the last 6 years with my friend and colleague Kaare Gundersen on this and related topics.

ISOLATION AND CHARACTERIZATION OF SOME DEEP-SEA BACTERIA

RITA R. COLWELL and RALPH C. KETTLING, JR.

Department of Microbiology
University of Maryland

Hydrostatic pressure tolerance and numerical taxonomy studies of deep-sea bacteria were carried out. Twenty-two strains of aerobic, heterotrophic marine bacteria were isolated from a sediment sample collected in the Virgin Island Basin of the Atlantic Ocean at a depth of 4000 m. The organisms, with one exception, were shown to have a high degree of pressure tolerance to the *in situ* pressure. Taxonomic analysis showed these strains to cluster into two groups, related at low S-values to a set of 38 deep-sea strains of bacteria from the Pacific Ocean. One of the clusters was identified as *Vibrio* and the other group resembled species of the genus *Acinetobacter*.

INTRODUCTION

Brunn (1956) reviewed the major physical, chemical, and biological parameters prevailing in the abyssal region of the oceans. Hydrostatic pressure is the most variable of these parameters because the rate of increase in hydrostatic pressure ranges from 0.099 to 0.105 atm/m, depending upon salinity, latitude, temperature, and water depth. The range of hydrostatic pressure in the sea is from 1 atm at the surface to 1140 atm at a depth of 10,860 m (Wüst, 1952). A useful rule for determining *in situ* hydrostatic pressure is that it increases 1 atm for every 10-m increase in depth.

Observations on deep-sea bacteria began with the voyage of the *Talisman* (1882–1883) and Certes' report of the presence of bacteria in sediment samples collected from depths as great at 5100 m (Certes, 1884*a*). He also reported that the microorganisms tolerated hydrostatic pressure isobaric with the *in situ* conditions (Certes, 1884*b*). Bacteria were found to be present in deep-sea sediments collected at depths of up to 5280 m in the West Indies during the Humboldt Plankton Expedition (Fischer, 1894).

Rittenberg (1940) isolated bacteria from cores of marine sediments, observing that large populations of bacteria were present near the mud-water interface and noting that these populations were drastically reduced with core depth. However, bacteria were never found to be absent from any of the cores examined. Rittenberg also noted a number of aerobic bacteria that occurred with greater frequency than anaerobic forms, even in deeper portions of the cores where anoxic conditions prevailed.

ZoBell (1952*a,b*) reported that in the period 1948–1950, bacteria were isolated from sediment samples taken in the mid-Pacific Ocean and on the floor of the Atlantic Ocean at depths of 5300 and 5800 m, respectively. Morita and ZoBell (1955) found bacteria in sediment samples taken during the Mid-Pacific Expedition at depths of 5942 m. Proof of the existence of bacteria in the deepest parts of the ocean was obtained during the Galathea Expedition (1950–1952) through the efforts of ZoBell and Morita (ZoBell, 1952*a,b*, 1954; ZoBell and Morita, 1957, 1959). These studies indicated that there were large populations of bacteria present in sediment at depths of 7000 to greater than 10,000 m. Unlike bacteria collected from shallow depths, these bacteria grew preferentially, or exclusively, at high pressures and at refrigerator temperatures. Ten to 1000 times as many bacteria grew in nutrient media incubated at 1000 atm and at 2.5°C than in replicate cultures incubated at 1 atm at 2.5°C. The difference was attributed to the presence of pressure-loving microorganisms, termed barophiles (ZoBell and Johnson, 1949). The observations of ZoBell (1952*a*) concerning numbers of bacteria in core samples confirmed the results of Rittenberg (1940).

Morita and ZoBell (1955) examined cores from the Mid-Pacific Expedition and found bacteria in the sediments to depths of 8 m in material which may have been deposited more than a million years ago.

Bartholomew and Rittenberg (1949) reported the isolation of gram-positive, spore-forming, thermophilic bacteria from deep-ocean bottom cores. The possibility thus exists that organisms in deep-ocean sediments may be in a dormant state, preserved by conditions of high pressure and low temperature, presumably not metabolizing actively under these conditions.

Only a single taxonomic analysis of deep-sea bacteria has been done (Quigley, 1967; Quigley and Colwell, 1968*a,b*). In this study, the properties of bacteria isolated from sediment samples of the Philippine Trench and the Challenger Deep, in the Marianas Trench of the Pacific Ocean, were studied. Hogan and Colwell (1969) studied the DNA base composition and esterase patterns of the organisms.

Thus, evidence from the literature suggests that a significant percentage of a given deep-sea sediment bacterial population is very likely not actively metabolizing under the conditions of high hydrostatic pressure and low temperature found in the deep-sea environment (Bartholomew and Rittenberg, 1949). A goal of the present study was to combine the methods of numerical taxonomy (Sneath, 1957; Colwell and Liston, 1961; Sokal and Sneath, 1963; Quigley and Colwell, 1968a,b) and pressure-tolerance measurement (Oppenheimer and ZoBell, 1954; ZoBell and Cobet, 1962) to analysis of deep-sea isolates to determine differences, if any, between the active, indigenous microorganisms of the deep sea and the metabolically "inert", opportunistic microflora present in deep-sea sediments.

Also, ZoBell and Morita (1957) stated that hydrostatic pressure is probably not a sufficiently distinctive characteristic to delineate a species. This may be true, but members of a species having a greater tolerance to pressure may compete selectively in deep-sea sediments. Thus, it was an objective of this study, also, to determine whether certain taxonomic groups of bacteria are favored by high pressure, becoming predominant in sediment bacterial populations.

MATERIALS AND METHODS

The cultures included in the study are listed in Table 1. *Pseudomonas bathycetes* strain C2M#2 (ATCC no. 23597) was isolated from sediment obtained during the Dodo Expedition on November 24, 1964 (isolated by Morita and ZoBell). This organism was isolated from a core taken in the Challenger Deep, in the Marianas Trench of the Pacific Ocean, at a depth of 10,373 m. The position of the station, no. 282 of the Dodo Expedition, was 11°20′ N and 142°19′ E.

The strains with prefix VI (Virgin Island) were isolated from a sediment sample obtained in the Virgin Island Basin of the Atlantic Ocean during a cruise of the USNS *Mizar,* Naval Research Laboratory (NRL), Washington, D.C., to the Caribbean Sea in November–December, 1969. The sample was taken at a depth of 3950 m on November 21, 1969, at 1321 hr and the station position was 17°52.5′ N and 64°49.5′ W. The sample was retrieved by a Shipek grab sampler during the first leg of lowering no. 4 and is identified as sediment sample no. 1. The *in situ* temperature at the point of sampling was 3.47°C and the salinity was 34.96°/$_{oo}$. The sample was taken within 7

Table 1. Cultures included in the study[a]

1. *Pseudomonas bathycetes* strain	
C2M#2 (ATCC strain no. 23597)	17. VI-18
2. VI-1	18. VI-19
3. VI-2	19. VI-20
4. VI-3	20. VI-21
5. VI-4	21. VI-22
6. VI-5	22. VI-23
7. VI-6	23. VI-24
8. VI-7	24. *Micrococcus* sp. strain CB-234
9. VI-8	25. *Escherichia coli* strain 11303
10. VI-9	26. *Achromobacter fischeri*
11. VI-10	27. *Achromobacter* sp. strain
12. VI-11	CB-263
13. VI-12	28. *Vibrio marinus* strain PS-207
14. VI-13	29. *Pseudomonas fluorescens*
15. VI-14	strain Ox Sawyer
16. VI-16	30. *Vibrio parahaemolyticus*
	strain Sak-3

[a] For details of isolation and source see text.

nautical miles of the island of St. Croix near the base of the Virgin Island Basin.

Immediately after the sampler was brought aboard, 10 g of sediment were removed aseptically from the center of the sediment sample and transferred to a sterile bottle. The sample was stored at 5°C for 30 days, i.e., until the end of the cruise, at which time the sample was received at the laboratory and immediately processed.

The sample was diluted with 50 ml MSWYE broth (*vide infra*), and 0.1 ml of the suspension was transferred to MSWYE agar. The plates were incubated at 25°C. Colonies were picked from the plates and streaked out at least three times, or until pure cultures were obtained. Twenty-four pure cultures were isolated (see Table 1, strains VI-1 through VI-24). Strains VI-15 and VI-17 failed to transfer, and hence were discarded. Stock cultures were stored on MSWYE slants under sterile mineral oil.

The other strains included in the taxonomic analysis (Table 1) served as reference strains, since they have been studied extensively in early numerical taxonomy studies. The deep-sea bacterial strains reported by Quigley and Colwell (1968a) were also compared with the Virgin Island isolates. The deep-sea bacteria were isolated from the top of sediment cores collected during the Dodo Expedition to the Philippine Trench. The cores were taken at depths of 9854 and 9443 m. Strains from the Challenger Deep of the Marianas Trench of the Pacific Ocean at 10,373 m, including *Pseudomonas bathycetes* strain C2M#2, were also included in the analysis.

Maintenance media used in the study were as follows: MSWYE, consisting of yeast extract (Difco), 0.1%; proteose-peptone (Difco), 0.1%; four-salts solution, the composition of which was sodium chloride, 2.4%, potassium chloride, 0.07%, magnesium chloride, 0.53%, and magnesium sulfate, 0.7%, pH 7.2−7.4 (adjusted with 0.1 N sodium hydroxide). The SWYE medium used was identical to MSWYE except that the concentration of yeast extract (Difco) was 0.3% and that of proteose-peptone (Difco) was 1.0%. Solid medium was prepared by the addition of 2.0% Bacto-agar (Difco).

Serial dilutions, where necessary, were done using a four-salts solution buffered with 0.20 M Tris-HCl buffer, to a final pH of 7.2−7.4.

All media for the determinative tests performed in the numerical taxonomy analysis were according to Colwell and Wiebe (1970), except for tests for salt requirement, in which Ionagar (Difco) was omitted because it contains salts, which interfere with the test.

Pressure studies were done using a hydrostatic pressure apparatus of the type described by ZoBell and Oppenheimer (1950), ZoBell and Morita (1956), and Morita (1970). Details of the experiments are published elsewhere (Kettling, 1972).

In Table 2 the taxonomic features employed in the taxonomic analyses are listed.

An IBM 360 system with 2314 disk storage, printer, plotter, and card punch was used in the data analysis. The program used was Georgetown Taxonomy Program 2 (GTP-2) written in PL/1 for the IBM 360 system. The GTP-2 program has several sub-programs written into it for the performance of specialized tasks. The sub-programs used in this analysis were GTP 2-2, GTP 2-CHART, and sub-program OLVER, the latter consisting of several sub-routines. GTP 2-2 is a clustering program which uses, as the basis for forming groups, the computed similarity values (S-values) of all the strains, each against the other (Colwell and Liston, 1961).

GTP 2-CHART is a program which takes the computed S-values and arranges them in tabular form as a triangle of similarities. Program OLVER sub-routine GRPICK scans the triangle of similarities to identify groupings of strains (two or more organisms per group) having S-values greater than a pre-selected cut-off value, in this case 71%. The output consists of a table of the groups of similar strains, assigned group numbers, strain numbers, and strain labels. Details of this program will be published elsewhere (Oliver and Colwell, 1974).

Table 2. List of taxonomic features

1. Length, <0.5 µm
2. Length, 0.5–1.2 µm
3. Length, 1.2–3.0 µm
4. Length, variable
5. Width, L × W 3:1–slender
6. Width, L × W 2–3:1–short
7. Width, L × W 1:1–oval[a]
8. Width, variable
9. Cell shape, straight[a]
10. Cell shape, curved[a]
11. Cell shape, spiral[a]
12. Cell shape, coccal-bacillary
13. Cell shape, coccus
14. Cell shape, branching
15. Cell shape, pleomorphic
16. Square end
17. Round end[a]
18. Tapered end[a]
19. Fusiform end
20. Spore present
21. Sporangium swollen
22. Sporangium normal
23. Terminal spore
24. Central spore
25. Gram-negative[a]
26. Gram-positive
27. Gram-variable[a]
28. Motile[a]
29. Gliding motility
30. Rotatory motility
31. Polar flagella[a]
32. Peritrichous flagella[a]
33. Lateral or mixed flagella
34. Fat (Burdon's stain)
35. Metachromatic granules

Cell arrangement

36. Single[a]
37. Pair[a]
38. Chain short < 5 cells
39. Chain long > 5 cells
40. Coccal clumping
41. Coccal packet
42. Filaments 8 µ L × W 20:1[a]
43. Variable
44. Colony size < 2 mm[a]

45. Colony size 2–5 mm[a]
46. Colony size > 5 mm
47. Colony shape, effuse
48. Colony shape, raised
49. Colony shape, convex[a]
50. Colony shape, umbonate
51. Entire edge[a]
52. Crenated edge
53. Erose edge
54. Fimbriated edge
55. Spreading (cytophaga)[a]
56. Transparent
57. Translucent[a]
58. Opaque[a]

Non-diffusible pigment

59. White[a]
60. Off-white to gray[a]
61. Red
62. Pink
63. Orange
64. Yellow
65. Green
66. Violet purple
67. Brown
68. Black
69. Blue
70. Yellow–diffusible pigment
71. Green–diffusible pigment
72. Brown–diffusible pigment
73. Black–diffusible pigment
74. Fluorescent–UV/any condition[a]
75. Luminous–any condition
76. Slight turbidity[a]
77. Moderate turbidity[a]
78. Heavy turbidity[a]
79. Even turbidity
80. Flocculent turbidity
81. Pellicle
82. Ring

Temperature range of growth °C

83. 0
84. 4
85. 7

Table 2. *Continued*

86. 10	123. Sucrose → gas
87. 15^a	124. Lactose → acid^a
88. 20	125. Lactose → gas
89. 25^a	126. Iodoacetate + glucose → growth^a
90. 30^a	127. Iodoacetate + glucose → acid^a
91. 35–37^a	
92. 40	*Carbohydrate compounds as sole*
93. 42–45	*carbon sources for growth*

pH Range of growth

128. Glucose
129. Ribose

94. 4.0

130. Sodium acetate

95. 4.5

131. Sodium citrate

96. 5.0

132. Starch hydrolysis^a

97. 5.5

133. Agar digestion

98. 6.0

134. Lysine decarboxylase^a

99. 6.5

135. Ornithine decarboxylase^a

100. 7.0

136. Arginine decarboxylase^a

101. 7.5

137. Nitrate to nitrite^a

102. 8.0

138. Nitrite to inorganic nitrogen

103. 8.5

139. Gelatin liquefaction^a

104. 9.0

140. Ammonia from peptone

Salt requirements in broth

Mixed amino acid requirements
for growth

105. Sodium
106. Calcium
107. Potassium
108. Magnesium
109. No salts

141. Medium no. 1
142. Medium no. 2
143. Medium no. 3
144. Medium no. 4
145. Medium no. 5
146. Medium no. 6
147. Medium no. 7
148. Medium no. 8
149. Medium no. 9

MOF-glucose

110. Aerobic   → acid^a
111. Aerobic   → gas^a
112. Anaerobic   → acid^a
113. Anaerobic   → gas^a

Amino acids as sole carbon and nitrogen
sources for growth

114. Glucose   → acid
115. Glucose   → gas
116. Galactose   → acid^a
117. Galactose   → gas
118. Mannitol   → acid^a
119. Mannitol   → gas
120. Ribose   → acid
121. Ribose   → gas
122. Sucrose   → acid

150. Alanine
151. Proline
152. Glutamic acid
153. Methionine
154. Skim milk, growth
155. Skim milk, casein hydrolysis
156. Lecithinase^a

continued

Table 2. *Continued*

157. Lipase (Tween 20)[a]	175. Voges–Proskauer test positive
158. Lipase (Tween 40)[a]	176. Simmon's citrate[a]
159. Lipase (Tween 60)[a]	177. MOF–galactose–aerobic→acid
160. Lipase (Tween 80)[a]	178. MOF–galactose–aerobic→gas
161. Kovac's oxidase	179. MOF–galactose–anaerobic→acid[a]
162. Catalase	180. MOF–galactose–anaerobic→gas
163. Cysteine HCl → H_2S	181. MOF–mannitol–aerobic→acid
164. Na thiosulphate → H_2S	182. MOF–mannitol–aerobic→gas
165. Phosphatase[a]	183. MOF–mannitol–anaerobic→acid[a]
166. Penicillin, 10 units	184. MOF–mannitol–anaerobic→gas
167. Chloromycetin, 30 μg[a]	185. MOF–ribose–aerobic→acid
168. Tetracycline, 30 μg	186. MOF–ribose–aerobic→gas
169. Dihydrostreptomycin, 10 μg	187. MOF–ribose–anaerobic→acid
170. Colimycin, 10 μg	188. MOF–ribose–anaerobic→gas
171. Pteridine, 0/129[a]	189. MOF–sucrose–aerobic→acid
172. Penicillinase	190. MOF–sucrose–aerobic→gas
173. Indole	191. MOF–sucrose–anaerobic→acid[a]
174. Methyl red test positive[a]	192. MOF–sucrose–anaerobic→gas

[a] Features which were also examined by Quigley (1967) and used in the comparative taxonomic study of strains of deep-sea bacteria isolated from the Pacific and isolates from the Virgin Island Basin of the Atlantic.

RESULTS AND DISCUSSION

The Virgin Island isolates were examined for the effect of hydrostatic pressure on the growth and death rate. The data are presented in Table 3. The experiments were carried out at 1 atm and 400 atm, at 25°C. An inoculum of $10^3 - 10^5$ cells/ml was used in all cases, except for two strains, VI-8 and VI-10.

The response of the Virgin Island isolates to pressure was variable. All but one strain, VI-12, showed marked barotolerance to *in situ* conditions of pressure. Some of the isolates were clearly capable of growth under pressure.

All strains listed in Table 1 have been analyzed by numerical taxonomy using the 192 phenotypic tests listed in Table 2. Similarity values were calculated using the GTP 2-2 program, which also clustered the organisms (see Table 4).

All 29 strains of the analysis clustered at the 46% level of similarity. Two major clusters within the Virgin Island isolates were detected and were identified as Groups I and II. Group I comprised strains VI-1, -2, -3, -4, -8, -10, -12, -13, -14, -16, -20, and -22, and Group II comprised strains VI-5, -6, -7, -9, -11, -18, -19, -23, and -24.

The strains of Group I were gram-negative, short, straight or curved rods and appeared in early culture as coccobacillary forms. They produced acid from glucose, both aerobically and anaerobically in MOF medium. Colonies on agar were off-white and strains VI-1, VI-2, VI-3, VI-4, VI-14, and VI-20 showed a strong blue-green luminescence, which persisted in these strains 2 yr after their initial isolation. The Group I strains were motile, by means of a single polar flagellum, and the cells occurred singly or in pairs. The strains required seawater for growth. They were all sensitive to iodoacetate and to chloromycetin, but not to penicillin. They were oxidase- and catalase-positive and produced H_2S from sodium thiosulfate and cysteine. They grew with proline as a sole carbon and nitrogen source and were capable of utilizing ribose, sodium acetate, and sodium citrate as sole carbon sources. They produced nitrite from nitrate, but not molecular nitrogen from nitrite. They were also capable of gelatin liquefaction, starch hydrolysis, and production of ammonia from peptone. Most of the strains were sensitive to 0/129 (vibrio-static agent).

All of the 13 organisms in Group I clustered at the 86% level of similarity, indicating a degree of relatedness at the species level. *Achromobacter fischeri* was found to cluster with Group I organisms at a similarity value of 69%. *Vibrio parahaemolyticus* strain Sak-3 clustered at the 68% level. This indicates a relationship at or near the generic level. *Vibrio marinus* strain PS-207 was found to cluster only at the 56% level with Group I organisms.

The Group I organisms showed significant similarity (68%) to *Vibrio parahaemolyticus*. They also were closely related to *Achromobacter fischeri* (69%), redefined by Colwell (1970) as *Vibrio*. The organisms of Group I are considered to be a new and separate species of *Vibrio*.

The Group II organisms clustered at a level of 75% similarity, with the exception of VI-18, which did not cluster with this group above 60% similarity. At the 60% level of similarity, members of Group I clustered with Group II organisms. These organisms did not show significant similarity to any of the control strains used in the study.

The Group II strains were gram negative (VI-18, VI-23, and VI-24 showed gram variability in older culture), rod-shaped bacteria, motile by means of one or two polar flagella. Strains VI-5, VI-6, VI-7, VI-9, VI-11, and VI-19 produced yellow-orange pigmented colonies. VI-18 produced a distinct rose-colored pigment and strains VI-23 and VI-24 were off-white. The pigments were not soluble. The organisms were proteolytic and produced ammonia from peptone. They produced acid from glucose in MOF medium, both aerobically and anaerobically, but gas was not produced. Action of these organisms in MOF was much slower than that observed for Group I organisms. All were sensitive to iodoacetate, penicillin, Chloromycetin, tetra-

Table 3. Pressure response of viability of VI-5 through VI-24 at pressures of 1 atm and 400 atm and 25°C

Strain	Pressure (atm)	Inoculum size	Total viable counts in cells/ml					
			24 hr	40 hr	70 hr	120 hr	216 hr	264 hr
VI-5	1	2.6×10^4		8.8×10^5	1.1×10^5		2.0×10^3	1.2×10^3
	400	2.6×10^4		2.5×10^5	1.3×10^5		4.3×10^4	
VI-6	1	1.9×10^3		3.6×10^6	1.1×10^5		$<10/ml$	$<10/ml$
	400	1.9×10^3		2.7×10^6	1.6×10^5		$<10/ml$	$<10/ml$
VI-7	1	2.5×10^4		1.6×10^6	1.1×10^5		$<10/ml$	$<10/ml$
	400	2.5×10^4		5.0×10^3	1.6×10^5		$<10/ml$	$<10/ml$
VI-8	1	1.6×10^7	2.9×10^7		6.9×10^6	9.0×10^6		
	400	1.6×10^7	6.5×10^7		3.0×10^6	2.1×10^5		
VI-9	1	8.2×10^4	5.3×10^6		8.2×10^4	5.5×10^4	$<10/ml$	
	400	8.2×10^4	1.3×10^6		4.5×10^5	1.5×10^3		
VI-10	1	8.0×10^7	2.5×10^7		1.4×10^7	1.3×10^7		
	400	8.0×10^7	3.2×10^7		3.3×10^6	2.3×10^5		
VI-11	1	1.3×10^5	3.2×10^6		1.4×10^5	7.0×10^5		
	400	1.3×10^5	5.8×10^5		4.5×10^5	2.0×10^4		
VI-12	1	7.5×10^4		1.0×10^7	4.6×10^6		7.0×10^5	2.2×10^6
	400	7.5×10^4			3.0×10^4		$<10/ml$	$<10/ml$

Sample	Dilution							
VI-13	1	7.3×10^4	2.6×10^7		8.3×10^6	1.1×10^7	1.2×10^6	
	400	7.3×10^4	3.4×10^6		8.5×10^5	5.0×10^5	9.8×10^4	
VI-14	1	7.4×10^4	1.8×10^7			2.3×10^6	5.5×10^5	
	400	7.4×10^4	1.6×10^5			1.0×10^4	1.8×10^4	
VI-16	1	3.9×10^4	2.5×10^7		8.2×10^6	4.4×10^6	2.0×10^6	
	400	3.9×10^4	4.8×10^6		3.0×10^5		1.4×10^5	
VI-18	1	1.7×10^4	6.2×10^5		1.4×10^4	7.0×10^2	1.6×10^2	
	400	1.7×10^4	5.6×10^3		1.4×10^4	1.7×10^5	7.0×10^3	
VI-19	1	3.5×10^4	1.1×10^7		4.8×10^5	5.1×10^4	1.8×10^5	
	400	3.5×10^4	1.6×10^4		1.8×10^4	8.0×10^5	3.0×10^1	
VI-20	1	1.0×10^4	2.2×10^5		1.8×10^6		8.0×10^4	6.0×10^3
	400	1.0×10^4	5.5×10^3		3.2×10^4		5.7×10^3	
VI-21	1	2.7×10^4	3.1×10^7		1.4×10^7	9.6×10^6	2.8×10^6	
	400	2.7×10^4	2.2×10^6		9.5×10^5	7.5×10^6	2.0×10^6	
VI-22	1	9.8×10^4	2.3×10^7		1.4×10^7	8.8×10^6	2.6×10^6	
	400	9.8×10^4	4.4×10^6		1.7×10^6	5.0×10^5	1.0×10^5	
VI-23	1	1.3×10^4		7.5×10^5	1.2×10^5		$< 10/ml$	$< 10/ml$
	400	1.3×10^4		2.5×10^3			8.6×10^4	2.0×10^1
VI-24	1	3.4×10^3		8.9×10^5	7.4×10^4		9.2×10^4	2.5×10^5
	400	3.4×10^3		2.7×10^6	5.0×10^5		2.2×10^3	3.4×10^3

Table 4. Similarity triangle for Virgin Island isolates

KEY

41–50 –
51–60 =
61–70 +
71–75 $
76–80 #
81–90 &
91–100 *

```
Pseudomonas fluorescens strain Ox Sawyer  1
Vibrio marinus strain PS-207              2 –
Escherichia coli strain 11303            3 – –
Vibrio parahaemolyticus strain Sak-3     4 = – – =
VI-20                                    5 = = + +
VI-1                                     6 = = + + *
VI-2                                     7 = = + & *
VI-4                                     8 – = + & *
VI-3                                     9 – = + & *
VI-14                                   10 = = = + & & & &
VI-13                                   11 = = = + & & # # & &
VI-16                                   12 = = = + # $ # # & & & &
VI-8                                    13 = = = + # # & & & & & *
VI-22                                   14 = = = + # # & & & & & * * &
VI-21                                   15 = = = + & & & & & & & * * * *
VI-12                                   16 = – = + # # & & & & & & # * * & *
VI-10                                   17 – – + $ $ # # & & * * * * & & & *
Achromobacter fischeri                  18 – = = = + + = = + + + + + + + + + +
VI-9                                    19 @ – – = = = = = = = + + + + + + + + +
VI-11                                   20 @ – – = = = = = = = = = = + + = + + + +
VI-7                                    21 @ – – = = = = = = = = = = = + = + + = & &
VI-19                                   22 – – – = = = = = = = = + = = = @ # & & &
VI-23                                   23 – – – = = = = = = = + = = = – $ $ $ $ &
VI-24                                   24 – – – = = = = = = = = = = – – $ $ $ # *
VI-5                                    25 @ @ @ – – – – – – – – – – – – @ + + + $ # # $
VI-6                                    26 @ @ – @ – – – – – – – – – – – @ + = + $ $ $ $ *
VI-18                                   27 @ – – @ – – = – – – – – – – – @ + = + $ $ $ $ # =
Achromobacter strain CB-263             28 = – – – – – – = – = – – – – – @ @ @ @ @ – – = = = =
Micrococcus strain CB-234               29 @ @ ; @ @ @ @ @ @ @ @ @ @ @ @ @ @ @ @ @ – – @ – – @ –
                                           1 2 3 4 5 6 7 8 9 0 1 2 3 4 5 6 7 8 9 0 1 2 3 4 5 6 7 8 9
                                                             1                   2
```

cycline, and dihydrostreptomycin. All except for strain VI-18 were sensitive to the vibriostatic agent (0/129). All members of the group required seawater for growth, except strains VI-5 and VI-6. Group II strains were not lipolytic and a majority utilized proline and alanine as sole carbon and nitrogen sources. These organisms could be classified as *Aeromonas, Vibrio,* or *Flavobacterium.* The final decision will require additional study, including determination of mole per cent guanine plus cytosine (% G + C) content of the DNA.

The original data of Quigley (Quigley, 1967; Quigley and Colwell, 1968*a*) were combined with the data for the Virgin Island isolates, using the computer program GTP-2 and the sub-programs OLVER, GRPICK, SEARCH, and PLOT. The cut-off value for sub-routine GRPICK was set at 71% S.

Quigley (1967) selected six clusters from his strain data on the basis of the similarity values generated by the computer analysis. Although some differences in placement of cultures into groups were noted, the major groupings detected by Quigley were also observed in this analysis. Quigley did not use a specified S-value as a cut-off, but clustered organisms at similarity values which applied best to that group. Grouping of peripheral strains in numerical taxonomy remains a difficult problem, even with the most recently developed computer programs.

The combined Virgin Island and Pacific Ocean data sets showed, in general, that the Pacific Ocean sediment isolates were not closely related to the Virgin Island isolates.

The median organisms computed from the Quigley data confirmed the low level of phenetic similarity between the two sets of organisms. The Pacific Ocean sediment samples were predominantly *Pseudomonas* spp. (Quigley and Colwell, 1968*a*), whereas the majority of the Virgin Island sediment isolates were *Vibrio* spp. It is interesting to speculate that, since the character of the sediment in the Virgin Island Basin receives a more direct influence from the coastal region, the microorganisms found in those sediments are therefore reflective of sediments of the coastal region. The difference of *in situ* pressure, 400 atm *versus* 600–700 atm for the Virgin Island and Pacific Ocean sediment samples, respectively, may also act as a selective factor in determining the autochthonous microbial flora.

ACKNOWLEDGMENTS

The assistance and advice of Miss T. Elizabeth Lovelace and Dr. James D. Oliver in the taxonomic aspects and use of the computer programs applied to analysis of the taxonomic data are gratefully acknowledged. This work was supported, in part, by Contract No. N00014-67-A-0239-0027 with the Office of Naval Research and Grant No. GB-35261X from the National Science Foundation. The authors acknowledge the support of the Oceanographic Program of Duke for use of the R/V *Eastward*. The Oceanographic Program is supported by National Science Foundation Grant GA-27725.

LITERATURE CITED

Bartholomew, J. W., and S. C. Rittenberg. 1949. Thermophilic bacteria from deep ocean bottom cores. J. Bacteriol. 57:658.

Bruun, A. F. 1956. The abyssal fauna: Its ecology, distribution, and origin. Nature (London) 177:1105–1108.

Certes, A. 1884*a*. Sur la culture, a l'abri des germes atmospheriques, des eaux et des sediments rapportes par les expeditions du "Travailleur" et du "Talisman", 1882–1883. C. R. Acad. Sci. Paris 98:690. Cited in ZoBell (1952*a*).

Certes, A. 1884*b*. Note relative a l'action des hautes pressions sur la vitalité des microorganismes d'eau douce et d'eau de mer. C. R. Soc. Biol. 36:220–222. Cited in ZoBell (1952*a*).

Colwell, R. R. 1970. Polyphasic taxonomy of the genus *Vibrio*: Numerical taxonomy of *Vibrio cholerae, Vibrio parahaemolyticus* and related *Vibrio* species. J. Bacteriol. 104:410–433.

Colwell, R. R., and J. Liston. 1961. Taxonomy of *Xanthomonas* and *Pseudomonas*. J. Gen. Microbiol. 37:617–619.

Colwell, R. R., and W. J. Wiebe. 1970. Methods for characterizing aerobic, heterotrophic, marine and estuarine bacteria. J. Georgia Acad. Sci. 28:165–185.

Fischer, B. 1894. Ergebnisse Planckton-Expedition Humboldt Stiftung, Vol. 4, p. 1. Cited in ZoBell (1952*a*).

Hogan, M. A., and R. R. Colwell. 1969. DNA base composition and esterase patterns of bacteria isolated from deep sea sediments. J. Appl. Bacteriol. 32:103–111.

Kettling, R. C., Jr. 1972. Properties and hydrostatic pressure studies of some deep-sea bacteria. M.S. Thesis, Georgetown Univ., Washington, D. C.

Morita, R. Y. 1970. Application of hydrostatic pressure to microbial cultures. *In* D. W. Robbins and J. R. Norris (eds), Methods in Microbiology, Vol. 2, pp. 243–257. Academic Press, New York.

Morita, R. Y., and C. E. ZoBell. 1955. Occurrence of bacteria in pelagic sediments collected during the mid-Pacific expedition. Deep-Sea Res. 3:66–73.

Oliver, J. D., and Colwell, R. R. 1974. A computer program designed to follow fluctuations in natural bacterial populations and its application in a study of Chesapeake Bay microflora. Appl. Microbiol. (In press.)

Oppenheimer, C. H., and C. E. ZoBell. 1954. The growth and viability of sixty-three species of marine bacteria as influenced by hydrostatic pressure. J. Marine Res. 11:10–18.

Quigley, M. M. 1967. Properties of bacteria isolated from deep-sea sediments. B.S. Thesis, Georgetown Univ., Washington, D. C.

Quigley, M. M., and R. R. Colwell. 1968a. Properties of bacteria isolated from deep-sea sediments. J. Bacteriol. 95:211.

Quigley, M. M., and R. R. Colwell. 1968b. Proposal of a new species, *Pseudomonas bathycetes*. Internat. J. System. Bacteriol. 18:214.

Rittenberg, S. C. 1940. Bacteriological analysis of some long cores of marine sediment. J. Marine Res. 3:191–201.

Sneath, P. H. A. 1957. The application of computers to taxonomy. J. Gen. Microbiol. 17:201–226.

Sokal, R. R., and P. H. A. Sneath. 1963. Principles of Numerical Taxonomy. W. H. Freeman, San Francisco.

Wüst, G. 1952. Neue Rekordtiefen des Wellmeres im Marianen-Graben. Die Erde Jahresheft 1952, pp. 114–115. Cited in C. E. ZoBell and K. M. Budge. 1965. Nitrate reduction by marine bacteria at increased hydrostatic pressure. Limnol. Oceanogr. 10:207–214.

ZoBell, C. E. 1952a. Bacterial life at the bottom of the Philippine Trench. Science 115:507–508.

ZoBell, C. E. 1952b. Dredging life from the bottom of the sea. Res. Rev. Office Naval Res. 14–20.

ZoBell, C. E. 1954. The occurrence of bacteria in the deep sea and their significance for animal life. Internat. Union Biol. Sci. Ser. B 16:20–26.

ZoBell, C. E., and A. B. Cobet. 1962. Growth, respiration and death rates of *Escherichia coli* at increased hydrostatic pressure. J. Bacteriol. 84:1228–1236.

ZoBell, C. E., and F. H. Johnson. 1949. The influence of hydrostatic pressure on the growth and viability of terrestrial and marine bacteria. J. Bacteriol. 57:179–189.

ZoBell, C. E., and R. Y. Morita. 1956. Apparatus for studying effect of hydrostatic pressure on microbial activities. Bacteriol. Proc. 21 (abstr.).

ZoBell, C. E., and R. Y. Morita. 1957. Barophilic bacteria in some deep-sea sediments. J. Bacteriol. 73:563–568.

ZoBell, C. E., and R. Y. Morita. 1959. Deep-sea bacteria. Galathea Rep. 1:139–154.

ZoBell, C. E., and C. H. Oppenheimer. 1950. Some effects of hydrostatic pressure on the multiplication and morphology of marine bacteria. J. Bacteriol. 60:771–781.

IV
NUTRIENT EFFECTS

NUTRIENTS AND MICROBIAL RESPONSE
TO NUTRIENTS IN SEAWATER

A. F. CARLUCCI

Institute of Marine Resources
University of California, San Diego

In order for marine bacteria to grow in the sea or in laboratory culture, nutrients must be synthesized by the organisms or must be provided externally to the cell. For heterotrophic bacteria some of these nutrients provide an energy source when they are utilized as organic compounds. These types of compounds include simple sugars, amino acids, oils, and the more complex molecules. Autotrophic bacteria may utilize the inorganic forms of many compounds as an energy source. Both heterotrophic and autotrophic bacteria require nutrients which are not involved directly in energy coupling, but are part of the cell constituents. Bacterial cells require both macronutrients, those substances such as glucose, in relatively large concentrations, and micronutrients, such as vitamins and trace metals, in small amounts. Some micronutrients, for example, the metals, may be toxic when present in high concentrations. For discussions on nutrients and bacteria the reader should consult recent textbooks on general microbiology (e.g., Brock, 1970; Alexander, 1971; Lechevalier and Pramer, 1971). MacLeod (1965) summarizes the reports dealing with nutrients and marine bacteria.

One of the important factors governing the distribution of bacteria in the sea is the availability of nutrients (ZoBell, 1946, 1968). In addition to the possible unfavorable effects of temperature and hydrostatic pressures (Jannasch *et al.,* 1971), the lack of nutrients is a major factor responsible for the few bacteria that are found in deep waters. In upper waters of the sea, and in sediments, there are relatively more organic nutrients which bacteria may utilize than in intermediate waters; hence, their numbers are proportionally higher. Inorganic nutrients (nitrate, phosphate, silicate), however, increase with depth. Even in relatively cold waters psychrophilic bacteria are influenced by the availability of nutrients. Bacterial numbers are higher in coastal waters, reflecting a high concentration of nutrients which results from land run-off and upwelling. Readily available dissolved organic matter and other nutrients are found in high concentrations near sewage outfalls; these waters also show elevated bacterial numbers and activities.

Earlier work in enumerating bacteria from the marine environment employed media rich in organic nutrients. ZoBell (1946) summarizes the various media used by these early workers. ZoBell (1946) employed a successful medium for assessing bacterial numbers which contained 5 g peptone, 0.1 g ferric phosphate, and 15 g agar in 1000 ml of "aged" seawater. This medium was referred to as 2216 and was the best medium available for maximum counts and reproducibility. It was subsequently improved when 1 g/liter yeast extract was included. The improved medium is referred to as 2216E and is employed by marine microbiologists throughout the world in enumerating marine bacteria. Kriss (1963) used a fish peptone nutrient agar in studies described in his monograph. More recent laboratory manuals describing techniques in marine microbiology also give various media for culturing marine bacteria (Aaronson, 1970; Rodina, 1972). All these media contain relatively high concentrations of nutrients and, hence, select for those bacteria which can utilize the nutrients at these high levels. The argument in favor of using highly enriched media is that maximum numbers of bacteria will grow, and that most bacteria that develop on the enriched media simply will not grow in dilute media. Studies of marine bacterial growth in these various media, however, are difficult to interpret with respect to microbial activity in the sea since natural seawater contains extremely low concentrations of nutrients (Jannasch, 1968).

Jannasch (1967, 1970) states there is a threshold level of nutrients needed before marine bacteria will grow in the laboratory, and that this level is generally higher than what is found in seawater. Even though these bacteria do not develop in a minimal medium they are found in the sea. Jannasch (1967) differentiated between those bacteria which are surviving and those which are growing in the marine environment.

Menzel and Ryther (1970) report that the sea below 200–300 m has little biological activity because of low concentrations of available organic matter. Jannasch *et al.* (1971) also found greatly reduced microbial activity in the deep sea. These authors observed rates of microbial degradation to be 10–100 times slower in the deep sea than in laboratory controls. They felt that hydrostatic pressure exerted an effect on bacterial cells, raising the minimal growth temperature.

Where substances like glucose have been determined in seawater their concentrations have been found to be low, in microgram per liter quantities (Vaccaro *et al.*, 1968; Vaccaro and Jannasch, 1966, 1967; Hamilton and Preslan, 1970; Andrews and Williams, 1971; and references cited therein).

Sediments are much richer in nutrients than the water above (ZoBell, 1946). The sea surface also contains considerably more nutrients than the water below. This high level of nutrients is reflected in the abundant number of bacteria found in the surface film (Sieburth, 1965; Bezdek and Carlucci, 1972).

In order to be near concentrated nutrient sources and other favorable conditions certain bacteria grow in association with other organisms or attached to surfaces (Wood, 1965; Brock, 1966).

The following section includes papers discussing nutrients and microbial growth in the marine environment. The toxicity of some pollutants to marine bacteria is also presented. Papers on film-forming bacteria and on nitrogen fixation are included.

LITERATURE CITED

Aaronson, S. 1970. Experimental Microbial Ecology. Academic Press, New York.

Alexander, M. 1971. Microbial Ecology. John Wiley, New York.

Andrews, P., and P. J. LeB. Williams. 1971. Heterotrophic utilization of dissolved organic compounds in the sea. III. Measurement of the oxidation rates and concentrations of glucose and amino acids in seawater. J. Mar. Biol. Ass. U.K. 51:111–125.

Bezdek, H. F., and A. F. Carlucci. 1972. Surface concentration of marine bacteria. Limnol. Oceanogr. 17:566–569.

Brock, T. D. 1966. Principles of Microbial Ecology. Prentice-Hall, Englewood Cliffs, New Jersey.

Brock, T. D. 1970. Biology of Microorganisms. Prentice-Hall, Englewood Cliffs, New Jersey.

Hamilton, R. D., and J. E. Preslan. 1970. Observations of heterotrophic activity in the eastern tropical Pacific. Limnol. Oceanogr. 15:395–401.

Jannasch, H. W. 1967. Growth of marine bacteria at limiting concentrations of organic carbon in seawater. Limnol. Oceanogr. 12:264–271.

Jannasch, H. W. 1968. Growth characteristics of heterotrophic bacteria in seawater. J. Bacteriol. 95:722–723.

Jannasch, H. W. 1970. Threshold concentrations of carbon sources limiting growth in seawater. *In* D. W. Hood (ed.), Organic Matter in Natural Waters, pp. 321–330. Inst. Mar. Sci., College, Alaska.

Jannasch, H. W., K. Eimhjellen, C. O. Wirsen, and A. Farmanfarmaian. 1971. Microbial degradation of organic matter in the deep sea. Science 171: 672–685.

Kriss, A. E. 1963. Marine Microbiology (Deep-Sea). (Trans. by J. M. Shewan and Z. Kabata.) Oliver and Boyd, London.

Lechevalier, H. A., and D. Pramer. 1971. The microbes. Lippencott, Philadelphia, Pennsylvania.

MacLeod, R. A. 1965. The question of the existence of specific marine bacteria. Bacteriol. Rev. 29:9–23.

Menzel, D. W., and J. H. Ryther. 1970. Distribution and cycling of organic matter in the oceans. *In* D. W. Hood (ed.), Organic Matter in Natural Waters, pp. 31–54. Inst. Mar. Sci., College, Alaska.

Rodina, A. G. 1972. Methods in Aquatic Microbiology. University Park Press, Baltimore, Maryland.

Sieburth, J. McN. 1965. Bacteriological sampler for air-water and water-sediment interfaces, pp. 1064–1067. Trans. Jt. Conf. Ocean Sci. Ocean Eng. Mar. Technol. Soc. Amer. Soc. Limnol. Oceanogr.

Vaccaro, R. F., S. E. Hicks, H. W. Jannasch, and F. G. Cary. 1968. The occurrence and role of glucose in seawater. Limnol. Oceanogr. 13:356–360.

Vaccaro, R. F., and H. W. Jannasch. 1966. Studies on heterotrophic activity in seawater based on glucose assimilation. Limnol. Oceanogr. 11:596–607.

Vaccaro, R. F., and H. W. Jannasch. 1967. Variations in uptake for glucose by natural populations in seawater. Limnol. Oceanogr. 12:540–542.

Wood, E. J. F. 1965. Marine Microbial Ecology. Reinhold, New York.

ZoBell, C. E. 1946. Marine Microbiology. Chronica Botanica, Waltham, Massachusetts.

ZoBell, C. E. 1968. Bacterial life in the deep sea. Bull. Misaki Mar. Biol. Inst. Kyoto Univ. 12:77–96.

IMPROVEMENT OF MEDIA FOR ENUMERATION AND ISOLATION OF HETEROTROPHIC BACTERIA IN SEAWATER

USIO SIMIDU*

Institute of Food Microbiology
Chiba University

An attempt was made to develop media on which the highest viable counts of seawater bacteria could be obtained. The enrichment of a chemically defined medium with artificial seawater as a basal diluent with a vitamin mixture had no stimulating effect, while a mixture of 17 inorganic ions, added with small amounts of organic nitrogen to the medium, greatly improved the growth of the bacteria in seawater.

Among the different carbon and nitrogen sources tested, the following gave the highest viable counts when incorporated into the medium: malate and lactate as organic acids, mannitol and sucrose as carbohydrates, Bacto-peptone as peptone, and N-tris (hydroxymethyl) methyl-2-aminoethane-sulfonic acid (TES) and N-2-hydroxyethyl-piperazine-N-ethane-sulfonic acid (HEPES) as buffers. The formation of precipitate during autoclaving could be avoided without any significant change in viable counts by reducing the concentrations of Ca, Mg, and Sr in the media to one-fourth that in seawater.

Although there remain further problems to be considered, viable counts equal to, or more than, those obtained with ZoBell's 2216E medium were attained by a medium of the following composition: 0.06% $NaNO_3$, 0.06% $(NH_4)_2SO_4$, 0.04% malic acid, 0.04% lactic acid, 0.03% sucrose, 0.03% mannitol, 0.04% Bacto-peptone, 0.02% yeast extract, 0.01% Lab-Lemco, 0.01% Ca-glycerophosphate, 0.003% Tween 80, 0.001% ferric citrate, and 1.5% agar in artificial seawater which was supplemented with minute amounts of inorganic ions.

* Present address: Ocean Research Institute, University of Tokyo.

The selection of media is, of course, one of the crucial factors in investigating the bacterial flora of natural environments. It is especially true for the study of seawater, since the bacterial numbers that can be obtained by cultural methods from seawater are at times only one-thousandth of those obtained by direct microscopic methods.

After ZoBell's (1941) detailed study on the media for seawater bacteria, little advance has been achieved in this field, and many workers have used ZoBell's medium 2216 with some modifications, such as addition of yeast extract at concentrations of 0.01% (Carlucci and Pramer, 1957), 0.1% (Gunkel and Oppenheimer, 1963), or 0.25% (Anderson, 1962). Taga (1968) also improved ZoBell's medium by varying the composition of peptones and incorporating marine-mud extract. Some workers used different media including fish peptone (Kriss, 1959) and fish extract (Simidu and Aiso, 1962). Although the incorporation of fish peptone and fish extract might have some advantages over ZoBell's medium or its modifications, the exact experimental bases for these media were not given.

Most of these media for marine bacteria are composed of natural seawater. The composition of seawater, however, varies from place to place. Recent studies have revealed the significance of minute organic and inorganic substances in seawater for the nutrition of marine planktonic microorganisms, and it is the quantities of these substances that show particularly great deviation in different areas and different seasons.

The media based on natural seawater also tend to produce copious precipitates, especially at the higher pH range of pH 7.6 or above, where the growth of marine bacteria is optimal.

We have been carrying out a series of experiments to develop media which, being based on artificial seawater, would give the highest possible bacterial counts.

MATERIALS AND METHODS

Seawater samples were taken either in Tokyo Bay or off the Pacific coast of Boso Peninsula, 2 km offshore, using Hiroth water samplers at a 5-m depth.

The viable counts were made by a surface plate method in which 0.05 ml aliquots of serially diluted seawater were placed on agar plates and spread by a bent nichrome wire. The colonies developed were counted after 12 days incubation at $20°C$. In some experiments the results of the colony counts were compared with the total bacterial counts measured by the direct microscopic method of Jannasch (1958).

Effect of Trace Elements, Vitamins, and Organic Nitrogen

A basal, chemically defined medium (medium A in Table 1) was made according to the formula given in Table 1.

Throughout the present investigation, the concentrations of nutrients in the media were kept at rather low levels for the following reason. Most nutrients are known to exhibit adverse effects on the growth of bacteria at certain levels of concentrations, and apparently the levels that inhibit the growth differ even among a crop of bacteria in a small volume of seawater. Since we have only restricted knowledge of this aspect of suppression of the growth of a marine bacterial population, the levels of nutrient content in the media should be kept minimal. The total concentration of the soluble organic substances in natural seawater is generally between 1 and 10 mg/liter. We can reasonably postulate that a concentration one order higher than this upper value would be enough to support most of the planktonic bacteria, although the actual nutrient concentrations in the niche where marine bacteria grow will be much greater than the levels in seawater.

Mixtures of (1) 17 minor elements found in seawater, (2) water-soluble vitamins, and (3) organic nitrogenous nutrients were added to the basal

Table 1. Basal media

| | Amount[a] | |
Component	Medium A	Medium B
Bacto-yeast extract	—	0.1
Lab-Lemco beef extract	—	0.1
$NaNO_3$	0.3	0.3
$(NH_4)_2SO_4$	0.3	0.3
Na-acetate	0.2	—
Na-citrate	0.2	—
Glucose	0.4	—
Tween 80	0.025	0.025
Ca-glycerophosphate	0.05	0.05
$NaH_2PO_4 \cdot 2H_2O$	0.03	0.03
Ferric citrate	0.01	0.01
Agar	15	15
Artificial seawater	1000 ml	1000 ml [b]

[a] In grams, unless otherwise noted.

[b] Enriched with the minor elements listed in Table 2.

medium A at two different concentrations, one being 5 times as high as the other. The compositions of these mixtures are given in Tables 2–4. The lower concentration levels are shown in these tables. In these media, the surface-colony counts were made with the seawater samples from Tokyo Bay.

Table 2. Amount of minor elements added to the media (μg/liter)

Element	Amount in seawater[a]	Amount in media[b]	Compounds added
Si	3000	3000	$Na_2SiO_3 \cdot 9H_2O$
Li	170	170	$LiSO_4 \cdot H_2O$
I	60	60	KI
Ba	30	60	$BaCl_2 \cdot 2H_2O$
In	20	40	$In_2(SO_4)_3 \cdot 9H_2O$
Al	10	20	$K_2Al_2(SO_4)_4 \cdot 24H_2O$
Zn	10	50	$ZnCl_2$
Mo	10	50	$Na_2MoO_4 \cdot 2H_2O$
Se	4	40	H_2SeO_3
Cu	3	30	$CuSO_4$
Sn	3	30	$SnCl_2 \cdot 2H_2O$
Mn	2	20	$MnCl_2 \cdot 2H_2O$
Ni	2	20	$NiSO_4 \cdot 7H_2O$
V	2	20	VCl_3
Ti	1	10	$TiCl_3$
Co	0.5	10	$CoSO_4 \cdot 7H_2O$
Cr	0.05	3	K_2CrO_4

[a] From Hill (1966).

[b] The stock solution of 150 times these concentrations was made in 0.1% citric acid solution.

Table 3. Amounts of vitamins added to the media (μg/liter)

Thiamine	200
Riboflavin	200
Pyridoxamine	200
Nicotinic acid	200
Pantothenate	200
Choline	200
Folic acid	10
p-Aminobenzoic acid	10
Biotin	1
Cyanocobalamin	1

Table 4. Amounts of organic
nitrogen added to the media (%)

Polypepton (Daigo, Osaka)	0.02
Bacto-yeast extract (Difco)	0.01
Lab-Lemco beef extract (Oxoid)	0.01

The result (Table 5) showed that vitamins had no stimulative effect at these concentrations either with or without other additive nutrients. The lower level of inorganic ions improved the development of colonies when they were added with the mixture of peptone and extracts, although when added alone they seemed to have adverse effects. The mixture of peptone and extracts stimulated the growth greatly even in rather low concentrations as in the present experiment.

Table 5. Effect of organic
nitrogen and minor nutrients

	Numbers of colonies[b]	
Component added[a]	Concentrations in Tables 1−3	5 times the concentrations
M	47	70
V	70	63
ON	156	139
M+V	61	76
M+ON	213	150
V+ON	125	144
M+V+ON	135	154
Control	83	102

[a] M, minor elements; V, vitamins; ON, organic nitrogen.

[b] Average of duplicate plates.

Comparison of Different Organic Nutrients and Buffers

As the next step we compared the efficiency of different organic acids, carbohydrates, nitrogen sources, and buffers as the components of the medium. For each category of components, five different compounds or substances were compared. The experiment was carried out according to a Greco-Latin square design (Cochran and Cox, 1950). Carbohydrate (0.04%),

Table 6. Effect of different organic nutrients and buffers

<table>
<tr><td colspan="6" align="center">Numbers of colonies[a]</td></tr>
<tr><td></td><td>1</td><td>2</td><td>3</td><td>4</td><td>5</td></tr>
<tr><td>α</td><td>A a : 93</td><td>B c : 181</td><td>C e : 61</td><td>D b : 145</td><td>E d : 145</td></tr>
<tr><td>β</td><td>B b : 223</td><td>C d : 253</td><td>D a : 243</td><td>E c : 249</td><td>A e : 209</td></tr>
<tr><td>γ</td><td>C c : 226</td><td>D e : 352</td><td>E b : 236</td><td>A d : 248</td><td>B a : 248</td></tr>
<tr><td>δ</td><td>D d : 300</td><td>E a : 267</td><td>A c : 242</td><td>B e : 277</td><td>C b : 165</td></tr>
<tr><td>ϵ</td><td>E e : 152</td><td>A b : 203</td><td>B d : 208</td><td>C a : 164</td><td>D c : 176</td></tr>
</table>

Peptone

1:Polypeptone	2:Bacto-peptone	3:Phytone	4:Proteosepeptone no. 3	5:Casamino acid
994	1256	990	1083	943

Buffer

a:Ca-glycerophosphate β:Tris γ:TES δ:HEPES ϵ:Tris+glycerophosphate

625	1177	1310	1251	903

Carbohydrate

A:Glucose	B:Sucrose	C:Maltose	D:Mannitol	E:Glycerol
995	1137	869	1216	1049

Organic acid

a:Acetate	b:Citrate	c:Malate	d:Lactate	e:Succinate
1015	972	1074	1154	1051

Analysis of variance

	Degrees of freedom	Sum of squares	Mean square
Total	24	99,360	—
Peptone	4	12,332	3083[b]
Buffer	4	65,263	16,316[b]
Carbohydrate	4	14,172	3543[b]
Organic acid	4	3730	933
Error	8	3863	483

[a] Mean of 3 plates.

[b] Significant at 1% level.

an organic acid (0.04%), a peptone (0.03%), and a buffer (0.01M) were added to a basal medium (medium B in Table 1). Seawater samples used were from Tokyo Bay. The results of the surface-colony counts are shown in Table 6 with an analysis of variance.

The results showed the superiority of malate and lactate among organic acids, of mannitol and sucrose among carbohydrates, of Bacto-peptone among peptones, and of TES and HEPES among buffers, although the differences were not statistically significant for the organic acids.

Prevention of Precipitate

The media in this investigation contained a small amount of citrate as a chelating agent, and their peptone content was rather low. Hence the amount of precipitate in sterilization was much less when compared with other media for marine bacteria. However, they formed small but varying amounts of precipitates on autoclaving, especially at a higher pH range of 7.8–8.4.

In order to avoid precipitation as far as possible, we reduced the concentration of the major divalent cations, Ca, Mg, and Sr, to one-fourth the concentrations in seawater, and substituted inorganic phosphate with organic glycerophosphate. These modifications greatly diminished the formation of the precipitates without any significant change in colony counts. The reduction of I, Br, Al, and Si in the medium to one-half of their concentrations in seawater also did not seem to affect the colony counts.

Concentration of Carbon and Nitrogen Sources

In our media, carbon and nitrogen sources were composed of nitrate, ammonium salt, organic acids, carbohydrates, Tween 80, peptone, and extracts. Obviously, it is a laborious task to determine the most suitable concentrations of this many substances in their varying combinations. To date we have only tested the effect of varying concentrations of these nutrients in the proportions shown in Table 7.

The carbon and nitrogen sources were incorporated into the medium at ½, 1, 2, and 4 times the strength of the concentrations given in Table 7. The seawater samples used were taken off the coast of Boso Peninsula. The surface counts were made on these media and also on ZoBell's 2216E medium. The latter medium was prepared from both seawater and an artificial seawater. The media were adjusted to pH 7.8.

The results (Table 8) showed that the highest counts were obtained on the media containing nutrients at twice the concentration of those shown in Table 7.

Beside the experiment given here, we compared our medium several times with ZoBell's 2216E medium. In some experiments our medium gave twice as high surface counts as given by ZoBell's medium, but in other

Table 7. Composition of medium C

Component	Amount[a]
Carbon and nitrogen source	
$NaNO_3$	0.3
$(NH_4)_2 \cdot SO_4$	0.3
Na-malate	0.23
Ca-lactate	0.34
Mannitol	0.15
Sucrose	0.15
Bacto-peptone	0.2
Bacto-yeast extract	0.1
Lab-Lemco beef extract	0.05
Other components	
TES	1.15
Ca-glycerophosphate	0.1
Tween 80	0.025
Ferric citrate	0.01
Modified artificial seawater[b]	900 ml
Distilled water	100 ml

[a] In grams, unless otherwise noted.

[b] Divalent ions reduced to one-quarter and enriched with minor elements in Table 2.

Table 8. Effect of nutrient concentration

Medium	Concentration of nutrient	Number of colonies[a]	Standard deviation
Medium C	X 4	93	24.4
	X 2	144	10.6
	X 1	109	14.9
	X ½	104	15.9
ZoBell 2216E	—	134	17.3
	—[b]	102	6.9

[a] Mean of 4 or 5 plates.

[b] Composed of artificial seawater.

experiments counts on both the media were approximately equal. The total bacterial numbers in the seawater from Tokyo Bay and off the coast of Boso Peninsula, which were counted by the direct microscopic method, were 35–63 times greater than the surface-colony counts on our medium.

We realize that there are still a number of factors to be considered for further improvement of the medium. For instance, the balance among various organic nutrients as well as among inorganic ions, the effect of pH, the selection of suitable chelating agents, and the purity of the agar all demand extensive studies in future. We should like to point out that the media which we use in microbiological studies of environments reflect the state of our knowledge of microbial physiology and ecology, and therefore will be improved from time to time according to the progress in our studies in this field.

LITERATURE CITED

Anderson, J. I. W. 1962. *Cited in* R. B. Scholes and J. M. Shewan. 1964. The present status of some aspects of marine microbiology. Advan. Mar. Biol. 12:133–170.

Carlucci, A. F., and D. Pramer. 1957. Factors influencing the plate method for determining abundance of bacteria in sea water. Proc. Soc. Exp. Biol. Med. 96:392–394.

Cochran, W. G., and G. M. Cox. 1950. Experimental Designs. Mathematical Statistics Series. John Wiley, New York.

Gunkel, W., and C. H. Oppenheimer. 1963. Experiments regarding the sulfide formation in sediments of the Texas Gulf Coast. *In* C. H. Oppenheimer (ed.), Symposium on Marine Microbiology, pp. 674–684. Charles C Thomas, Springfield, Illinois.

Hill, M. N. (ed.), 1966. The Sea, Vol. 2, p. 4. John Wiley (Interscience), New York.

Jannasch, H. W. 1958. Studies on planktonic bacteria by means of a direct membrane filter method. J. Gen. Microbiol. 18:609–620.

Kriss, A. E. 1959. Marine Microbiology (Deep Sea) Izv. Akad. Nauk, Moscow. (Trans. by J. M. Shewan and Z. Kabata, 1963.) Oliver and Boyd, London.

Simidu, U., and K. Aiso. 1962. Occurrence and distribution of heterotrophic bacteria in seawater from the Kamogawa Bay. Bull. Jap. Soc. Sci. Fish. 28:1133–1141.

Taga, N. 1968. Some ecological aspects of marine bacteria in the Kuroshio current. *In* H. Kadota and N. Taga (eds), Proc. U.S. Jap. Sem. Mar. Microbiol. Bull. Misaki Mar. Biol. Inst. Kyoto Univ. No. 12:65–76.

ZoBell, C. E. 1941. Studies on marine bacteria. I. The cultural requirements of heterotrophic aerobes. J. Mar. Res. 4:42–75.

SELECTIVE MEDIA FOR CHARACTERIZING MARINE BACTERIAL POPULATIONS

DARRELL PRATT and JOHN REYNOLDS

Department of Microbiology
University of Maine

Methylene blue and polymyxin B were studies as potentially effective agents for characterizing marine bacterial populations. Methylene blue selected for oxidative, oxidase positive, motile rods, and against oxidase positive, fermentative rods. Samples of seawater from differing localities were observed to differ in the percentage of methylene-blue-resistant cells present. Polymyxin B selected for some non-pigmented fermentative bacteria and for pigmented (yellow, orange, pink) non-fermentative gram-negative rods. The organisms producing pigmented colonies failed to grow on media without polymyxin and were inhibited by the growth of polymyxin-sensitive non-pigmented rods. Pigmented colonies isolated in the absence of polymyxin were found to be polymyxin resistant. Samples varied considerably with respect to the number of polymyxin-resistant cells present.

Environmental microbiologists need more rapid and convenient methods for analyzing natural populations of bacteria. The bacterial population is, like other populations, a function of the environment, but, because bacteria have relatively short generation times, we can expect that changes in the environment will be rapidly reflected in the kinds and numbers of bacteria present. Methods which involve the random isolation of bacterial cultures from samples and their careful identification are essential since they tell us what possibilities to expect, but these procedures are too slow to allow the extensive sampling needed to study bacterial population dynamics. Obviously, if we are to maximize our use of such information, large numbers of sample analyses will be needed to establish meaningful correlations with environmental parameters.

The marine bacterial population consists of a wide variety of species and presents real problems, some perhaps without solution, to the analytical microbiologist. The methods used by most of us have been directed to heterotrophs capable of growing aerobically on complex media. Except for a few particularly directed investigations, obligate anaerobes and autotrophs are overlooked. Very often obligate psychrophiles and thermophiles have not been considered in the methods being used. In addition, some highly exacting strains may fail to grow because of a lack of a particular growth factor and others may fail to grow because of the presence of an inhibitory substance. Always in the background are those bacteria which fail to grow in the high nutrient concentrations required to produce visible colonies. Our methods usually do not protect the slowly growing indigenous cells from the adverse effects of rapidly growing zymogenous bacteria. Ultimately any set of procedures for a thorough analysis of marine bacterial populations must consider these many problems.

Many studies have been reported for marine as well as other environments in which analyses of bacterial populations have been made. Often separations of yeasts, molds, actinomycetes, and bacteria have been made using selective media and morphology. Protein hydrolysis and starch hydrolysis have frequently been used as differential qualities. A wider variety of physiological groups has been estimated by means of limiting-dilution (MPN) methods. Quite possibly these should be exploited more fully since an infinite number of substrates can be introduced. A logical difficulty exists here because the experimenter has to guess which are the physiological activities of importance. These methods should be considered as potential components of any scheme for population analysis.

The possibility of using media containing selective agents for the characterization of a marine population was prompted by their widespread use in microbiology for the isolation of particular bacterial species. The initial hypothesis was simply that a given selective agent would divide the population into two groups, resistant and sensitive. With two such agents it would be possible to create four groups: (1) resistant to both; (2) resistant to neither; (3),(4) resistant to one or the other. With three agents this becomes eight groups and would offer an empirical approach to characterizing populations. A wide range of selective agents, both chemical and physical, is available. For example, dyes, antibiotics, antibacterial chemotherapeutics, antivitamins, carbon source, salts, pH, anaerobiosis, and incubation temperatures are all possible selective agents; combinations of these greatly increase the various possibilities. The best anticipated results of these studies would be to produce a series of media which would estimate the numbers of various taxonomic groups present. Perhaps the least satisfying result would be that media would be developed which would allow us to make an empirical characterization of the population without reference to taxonomic groups. The usefulness of the procedure in any case will depend on the correlation of population composition with identifiable environmental factors. We plan this to be a long-range program and our present report is a preliminary one.

Our approach to this project was to begin by studying the sensitivity of stock cultures of marine bacteria to a series of dyes and to antibiotics. The idea was that if we could make meaningful separations among these cultures, based on their resistance to these agents, we might be able to apply them to natural populations of marine bacteria. We have considered thus far only the aerobic heterotrophs since technically these are the easiest with which to work and would allow us to learn more about the basic problems to be solved.

MATERIALS AND METHODS

The medium employed in these studies was Marine Broth (Difco), with and without 2% agar (Difco) according to the experimental circumstance. This medium, a peptone and yeast extract combination, was chosen because it was commercially available and easy to use.

The stock cultures employed had been isolated a number of years ago from samples of seawater taken from the Atlantic Ocean off the Florida east coast. These were characterized and described earlier (Tyler *et al.*, 1960). All appeared to be either in the genus *Vibrio* or *Pseudomonas* with the exception of two which were tentatively *Vibrio*. The cultures have been maintained under a mineral-oil seal in a Trypticase (BBL)-seawater medium.

Samples of seawater were obtained at several locations near Schoodic Point, Maine. These were taken in the surf at points well away from any source of terrestrial pollution. Several samples were obtained from a large tidal lagoon. The condition of the tides was noted and the temperature of the water was recorded. The samples were packed in ice and transported to the laboratory. All media were inoculated within 3 hr after the samples were taken.

Agar media were inoculated by spreading 0.1 ml of the sample over the surface of a dried and chilled plate. The inoculations were made with sterile curved glass rods.

All plates were incubated at 23°C. They were observed and counted periodically. Methylene blue hydrochloride (Fisher) and polymyxin B sulfate (Sigma) were added to the basal medium in the desired amounts. Polymyxin B sulfate was sterilized by filtration and added aseptically.

RESULTS

Preliminary studies (Pratt and Reynolds, 1970) indicated that, of several promising selective agents, methylene blue and polymyxin B were perhaps the best with which to begin. Basic dyes as a group were found to be far more inhibitory to marine bacteria than could have been anticipated from their behavior toward non-marine gram-negative bacteria. Methylene blue was slightly less inhibitory than crystal violet or malachite green, and among the stock cultures the vibrios were markedly susceptible. The sensitive strains, in addition to being fermentative, had very low catalase content. Antibiotic sensitivity tests revealed that polymyxin B was inhibitory to the marine pseudomonads being tested but not to the vibrios. Thus these two agents had the potential of separating two important groups of marine bacteria. The 20 Florida strains of marine bacteria were streaked on plates containing several concentrations of either methylene blue or polymyxin B. The two selective agents divided the cultures into four groups. The polymyxin-sensitive-methylene-blue-resistant group (Group A) had eight cultures, as did the polymyxin-resistant-methylene-blue-sensitive group (Group B). The former were all pseudomonads and the latter were all vibrios. Group C appeared to be pseudomonads which were sensitive to methylene blue, and Group D were possibly vibrios which were insensitive to both (Table 1).

Table 1. Response of 20 cultures of marine bacteria (Florida series) to polymyxin and methylene blue[a]

| | | Polymyxin units/ml | | | Methylene blue μg/ml | |
Fermentation	0	50	100	2	8
A (8)[b]	−	+	−	−	+	+
b (8)	+	+	+	+	−	−
C (2)	−	+	−	−	−	−
D (2)	+	+	+	+	+	+

[a] + = growth, − = no growth.

[b] Number of cultures in group.

Methylene blue was very inhibitory to colony formation using samples of seawater as the inoculum. As little as 2 μg/ml caused a substantial reduction in colony numbers in all cases, and occasionally was nearly as effective as 20 μg/ml (Table 2). The three samples of seawater could be distinguished on the basis of the fraction of colonies capable of growing in 2 μg/ml of methylene blue. The presence of methylene blue in the medium resulted in the selection of non-fermentative organisms. When colonies were selected from the control plate by a method of random selection and their ability to ferment glucose was determined (Table 3), 62% were fermentative. If the colonies were picked from media containing 2 μg/ml of methylene blue, only 16% were fermentative. Thus the presence of methylene blue in the medium restricted colony formation by those bacteria capable of fermenting glucose. This observation was in keeping with the predicted results.

Table 2. Comparative methylene-blue sensitivity of the bacterial populations in three samples of seawater

Sample no.	Methylene blue (g/ml)	Colonies (no./0.1 ml)	Inhibition (%)
1	0	1530	—
1	2	510	67
1	20	170	89
2	0	240	—
2	2	147	39
2	20	45	8[1]
3	0	112	—
3	2	11	90
3	20	3	97

Table 3. Selective inhibition of marine bacterial populations by methylene blue

Methylene blue (g/ml)	Colonies (no./0.1 ml)	Inhibition (%)	Fermentative[a]
0	293	0	25/40
0.5	172	42	7/20
1.0	153	48	6/20
2.0	116	60	4/24
4.0	89	70	3/24
8.0	66	77	4/27

[a] Data given as ratio of the number capable of fermenting glucose to the number tested.

Table 4. Selective inhibition of a
marine bacterial population by polymyxin

Polymyxin (units/ml	Colonies (no./0.1 ml)	Inhibition (%)	Fermentative[a]
0	287	0	25/40
2	249	13	18/22
20	86	70	4/20
50	74	74	6/20
100	80	72	7/20
200	78	73	3/20

[a] Data given as ratio of the number capable of fermenting glucose to the number tested.

Table 5. Influence of polymyxin B
on colony numbers and pigmentation

Polymyxin (units/ml)	Colonies (no./0.1 ml)			
	Sample 1		Sample 2	
	Total	Yellow	Total	Yellow
0	240	1	106	2
10	110	36	23	9
20	87	41	18	15
50	98	44	18	10
100	81	46	12	6

The results with polymyxin were rather different. With our first sample of seawater the colony formation was inhibited only 13% with 2 units/ml and 70% with 20 units. Higher concentrations did not inhibit colony formation to a significantly greater degree. These organisms behaved as though they were completely indifferent to the action of polymyxin. With media containing 2 units of polymyxin the proportion of fermenters among randomly sampled colonies increased to nearly 100%; however, the fraction dropped markedly with concentrations above 20 units/ml (Table 4). On the polymyxin agar the colonies were predominantly yellow and orange, with some pink and some white. The pigmented colonies when streaked to a medium without polymyxin gave pigmented growth.

With two subsequent samples the proportion of yellow colonies forming in media containing several concentrations of polymyxin was estimated. The two samples were quite different with respect to absolute numbers, but in each case not only did a greater proportion of pigmented colonies develop in the presence of polymyxin, but the actual number was greater (Table 5). In the absence of polymyxin essentially no pigmented colonies were observed. Since the pigmented colonies appeared on plates where the total number of colonies was less, the possibility that crowding prevented their development was considered. A sample of seawater was taken from a rather sizable tidal pool where one might suspect a relatively high count. This pool did not drain completely at low tide and filled vigorously on the incoming tide. At the time of sampling the pool was beginning to fill, and the water temperature was 20°C as compared with 12°C for the open water. Dilutions of this sample were plated on marine agar and on media containing 50 units/ml of polymyxin. Although the proportion of pigmented colonies was still rather low, more were observed when the growth was less crowded. The unexpected result with this sample was that the counts on polymyxin agar were high, being nearly equal to those on marine agar, and approximately 85% of the colonies were pigmented (Table 6). There was then a cryptic population of pigmented organisms which grew only in the polymyxin agar on primary isolation.

A number of pigmented colonies and a number of white colonies were isolated. These were streaked separately on marine agar and on polymyxin agar and then in combination. The inocula were prepared by suspending in artificial seawater growth taken from a slant. In every case of six different combinations the pigmented organism failed to grow in the presence of the non-pigmented organism when inoculated together on marine agar but did

Table 6. Influence of inoculum size on numbers of pigmented colonies observed from a natural source

Inoculum (ml)	Colonies on marine agar		Colonies on polymyxin agar	
	Total	Pigmented	Total	Pigmented
0.01	92	14	61	54
0.02	137	7	124	105
0.04	322	9	196	172
0.06	447	8	278	255
0.08	>500	3	>500	—[a]
0.10	>500	0	>500	—[a]

[a] Approximately 30 white colonies per plate.

Table 7. Characteristics of organisms
randomly isolated from selective media[a]

Source medium	N	M	F	Cat	Ox
Marine agar	40	35	20	28	30
Methylene blue agar	50	50	10	18	38
Polymyxin agar	37	1	0	11	6
Polymyxin agar (white)	13	9	11	6	12
Marine agar (pigmented)	10	2	0	3	2
Methylene blue agar (blue)	10	0	0	0	0

[a] N, number of cultures; M, motility; F, fermentative; Cat, catalase;
  Ox, oxidase. The number of cultures positive for each property
  are indicated.

grow when the mixture was inoculated onto polymyxin agar. The non-pigmented strains grew rapidly and formed large colonies within 24 hr; the pigmented strains grew more slowly and required 48 hr to produce visible colonies. The evidence suggested that the rapidly growing zymogenous bacteria inhibited the development of the slower growing pigmented colonies. These interactions are being investigated further; the inhibitions could result from the presence of antibiotic substances or from metabolic effects such as the production of amines or the alteration of the pH.

Over the period of several experiments a number of colonies were selected from the various media. For the most part these were selected by a procedure designed to make the selection random; some were selected because of a particular attribute. Each was characterized with respect to motility (soft agar), fermentation of glucose (Leifson, 1963), catalase, and oxidase (Kovacs, 1956) (Table 7). Of the 40 colonies selected from the control agar half were fermentative and 75% were oxidase positive. Methylene blue as before selected non-fermenters, although it was not completely efficient. Among the colonies growing on media containing methylene blue were some which were stained deeply blue; these were characterized by being negative in each of the characteristics tested. The pigmented organisms from polymyxin agar were non-fermentative and oxidase positive, but the white colonies from polymyxin agar were mostly fermentative and oxidase positive. This was in keeping with the original expectation that polymyxin would select for fermentative cultures. In addition, it is consistent with the observation that media containing only 2 units of polymyxin selected for fermentative organisms.

CONCLUSIONS

Methylene blue when added to culture media served as a selective agent and inhibited a considerable fraction of the fermentative bacteria present in the samples of seawater investigated. Different samples had differing proportions of methylene-blue-resistant bacteria present and it would be possible with this medium to characterize a given sample. Such information would only be of value if a low or high proportion of methylene-blue-resistant cells could be correlated with some pertinent environmental parameter. More interesting was the selection by polymyxin B of pigmented organisms which failed to grow on media in which the general population was not inhibited. These gram-negative, non-motile, rods producing yellow, orange, or pink colonies were probably "cytophagas" and will require further taxonomic study to resolve their identity (Hendrie *et al.,* 1968; Hayes, 1963; Lewin, 1969; Colwell, 1969). Polymyxin-containing media must be compared further with that of Warke and Dhala (1968) which contained penicillin and Chloromycetin and was reported to be effective for enumerating cytophaga in soils. The unmasking of the cryptic population of pigment-producing bacteria by polymyxin was of interest because very probably these organisms are among the types not usually estimated by viable counts (Jannasch and Jones, 1959). These were then bacteria which fail to compete with rapidly growing cells on the standard, rich, plating medium and thus satisfy one of the speculations often advanced as one reason that the viable count is often much smaller than the microscopic count of cells in marine samples.

LITERATURE CITED

Colwell, R. 1969. Numerical taxonomy of the flexibacteria. J. Gen. Microbiol. 58:207–215.

Hayes, P. R. 1963. Studies on marine flavobacteria. J. Gen. Microbiol. 30:1–19.

Hendrie, M. S., T. G. Mitchell, and J. M. Shewan. 1968. The identification of yellow-pigmented rods. *In* B. M. Gibbs and D. A. Shapton (eds), Identification Methods for Microbiologists, Part B, pp. 67–78. Academic Press, New York.

Jannasch, H. W., and G. E. Jones. 1959. Bacterial populations in seawater as determined by different methods of enumeration. Limnol. Oceanogr. 2:128–139.

Kovacs, N. 1956. Identification of *Pseudomonas pyocyanea* by the oxidase reaction. Nature (London) 178:703.

Leifson, E. 1963. Determination of carbohydrate metabolism of marine bacteria. J. Bacteriol. 85:1183–1184.

Lewin, R. A. 1969. A classification of flexibacteria. J. Gen. Microbiol. 58:189–206.

Pratt, D. B., and J. Reynolds. 1971. Growth inhibition of marine bacteria by methylene blue. Bacteriol. Proc. 71:57.

Tyler, M. E., M. C. Bielling, and D. B. Pratt. 1960. Mineral requirements and other characters of selected marine bacteria. J. Gen. Microbiol. 23:153–161.

Warke, G. M., and S. A. Dhala. 1968. Use of inhibitors for selective isolation and enumeration of cytophagas from natural substrates. J. Gen. Microbiol. 51:43–48.

AEROBIC, HETEROTROPHIC BACTERIAL POPULATIONS IN ESTUARINE WATER AND SEDIMENTS*

L. HAROLD STEVENSON, CHARLES E. MILLWOOD,
and BRUCE H. HEBELER

Department of Biology and
Belle W. Baruch Coastal Research Institute
University of South Carolina

INTRODUCTION

Marshes and estuaries of various types comprise approximately 80% of the Atlantic and Gulf Coasts and 10–20% of the Pacific Coast of the United States. These areas are significant to human welfare because of their importance in transportation, food production, waste disposal, and recreational activities. Despite this importance, marine bacteriologists have largely neglected these areas in favor of "blue-ocean" bacteriology. Most of the near-shore studies have centered in areas with a river-ocean interface.

The Belle W. Baruch Coastal Research Institute of the University of South Carolina, in conjunction with the Departments of Biology and Geology, has undertaken a multidisciplinary analysis of a salt-marsh ecosystem. The preliminary bacteriological study of marsh water and sediment is the subject of this presentation. The objectives of the work reported here were to develop techniques for the isolation and enumeration of common heterotrophic bacteria, to estimate the bacterial populations of estuarine samples,

* Contribution No. 84 of the Belle W. Baruch Library in Marine Science.

and to make a preliminary taxonomic evaluation of the bacteria using numerical methods.

MATERIALS AND METHODS

Study Area

The estuary is located on the Southeast coast of the United States near Georgetown, South Carolina (33°20'N 79°10'W). The area, referred to as the North Inlet Estuary, is confined chiefly within the property lines of the Belle W. Baruch Plantation, which is maintained for environmental research. The North Inlet Estuary is a semitropical saltwater marsh. There is little or no freshwater inflow with the exception of rainwater run-off. Typically, the salinity is greater than 30 °/oo. The water level in the system is dependent upon the tides, which have a normal amplitude of about 6 ft. The mouth of the estuary, North Inlet, is the only major connection between the ocean and marsh. A complete description of the area has been published elsewhere (Vernberg, 1973).

Sampling, Isolation, and Storage

Water samples were obtained from the center of Debidue Creek and sediment was collected on the inland side of Debidue Island (Fig. 1). Water was collected using a Johnson-ZoBell sampler (ZoBell, 1941) modified for use in shallow water (Fig. 2). The sampler, as modified by Millwood, consists of a stainless-steel holder and interchangeable 1-liter glass bottles. The changes in design involve the activating mechanism. The device is not activated by a messenger; rather, the operator initiates water collection by pulling a nylon cord attached to a lever. This breaks the collection tube allowing water to enter the container. Sediment cores were obtained with 1.5-cm sterile glass cylinders. After collection, both water and sediment samples were stored in an ice chest and processed within 1 hr.

Viable counts were determined by first diluting the specimens with four-salts solution (Quigley and Colwell, 1968) and then plating on ZoBell's marine agar 2216 (Difco) using the spread-plate technique. The populations of sediment organisms were expressed on the basis of number per gram dry weight. Two cores were obtained from each location. One was used to estimate the number of bacteria and the other was placed in tared weighing jars for determination of dry weight. The organisms used for numerical analysis were selected from the spread-plates and stored in 1-ounce prescription bottles on the same medium.

Numerical Taxonomy

The organisms obtained from the estuarine materials were subjected to determination of 44 cell morphology, 39 cultural, 27 physiological, 53 biochemical, and 14 antibiotic sensitivity characteristics. The characters, methods, and media used were primarily those outlined by Colwell and Wiebe (1970) and Quigley and Colwell (1968). The data were analyzed using Georgetown Taxonomy Programs 2, 3, 4, and 5 (Quigley and Colwell, 1968). Groups were determined by visual observation of sorted triangles obtained by computer analysis and no attempt was made to establish rigorous mathematical verification of their homogeneity.

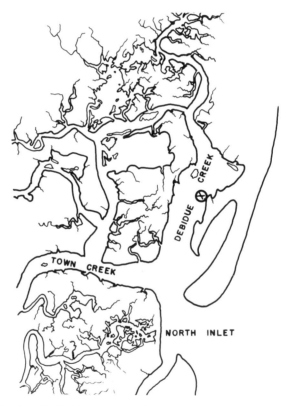

Fig. 1. Map showing North Inlet estuary and the sampling site, which is designated by the circled ×.

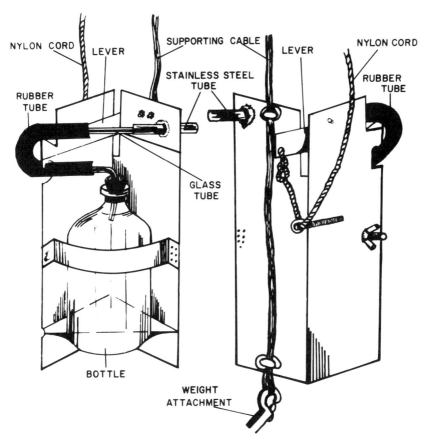

Fig. 2. Diagram of a modified Johnson-ZoBell water-sampling device.

RESULTS AND DISCUSSION

The use of viable plate counts for determining bacterial populations in natural materials has been criticized in recent years (Jannasch, 1965; Jannasch and Jones, 1959). Direct microscopic counts and the technique of ATP analysis (Holm-Hansen, 1969) have yielded data that indicate the population may be from 2 to 1000 times greater than previously suspected. Yet, if an investigator desires to examine the properties and types of organisms present, then the normal plating procedures offer a variety of advantages. Since our long-range goals involved the use of numerical taxonomy to characterize the populations in the estuary, we chose the viable plate count as the vehicle to begin our studies. The dependability of the sampling and plating procedures was tested by examining up to six replicate samples. The standard percentage error was always less than 2% when replicate water samples were tested and from 7 to 10% when sediment was examined.

The recovery of bacteria from aquatic sources is frequently complicated by the tendency of organisms to clump or form microcolonies in association with detritus or sediment particles (Meadows and Anderson, 1968). Consequently, the disruption of these clusters could result in the recovery of a higher number of organisms from a given amount of water or sediment (Jannasch, 1965; Skerman, 1963). Two methods were attempted in an effort to disrupt clumps: blending with a Waring Blendor and treatment with Tween 80. The effect of blending on the isolation of bacteria from water is illustrated in Fig. 3. In this, and other blending experiments, the container was precooled in an ice bath and cooled between each time interval. The number of bacteria recovered per milliliter of water decreased from an initial value of 25×10^2 to 15×10^2 after 5 min of agitation in the first experiment and a similar decrease was noted in the second. The results obtained when water was shaken and blended with Tween 80 are illustrated in Fig. 4. Shaking for the indicated time had little effect on the number of organisms recovered and blending again resulted in a decrease in the number obtained.

On the other hand, blending of sediment samples resulted in an increase in the number of organisms recovered. Figure 5 illustrates the results obtained from one experiment. Blending of the sample for 4–5 min gave a 1.5- to

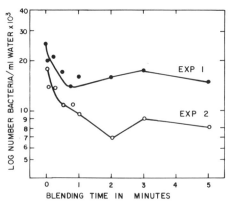

Fig. 3. The effect of blending on the recovery of heterotrophic bacteria from estuarine water. The number of bacteria per milliliter obtained in two experiments is plotted against blending time.

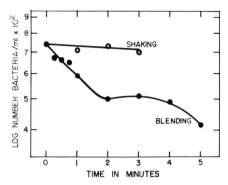

Fig. 4. The relationship observed between the number of bacteria recovered from the sample after shaking and blending with Tween 80.

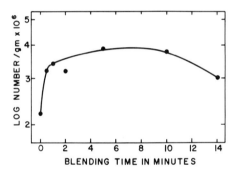

Fig. 5. The effect of blending on the recovery of bacteria from estuarine sediments.

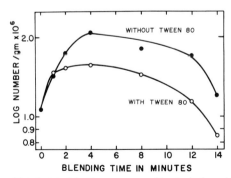

Fig. 6. The relationship observed in the recovery of bacteria from blended samples with and without the addition of Tween 80.

2-fold increase in colony forming units. Continued blending after 5 min resulted in a decrease. Shaking of sediment by hand in the presence of Tween 80 had no effect on the number of viable cells recovered, and the number of colony-forming units obtained from sediment samples blended in the presence of Tween 80 was consistently lower than the number recovered from duplicate samples blended without the addition of the agent (Fig. 6).

Several natural factors, such as location and depth, possibly associated with the recovery of bacteria from the estuary were examined. When water was collected from several locations within the estuary, higher populations were obtained from the stations that were farther inland; however, the differences were not marked. The number of colony-forming units per milliliter of water from various depths within 3- to 4-m water columns in the area was also determined. Slightly greater populations were obtained near the surface than near the bottom. All subsequent samples were collected 0.5 m from the surface.

The tides proved to have a marked effect on the isolation of bacteria from the water. Figure 7 illustrates data obtained during a falling tide. Approximately 2×10^3 bacteria per milliliter were recovered from water collected at maximum high tide. As the level of water fell in the estuary, there was an increase in the bacterial density until a population of 2.2×10^4 per milliliter obtained. Data obtained from samples collected on a later occasion during a rising tide yielded similar results (Fig. 8). The highest population density was obtained from the low-tide sample, which was collected first. The number gradually decreased as the tide flooded the marsh.

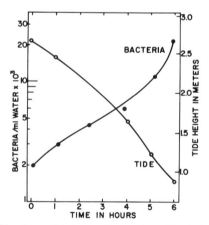

Fig. 7. The relationship observed between the number of bacteria per milliliter of water and the level of an ebb tide. The samples were collected 19 August 1970.

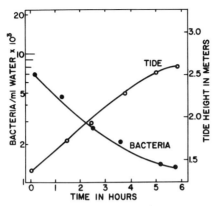

Fig. 8. The relationship observed between the number of bacteria per milliliter of water and the level of a flood tide. The samples were collected 9 October 1970.

The increase in population observed in water leaving the marsh may reflect the high productivity of the estuary relative to the open ocean. However, a factor of greater significance may be the transport of bacteria from, or in association with, marsh detritus and sediment. The water that flooded the marsh at high tide probably contained a relatively low number of bacteria. Then, as the tide receded, the marsh was "flushed" and the sediment bacteria were swept into the water system. This would not only increase the population but could also alter the nature of the population recovered. Organisms obtained from estuarine water at high tide may be more representative of "marine bacteria" and those obtained at low tide may reflect properties of sediment organisms.

Sediments of various types and depths were also examined for bacterial density. The number of colony-forming units obtained from different locations at the Debidue Island station are illustrated in Table 1. The data were collected from 5- to 6-cm cores and none of the samples was blended. The highest number of bacteria recovered from any location was 7.6×10^6 colony-forming units per gram dry weight of sediment taken from a mixed area of *Spartina* grass and oyster community. Sediment taken from a dry sand dune had the lowest population density.

The estuarine sediment was examined to determine the variance in population with depth in the upper 10 cm. Core samples were aseptically divided into 2-cm sections. The sections were first agitated by shaking and a plating series prepared. Each sample was then blended for 4 min and plated a second time. The results, illustrated in Fig. 9, show that in all cases the blended samples yielded a higher number of viable colonies than the shaken

Table 1. Distribution of
bacteria in various types of sediment

Sediment type	Bacteria/g ($\times 10^3$)		
	July[a]	August	October
Soft silt deposit[b]	610	600	550
Firm sandy mud[b]	900	—[c]	180
Spartina stand; sandy mud[b]	690	—[c]	290
Mixed *Spartina* stand and oyster bank; sandy mud[b]	7600	1000	390
Sand dunes	32	38	11

[a] All samples were collected during 1970.

[b] Area covered by water at high tide.

[c] Sample not obtained.

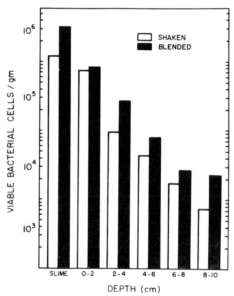

Fig. 9. Comparison between the number of aerobic heterotrophic bacteria in the vertical stratum of sediment cores that were treated by shaking and blending.

Table 2. Selected features describing the organisms from each source

Feature	Water isolates	Sediment isolates	Feature	Water isolates	Sediment isolates
Cells straight rods	55[a]	42	Lactose fermented	14	8
Cells branching	0	5	Sucrose fermented	20	15
Cells 1–3 μm	67	30	Glucose used as sole carbon source	14	32
Cells >3 μm	17	47	Acetate used as sole carbon source	32	21
Gram negative	83	68	Glutamate used as sole carbon source	58	71
Motile	46	40	H_2S from cysteine	6	26
Polar flagella	40	33	NH_4 from peptone	90	88
Endospores present	<1	4	Proteolytic on gelatin	55	67
Colony less than 2 mm	90	60	Proteolytic on casein	47	52
White-off white colonies	42	49	Lipolytic activity on Tween 20, 40, and 60	60–67	69–76
Yellow colonies	30	19	Lipolytic activity on tributyrin	90	90
Orange colonies	14	20	Sensitivity to chloromycetin	95	94
Growth without added mineral salts	50	49	Erythromycin	90	94
Growth without added Na^+	54	50	Penicillin	57	71
Growth without added K^+, Ca^{2+}, or Mg^{2+}	>90	>95	Vibriostat	47	45
MOF aerobic acid	61	62			
MOF anaerobic acid	84	78			

[a] Percentage of organisms tested showing positive reactions.

sample. The surface slime layer contained the highest number of bacteria per gram dry weight with a progressive decrease in the population at lower depths.

One hundred ten isolates from water and 165 isolates from sediment were selected from dilution plates and subjected to the battery of tests required for numerical analysis. A list of features selected from the 178 characters used to describe each organism is shown in Table 2.

Greater than 80% of the water isolates were gram-negative rods; 46% were motile, most by polar flagella. Most of the isolates produced entire raised colonies less than 2 mm in diameter when grown on agar plates and produced very little turbidity in broth culture. Approximately one-half of the isolates did not require the addition of sodium ions for growth and greater than 90% did not require the addition of divalent metal ions. Only 6% of the isolates produced hydrogen sulfide from cysteine, while 90% produce ammonia from peptone. About one-half of the organisms were proteolytic. It is interesting to note that 60–67% demonstrated lipolytic activity when tested on Tweens and 90% produced zones of clearing on tributyrin agar.

The organisms obtained from sediment, as a group, were generally similar to the water organisms in antibiotic sensitivities, degradative and fermentative capabilities, and cation requirements. However, they differed in several respects: ability to use compounds as the sole carbon source, production of H_2S from cysteine, and several cellular and cultural characteristics. Only 68% of the isolates were gram negative and 4% formed endospores. Almost 50% produced cells that were greater than 3 μm in length as compared with 17% of the water isolates. Branching was exhibited by 5% of the sediment cultures and only 60% produced colonies that were less than 2 mm in size, whereas 90% of the water isolates produced small colonies.

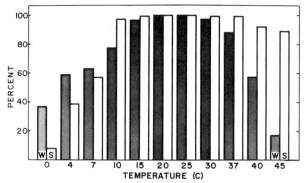

Fig. 10. The effect of temperature of incubation on the percentage of isolates from water (W) and sediment (S) that developed into visible colonies.

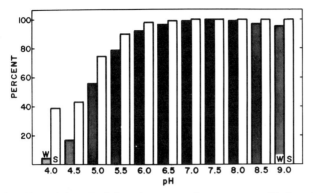

Fig. 11. The effect of the pH of the substrate on the percentage of isolates from water (W) and sediment (S) that exhibited growth.

The temperature and pH tolerances of each set of organisms revealed some interesting differences. Figure 10 illustrates the percentage of each set of cultures that grew at different temperatures of incubation. The sediment cultures demonstrated a distinct preference for the higher temperatures. Only 7% of the sediment isolates grew at 0°C compared with 37% of the water isolates, while 89% of the sediment bacteria grew at 45°C, compared with 17% of the water isolates. The sediment organisms also demonstrated a greater tolerance of pH (Fig. 11). Only 4% of the water isolates grew on media which had a pH of 4, whereas 39% of the sediment organisms grew at that pH.

The degree of similarity between each of the possible pairs of organisms obtained from the water was computed and the organisms re-sorted using the single-linkage cluster technique (Sokal and Sneath, 1963) (Fig. 12). The degree of similarity among the isolates proved to be low indicating a wide diversity among the organisms tested. The highest S value was 89. When the organisms were grouped by proceeding through decreasing values to 75 S, the phenon level conventionally used to designate species, only six groups representing a total of 16 organisms had been established. The other 87 isolates remained as singlets. Most of the clustering occurred at the 62–69 S levels. All but one of the strains were grouped at 58 S.

The triangular print-out of the similarity matrix (Fig. 12) indicated that approximately one-half of the strains were clustered into five loosely defined groups. It was noted that very few of the strain pairs had similarity values greater than 80%, symbols A and B, and most of the pairs were only 60–70% similar, symbols E and F.

Conceding that speciation using numerical analysis is, at best, a risky proposition, some generalities can, nonetheless, be offered regarding the

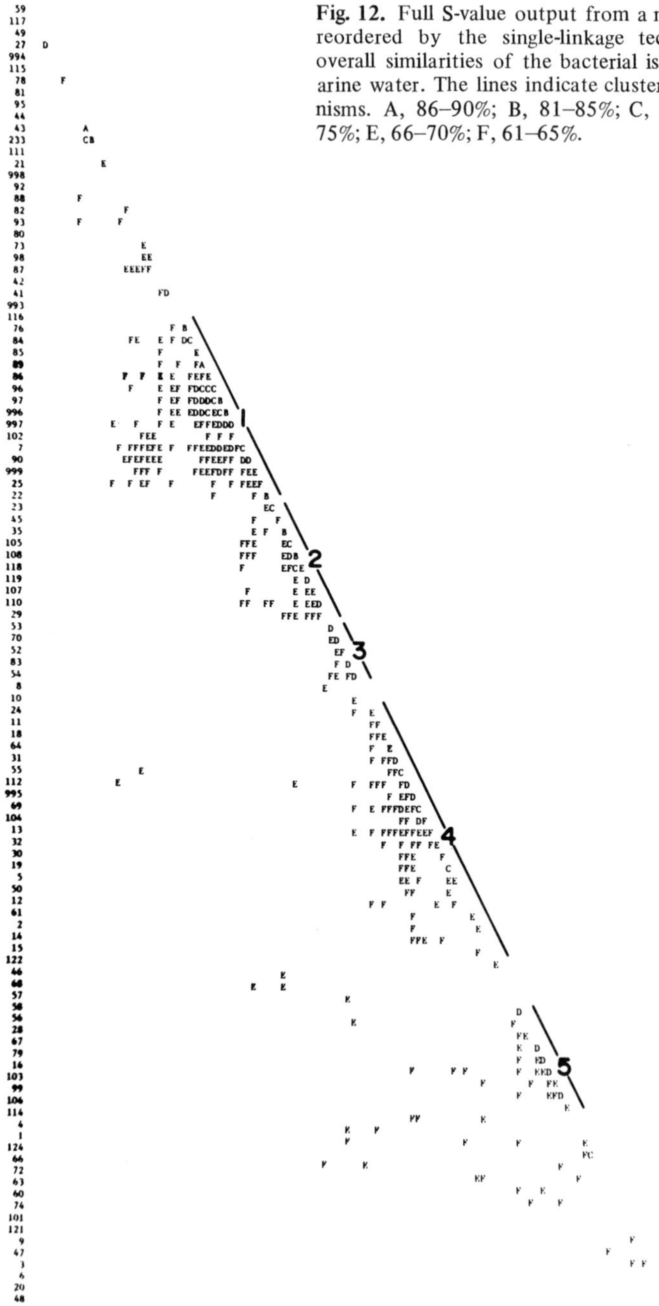

Fig. 12. Full S-value output from a matrix sorted and reordered by the single-linkage technique showing overall similarities of the bacterial isolates from estuarine water. The lines indicate clusters of related organisms. A, 86–90%; B, 81–85%; C, 76–80%; D, 71–75%; E, 66–70%; F, 61–65%.

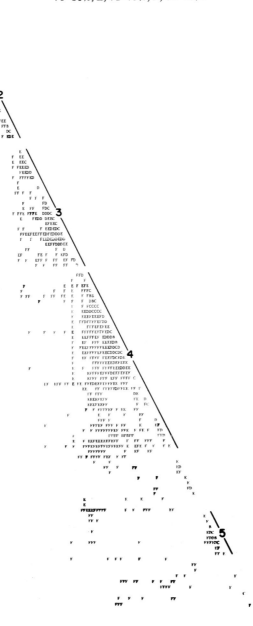

Fig. 13. Full S-value output from a matrix sorted and reordered by the single-linkage technique showing overall similarities of the bacterial isolates from estuarine sediment. The lines indicate clusters of related organisms. A, 91–95%; B, 86–90%; C, 81–85%; D, 76–80%; E, 71–75%; F, 66–70%.

groups. Probable genus names were assigned to each group using the schemes proposed by Murchelano and Brown (1970) and Shewan (1963). The properties given represent a majority of each group and not necessarily every individual organism.

The most compact cluster, no. 1, apparently represented the genera *Vibrio* and *Pseudomonas.* The organisms were all gram-negative, polar-flagellated rods and produced off-white colonies. Only one-third of the group was sensitive to Vibriostat, and members of the group demonstrated mixed response to the oxidase and oxidation/fermentation tests. Most of the isolates required the addition of Na^+ for growth. Group 2 consisted of a combination of organisms resembling *Flavobacterium* and *Achromobacter* species. All were non-motile, gram-negative, small rods, and about one-half were pigmented. Sixty-five per cent required Na^+ for growth. Group 3 was a small cluster of *Pseudomonas*-type organisms that did not require the addition of any mineral salts for growth.

The gram-positive bacteria obtained from the water were sorted into group 4. About 75% of the strains were non-motile cocci that were sensitive to penicillin. They produced H_2S from cysteine and did not require the addition of sodium for growth. Group 5 was apparently comprised of pigmented, non-motile gram-negative rods similar to the genus *Flavobacterium.* Most required the addition of sodium for growth. A few *Cytophaga* strains were encountered but they did not appear in any of the major clusters.

The overall similarity of the sediment organisms was higher than that observed among the water isolates (Fig. 13). However, a degree of heterogeneity was still evident. Only three pairs of strains had S values above 90%, with the highest being 94%. When the strains were clustered by proceeding through decreasing S values, 14 groups, representing 98 of the 165 strains tested, were established. Because of space limitations, only 128 of the isolates were included in the triangular print-out of the re-sorted organisms. The upper and lower extremes of what would have been a larger triangle were eliminated. Three large and two smaller clusters were evident. Sub-groupings could be established within some clusters, with cluster no. 2 showing the obvious presence of two subgroups.

Group 1 was a cluster of organisms typical of the genus *Vibrio,* none of which required the addition of Na^+. Most of the organisms in group 2 were isolated from sandy sediments. They were gram-negative rods, non-motile and nonfermentative. Almost all required Na^+ for growth and produced H_2S for cysteine. All of the strains required complex media for growth. In many respects they appeared to be *Achromobacter* or *Alkaligenes* type organisms.

The isolates obtained from the soft, silty mud were clustered into two groups, nos. 3 and 4. Group 3 was composed of long, gram-negative rods, most producing large colonies. More than half of the organisms, all of the larger subgroup, were similar to the genus *Cytophaga.* All demonstrated the typical spreading on agar plates. No flagella or motility in liquid media were evident. Most required Na^+ for growth. The organisms in group 4 were heterogeneous with *Pseudomonas* appearing to be the dominant type of gram-negative organism present. However, some gram-positive organisms were also included in the lower portion of the cluster. None of the organisms required mineral salts for growth.

The organisms in group 5 were all actinomycetes obtained from the sand-dune area, the only collection point that was not subject to flooding at high tide. Fifty to 80% of the organisms obtained on spread plates prepared from the dry sand were actinomycetes-type cultures. None was routinely found in other areas.

The preceding investigation was undertaken for two primary reasons: first, to determine the nature and distribution of bacteria within the estuary; second, and perhaps more importantly, to uncover some of the problems associated with the use of numerical methods in describing the bacterial flora in such a natural system. The data obtained in response to the first question were straightforward and basically what were expected. However, the questions relating to the use of numerical methods were varied and defied simplified answers.

A problem frequently encountered early in the investigation related to the availability of suitable, well-characterized marker strains from the marine environment. Some of the available ATCC cultures bear little resemblance to the cultures isolated from the estuary. Marker strains would not only aid in attempts to compare data from different sets of isolates or different investigators, but would also serve as standards for gauging the many laboratory tests involved in screening.

Although not reported in the "Results" section, our data indicate that it is imperative to process samples as soon as possible after collection. A delay of only a few hours results in alterations in the type of population observed. Selection of the organisms presented another problem. The organisms used in this study were picked without deliberate bias, but not in truly random fashion. Subsequent findings have indicated that the use of a table of random numbers to select isolates may be a necessity. One of the most difficult questions concerned the number of organisms one must use in the cluster analysis.

The effect of the tides appeared to be one of the major factors associated with the qualitative and quantitative isolation of bacteria in the system studied. This is neither a new nor a revolutionary concept. Indeed, one would expect tide fluctuations to influence the bacterial populations in marsh areas.

ZoBell and Feltham (1942) have shown that the water leaving a shallow mud flat at low tide has a higher population of bacteria than that found in incoming water during a rising tide. Similarly, in a report on a salt-marsh ecosystem, Odum and de la Cruz (1967) noted that the amounts of detritus, phytoplankton, and zooplankton were higher at mid-ebb tides than during mid-flood tides. The tides have also been shown to affect chemical and physical parameters of the water in coastal areas (Alexander *et al.*, 1973). However, even a cursory examination of the published data on bacterial populations in near-shore environments reveals an almost complete disregard of the warning issued by ZoBell and Feltham (1942). The influence of tidal flux is of particular importance in investigations designed to study bacterial variations on a daily or seasonal basis. A series of samples taken at one specific time cannot be considered as representative of the bacteria present in a salt marsh on a given day.

The multiplicity of microenvironments and colonization of particles in the sediment represent an even more serious factor that must be considered. The variation in numbers of bacteria with depth of sediment is indeed not a new finding (ZoBell and Anderson, 1936); however, there is evidence that the types of aerobic heterotrophs also varies with depth (Sizemore *et al.*, 1973). When multiple samples were taken within 1 cm of each other, the number per gram of sediment varied from 7 to 10% at the most, but even casual examination indicated that the nature of the organisms varied extensively. Likewise, the blending of sediment not only changes the number but also the kinds of bacteria obtained.

ACKNOWLEDGMENTS

This research was supported in part by a grant from the Belle W. Baruch Foundation. We express our gratitude to R. R. Colwell for her generous assistance in the computer analysis of the numerical data.

LITERATURE CITED

Alexander, J. E., R. Hollman, and S. A. Fisher. 1973. The oceanography of Block Island Sound. Part I, Sampling. *In* L. H. Stevenson and R. R. Colwell (eds), Belle W. Baruch Library in Marine Science, Vol. I: Estuarine Microbial Ecology. Univ. South Carolina Press, Columbia, South Carolina.

Colwell, R. R., and W. J. Wiebe. 1970. "Core" characteristics for use in classifying aerobic, heterotrophic bacteria by numerical taxonomy. Bull. Ga. Acad. Sci. 28:165–185.

Holm-Hansen, O. 1969. Determination of microbial biomass in ocean profiles. Limnol. Oceanogr. 14:740–747.

Jannasch, H. W. 1965. Biological significance of bacterial counts in aquatic environments. Proc. Atmos. Biol. Conf., pp. 127–131.

Jannasch, H. W., and G. E. Jones. 1959. Bacterial populations in sea water as determined by different methods of enumeration. Limnol. Oceanogr. 4:128–139.

Meadows, P. S., and J. G. Anderson. 1968. Microorganisms attached to marine sand grains. J. Mar. Biol. Ass. U.K. 48:161–175.

Murchelano, R. A., and C. Brown. 1970. Heterotrophic bacteria in Long Island Sound. Mar. Biol. 7:1–6.

Odum, E. P., and A. A. de la Cruz. 1967. Particulate organic detritus in a Georgia salt marsh-estuarine ecosystem. *In* G. H. Lauff (ed.), Estuaries, pp. 383–388. Amer. Ass. Advan. Sci., Washington, D. C.

Quigley, M. M., and R. R. Colwell. 1968. Properties of bacteria isolated from deep-sea sediments. J. Bacteriol. 95:211–220.

Shewan, J. M. 1963. The differentiation of certain genera of gram-negative bacteria frequently encountered in marine environments. *In* C. H. Oppenheimer (ed.), Symposium on Marine Microbiology, pp. 499–521. Charles C Thomas, Springfield, Illinois.

Sizemore, R. K., L. H. Stevenson, and B. H. Hebeler. 1973. Distribution and activity of proteolytic bacteria in estuarine sediments. *In* L. H. Stevenson and R. R. Colwell (eds), Belle W. Baruch Library in Marine Science, Vol. 1: Estuarine Microbial Ecology. Univ. South Carolina Press, Columbia, South Carolina.

Skerman, T. M. 1963. Nutritional patterns in marine bacterial populations. *In* C. H. Oppenheimer (ed.), Symposium on Marine Microbiology, pp. 685–698. Charles C Thomas, Springfield, Illinois.

Sokal, R. R., and P. H. A. Sneath. 1963. Principles of Numerical Taxonomy. W. H. Freeman, San Francisco, California.

Vernberg, F. J. 1973. Introduction. The estuarine ecosystem. *In* L. H. Stevenson and R. R. Colwell (eds), Belle W. Baruch Library in Marine Science, Vol. 1: Estuarine Microbial Ecology. Univ. South Carolina Press, Columbia, South Carolina.

ZoBell, C. E. 1941. Apparatus for collecting water samples from different depths for bacteriological analysis. J. Mar. Res. 4:173–188.

ZoBell, C. E., and D. Q. Anderson. 1936. Vertical distribution of bacteria in marine sediment. Bull. Amer. Ass. Petrol. Geol. 20:258–269.

ZoBell, C. E., and C. B. Feltham. 1942. The bacterial flora of a marine mud flat as an ecological factor. Ecology 23:69–78.

THE ROLE OF UREA IN MARINE MICROBIAL ECOLOGY*

CHARLES C. REMSEN, EDWARD J. CARPENTER,
and BRIAN W. SCHROEDER

Woods Hole Oceanographic Institution

INTRODUCTION

The availability of nutrients is of prime importance in the regulation of photosynthesis by phytoplankton. Nitrogen has been recognized as the most important nutrient limiting to phytoplankton growth in oceanic and coastal waters (Ryther and Dunstan, 1971). Estimates of the nitrogen available for growth of phytoplankton, however, have usually included measurements of only the inorganic nitrogen present as ammonia, nitrite, and nitrate. In fact, prior to the work of Vaccaro (1963), most measurements of available nitrogen included only nitrate and nitrite. Over the years, however, several workers (Harvey, 1940; Ryther, 1954; Degens *et al.*, 1964) have pointed out the significance of another nitrogen source, urea. Much of the work that we have been doing over the past 2–3 years indicates that urea may be a significant source of combined nitrogen for many marine microbes.

Since urea-decomposing bacteria have been shown to be present in both freshwater and marine environments (ZoBell, 1946), and marine and estuarine phytoplankters have also been shown to utilize urea as a source of nitrogen (Ryther, 1954; Droop, 1960; Syrett, 1962; Guillard, 1963; Carpenter *et al.*, 1972b), our major interests have been to examine the availability of

* Contribution No. 3266 of the Woods Hole Oceanographic Institution.

urea as a substrate for these microorganisms, as well as to elucidate the roles that both groups of microplankton play in the utilization of urea in the sea. In addition, our interests have included an evaluation of the ecological effects of urea enrichment in certain coastal and estuarine environments.

MATERIAL AND METHODS

General Methodology

As a means of examining the respective roles of bacteria and phytoplankton in the decomposition and/or utilization of urea, we developed a relatively simple method for qualitatively separating bacterial activity from phytoplankton activity. Typically, a water sample was treated in the following manner. Samples were collected with 6-liter PVC Niskin bottles and a small portion immediately frozen for nutrient analysis. Half of the remaining sample was then filtered through either a 10- or 20-μm aperture Nitex net, and a sample of this was frozen for nutrient analysis. One-hundred-ml aliquots of the filtered and unfiltered sample were then separately placed in 250-ml Erlenmeyer flasks for urea-decomposition experiments. Other analyses carried out on both filtered and unfiltered fractions included cell counts of urea-decomposing bacteria and counts of phytoplankton species and abundance in preserved samples.

The filter mesh used (10–20 μm) was large enough to allow bacteria to pass, but small enough to capture the larger or chained phytoplankton. Using this simple fractionation technique, differences in the decomposition of urea in filtered and unfiltered aliquots could be ascribed to either smaller microplankton or the larger phytoplankton.

To determine whether the filter did selectively separate some phytoplankton from bacteria, measurements were made of the phytoplankton and urea-decomposing bacterial populations in both filtered and unfiltered samples. Since it is generally felt that healthy, photosynthesizing diatoms are relatively free of bacteria (Droop and Elson, 1966), differences in the rate of urea decomposition between filtered and unfiltered samples could be assigned to differences in the phytoplankton populations (cell and chlorophyll concentrations), providing that the bacterial populations were the same.

Enumeration of Urea-Decomposing Bacteria

One-tenth ml of sample (for both filtered and unfiltered water) was aseptically added to each of five tubes containing 10 ml of sterile nutrient media containing urea. A distilled water medium was arbitrarily used for samples having a salinity of less than 5°/oo , 50% seawater for samples having salinities between 5°/oo and 25°/oo , and full-strength seawater for all samples having salinities greater than 25°/oo. One ml from each of these tubes, after mixing on a vortex mixer, was transferred to a second set of five tubes containing 9 ml of medium, and likewise from the second to a third set. The tubes were allowed to incubate for at least 4–5 days before being examined. A change in color of the phenol red indicator in the medium, that resulted from an increase in pH, was considered a positive result. Most probable numbers (MPN) tables (Hoskins, 1934) were then consulted. According to Prescott *et al.* (1947), for three dilutions of five tubes each, 97% of the determinations will yield an MPN in the range of 0.3–3.6 times the true bacterial density. In other words, if an MPN of 100 bacteria per milliliter is obtained, then there is a 97% probability that the true density is within the range of 28–333 bacteria per milliliter.

Enumeration of Phytoplankton

Samples (filtered and unfiltered) were preserved in Lugol's solution and then settled in 10-ml volume and 25-mm diameter counting chambers. Phytoplankton were counted at 600 X magnification with an inverted microscope using the Utermöhl technique (Lund *et al.*, 1958). The confidence-limit values of Lund *et al.* (1958) were used to determine whether statistically significant differences (95% confidence level) exist between filtered and unfiltered samples for each species.

Determination of Urease Activity in Natural Waters

If the carbon in urea is labeled with ^{14}C, then $^{14}CO_2$ will be released by enzymic hydrolysis and can be measured. Thus, the following method was developed to measure the urease activity in natural waters.

One hundred ml of water (sample) is aseptically placed in a 250-ml wide-mouth Erlenmeyer flask, to which is added 0.1-ml urea solution (0.1 μmole cold urea + 0.4 μCi [^{14}C] urea). A scintillation vial containing a small piece of filter paper moistened with 0.2 ml 10% CO_2-free KOH is suspended in the flask by means of a piece of thin copper wire held in position by a tightly inserted rubber stopper. These flasks are then gently shaken for 4 hr on a laboratory rotator. At the end of this time, the hydrolysis is terminated by the addition of 0.4 ml 1 N HCl to each flask, after which the flask is shaken for an additional 4 hr to allow the KOH-moistened filter paper to absorb the CO_2 produced by the hydrolysis of urea and released on acidification. This CO_2 contains $^{14}CO_2$ from the labeled urea. The scintillation vials are then tightly capped and stored until ready for counting. When ready for counting, 15 ml of Aquasol scintillation fluid is added and the activity measured in a Packard Model 3320 liquid scintillation spectrometer or in a Nuclear Chicago Unilux II liquid scintillation counter. Controls for initial urea activity, for urea breakdown under sterile conditions, and for efficiency of CO_2 capture using $NaH^{14}CO_3$ (ca. 85%) are also run with each set of samples. From these samples and control counts, and from the *in situ* urea concentrations, the rate of urea decomposition is calculated (nmole/liter/hr).

Urea Determination

Urea concentrations in natural waters were determined by the diacetyl-monoxime method of Newell *et al.* (1967). Nitrite was determined by the method of Bendschneider and Robinson (1952), nitrate by the method of Wood *et al.* (1967) using cadmium-copper reduction to nitrite, and ammonia by the method of Solórzano (1969).

RESULTS

Urea Concentration in River, Estuarine, Coastal, and Oceanic Environments

Before we could reasonably investigate the role of urea in the marine environment, it was necessary to demonstrate the presence of urea in signifi-

Table 1. Urea concentration in river, estuarine, coastal and oceanic waters

Sampling site	Date	Urea conc. (μM)	Conc. range (μM)
Hudson River, mouth	Sept. 1969	5.60	—
Raritan River, mouth	Sept. 1969	4.65	—
Mississippi River at Pilottown	Dec. 1969	4.58	—
Savannah River	Mar. 1971	4.02	0.59–8.89
Savannah River, mouth	Mar. 1971	3.57	—
Mississippi River	Dec. 1969	2.81	1.16–7.50
Ogeechee River, Ga.	Mar. 1971	2.74	1.26–4.89
New York Bight	May 1971	2.29	1.00–3.50
Peru, coastal south of Callao	Mar. 1969	1.73	0.87–3.50
Wilmington River, Ga.	Mar. 1971	1.45	1.00–1.91
N. Africa, coastal off Ifni	Oct. 1971	1.29	1.00–1.72
New York Bight	Sept. 1970	1.24	0.53–2.49
S. America, coastal Panama to Ecuador	Mar. 1969	0.98	0.27–2.50
Peru, coastal north of Callao	Mar. 1969	0.92	0.46–1.53
N. Atlantic (48°N between 35° and 52°W)	June 1972	0.61	0.09–1.36
Louisiana, coastal	Dec. 1969	0.57	0.50–0.63
New Jersey, coastal	Sept. 1969	0.43	0.09–0.80
N. Atlantic (between 30° and 40°N and 30° and 13°W)	July 1972	0.28	0.06–0.77
Azores, coastal	July 1972	0.20	0.09–0.39
Sargasso Sea	May 1970	0.20	0.15–0.25

cant concentrations. Table 1 summarizes the data which have been accumulated. Sampling areas are listed in order of decreasing urea concentration; thus the mouth of the Hudson River showed the highest values (5.6 μM) and the Sargasso Sea the lowest (0.2 μM).

As might be expected, known polluted rivers such as the Hudson, Raritan, Mississippi, and Savannah show particularly high urea concentrations. This is probably due in large part to the untreated and primary-treated sewage outfalls which continually pour vast quantities of waste water into the rivers. In addition, drainage from agricultural regions also might contribute heavily if urea-based fertilizers are used. Conversely, the relatively high urea values found off the coast of Peru (1.7 μM) and off the coast of Ifni in North Africa reflect the highly productive upwelling areas. A major source of urea may be the anchovy that are present in high concentrations in these upwelling areas. According to Whitledge and Packard (1971), the Peruvian anchovy, *Engraulis rigens,* excretes urea at the rate of 3.29 μg-atom/g dry wt/hr. As might be expected, urea concentrations in the open North Atlantic and in the Sargasso Sea were low, 0.28 and 0.20 μM, respectively.

The most striking indication of the potential (or actual) importance of urea is shown in a comparison of its concentration with that of combined inorganic nitrogen in coastal water. Table 2 shows average concentrations at various depths between (1) Cape Cod and Cape May in September 1969, and (2) the Canary Islands and Madeira Island in September 1970. Both comparisons illustrate the relative importance of urea when this nitrogen source is compared with ammonia, nitrite, and nitrate, the inorganic sources of nitrogen. For inshore waters (100 fathoms or less) between Cape Code and Cape May, urea concentrations ranged from 10% at 100 m to 72% at 25 m. In off-shore waters, urea ranged from 4% at 1000 m to 62% at 25 m. Urea averaged 56% of the "available" nitrogen in surface waters at the inshore stations and 40% in the offshore stations (Table 2). This rather large reserve of nitrogen may be significant in affecting the growth rates of phytoplankters that have enzyme systems capable of using urea.

For waters around the Canary Islands and Madeira Island the urea concentration remains fairly constant around 1.25 μM. For inshore stations, the concentration was nearly constant along a vertical profile and averaged 1.28 μM. In deeper waters, the vertical distribution of urea is almost homogeneous (averaging 1.25 μM) to a depth of between 100 and 140 m. Below this depth there is an abrupt disappearance of urea, with only occasional low values being recorded. The cause of this abrupt decrease in urea concentration is as yet unknown.

Table 2. Comparison of urea-N concentration with inorganic N sources

		Amount of nitrogen at various depths								
		0 m		25 m		50 m		100 m		1000 m	
Location[a]		μg atom/liter	%	μg atom/liter	%	μg atom/liter	%	μg atom/liter	%	μg atom/liter	%
Site 1											
Inshore	Urea-N	1.32	56.3	1.46	72.2	—	—	1.48	10.6	—	—
	NH₃-N	0.89	38.2	0.45	22.2	—	—	0.49	3.5	—	—
	NO₂-N	0.03	1.6	0.02	0.9	—	—	0.08	0.5	—	—
	NO₃-N	0.09	3.8	0.09	4.4	—	—	11.69	85.4	—	—
Offshore	Urea-N	0.69	39.8	0.91	61.8	0.61	6.7	0.81	5.1	0.71	3.7
	NH₃-N	0.83	47.4	0.46	31.4	0.38	4.1	0.40	2.5	0.85	4.4
	NO₂-N	0.03	1.6	0.04	2.7	0.06	0.6	0.04	0.2	0.03	0.1
	NO₃-N	0.04	2.2	0.06	4.1	8.05	88.6	12.99	93.2	17.57	91.8
Site 2											
Inshore	Urea-N	2.57	57.1	2.66	58.3	2.56	45.6	2.48	24.1	—	—
	NH₃-N	1.60	35.5	1.54	33.8	1.27	22.6	1.16	11.3	—	—
	NO₂-N	0.04	0.8	0.05	1.1	0.11	1.9	0.09	0.9	—	—
	NO₃-N	0.29	6.6	0.31	6.8	1.68	29.9	6.55	63.7	—	—
Offshore	Urea-N	2.59	64.9	2.79	64.4	2.57	65.5	2.61	55.8	0.09	0.6
	NH₃-N	1.16	29.1	1.35	31.2	1.16	29.6	1.19	25.4	0.92	5.8
	NO₂-N	0.04	1.0	0.03	0.7	0.04	1.0	0.08	1.7	0.03	0.2
	NO₃-N	0.20	5.0	0.16	3.7	0.15	3.9	0.83	17.1	14.92	93.4

[a] Site 1, between Cape Cod and Cape May, Sept. 1969. Site 2, between the Canary Islands and Madeira, Sept. 1970.

Urease Activities of Microplankton
in River, Estuarine, Coastal, and Oceanic Waters

As indicated in Table 3, the rate of urea breakdown varies with the locality. Well over an order of magnitude difference in highest and lowest rates of urea decomposition and turnover time were observed. Most rapid urea decomposition (19.4 nM urea/liter/hr) was observed in the Savannah River plume. Slowest was in North Atlantic oceanic waters (0.31 nmole urea/liter/hr). It is probable that the variations in urea decomposition are related to the availability of other nitrogen sources, as well as phytoplankton and bacterial standing crops and other variables.

In the Savannah estuary, we found that about 40–60% of the total urea-decomposing activity could be assigned to a few phytoplankton species. The activity of the phytoplankton population appears to be responsible for a considerable amount of the variation in urease activity from place to place.

Table 3. Urease activity of microplankton in surface waters

Area sampled	Date of sampling	Number of samples	Average urea decomposition (nmole/liter/hr)[a]	Turnover time (days)[b]
Savannah River plume	Mar. 1970	7	19.37 (9.10 − 41.60)	3.2
Mississippi River (lower 100 mi)	Dec. 1969	28	17.70 (5.77 − 26.70)	6.6
Savannah River (lower 30 mi)	Mar. 1970	12	6. 86	24.4
Mississippi River plume	Dec. 1969	34	2. 70 (0.48 − 5.08)	14.4
Coastal Louisiana	Dec. 1969	10	1. 47 (0.88−2.45)	14.5
N. Africal, coastal off Ifni	Oct. 1970	15	0. 96 (0.06 − 2.50)	55.9
N. Atlantic (48°N between 35 and 52°W)	June 1972	3	1. 02 (0.67 − 1.64)	98
N. Atlantic (between 30–40°N and 30–13°W)	July 1972	5	0. 31 (0.20 − 0.52)	59.2

[a] Range of decomposition rates observed.

[b] Turnover times are based only on urea concentrations at stations where urea decomposition was measured. Thus, these data are not comparable with those in Table 1.

Table 4. Effect of filtration on urease activity

Site	Urea concentration	Activity (nmole liter^{-1} hr^{-1})		Activity remaining in 10 μm filtrate
		Unfiltered	Filtered (10 μm)	
River (Savannah-Ogeechee)	3.66	6.44	2.34	36.2
Estuary (Savannah-Ogeechee)	1.95	14.93	4.69	31.4
Coastal (Georgia)	1.45	23.29	3.28	14.1
Oceanic (North Atlantic)	0.45	0.45	0.23	51.1

By filtering portions of the water samples through 10 or 10 μm Nitex nets a varying amount of the activity could be removed. Table 4 shows the effect of filtration on urease activity present in selected waters. As can be seen, in coastal waters off Georgia 86% of the activity is removed by filtering through a 10–20 μm net. On the other hand, over 50% of the activity remains in similarly treated water of the open North Atlantic.

Table 5. Effect of filtration on MPN for urea-decomposing bacteria in river, estuarine, and coastal waters of Georgia

Station no.	Depth (m)	Unfiltered cells ($\times 10^5$)	Filtered cells ($\times 10^5$)
River (Savannah)			
2	0	3.5	3.5
	4		1.3
3	0	3.5	3.5
	4	5.4	3.5
4	0	1.4	2.3
	4	2.4	5.4
5	0	3.5	3.5
	4	5.4	5.4
	10	2.4	3.5
6	0	3.5	1.7
	4	5.4	5.4
Estuarine			
10	0	>16	>16
	10	9.2	9.2
13	0	>16	>16
	8	9.2	9.2
16	0	16	0.2
	4	2.4	5.4
Coastal			
32	0	1.4	2.4
33	0	1.1	5.4
34	0	9.2	5.4
55	0	0.02	0.35
River (Ogeechee)			
58	0	0.08	0.08
59	0	0.08	0.08
60	0	0.05	0.1
63	0	0.04	0.04

Filtration usually had little measurable effect on the cell concentration in freshwater, but removed about four to five species at each brackish water station. This is probably because species of diatoms with large cross-sectional areas (*Chaetoceros, Rhizosolenia,* etc.) were abundant in brackish water, whereas in freshwater small chlorophyte algae tended to dominate.

In the river, estuarine, and coastal waters studied, almost all of the species that were removed by filtration were diatoms. Twelve diatom species, belonging to eight genera, *Melosira, Asterionella, Thalassiosira, Stephanopyxis, Chaetoceros, Rhizosolenia, Bacteriastrum* and *Cyclotella,* were removed. In addition, the green algae *Ankistrodesmus falcatus* and *Dictyosphaerium* sp. were removed along with the dinoflagellate *Prorocentrum minimum.* From 38 to 100% of the cells of each of these species were removed by filtration, with an average of 88%. Details on numbers and species can be found in another paper (Remsen *et al.,* 1972).

Urea-Decomposing Bacteria

There was no significant difference in the concentration of urea-decomposing bacteria between filtered and unfiltered samples of water (Table 5); thus any differences in the rate of urea decomposition between filtered and unfiltered samples could be assigned to differences in the phytoplankton populations.

The highest concentration of urea-decomposing bacteria was found in estuarine waters where counts of about 10^6 cells/liter were common. In river water, the concentration of these bacteria ranged from a low of 4×10^3 cells/liter in the Ogeechee River to a high of 1.6×10^6 cells/liter in the Savannah River, the latter being quite obviously polluted.

The concentration of urea-decomposing bacteria in waters between the Canary Islands and Madeira Island remained fairly constant around $3-4 \times 10^3$ cells/liter (Table 6). Little variation was found in these numbers with increased depth, at least in the upper 100 m.

DISCUSSION

The data we have presented, especially in Table 1, show quite clearly that urea is present in natural waters in biologically significant quantities. Where does this urea originate?

Source of Urea in Natural Waters

Urea is excreted into the marine environment as an end product of nitrogen metabolism by many higher organisms and as a microbial degradation product of amino acids, purines, and pyrimidines. The organisms responsible for the excretion of urea may live in the ocean, rivers, or on land.

Table 6. Urea-decomposing bacteria in waters between the Canary Islands and Madeira Island

Depth (m)	No. of urea-decomposing bacteria ($\times 10^3$/liter) Station No.												
	1741	1742	1743	1744	1745	1746	1747	1748	1749	1750	1751	1752	1753
0	3.5	2.3	5.4	0.6	5.4	3.5	—	5.4	16	3.5	9.2	9.2	0.5
10	6.4	1.7	9.2	0.3	9.2	1.7	16	16	1.4	9.2	9.2	3.5	0.5
25	0.5	0.3	—	0.6	1.1	0.9	9.2	5.4	1.5	3.5	1.6	2.9	0.6
50	3.5	2.4	—	2.2	1.8	2.2	—	11.0	5.4	5.4	1.1	3.5	0.5
75	5.4	0.5	3.5	0.2	0.3	2.1	5.4	5.4	5.0	2.4	3.6	0.7	1.9
100	9.2	0.2	2.4	0.5	2.4	5.4	9.2	5.4	3.5	0.4	0.8	0.7	7.9

In the ocean up to 14% of the total nitrogen excreted by shrimp, crayfish, and mussels is as urea (Parry, 1960). Corner and Newell (1967) have reported that as much as 10% of the total nitrogen excreted by *Calanus helgolandicus* may be in the form of urea. Marine crustaceans also excrete dissolved amino acids such as arginine which are capable of being utilized by other organisms and being converted to urea (Webb and Johannes, 1969). The bony or teleostean fish excrete 5–10% of their total nitrogen as urea, while the cartilaginous fish (elasmobrachs) excrete up to 70–80% of their total nitrogen in the form of urea. This difference is reflected in the amount of urea present in the body tissue, 2% in the elasmobranch fish, and 0.2% in the teleostean fish (cf. Huggins *et al.,* 1969).

Similarly, terrestrial amphipods excrete significant quantities of urea which may eventually work its way into rivers. Amphibians, such as Anura (frogs) and Urodela (salamanders) excrete urea; Apoda (caecilians) excrete 66% of their total nitrogen in the form of urea (Carlisky *et al.,* 1969).

Finally, and probably quite important ecologically, is the fact that man is a prime source of urea along with most of the terrestrial vertebrates. Man can excrete up to 35 g urea per day. A considerable amount of this reaches the sea by domestic sewage discharge.

A large number of other organic nitrogen compounds is present in oceanic water. In localized areas the concentrations of these compounds may be quite high. For example, in the rich upwelling waters off the coast of Peru there are huge quantities of bony fish (anchovies) and a large number of guano-producing birds. Both the fish and the birds excrete large amounts of uric acid as an end product in nitrogen metabolism. By the same token, the duck farms located around Great South Bay, Long Island, have resulted in a very high organic nitrogen content, particularly uric acid, in the bay water from the duck wastes. Through the action of several enzyme systems, uric acid may be broken down to urea.

(a) Cellular decomposition:

xanthine $\longrightarrow$ uric acid $\longrightarrow$ allantoin $\longrightarrow$ allantoic acid $\longrightarrow$ urea
      (xanthine   (uricase)   (allantoinase)   (allantoicase)
      oxidase)

(b) excretion:

urea $\longrightarrow$ ammonia + CO_2
   (urease)

uric acid ⟶ allantoin ⟶   allantoic acid ⟶   urea
  (uricase)   (allantoinase)   (allantoicase)

arginine ⟶   urea + ornithine.
  (arginase)

Most algae are capable of attacking uric acid and producing urea from it; some algae carry this activity to its conclusion and are able to use urea as a source of nitrogen (McKee, 1962). However, it is reasonable to assume that urea levels will be higher in areas of uric-acid excretion.

Some amino acids can be converted to other compounds with the release of urea. Marine plankton are notable high arginine excretors (Smith and Young, 1955) due largely to the fact that they contain high concentrations of arginine (Degens et al., 1968) in their tissues. Arginine may be converted to ornithine plus urea by the action of arginase, an enzyme present in mammalian tissue, invertebrates, most higher plants, and in a number of different marine algae (Smith and Young, 1955).

Results of Siegel and Degens (1966) showing high arginine and low ornithine concentrations in particulate matter, and high ornithine plus urea in the dissolved material, suggest that the production of urea by this mechanism is plausible. Other amino acids can be similarly acted upon. For example, canavanine can be converted to canaline with the release of urea. Finally, the action of bacteria, fungi, and yeasts on detrital material eventually leads to the production of ammonia and/or in many cases urea.

Fate of Urea in Natural Waters and the Relative Role of Bacteria and Phytoplankton

From the data presented, urea is (a) an available source of nitrogen for phytoplankton growth, and (b) is decomposed to ammonia and CO_2 by urea-decomposing bacteria. The specific role(s) of bacteria and phytoplankton in the utilization and/or decomposition of urea varies with the water area being investigated and can be determined when the previously described filtration method is used.

In river, estuarine, and coastal waters of Georgia, phytoplankton were responsible for the major part of urea being broken down to ammonia and CO_2. Filtration of this water through a 10- to 20-μm net significantly altered the phytoplankton cell concentrations and urea decomposition rates. Approximately 73% of the activity (urease, urea $\rightarrow$ ammonia + CO_2) was removed, indicating that most of the activity was associated with the larger centric and pennate diatoms.

The diatom species removed by filtration are commonly found in other east-coast estuaries (Conover, 1956; Griffith, 1961). One of the species, *Stephanopyxis costata* (*Skeletonema costatum*) (Hustedt, 1956), is the most abundant diatom in the coastal waters of the eastern United States (Riley, 1967), and it has previously been shown to break down urea in laboratory studies (Guillard, 1963; Carpenter *et al.*, 1972*a*).

It might be argued that urea-decomposing bacteria were on the surfaces of the phytoplankton, thereby negating the basis for the filtration. However, if this were true we should have seen differences in the MPN analyses. Furthermore, previous data indicate that healthy pelagic diatoms, living in coastal marine waters, are virtually free of bacteria (Droop and Elson, 1966).

There was considerable variation in the urea decomposition rates from one area to another. While there was an overall range of only 16-fold in the *in situ* concentration of urea, there was a 200-fold range in decomposition rates. The only parameter to which these rates correlated was salinity. The average rate of urea decomposition in brackish waters was almost three times that in freshwater. Since the *in situ* concentration of urea in brackish waters was half that in freshwater, this meant that the turnover time for urea in brackish water was six times that in freshwater. As a result, one might surmise that the efficacy of urea-decomposition products (as a potential substrate) are more significant for brackish water and marine phytoplankton than for freshwater species.

In the case of water samples from the North Atlantic, bacteria seem to play a much more important part. While the absolute activity is comparatively low (0.45 nmole/liter/hr *versus* 23.99 nmole/liter/hr) the amount of activity remaining in the 10-μm filtrate is comparatively higher (51% *versus* 14%).

The relative importance of bacteria in the utilization of dissolved organic compounds in the open ocean is not surprising, albeit the actual levels of activity are quite low. What was surprising was the relative unimportance of bacteria in the estuarine and coastal waters. Our data appear to be contrary to the conclusions of a number of other investigators (Fred *et al*, 1924; Henrici, 1937; Waksman, 1941; ZoBell, 1946; Potter and Baker, 1961) that bacteria in both freshwater and marine environments are the most important agents in the utilization of dissolved organic compounds.

Hobbie and Wright (1965) have commented that "any competition between algae and bacteria for dissolved organic compounds appear to favor the bacteria, at least at the low substrate concentrations found in nature." The decomposition of urea in an estuarine and coastal environment would appear to be an exception to this general rule.

Urea as a Sole Source of Nitrogen for Marine Phytoplankton

It has been shown that some marine phytoplankters can grow on urea as a sole source of nitrogen (Ryther, 1954; Droop, 1960; Syrett, 1962; Guillard, 1963). Of the twelve coastal and estuarine species that Guillard (1963) tested, six could grow on urea and six could not. It is possible that changes in the concentration of urea can alter the species composition of phytoplankton. It has already been shown that some phytoplankters can grow on urea as a sole nitrogen source while others cannot (Table 7). Obviously, this alone can influence species composition since, if nitrogen is limiting, the presence of urea in sufficient quantities will increase the growth rates of those species that can use it while not affecting those that cannot. Remsen (1971) observed that, in September, urea constitutes up to 72% of the available nitrogen (NO_2, NO_3, NH_3, urea) in New York Bight waters, and this suggests that its influence on species composition could be considerable.

It is already known that *Stephanopyxis costata* can grow as well on urea as on NO_3 and NH_3 (Guillard, 1963; Carpenter *et al.,* 1972b). However, these growth measurements were made at very high concentrations of urea, NO_3,

Table 7. Algae capable of using urea as a sole N source

Haematococcus plurialis (Droop, 1960)
Chlamydomonas pulsatilla (Droop, 1960)
Chlamydomonas reinhardii (Sager and Granik, 1953)
Chlorella ellipsoida (Hattori, 1958)
Chlorella vulgaris (Arnow, Oleson, and Williams, 1953)
Chlorella pyrenoidosa (Samejima and Myers, 1958)
Scenedesmus obliques (Ichioka and Arnon, 1955)
Nannochloris atomus (Ryther, 1954)
Nannochloris oculata (Droop, 1955)
Nannochloris sp. (Thomas, 1966)
Gymnodinium simplex (Thomas, 1966)
Chaetoceros gracilis (Thomas, 1966)
Stichococcus sp. (Ryther, 1954)
Cyclotella nana (Guillard, 1963)
Cyclotella caspia (Guillard, 1963)
Phaeodactylum tricornutum (Hayward, 1965)
Skeletonema sp. (Guillard, 1963)
Stephanopyxis costata (*Skeletonema costatum*) (Guillard, 1963)
Thalassiosira nordenskioldii (Guillard, 1963)
Anabaena variabilis (Kratz and Myers, 1955)
Tribonema aequale (Belcher and Fogg, 1958)
Rhizosolenia setigera (Guillard, 1963)

and NH_3. Data from a urea uptake kinetics study (Carpenter et al., 1972) indicated that, at the average urea concentrations in surface water between Cape Cod and Cape May in September (0.65 μM), *Stephanopyxis costata* could take up enough urea to double its nitrogen content, and thus divide, every 3.3 days. This division rate is approximately the same as that reported for average division rates of coastal phytoplankters by McAllister et al. (1961).

In summary, the data presented here indicate that (1) urea is present in biologically significant amounts in most natural waters, (2) phytoplankton play a more dominant role than bacteria in the utilization and/or decomposition of urea in coastal waters, while bacteria play a dominant role in the breakdown of urea in the open ocean, (3) the rates of urea breakdown are far greater in coastal waters than in oceanic waters, and (4) since the ability to use urea is not present in all phytoplankton species, it is possible that the presence of urea can exert considerable influence on phytoplankton species composition.

ACKNOWLEDGMENTS

The authors wish to express their appreciation to Mrs. Linda Graham and Mrs. Frederica Valois for their technical assistance in portions of the research summarized in this paper, and to Dr. Stanley Watson for his encouragement and interest over the years. This work was partly supported by a Public Health Service Grant GM16754 from the National Institute of General Medical Sciences, and by Contract N00014-69-0184 of the Office of Naval Research.

We are particularly indebted to the masters and crews of the research vessels *Atlantis II, Chain,* and *Gosnold* for their enthusiastic support of our research efforts.

LITERATURE CITED

Arnow, P., J. J. Oleson, and J. H. Williams. 1953. The effects of arginine on the nutrition of *Chlorella vulgans.* Amer. J. Bot. 40:100–103.

Belcher, J. H., and G. E. Fogg. 1958. Studies on the growth of Xanthophyceae in pure culture. III. *Tribonema aequale* Pascher. Arch. Microbiol. 30:17–22.

Bendschneider, K., and R. J. Robinson. 1952. A new spectrophotometric determination of nitrite in sea water. J. Mar. Res. 11:87–96.

Carlisky, N. J., A. Bamio, and L. I. Sadnik. 1969. Urea biosynthesis and excretion in the legless amphibian *Chtonerpeton indestructum* (Apoda). Comp. Biochem. Physiol. 29:1259–1262.

Carpenter, E. J., C. C. Remsen, and B. W. Schroeder. 1972a. Comparison of laboratory and *in situ* measurements of urea decomposition by a marine diatom. J. Exp. Mar. Biol. Ecol. 8: 259–264.

Carpenter, E. J., C. C. Remsen, and S. W. Watson. 1972b. Utilization of urea by some marine phytoplankters. Limnol. Oceanogr. 17:265–269.

Conover, S. A. M. 1956. Oceanography of Long Island, 1952–1954. Bull. Bingham Oceanogr. Coll. XV: 62–111.

Corner, E. D. S., and B. S. Newell. 1967. On the nutrition and metabolism of zooplankton. IV. The forms of nitrogen excreted by *Calanus*. J. Mar. Biol. Ass. U.K. 47:113–120.

Degens, E. T., M. Behrendt, B. Gotthardt, and E. Reppmann. 1968. Metabolic fractionation of carbon isotopes in marine plankton. II. Data on samples collected off the coasts of Peru and Ecuador. Deep-Sea Res. 18:11–20.

Degens, E. T., J. H. Reuter, and K. T. Shaw. 1964. Biochemical compounds in offshore California sediments and seawater. Geochim. Cosmochim. Acta 28:45–65.

Droop, M. R. 1955. Some new supra-littoral protista. J. Mar. Biol. Ass. U.K. 34:233–245.

Droop, M. R. 1960. *Haematococcus plurialis* and its allies. III. Organic nutrition. Rev. Algol. 5:247–259.

Droop, M. R., and K. G. R. Elson. 1966. Are pelagic diatoms free from bacteria? Nature (London) 211:1096–1097.

Fred, E. B., F. C. Wilson, and A. Davenport. 1924. The distribution and significance of bacteria in Lake Mendota. Ecology 5:322–339.

Griffith, R. E. 1961. Phytoplankton of Chesapeake Bay. Hood College Monograph No. 1, Frederick, Maryland.

Guillard, R. R. L. 1963. Organic sources of nitrogen for marine centric diatoms. *In* C. H. Oppenheimer (ed.), Symposium on Marine Microbiology, pp. 93–104. Charles C Thomas, Springfield, Illinois.

Harvey, H. W. 1940. Nitrogen and phosphorus required for the growth of phytoplankton. J. Mar. Biol. Ass. U.K. 24:115–123.

Hattori, A. 1958. Studies on the metabolism of urea and other nitrogenous compounds in *Chlorella allipsoidea*. II. Changes in levels of amino acids and amides during the assimilation of ammonia and urea by nitrogen starved cells. J. Biochem. Tokyo 45:57–64.

Hayward, J. 1965. Studies on the growth of *Phaeodactylum tricornutum* (Bohlin). 1. The effect of certain organic nitrogenous substances on growth. Physiol. Plant. 18:201–207.

Henrici, A. T. 1937. Studies of freshwater bacteria. IV. Seasonal fluctuations of lake bacteria in relation to plankton production. J. Bacteriol. 35:129–139.

Hobbie, J. E., and R. T. Wright. 1965. Competition between planktonic bacteria and algae for organic solutes. Mem. Ist. Ital. Idrobiol. 18 Suppl.: 175–185.

Hoskins, J. K. 1934. Most probable numbers for evaluation of coliaerogenes tests by fermentation tube method. Pub. Health Rep. 49:393–405.

Huggins, A. K., G. Skutsch, and E. Baldwin. 1969. Ornithine-urea cycle enzymes in teleostean fish. Comp. Biochem. Physiol. 28:587–602.

Hustedt, F. 1956. Diatomeen aus dem Lago de Maracaibo in Venezuela. *In* F. Gessen and V. Vareschi (eds), Ergebnisse der Deutschen Limnologischen Venezúela–Expedition 1952. Vol. 1, pp. 93–140. V.E.B. Deutscher Verlag der Wissen, Berlin.

Ichioka, P., and D. Arnon. 1955. Molybdenum relation to nitrogen metabolism. II. Assimilation of ammonia and urea without molybdenum by *Scenedesmus*. Physiol. Plant. 8:552–560.

Kratz, W. A., and J. Myers. 1955. Nutrition and growth of several blue-green algae. Amer. J. Bot. 42:282–287.

Lund, J. W. G., C. Kipling, and E. D. LeCren. 1958. The inverted microscope method of estimating algal numbers and the statistical basis of estimations by counting. Hydrobiology 11:143–170.

McAllister, C. D., T. R. Parsons, K. Stephens, and J. D. H. Strickland. 1961. Measurements of primary production in coastal sea water using a large-volume plastic sphere. Limnol. Oceanogr. 6(3):237–258.

McKee, H. S. 1962. Nitrogen Metabolism in Plants. Clarendon Press, Oxford.

Newell, B. S., B. Morgan, and J. Cundy. 1967. The determination of urea in sea water. J. Mar. Res. 25:201–202.

Parry, G. 1960. Physiology of Crustacea. Academic Press, New York.

Potter, L. F., and G. E. Baker. 1961. The microbiology of Flathead and Rogers Lakes, Montana. II. Vertical distribution of the microbial populations and chemical analyses of their environments. Ecology 42:338–348.

Prescott, S. C., C-E. A. Winslow, and M. H. McCrady. 1947. Water Bacteriology, 6th edn., Chap. VII, pp. 134–138. John Wiley, New York.

Remsen, C. C. 1971. The distribution of urea in coastal and oceanic waters. Limnol. Oceanogr. 16:732–740.

Remsen, C. C., E. J. Carpenter, and B. W. Schroeder. 1972. Competition for urea among estuarine microorganisms. Ecology 53:921–926.

Riley, G. A. 1967. The plankton of estuaries. *In* G. Lauf (ed.), Estuaries, pp. 316–326. Amer. Ass. Advan. Sci., Washington, D. C.

Ryther, J. H. 1954. The ecology of phytoplankton blooms in Moriches Bay and Great South Bay, Long Island, New York. Biol. Bull 06:198–209.

Ryther, J. H., and W. M. Dunstan. 1971. Nitrogen, phosphorus and eutrophication in the coastal marine environment. Science 171:1008–1013.

Sager, R., and S. Granik. 1953. Nutritional studies with *Chlamydomonas reinhardii*. Ann. N.Y. Acad. Sci. 56:831–838.

Samejima, H., and J. Myers. 1958. On the heterotrophic growth of *Chlorella pyrenoidosa*. J. Gen. Microbiol. 18:107–117.

Siegel, A., and E. T. Degens. 1966. Concentration of dissolved amino acids from saline waters by ligand-exchange chromatography. Science 151:1098–1101.

Smith, D. G., and E. G. Young. 1955. The combined amino acids in several species of marine algae. J. Biol. Chem. 217:845–853.

Solórzano, L. 1969. Determination of ammonia in natural waters by the phenol-hypochlorite method. Limnol. Oceanogr. 14:799–801.

Syrett, P. J. 1962. Nitrogen assimilation. *In* R. A. Lewin (ed.), Physiology and Biochemistry of Algae, pp. 171–188. Academic Press, London.

Thomas, W. H. 1966. Surface nitrogenous nutrients and phytoplankton in the northeastern tropical Pacific Ocean. Limnol. Oceanogr. 11:393–400.

Vaccaro, R. F. 1963. Available nitrogen and phosphorus and the biochemical cycle in the Atlantic off New England. J. Mar. Res. 21:284–301.

Waksman, S. A. 1941. Aquatic bacteria in relation to the cycle of organic matter in lakes. *In* A Symposium on Hydrobiology, pp. 86–105. University of Wisconsin Press, Madison, Wisconsin.

Webb, K. L., and R. E. Johannes. 1969. Do marine crustaceans release dissolved amino acids? Comp. Biochem. Physiol. 29:875–878.

Whitledge, T. E., and T. T. Packard. 1971. Nutrient excretion by anchovies and zooplankton in Pacific upwelling regions. Invest. Pesq. 35:243–250.

Wood, E. D., F. A. J. Armstrong, and F. A. Richards. 1967. Determination of nitrate in sea water by cadmium-copper reduction to nitrite. J. Mar. Biol. Ass. U.K. 47:23–31.

ZoBell, C. E. 1946. Marine Microbiology. Chronica Botanica, Waltham, Massachusetts.

RESPONSES BY OPEN-OCEAN MICROORGANISMS TO ENVIRONMENTAL POLLUTION*

CHARLES C. REMSEN, V. T. BOWEN, and S. HONJO

Woods Hole Oceanographic Institution

INTRODUCTION

It has been generally concluded that many pollutants, notably DDT and the polychlorinated biphenyls (PCB'S), have major pathways of distribution via the atmosphere. This conclusion is based on data from the distribution of their concentrations among a wide variety of open-ocean biota, from the way these distributions change geographically, and from our knowledge of their chemistry and geochemistry. This fact suggests to us that such pollutants as DDT and PCB's may reach levels that are damaging to the open-ocean biota without necessarily first affecting the biota of near-shore waters. This prediction is a source of great concern with respect to the future status of most of the organisms of the open ocean, and of those others whose life histories include pelagic larval stages. Furthermore, we believe that special concern should be felt for the marine bacteria, phytoplankton and planktonic protozoa that form the basis of the food webs of the oceans. The reasons for this special concern are as follows.

a. If these populations were seriously interfered with, the ocean as a biological-geochemical system would almost certainly change in all its properties.

* Contribution No. 3267 of the Woods Hole Oceanographic Institution.

b. The physical-chemical nature of DDT and PCB's, coupled with the surface-active properites of the majority of these microorganisms, leads us to suppose that the microorganisms must suffer high exposures to pollutants.

c. Many of these organisms are difficult to maintain in culture, and hence to study experimentally. Their small size also appears to make impossible analytical assessments of their body burdens of chemical pollutants. Because of the difficulty in assessing the impact of pollutants in these species, the problem is in danger of suffering from neglect in spite of its importance.

We have recently obtained funds from the Sarah Mellon Scaife Foundation to begin a study on the responses of open-ocean microorganisms to environmental pollution. The bulk of this presentation will, by necessity, be concerned with how we intend to approach the problem; however, some preliminary data are available and will be discussed.

GENERAL APPROACH TO THE PROBLEM

By combining the approaches of population ecology, physiology, and cyto-chemistry along with cytomorphology, it will be possible to establish the present status of selected populations of open-ocean microorganisms in areas of the North and South Atlantic and Mediterranean, selected for measurable or predictable differences in rates of delivery of atmospherically distributed chemical pollutants. It will be necessary to study both field populations and cultured species whenever possible, to confirm the relationship between specific chemical, physiological, or morphological lesions and specific sub-lethal levels of pollutants. The microorganisms that we will study and the reasons for selecting them are as follows.

a. The film- and floc-forming marine bacteria. Since these can be easily collected and since their association with surfaces should make them subject to the effects of surface-active pollutants, bacteria and fungi associated with surface films will be studied in great detail.

b. The smaller marine phytoplankton. Because of their over-riding importance to the biological systems of the open ocean, the availability of many species in culture, and the large amount of data available concerning their population dynamics in many ocean areas, diatoms and cocco-lithophorids will be primary targets for investigation.

c. The Acantharia. These strontium sulfate skeleton forming planktonic protozoa will be studied because of their very strong surface-active properties, their general abundance in the open ocean, and the large amount of data available concerning their population dynamics.

d. The Foraminifera. The planktonic protozoa forming calcium carbonate skeletons will also be examined because of the interest in their apparently quite different surface properties, because they are basic to the Ca^{2+} and Mg^{2+} cycles of the oceans, and because they consist of populations of relatively few species of extensive geographic range, and have been very extensively studied in other contexts.

RESEARCH PLAN AND PRELIMINARY OBSERVATIONS

In terms of our own research, the studies we plan fall into three main categories: (a) population ecology; (b) cell chemistry and structure of natural populations of microorganisms; (c) experimental populations exposed to sublethal levels of pollutants such as PCB's and DDT, uptake studies using isotope-labeled compounds, and study of the changes in cell chemistry and structure of exposed populations.

It should be emphasized that the interpretation of the data from the first two categories of study, and the selection of exposure levels in the third category, all depend on information now at hand, or to be collected, as results of analytical studies of chemical pollutants in open-ocean organisms. From these studies (Harvey et al., 1972; Grice et al., 1972; Harvey et al., 1974) we can with some assurance select sampling areas that should yield populations whose exposure histories to DDT, PCB's or other atmospherically borne pollutants differ over a range of an order of magnitude or more. The program that we have begun would be quite impossible in the absence of this sort of data on exposure levels.

Population Ecology

We know a good deal about the history of the sizes and the species composition of populations of phytoplankton, planktonic *Foraminifera*, and *Acantharia* in many parts of the Atlantic Ocean and Caribbean and Mediterranean Seas; in some areas our information extends back to the late 1950's. It seems to us that collection of comparable information about the marine film-forming bacteria—as the group of bacteria most easily collected by standardizable methods—would be very useful. We wish especially to examine each of these groups of populations in areas we know to be of high, or low, exposure to air-borne pollutants to see whether changes are appearing either in total number of individuals or in the relative abundance of various species. Such data are also needed as the best means of monitoring the open ocean for small changes as a result of man's activities.

Cell Chemistry and Structure of Natural Populations

A large fraction of this phase of our study, already started with a cruise in the North Atlantic this past summer (1972), will be done at sea. Immediately after the collection of organisms, the living specimens—identified as to species—will be fixed and mounted for examination by the electron microscope, both scanning and transmission, and for chemical analysis by non-dispersive X-ray emission spectrometry. Specimens will also be embedded for sectioning, or sectioned while frozen, for the same purposes. Other specimens will be treated histochemically to reveal the intracellular distributions of various enzyme activities. Since, as noted above, these specimens will be drawn from populations having very different exposure histories to atmospherically borne pollutants, it will be possible to correlate any differences found in their cytomorphology or cytophysiology with the differences in pollutant exposure, as well as the effects of such pollutants observed in experimental populations.

Experimental Populations Exposed to Sub-lethal Levels of Pollutants

Experimental studies, as we noted above, are possible at present only on a small fraction of the species in which we are interested. These species fall into two classes: those that can be maintained in laboratory culture (a group that at present includes *only some* of the open-ocean phytoplankton and some of the film- and floc-forming bacteria), and those that can be maintained in healthy condition for short periods (a group that includes some species of each of the kinds of organisms we wish to study). In the latter case we will have to undertake ship-board experiments on freshly collected specimens.

In general we plan experiments of two kinds: first, the localization of environmental pollutants in microorganisms; second, to determine the effect(s) of the pollutant on the life processes of these microorganisms.

Localization of Chlorinated Hydrocarbons in Plant Microorganisms. By exposing a variety of marine microorganisms to isotopically labeled DDT, dieldrin, aldrin, and PCB (either ^{3}H or ^{14}C), we hope to be able to determine the *in vivo* localization of these pollutants. Once exposed to the isotope, the organism will be chemically stabilized and prepared for autoradiography. Ultrathin sections are prepared by standard techniques; a film emulsion is placed over the section and exposed to the decays of the isotope. After an appropriate exposure time, normally 2–4 weeks at 4°C, the film is developed (while still remaining attached to the thin section) and the entire specimen is examined in the TEM. Silver grains appear over that portion of the cell in which the isotope is concentrated (Yamamoto and Shahrabadi, 1971).

Localization and Identification of Heavy-Metal Concentration in Micro-organisms. A similar experiment to that just described will be carried out in order to determine the *in vivo* localization of heavy metals. Thin section microanalysis can detect and locate intracellular elements with increased sensitivity over cytochemical and histochemical techniques. Abnormal concentrations of toxic elements such as lead, cadmium, nickel, mercury, beryllium, bismuth, antimony, and tin present in phytoplankton and zooplankton can be determined in a SEM equipped with a wavelength-dispersive X-ray spectrometer.

The Effect of Pollutants on the Life Processes of Organisms of Different Trophic Levels. Recently it has been shown that DDT binds to lecithin, a major component of cellular membrane systems (Tinsley *et al.*, 1971). The interactions between such membrane systems as mitochondria and chloroplasts and chlorinated hydrocarbons, can be carefully examined by a variety of electron-microscopic techniques.

Of particular interest will be the interaction of DDT and the cell membrane. Some evidence has been given to show that DDT inhibits K^+-, Na^+-, Mg^{2+}-adenosine triphosphatase (ATPase) (Janicki and Kenter, 1971). Since these enzymes appear to function in osmoregulatory processes, one effect of DDT and other chlorinated hydrocarbons may be to inhibit sodium transport across membranes. If one of the results of this inhibition is the accumulation of cations within a cellular organelle, the analytical electron microscope can detect both the binding sites and concentration of cation.

The chlorinated hydrocarbons have also been implicated as inhibitors of SH-containing enzymes. A variety of cytochemical methods are available to the electron microscopists to determine the presence and activities of many of these enzymes in marine microorganisms.

The action of dieldrin, other chlorinated hydrocarbons, and heavy metals on photosynthesis in phytoplankton will be another subject of investigation (Harriss *et al.*, 1970). Dieldrin is particularly interesting because of its relatively high solubility in water (DDT, 1.2 ppb; aldrin, 27 ppb; dieldrin, 186 ppb). Laboratory experiments will be conducted to see what effects these pollutants have on the function and structure of the photosynthetic membranes. Since inhibition of the electron-transport system has been implicated as an effect of both DDT and DDE, cytochemical methods will be used to measure the presence and activities of those enzymes involved in electron transport (Bowes and Gee, 1971).

MacFarlane *et al.* (1972) studied the interaction of light intensity and DDT concentration upon the marine diatom *Nitzchia delicatessima.* A consistent reduction in carbon fixation and chlorophyll-*a* per cell, over controls, in a 24-hr period was observed with increasing DDT concentrations between 9.4 and 1000 ppb. At 100 ppb, carbon fixation per cell was reduced as much as 94%. Distortion of the chloroplasts in the cells exposed to DDT was observed even in cells exposed to as little as 9.4 ppb. This is particularly significant since Seba and Corcoran (1969) have shown that surface slicks contain as much as 3.46 ppb DDT, or slightly less than one-third the concentration used to demonstrate drastic morphological distortion of chloroplasts.

THE CONCENTRATION OF CHLORINATED
HYDROCARBON PESTICIDES IN THE OPEN OCEAN

In their report to the National Academy of Sciences in 1971, the Panel* on Monitoring Persistent Pesticides in the Marine Environment concluded that "the oceans are an ultimate accumulation site for the persistent, chlorinated hydrocarbons" and that "as much as 25 percent of the DDT compounds produced to date may have been transferred to the sea". According to Seba and Corcoran (1969) the concentration of DDT in surface slicks may reach as high as 3.4 ppb. During a recent cruise in the North Atlantic (R/V *Chain* No. 105) concentration of polychlorinated biphenyls (PCB's) as high as 150 ppt were recorded in surface waters of the open ocean (George Harvey, personal communication).

Certainly these levels of concentration are sufficiently high to expect some effect on the population of microorganisms associated with the surface waters of the open ocean.

ROLE OF BACTERIA AND FUNGI IN THE UPTAKE
AND TRANSFER OF ENVIRONMENTAL POLLUTANTS

Bacteria and fungi may play prominent roles in the uptake and concentration of heavy metals and chlorinated hydrocarbons in the oceanic environment. In our work we plan to study the floc-forming bacteria indigenous to natural waters as well as those other bacteria and fungi associated with the surface slick, or film, on water. Surface slicks are notably high in organic material and act as concentrators of pesticides and other environmental pollutants (Seba and Corcoran, 1969).

One physical-biological relationship which we believe to be quite significant is the association of pesticides and PCB's with microscopic particulate

* Edward D. Goldberg, Philip Butler, Paul Meier, David Menzel, Robert W. Risebrough, and Lucille F. Stickel.

matter in oceanic water. Pfister *et al.* (1969) were able to show that pesticides are held in suspension via adsorption to very small particles, and that these particles can, in turn, be adsorbed by microorganisms such as bacteria and algae.

Studies on the floc-forming bacteria have shown that the exocellular polymers formed by these bacteria possess polyelectrolyte properties and are able to bind high concentrations of metallic ions (Friedman and Dugan, 1968; Friedman *et al.*, 1969). Leshniowsky *et al.* (1970) at Ohio State University have shown that floc-forming bacteria isolated from Lake Erie adsorb and concentrate aldrin from colloidal dispersion so that the settling of the bacterial flocs removes aldrin from the water phase. The microparticulate fraction which has been shown to be relatively high in chlorinated pesticides, including DDT, may consist, to a large extent, of floc-forming bacteria (Pfister *et al.*, 1969).

SENSITIVITY OF OPEN-OCEAN
MICROORGANISMS TO ENVIRONMENTAL POLLUTION

We have recently completed a study comparing the sensitivity of diatoms to PCB's. Mr. Nicholas Fisher, a graduate student at Stony Brook in residence with us at Woods Hole, has compared the sensitivity of oceanic and coastal clones of the same species of diatoms. Cells were grown in batch culture starting with low cell densities (5000 cells per milliliter) and grown to the stationary phase, approximately 5 days later. Each experiment was replicated three times. Table 1 compares the results for 3 clones of *Thalassiosira pseudonana,* 2 clones of *Fragilaria pinnata,* and 2 clones of *Bellerochia* sp.

As seen in Table 1, in all three cases the oceanic clone was more sensitive to PCB's than the coastal clone of the same species. In the case of *Thalassiosira pseudonana,* a clone isolated from the continental shelf exhibited a sensitivity somewhere between the coastal and oceanic clones. These data illustrate quite nicely the relationship between environment and sensitivity to environmental stresses or perturbations. In the case of the clones isolated from coastal waters, the environment in which they were living was constantly subjected to a variety of stresses and fluxes in abiotic factors such as temperature, chemical pollution, etc. The microorganisms living there, in this case diatoms, have, therefore, developed a resistance to stress in general, and this is reflected in their growth pattern when exposed to 10 ppb PCB's. The oceanic clones, on the other hand, have been accustomed to the stability of the Sargasso Sea and, as a result, have become quite specialized in terms of strain variation. When suddenly placed in a stress situation, as illustrated in

Table 1. Ratios of cell densities from
treated (10 ppb PCB's) to control cultures

Clones	Ratio[a]				
	0 hr	47.5 hr	70 hr	95 hr	118 hr
Thalassiosira pseudonana[b]					
3H	100	69.4	36.5	67.6	97.6
7-15	100	54.4	36.8	52.0	84.4
13-1	100	42.2	24.1	17.2	34.1
Fragilaria pinnata[c]					
0-12	100	82.4	90.9	93.2	59.0
13-3	100	68.8	70.7	26.9	38.7
Bellerochia[d]					
Say-7	100	20.7	6.4	3.6	3.5
SD	100	7.4	2.0	1.3	0.5

[a] Data given are means of three replicates. The higher the ratio, the less sensitive the cells, with 100 = no sensitivity.

[b] 3 H, Moriches Bay, Long Island; 7-15, edge of the continental shelf; 13-1, Sargasso Sea.

[c] 0-12, Oyster Pond, Massachusetts; 13-3, Sargasso Sea.

[d] Say-7, Great South Bay, Long Island; SD, Sargasso Sea.

Table 1, the result is a clear-cut sensitivity. These results show quite clearly one of the major reasons why a study of open-ocean microorganisms at this time is quite important.

The coccolithophorids represent another group of oceanic microorganisms which we will study in detail in terms of their response to environmental pollutants. Coccolithophorids are biflagellate, golden-brown algae belonging to the class Haptophyceae. They inhabit the euphotic zone in marine waters and together with the diatoms and dinoflagellates comprise the bulk of the marine phytoplankton. Coccolithophorids have the ability to secrete internally minute calcite skeletal elements called coccoliths, which are extruded to form a coating on the cell surface. The taxonomy and systematics of this group of algae are based upon the complexities of coccolith structure.

Recent evidence suggests that deformities in coccolith structure may be caused by environmental stress conditions. Dr. Hisatake Okada of the Department of Geology and Geophysics, Woods Hole Oceanographic Institution, has spent the past several years studying plankton samples from the Pacific Ocean in an attempt to compare modern coccolithophorids present in the euphotic zone with fossil coccoliths present in deep-sea sediment. Hopefully this study will provide some insight into the evolutionary development of this interesting group of phytoplankton. Of interest to our work is the fact that Dr. Okada has been able to demonstrate deformities in coccolith structure which may be a response to environmental stress. For the purpose of this presentation, two examples are given. The first example is illustrated in Fig. 1. Figure 1, *A* and *B*, shows the normal variation in coccolith structure in the coccolithophore *Gephyrocapsa oceanica*. The species was found in water samples off the south coast of the Island of Hawaii. Deformed examples of the same

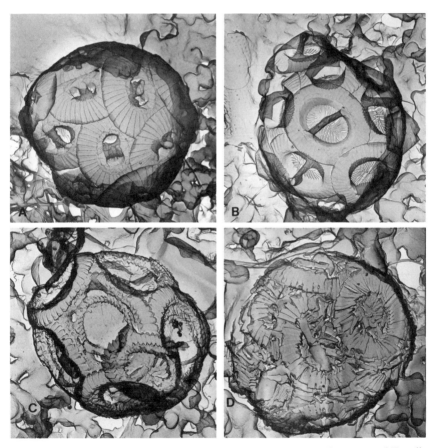

Fig. 1. Carbon replica of *Gephyrocapsa oceanica* isolated from southern coast of Hawaii. *A*, normal specimen × 4500; *B*, normal specimen showing variation in coccolith structure × 6000; *C*, deformed specimen × 4500; *D*, deformed specimen showing variation in coccolith structure × 4500.

species are shown in Fig. 1, *C* and *D*. The frequency of deformed specimens of this species is quite high, often approaching 50%; however, this situation is quite rare among the coccolithophorids in general. The second example is illustrated in Fig. 2. Figure 2, *A* and *B*, shows the normal variation in coccolith structure in *Emiliana huxleyi*. A warm variety (Fig. 2*A*) was found in waters off the coast of Hawaii, while a cold variety was found in waters off the south coast of Japan. Figure 2, *C* and *D*, shows examples of deformed coccolith structure. According to McIntyre (unpublished data) these deformities are due to a deficiency in vitamin B_{12}, and in subarctic waters ($\sim 50°$N) a high percentage of *Emiliana huxleyi* are deformed.

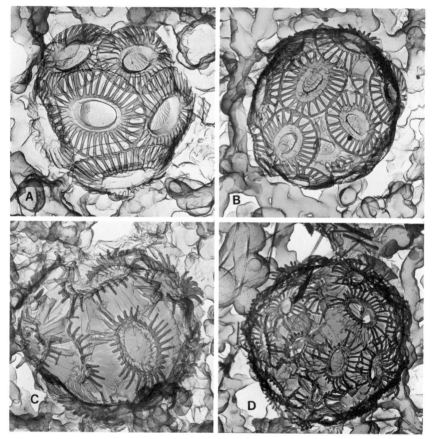

Fig. 2. Carbon replica of *Emiliana huxleyi*. *A,* normal specimen from southern coast of Hawaii $\times$ 6000. *B–D,* from southern coast of Japan; *B,* normal specimen showing variation in coccolith structure $\times$ 6000; *C,* deformed specimen $\times$ 6000; *D,* deformed specimen showing variation in coccolith structure $\times$ 7500.

Fig. 3. Scanning electron micrograph of a species of *Acantharia* × 3750.

In general, however, when all of the coccolithophorids are considered together, a definite trend emerges with respect to deformed *versus* normal coccoliths. In pelagic waters, the frequency of deformed coccoliths among all coccolithophores is less than 1%. In coastal waters, or inland seas, the frequency can be as high as 70% with the average being greater than 10%.

While there is no evidence to suggest that the deformities in coccolith structure are due to chemical pollution, the general feeling is that they do represent a response to a stress situation. It is our feeling that pollution of oceanic waters with environmental pollutants may contribute to the deformities seen in the coccolithophorids.

We also intend to examine the effects of chemical pollution on the skeletal structure of the marine planktonic protozoa. Figure 3 shows the skeleton of a species of Acantharia isolated from the open ocean. It is our feeling that environmental stress situations will produce a noticeable change or deformity on the Acantharia spicules in much the same way that stress appears to effect coccolith structure on the coccolithophorids.

ACKNOWLEDGMENT

This research is being supported by a grant from the Sarah Mellon Scaife Foundation to the Woods Hole Oceanographic Institution.

LITERATURE CITED

Bowes, G. W., and R. W. Gee. 1971. Inhibition of photosynthetic electron transport by DDT and DDE. Bioenergetics 2:47–60.

Chacko, C. I., J. L. Lockwood, and M. Zabik. 1966. Chlorinated hydrocarbon pesticides: Degradation by microbes. Science 154:893–895.

Friedman, B. A., and P. R. Dugan. 1968. Identification of *Zoogloea* species and the relationship to zoogloeal matrix and floc formation. J. Bacteriol. 95:1903–1909.

Friedman, B. A., P. R. Dugan, R. M. Pfister, and C. C. Remsen. 1969. Structure of exocellular polymers and their relationship to bacterial flocculation. J. Bacteriol. 98:1328–1334.

Grice, G. D., G. R. Harvey, V. T. Bowen, and R. H. Backus. 1972. The collection and preservation of open ocean marine organisms for pollutant analysis. Bull. Environ. Contam. Toxicol. 7:125–132.

Harriss, R. C., D. B. White, and R. B. MacFarlane. 1970. Mercury compounds reduce photosynthesis by plankton. Science 170:736–737.

Harvey, G. R., V. T. Bowen, R. H. Backus, and G. D. Grice. 1972. Chlorinated hydrocarbons in open-ocean Atlantic organisms. Nobel Symposium Proceedings 20, The Changing Chemistry of the Oceans, pp. 170–186.

Harvey, G. R., H. P. Miklas, V. T. Bowen, and W. G. Steinhauer. 1974. Observations on the distribution of chlorinated hydrocarbons in Atlantic Ocean organisms. J. Mar. Res. (In press.)

Janicki, R. H., and W. B. Kenter. 1971. DDT: Disrupted osmoregulatory events in the intestine of the eel *Angiulla rostrata* adapted to seawater. Science 173:1146–1148.

Leshniowsky, W. O., R. P. Dugan, R. M. Pfister, J. I. Frea, and C. I. Randles. 1970. Aldrin: Removal from lake water by flocculent bacteria. Science 169:993–995.

MacFarlane, R. B., W. A. Glooschenko, and R. C. Harriss. 1972. The interaction of light intensity and DDT concentration upon the marine diatom, *Nitzschia delicatissima* Cleve. Hydrobiologia 39:373–382.

Pfister, R. M., P. R. Dugan, and J. I. Frea. 1969. Microparticulates: Isolation from water and identification of associated chlorinated pesticides. Science 166:878–879.

Seba, D. B., and E. F. Corcoran. 1969. Surface slicks as concentrators of pesticides in the marine environment. Pestic. Monit. J. 3:190–193.

Tinsley, I. J., R. Hague, and D. Schmedding. 1971. Binding of DDT to lecithin. Science 174:145–147.

Tornabene, T. G., and H. W. Edwards. 1972. Microbial uptake of lead. Science 176:1334–1335.

Yamamoto, T., and M. S. Shahrabadi. 1971. Enzyme cytochemistry and autoradiography of adenovirus-infected cells as determined with the electron microscope. Can. J. Microbiol. 17:249–256.

BIODETERIORATION IN THE SEA AND ITS INHIBITION

JOHN McN. SIEBURTH and ALLAN S. DIETZ

Narragansett Marine Laboratory
University of Rhode Island
Graduate School of Oceanography

At 1–3°C food materials were consumed or decayed when held in perforated double containers at a depth of 2 m in inshore waters and at 5200 m in the deep sea for 2 and 10 weeks respectively. When identical food materials were held in triple enclosures which minimized water exchange and the passage of omnivorous scavenging animals, they were in a remarkable state of preservation. The latter conditions were apparently present to a very high degree in the well-preserved box lunches recovered from the research submersible *Alvin*. Such inhibition of biodeterioration must also occur when dumped organic wastes are compacted or allowed to accumulate in the sea, regardless of depth.

INTRODUCTION

Microorganisms are widely distributed in the deep sea (ZoBell, 1946; Kriss, 1963). Bacteria-free environments are relatively rare (Sieburth, 1961; Watson and Waterbury, 1969). Although bacteria can be inhibited or killed at pressures exceeding 200 atm (ZoBell and Oppenheimer, 1950) and enzymic reactions have been shown to be pressure sensitive (Morita, 1967), bacteria surviving ascent from depths of 10,000 m can grow at hydrostatic pressures of 1000 atm (ZoBell and Morita, 1957). Obligate marine psychrophiles readily develop at temperatures below 10°C in inshore waters (Sieburth, 1967) and occur in the deep sea (Sieburth, 1971). Scavenging omnivores are ubiquitous in the sea and are present even in the less productive abysses at depths of 5300 m (Hessler *et al.,* 1972). Biodeterioration of organic materials in the sea is a result of both consumption by scavenging omnivores such as the gammarid amphipods, which tear and triturate the materials thereby increasing surface area and the penetration of oxygen, and the activities of microorganisms which are limited by surface area and the degree of oxygenation. The food ingested by the macrofauna is also subject to the microbial activities of the gut flora during its passage of the gut as well as in the feces.

Since both scavengers and microorganisms are ubiquitous, one might assume, *a priori,* that at similar temperatures the biodeterioration of organic matter in the deep sea would occur in a manner and rate not too dissimilar from that in shallow waters. The report by Jannasch *et al.* (1971) of the remarkable state of preservation of the box lunch accidentally submerged with the research submersible *Alvin* for 10 months at 1540 m is paradoxically at odds with what one might expect. In order to see if the conditions of the deep sea do slow the rates of microbial activity, Jannasch *et al.* (1971) used microorganisms from shallow water and a depth of 200 m to inoculate soluble substrates held in syringes and bottles sealed with serum stoppers which were incubated at the *in situ* temperatures and pressures of the deep sea for 2 to 5 months. They found that these samples decomposed 666 to 8.6 times more slowly than the refrigerated controls. They concluded that the degree of preservation of the foodstuffs recovered from *Alvin* was no chance observation and that there was a general slow-down of life processes in the deep sea, presumably due to an interaction of low temperature and high pressure.

One thing shared in common by the box lunch in *Alvin* and the experiments reported by Jannasch *et al.* (1971) is that the materials were isolated from a free interchange with the waters of the deep sea and their biota. The lunch in *Alvin* was in a covered plastic box. It seemed possible to us that the protection offered by the enclosure might be sufficient to interfere with the natural biodeterioration near the ocean floor. To see if this is the case, experiments were designed in a manner similar to those of Payne (1965) in which scavenging organisms would be excluded.

MATERIALS AND METHODS

The box lunch in *Alvin* contained bologna sandwiches. Similar solid foodstuffs were selected to simplify the design and construction of the inexpensive "lunch box" test packages (Fig. 1*B*) used to hold the test materials under

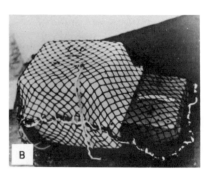

Fig. 1. The condition of the exposed, protected, and control test materials and the nature of the test packages. *A,* Current meter array with its test packages being recovered after 74 days immersion; *B,* a "lunch box" test package containing test materials protected in a triple enclosure (left) and exposed in a double perforated enclosure (right); *C,* the duplicate set of test materials in perforated containers from Station II (upper); *D–F;* a comparison of test materials from Station II (lower) showing frozen control (*D*), exposed (*E*), and protected (*F*) boxes.

both protected and exposed conditions. Polyethylene refrigerator boxes with a snug lid were used to hold 2.5 cm cubes of lean beef, suet, whole-wheat roll, 3% agar on a styrofoam core, a fresh crab claw from a Jonah crab, and corrugated cardboard which were strung and separated by sailmaker's Dacron twine to avoid protection of the surfaces of the foodstuff by the containers (Fig. 1D); the latter point is a critical feature of our experiments. Duplicate boxes were placed in a plastic lunch pail which was bolted to the inside of a plastic dishpan and enclosed with a matching dishpan held tightly lip to lip with bolts at each corner. These triple-enclosed test materials were intended to simulate the protected lunches in *Alvin*. In order to expose a replicate set of test materials to the environment, a second lunch pail and its enclosed food boxes were well perforated with 5-mm holes (Fig. 1C) and bolted to the outside of the dishpan enclosure. Plastic license-plate bolts were used throughout. Each completed test package was enclosed in a Dacron fish net bag (Fig. 1B) and frozen until used. On station, the bags were attached to the current meter arrays (Fig. 1A) which were submerged at the locations and depths and for the periods shown in Table 1. Test packages were also immersed in shallow water (5 m) from a dock in Narragansett Bay, Rhode Island, at a depth of 2 m for 2-week periods in February (1–2°C) and August (20–21°C). Frozen controls were kept for purposes of comparison. Upon recovery of the current meter arrays, the test packages were removed, the contents of the food boxes photographed to show gross changes and the materials were carefully examined to record their condition. Representative samples of the test materials were prepared for scanning electron microscopy (Gessner *et al.*, 1972) and kept frozen with Dry Ice.

RESULTS AND DISCUSSION

The results of the deep-sea tests are given in Table 1. The smallest differences in the condition of the test materials in the two types of containers were obtained with the upper packages held 1500 m off the ocean floor at depths of 3715–3875 m. Although all test materials were present in both types of containers, there was a detectably poorer quality in the color, texture, and feel of the more labile materials remaining in the perforated containers. The differences in the packages held some 5 m off the bottom were very striking as shown in Figs. 1D–F. Materials in the triple enclosure (Fig. 1F) were intact and of a good appearance while all readily decomposable or edible materials such as the beef, crabmeat, and bread, were absent from the perforated container (Fig. 1E). Several lysianassid gammarid amphipods were present in the perforated enclosures at Station II-lower. The crab claws in the perforated

Table 1. Station locations, incubation conditions, and state of the recovered "lunch box" test packages

	Station I		Station II				Station III			
Location	29°49.9'N, 70°22.3'W		28°54.3'N, 69°33.0'W				29°50.3'N, 60°34.2'W			
Date dropped	10/26/71		10/27/71				10/28/71			
Days incubation	73		74				72			
Incubation temp (°C)	2–3		2–3				2–3			
Package position	Upper[a]		Upper		Lower		Upper		Lower	
Depth (m)	3875		3815		5315		3715		5215	
Lunch box	Closed	Perf.[b]	Closed	Perf.	Closed	Perf.[c]	Closed	Perf.	Closed	Perf.[d]
Condition of food[e]										
Beef	G	P	G	P	G	Gone	G	P	G	Gone
Crabmeat	G	P	G	P	G	Gone	G	P	G	Gone
Cardboard	NC	NC	NC	NC	NC	NC	NC	NC	NC	NC
Agar	G	P	G	P	G	F	G	P	NC	F
Bread	NC	F	F	NC	NC	Gone	F	Gone	NC	NC
Fat	NC	F	NC	NC	F	F	NC	F	F	Gone

[a] Upper, 1500 m above ocean floor; lower, 5 m above ocean floor.

[b] Perf., perforated.

[c] Lysianassid gammarid amphipods present.

[d] Sediment present.

[e] G, good appearance, intact, some slime; P, poorer appearance, some sediment present, copious slime; NC, no apparent change; F, fragmented.

enclosures had a tendency to be brittle or soft while those in the protected enclosures were hard and difficult to crush.

Striking differences were also obtained with the shallow-water test packages held in Narragansett Bay at 1–2°C for 2 weeks. The triple-enclosed materials were in an unbelievable state of preservation with the beef still red inside while the partially consumed materials remaining in the perforated container were discolored, slimy, and highly putrid. After 2 weeks of immersion at 20–21°C in Narragansett Bay, differences in the rate of biodeterioration of test materials in the two types of containers were much less, but still detectable. This was the only test package fouled by pennate diatoms and barnacles. The readily decomposible food stuffs were absent from the perforated container over which gammarid amphipods were swarming, while in the triple-enclosed containers the cube of beef and the crabmeat were intact although in a highly putrid state. The exposed crab claw was soft and friable while the protected one was in a hard state.

The test materials in the triple enclosures, although intact and of good appearance, were apparently subject to microbial attack as the enclosed water was turbid and contained in excess of 10^7 bacteria per milliliter as determined by direct microscopic examination. The original purpose in fixing and freezing the test materials was to characterize microbial colonization by scanning electron microscopy. Results obtained with most of the materials were difficult to interpret. An exception was the surface of the corrugated cardboard. The cellulose fibers, which were free of apparent microorganisms at the time of immersion, became covered with forms suggestive of both bacteria and fungi in both the exposed and the protected packages. If microorganisms are present and active in the triple enclosures, why was not biodeterioration greater? Biodeterioration is not normally restricted to one class of organisms. The decomposition of dead flesh is greatly expedited by the disruption of tissues by carrion-feeding insects (Payne, 1965). At the shore's edge, kelp-fly larvae play a similar role in hastening the decomposition of storm-tossed seaweed through trituration and aeration (Bunt, 1955). In the sea, gammarid amphipods and other omnivores have the role of the fly larvae on land. In the deep sea, as well as in Narragansett Bay, gammarid amphipods were present in the perforated containers in which the readily edible and decomposible foodstuffs were absent.

The differences observed in the test materials do not seem to be due to just the presence of animals. In the deep-sea packages held 1500 m above the ocean floor, where gammarids were not trapped in the perforated containers and all materials were intact, there was still a visibly detectable difference between materials held in the two types of containers. Also the softening of the crab claws in only the perforated containers must have been due to chitinoclastic bacteria. The restriction of water flow may have a detectable effect on microbial activity. Microbial activities are greatly decreased by oxygen depletion. The strong putrefactive odors of our food packages indicated this was the case. This possibility was recognized but rejected by Jannasch *et al.* (1971) as there was no evidence in the *Alvin* lunch of reducing conditions or the lack of dissolved oxygen.

The carefully controlled and executed experiments of Jannasch *et al.* (1971) were designed to obtain quantitative data fundamental to an understanding of the effect of temperature and pressure on bacteria. Despite the qualitative and subjective nature of our observations, we present them here as we feel they indicate that labile foodstuffs held near the ocean floor may readily undergo biodeterioration. Although the metabolism of deep-sea microorganisms and scavenging animals are apparently much reduced in the deep sea, their accumulative activities may not be as slow as indicated by observations on individual components or in exposure experiments in which the bulk of the test material is protected by its enclosure. The process of biodeterioration appears to be easily inhibited by small changes such as a protective covering which shields a labile substrate from scavenging animals and dissolved oxygen.

As pointed out by Jannasch *et al.* (1971), the rates of biodeterioration on the deep-ocean floor have obvious implications with regard to the dumping of organic wastes in the deep sea. Municipal wastes are largely composed of paper, which is slow to degrade, and a small content of food wastes. In tests for ocean dumping, this material is being compacted into large cubes of approximately 1 m^3 which will sink. Laboratory studies have shown that biodeterioration is impeded within solid waste deposits in seawater after oxygen has been consumed and hydrogen sulfide has been produced (S. D. Pratt, personal communication). Such observations, as well as ours, emphasize the fact that biodegradation is a surface phenomenon involving the total benthic community. Conditions which enclose organic matter or lead to passive deposits, thereby excluding animals and microbial oxidation, are, therefore, conducive to storage rather than to the oxidation and mineralization of organic matter.

SUMMARY

1. Readily consumable foodstuffs such as lean beef, crabmeat in a claw and bread held in perforated double containers were consumed or decayed at 1–3°C within 2 weeks in inshore waters and within 10 weeks in the deep sea.
2. Triple enclosures did not prevent bacterial development, but the foodstuffs were in an excellent state of preservation.

3. The exclusive of omnivorous scavengers in the sea has exactly the same delaying effect on decomposition as the exclusion of carrion-feeding insects on land.
4. Enclosures apparently prevented animal consumption and microbial decay of the well-preserved box lunches recovered from the research submersible *Alvin*.

ACKNOWLEDGMENTS

The scientific parties of TR 104 and 109 who helped to drop and recover the test packages and T. A. Napora who identified the amphipods are gratefully acknowledged. This study was supported in part by NSF Grants GB 18000 and GA 28903.

LITERATURE CITED

Bunt, J. S. 1955. The importance of bacteria and other microorganisms in the seawater at MacQuarie Island, Austr. J. Freshw. Res. 6:60—65.

Gessner, R. V., R. D. Goos, and J. McN. Sieburth. 1972. The fungal microcosm of the internodes of *Spartina alterniflora*. Mar. Biol. 16(4):269—273.

Hessler, R. R., J. D. Isaacs, and L. Mills. 1972. Giant amphipod from the abyssal Pacific Ocean. Science 175:636—637.

Jannasch, H. W., K. Eimhjellen, C. O. Wirsen, and A. Farmanfarmaian. 1971. Microbial degradation of organic matter in the deep sea. Science 171:672—675.

Jannasch, H. W., and C. O. Wirsen. 1972. *Alvin* and the sandwich. Oceanus 16 (Dec.):20—22.

Jannasch, H. W., and C. O. Wirsen. 1973. Deep-sea microorganisms: *in situ* response to nutrient enrichment. Science 180:641—643.

Kriss, A. E. 1963. Marine Microbiology (Deep Sea). (Trans. by J. M. Shewan and Z. Kabata.) Oliver and Boyd, London.

Morita, R. Y. 1967. Effects of hydrostatic pressure on marine microorganisms. Oceanogr. Mar. Biol. Annu. Rev. 5:187—203.

Payne, J. A. 1965. A summer carrion study of the baby pig *Sus scrofa* Linnaeus. Ecology 46(5):592—602.

Sieburth, J. McN. 1961. Antibiotic properties of acrylic acid, a factor in the gastrointestinal antibiosis of polar marine animals. J. Bacteriol. 82:72—79.

Sieburth, J. McN. 1967. Seasonal selection of estuarine bacteria by water temperature. J. Exp. Mar. Biol. Ecol. 1:98—121.

Sieburth, J. McN. 1971. Distribution and activity of oceanic bacteria. Deep-Sea Res. 18:1111—1121.

Watson, S. W., and J. B. Waterbury. 1969. The sterile hot brines of the Red Sea. *In* E. T. Degens and D. A. Ross (eds), Hot Brines and Recent Heavy Metal Deposits in the Red Sea, pp. 272–281. Springer-Verlag, New York.

ZoBell, C. E. 1946. Marine Microbiology. Chronica Botanica, Waltham, Massachusetts.

ZoBell, C. E., and R. Y. Morita. 1957. Barophilic bacteria in some deep sea sediments. J. Bacteriol. 73:563–568.

ZoBell, C. E., and C. H. Oppenheimer. 1950. Some effects of hydrostatic pressure on the multiplication and morphology of marine bacteria. J. Bacteriol. 60:771–781.

Note added in proof. In discussing our paper with H. W. Jannasch and C. O. Wirsen, it was learned that similar experiments were conducted 2 years earlier as part of the original study by Jannasch *et al.* (1971), but results were not included as they were not quantitative. The subsequent description of the "BIO-PACK" experiment (Jannasch and Wirsen, 1972) indicates an entirely different design and objective of the experiments. The paper by Jannasch and Wirsen (1973) ably demonstrates that the *in situ* sediment microflora is also slow to convert organic matter, but points out the possible importance of animals and their microflora, the major premise of this paper.

DISTRIBUTION AND ACTIVITY OF NITROGEN-CYCLE BACTERIA IN WATER-SEDIMENT SYSTEMS WITH DIFFERENT CONCENTRATIONS OF OXYGEN

ISAO SUGAHARA, MOTOHIKO SUGIYAMA, and AKIRA KAWAI

Faculty of Fisheries
Prefectural University of Mie
and
Research Institute for Food Science
Kyoto University

In aquatic environments such as coastal regions, lakes, or inland waters, the production of the biological community depends largely upon the physical and chemical conditions of the water region. The environmental condition or water quality results mainly from the activities of the various kinds of microorganisms inherent in the water or bottom sediments. Conversely, the occurrence, abundance, and activity of the microorganisms are greatly affected by the environmental factors of the water region.

Of the elements of biological importance in the hydrosphere, nitrogen is the most important for living organisms, and the productivity of the water region is sometimes controlled by the form as well as the concentration of the nitrogenous compounds in the water. The dissolved oxygen concentration in water is one of the most important environmental factors which influence the nitrogen cycle.

METHODS

Enumeration of Bacterial Groups

Besides the total heterotrophic bacteria, the number of each group of bacteria which have the following biochemical activities was estimated in the bottom sediments by the most probable number (MPN) method: nitrification (ammonia oxidation $NH_4^+ \rightarrow NO_2^-$, nitrite oxidation $NO_2^- \rightarrow NO_3^-$), ammonification (organic nitrogenous compounds $\rightarrow NH_4^+$), nitrate reduction ($NO_3^- \rightarrow NO_2^-$), denitrification ($NO_3^- \rightarrow N_2$), proteolysis (gelatin liquefaction), nitrogen fixation ($N_2 \rightarrow NH_4^+$ or organic nitrogenous compounds) and sulfate reduction ($SO_4^{2-} \rightarrow S^{2-}$).

The composition of each culture medium used for estimating the bacterial populations is given in Table 1. The medium for total heterotrophic bacteria was also used to enumerate ammonifying bacteria (ammonia analysis), nitrate-reducing bacteria (nitrate analysis), and denitrifying bacteria (N_2 gas production as indicated in the Durham tube). After separate autoclaving, the KH_2PO_4 was added aseptically to tubes of each medium. Similarly a small amount of $CaCO_3$ was added to each culture tube of medium for both the ammonia oxidizers and the nitrite oxidizers.

Analysis

The pH value of the samples was measured by the use of a pH meter with glass electrode (Toa Electronics Ltd., Model HM-5A). Dissolved oxygen (DO) in the water was determined either according to the modified Winkler method or by the use of a DO meter (EIL, type 1520) and the COD value was estimated from the amount of $KMnO_4$ consumed under alkaline conditions. Nitrite was determined colorimetrically with Griess-Romijn reagent. Ammonia was determined by the method of Richards and Kletsch (1965), in which ammonia was oxidized with sodium hypochlorite to form nitrite which was then determined as above.

In some cases, the modified indophenol method was also employed (Manabe, 1969). Nitrate was reduced to nitrite by passing through a Cd-Cu column (Strickland and Parsons, 1965), and the nitrite formed was also determined by the Griess-Romijn method. Sulfide was determined colorimetrically by the use of p-aminodimethyl aniline reagent after distilling the sulfide according to the method of Tomiyama and Kanzaki (1951).

Assay of Bacterial Activities

The bacterial cells attached to the filter sand of the experimental aquaria were collected by shaking in an Erlenmeyer flask containing 300 ml of filter

Table 1. Composition of the culture media

Total heterotrophic bacteria, ammonifying bacteria, nitrate-
reducing bacteria, and denitrifying bacteria
   Bacto-casitone, 5 g; beef extract 3g; KH_2PO_4,
   0.1 g; Fe-EDTA, 6 mg; KNO_3, 0.5 g; artificial
   seawater, 1 liter
   Durham tube in each culture tube

Ammonia-oxidizing bacteria
   $(NH_4)_2SO_4$, 30 mg; KH_2PO_4, 0.1 g; Fe-EDTA,
   6 mg; artificial seawater, 1 liter

Nitrite-oxidizing bacteria
   KNO_2, 30 mg; KH_2PO_4, 0.1 g; Fe-EDTA, 6 mg;
   artificial seawater, 1 liter

Gelatin-liquefying bacteria
   Gelatin, 30 g; Bacto-casitone, 5 g; beef
   extract, 3 g; KH_2PO_4, 0.1 g; seawater,
   1 liter

Nitrogen-fixing bacteria
   Sucrose, 20 g; KH_2PO_4, 0.2 g; $Na_2MoO_4 \cdot 2H_2O$
   0.01 g; $FeSO_4 \cdot 7H_2O$, 0.001 g; $MnSO_4 \cdot 6H_2O$, 0.001 g;
   artificial seawater, 1 liter

Sulfate-reducing bacteria
   Peptone, 2 g; beef extract, 1 g; calcium
   lactate, 3 g; $FeSO_4 \cdot 7H_2O$, 0.2 g;
   $MgSO_4 \cdot 7H_2O$, 0.2 g; K_2HPO_4,
   0.05 g; Na-EDTA, 0.4 g; ascorbic acid, 0.1 g;
   agar, 3 g; distilled water, 1 liter

sand and sterilized seawater, which was then centrifuged at $0°C$ and $1000 \times g$ for 15 min. The precipitate obtained was washed by centrifugation and resuspended in sterile seawater. For the measurement of nitrifying activity, 50 ml of reaction mixture (bacterial suspension, substrate and Tris buffer, pH 8.2) in an Erlenmeyer flask were incubated for 4 hr under different degrees of aeration with air or argon gas. Thunberg tubes containing 10 ml of reaction mixture (bacterial suspension, 1 μM of substrate, 50 μM of Tris buffer, pH 8.2, 50 μM of sodium acetate, small amount of toluene) were used for measuring activities of nitrate reduction and denitrification. The reaction was carried out for a period of 4 hr under anaerobic conditions in an atmosphere of argon gas.

Experimental Aquaria

The experimental aquarium with circulating system which was employed in this study is shown in Fig. 1. The volume of the circulating water was about 2.7 liters, and about 0.43 liters of sand (2–3 mm in diameter) was used for a filter sand bed of about 3-cm thickness. In order to supply different amounts of oxygen continuously into the aquaria, three kinds of gas, i.e., air, air plus argon, and argon, were used for the circulation of the water. After inoculation with a small amount of coastal seawater, 100 mg of Bacto-casitone (Difco) were added to each aquarium. Each of the aquaria was set up in a somewhat darkened room at a definite temperature. The same amount of casitone was supplemented when the peak of nitrite disappeared after the mineralization of the organic nitrogen took place by the action of bacteria in the filter sand bed and circulating water. This procedure was repeated several times until the filter sand reached equilibrium.

RESULTS AND DISCUSSION

Microflora of Bottom Sediments

The bottom sediments were collected with a modified K-K core sampler (Kimata *et al.*, 1960) from a coastal region (Sagami Bay), inland bays

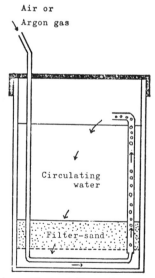

Fig. 1. Experimental aquarium with circulating system.

(Maizuru Bay and Kumihama Bay), and a seawater lake (Lake Suigetsu), all in Japan, and the occurrence of bacteria with different types of biochemical activities was determined.

Maizuru Bay is located in the northern part of Kyoto Prefecture in Japan. It has a surface area of some 26 km² and a maximum depth of about 25 m. Kumihama Bay is also in the northern part of Kyoto Prefecture and is connected with the Japan Sea through a shallow (3 m) and narrow (30 m) canal 300 m in length. It has a surface area of about 7 km², a coastal length of 22 km and a maximum depth of 22 m. Lake Suigetsu is one of five Mikata Lakes in Fukui Prefecture. It is connected with the other lakes and has a surface area of about 4.3 km² and a depth of about 34.0 m. The lake water is supplied from a freshwater lake (Lake Mikata) and flows out to the adjacent lake (Lake Hiruga) through an artificial channel. The seawater flows into Lake Suigetsu through Lake Hiruga only at high tide during autumn and winter.

Table 2 shows the analyses of the water and the bottom sediments of Maizuru Bay, Kumihama Bay, and Lake Suigetsu. There was no remarkable difference between the content of organic matter derived from land drainage

Table 2. Analyses of the water and the bottom sediments of Maizuru Bay, Kumihama Bay, and Lake Suigetsu

	Maizuru Bay[a]		Kumihama Bay[b]		Lake Suigetsu[c]	
	Bottom	Water (10 m)	Bottom	Water (10 m)	Bottom	Water (10 m)
Temperature (°C)	17.5	18.5	16.9	18.8	13.7	16.7
pH	7.8	8.0	7.6	7.9	7.3	7.6
DO (saturation %)	–	85.6	–	18.0	–	0
Ammonia-N (μg atom/g, liter)	0.03	1.19	1.38	5.13	8.28	240
Nitrite-N (μg atom/g, liter)	0	1.01	0	0.67	0	0.26
Nitrate-N (μg atom/g, liter)	0	1.92	0	1.37	0	0
Sulfide-S (μg atom/g, liter)	0.07	0	2.00	0.36	0.62	51.0
COD (O$_2$ mg/g, liter)	8.8	0.32	19.0	3.06	16.0	20.0

[a] Collected on 21 Nov. 1967; station depth, 15 m.

[b] Collected on 19 Nov. 1970; station depth, 19 m.

[c] Collected on 26 Nov. 1966; station depth, 20 m.

in the bottom sediments of the three water regions. The water at a depth of 10 m in Maizuru Bay was well oxygenated being at 86% saturation, whereas in Kumihama Bay the dissolved oxygen in the water was only 18% of saturation. The saline water in Lake Suigetsu is stagnant from a depth of 8 m to the bottom. No dissolved oxygen was found at 10 m. Matsuyama and Saijo (1971) suggested that the stagnation of the deep water of Lake Suigetsu is very stable and that the mixing of water between the upper aerobic and lower anaerobic layer is quite limited. In Kumihama Bay and Lake Suigetsu the exchange of water with the ocean is limited by a shallow entrance which probably prevents a horizontal exchange of water. The ratio of the content of ammonia to that of inorganic nitrogenous compounds in the water gradually increased with a lowering of the dissolved oxygen. This suggested that the nitrate ratio decreases with low oxygen supply, whereas the ammonia ratio increases from the viewpoint of the inorganic nitrogen balance.

Figure 2 shows the number of bacteria which are responsible for each process in the nitrogen cycle in the bottom sediments (Kawai, 1972). Heterotrophic bacteria were abundant in their population, i.e., 10^5 cells per gram in the bottom sediments of Lake Suigetsu, Kumihama Bay, Maizuru Bay, and Sagami Bay. The number of nitrifying bacteria in the bottom sediments was about 10^2-10^3 cells per gram in Maizuru Bay and $10-10^2$ cells per gram in Kumihama Bay. However, nitrifying bacteria were very scanty in Lake Suigetsu, i.e., 1 cell per gram or less.

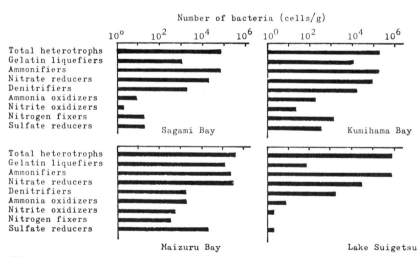

Fig. 2. Occurrence of bacteria with different types of biochemical activities in the bottom sediments of various water regions in Japan.

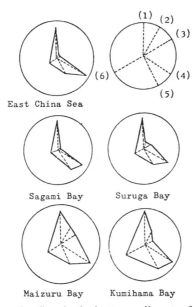

Fig. 3. Diagram of the microflora in the bottom sediments of various water regions in Japan: (1) ammonifying bacteria; (2) ammonia-oxidizing bacteria; (3) nitrite-oxidizing bacteria; (4) nitrate-reducing bacteria; (5) denitrifying bacteria; (6) nitrogen-fixing bacteria. The radius of the circle and the length of broken line show the number of total heterotrophic bacteria and each group of the bacteria in the nitrogen cycle in log scale, respectively.

The number of bacteria responsible for each process in the nitrogen cycle was plotted as a diagram. Figure 3 shows the microflora of the bottom sediments from five regions as diagrams (Kawai and Sugahara, 1972). The direction and the length from the center of the circle indicate the type and the number of the bacteria responsible for each biochemical reaction in the nitrogen cycle. The number of total heterotrophic bacteria is represented as the area of the circle. A characteristic pattern of the microflora was observed for each water region. The occurrence of nitrogen-fixing bacteria as well as nitrifying bacteria was very low, i.e., 1–10 cells per gram or less in offshore regions (Sagami Bay, Suruga Bay, and East China Sea), and the diagram for these areas was an L-shape. On the other hand, the ratio of the number of nitrogen-fixing bacteria as well as nitrifying bacteria to that of total heterotrophic bacteria was high in inland bays which were comparatively eutrophic, such as Maizuru Bay and Kumihama Bay, and their diagrams were triangular in form.

Microflora Differences under Differing Oxygen Supply

The experimental aquaria, which were kept under controlled conditions, were employed as a model system of coastal regions, since it might be quite difficult to analyze the results obtained from natural water regions in which the environmental conditions fluctuate with time and cannot be reproduced.

In order to make clear the effect of oxygen on the microflora of the filter sand, we used experimental aquaria with different concentrations of oxygen. Table 3 shows the oxygenation level of the circulating seawater of three aquaria which were supplied with oxygen at different rates (Kawai *et al.*, 1971). After 3 months of incubation, the saturation of dissolved oxygen in the water was 89%, 34%, and 6% for aquaria I, II and III, respectively. It can be seen in Table 3 that the pH increases as the DO decreases. The most remarkable differences between these three aquaria were found in the ammonium and nitrate nitrogen content of the water. The ammonia in the water was extremely high in the aquarium with DO 6% and very low in the other two, while nitrate content decreased with the lowering of oxygen tension. The amount of the organic nitrogen introduced into the three aquaria was not equal because the metabolic rate of the nitrogenous compounds decreases with oxygen tension. In the aquarium with DO 34%, the nitrate to total nitrogen ratio was nearly half that in the aquarium with DO 89%. This fact suggests that denitrification is associated with an anaerobic condition.

Figure 4 shows the occurrence and abundance of bacteria which are responsible for the metabolic processes in the nitrogen cycle (Kawai *et al.*,

Table 3. The effect of three levels of oxygenation on the water quality in seawater-sand aquaria

Aquarium	I	II	III
Oxygen tension (saturation %)	89	34	6
Circulation rate (liter/hr)	1.3	1.3	1.3
Water temperature ($^\circ$C)	22.0	22.0	22.0
Organic-N added (mg)	2250	2130	1330
pH	7.3	8.0	8.6
COD (O_2 mg/liter)	13.1	12.7	14.4
Ammonia-N (μg atom/liter)	43.5	33.2	12,400
Nitrite-N (μg atom/liter)	1.23	1.63	1.30
Nitrate-N (μg atom/liter)	17,700	9030	39.9
Sulfide-S (μg/liter)	0	0	0
(μg/g filter sand)	0	0.61	1.99

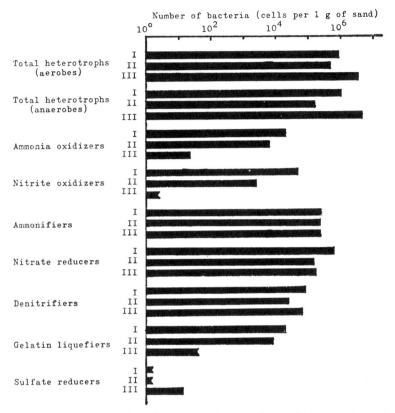

Fig. 4. Occurrence of bacteria with different biochemical activities in the filter sand of the seawater aquaria. Dissolved oxygen in saturation: I, 89%; II, 34%; III, 6%.

1971). It can be clearly observed that the lower the oxygen tension of the circulating water, the smaller the number of nitrifying bacteria, namely ammonia oxidizers and nitrite oxidizers, in the filter sand. On the other hand, there was an opposite correlation between the occurrence of sulfate-reducing bacteria and the DO value of the water. However, with the exception of the nitrifying bacteria and sulfate-reducing bacteria, many groups of bacteria having the ability to metabolize various nitrogenous compounds did not show any clear response to the change in oxygen supply.

Effect of Oxygen Supply on Bacterial Activities

Filter sand samples were collected from the experimental aquarium with air circulation and then the biochemical activities of the filter sand were determined at different concentrations of oxygen.

As shown in Fig. 5, the nitrifying activity of filter sand decreased with the rate of aeration, and the activity under strong aeration was less than half of that under non-aerated conditions (Kawai *et al.*, 1965). This shows that large amounts of oxygen have a tendency to suppress the nitrification process.

Figure 6 shows that the activity of nitrification also decreases under low oxygen tension maintained by bubbling argon gas into the reaction mixture (Kawai *et al.*, 1965). This indicates that the nitrification process may also be inhibited under low oxygen tension. However, nitrification still occurs even under low oxygen tension.

On the other hand, the activities of nitrate reduction and denitrification increased with a decrease in oxygen in the experimental aquaria supplied with oxygen at different rates. This is shown in Fig. 7.

Gundersen (1966) reported that nearly pure oxygen is toxic to *Nitrosocystis oceanus* growing on an agar medium. In liquid medium the organism was less affected by high partial pressures of oxygen than on agar medium. The formation of microcolonies took place at low partial pressures of oxygen. The lower limit of oxygen concentration which permitted oxidation of

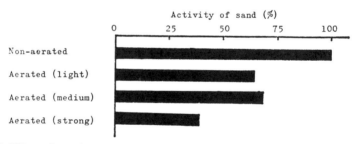

Fig. 5. Effect of aeration on the nitrifying activity of the filter sand in the seawater aquarium. Air was bubbled into the reaction mixture to a varying extent.

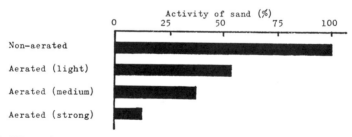

Fig. 6. Effect of argon aeration on the nitrifying activity of the filter sand in the seawater aquarium. Argon gas was bubbled into the reaction mixture to a varying extent.

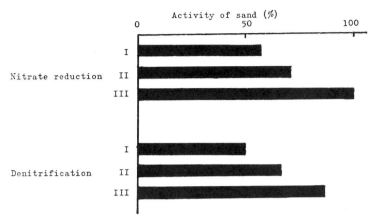

Fig. 7. Activities of nitrate reduction and denitrification of the filter sand of the seawater aquaria with different concentrations of oxygen. Dissolved oxygen in saturation: I, 89%; II, 34%; III, 6%.

ammonia in liquid medium corresponded to about 0.05 ml O_2 per liter. Carlucci and McNally (1969) observed that nitrifying bacteria oxidize low concentrations of substrate (5 μg atom NH_3-N or 20 μg atom NO_2-N per liter) in liquid medium containing less than 0.1 ml O_2 per liter and on solid medium under 0.002 atm O_2.

In this paper we have shown that nitrifying activity decreases under low oxygen tension. However, nitrification still occurs to a fair degree even under extremely low oxygen tension. This agrees with the results of Gundersen (1966) and Carlucci and McNally (1969).

On the other hand, oxygen is known to be an inhibitor for the processes of nitrate reduction as well as denitrification by virtue of its effective competition with nitrate as an electron acceptor during the metabolism of the cell (Delwiche, 1956). Goering (1968) reported that denitrification in seawater is affected by small changes in the bulk concentration of O_2, and at 0.02 ml O_2 (about 0.4% saturation with O_2) denitrification is about 1.7 times greater than at 0.19 ml O_2 (about 3.5% saturation). The amount of dissolved oxygen which limits denitrification is uncertain, but Goering believed it to be a few tenths ml O_2 per liter. Skerman and MacRae (1957) found no reduction of nitrate by *Pseudomonas denitrificans* at dissolved oxygen concentrations above 0.2 mg/liter. Under low oxygen concentrations (few tenths ml/liter) it is possible that some localized anoxic zones may exist in which active denitrification can take place.

In this paper, the filter sand in the aquaria with different concentrations of oxygen demonstrated nitrate reduction and denitrification under anaero-

Table 4. *In situ* rate of ammonia oxidation in water bodies and aquaria differing in oxygen content

	DO(%)	Ammonia-N (μg atom/liter)	Ammonia oxidation Velocity (μg atom/liter/hr)	Ammonia oxidation Turnover rate (%/hr)	Ammonia oxidizers (cells per milliliter)	Nitrite oxidizers (cells per milliliter)
Aquarium (aeration) 20 Sept. 1971	43	2316	6.04	0.26	9.2×10^2	9.3
Aquarium (non-aeration) 20 Sept. 1971	7	2085	3.10	0.15	$<1.8 \times 10^2$	7.9
Culture pond 12 June 1971	69.6	5.6	0.095	1.70	1.7×10^1	1.2×10^1
Pond I 15 Feb. 1972	96	2.5	0.053	2.12	$<1.8 \times 10^2$	$<1.8 \times 10^2$
Lake Biwa 5 Oct. 1970	102	2.9	0.026	0.96	9.1×10^{-1}	$<1.8 \times 10^{-1}$
Lake Biwa 2 Nov. 1971	89	4.5	0.11	2.44	1.7×10^1	1.7

bic conditions. Although it is not obvious whether denitrification and nitrate reduction are actually proceeding or not in the experimental aquaria with dissolved oxygen above a few tenths ml/liter, the results show that the enzyme systems which are responsible for the processes of nitrate reduction and denitrification may be formed in cells grown even under oxygenated conditions.

In aquatic environments, each of the reactions in the nitrogen cycle is mainly carried out by the action of microorganisms. However, the activity was not parallel to the biomass or the number of the microorganisms responsible for the reactions. In the experimental aquaria, the activity of nitrate reduction at 0.30 ml O_2 per liter (about 6% saturation) is about 1.7 times greater than at 4.59 ml O_2 per liter (about 89% saturation). On the contrary, the number of nitrate-reducing bacteria at 0.30 ml O_2 per liter is about one-fourth of the value at 4.59 ml O_2 per liter. Therefore, it is necessary to estimate the rate of the reactions which are actually proceeding *in situ* in order to demonstrate the role of microorganisms in the production of aquatic biocommunities. An attempt to estimate the *in situ* rate of each process in the nitrogen cycle was made using ^{15}N compounds as tracers in several water regions. Table 4 shows an example of the estimation of *in situ* rates of biochemical processes in the nitrogen cycle in some water regions. At this point, it seems probable that the range of activity per unit cell is not so great with each water region in view of the unavoidable error involved in the estimation of bacterial populations using the MPN method.

Further studies on this problem are necessary in order to demonstrate microbial nitrogen metabolism in aquatic environments.

LITERATURE CITED

Carlucci, A. F., and P. M. McNally. 1969. Nitrification by marine bacteria in low concentrations of substrate and oxygen. Limnol. Oceanogr. 14:736–739.

Delwiche, C. C. 1956. Denitrification. *In* W. D. McElroy and B. Glass (eds), Inorganic Nitrogen Metabolism, pp. 233–256. Johns Hopkins Press, Baltimore, Maryland.

Goering, J. J. 1968. Denitrification in the oxygen minimum layer of the eastern tropical Pacific Ocean. Deep-Sea Res. 15:157–164.

Gundersen, K. 1966. The growth and respiration of *Nitrosocystis oceanus* at different partial pressures of oxygen. J. Gen. Microbiol. 42:387–396.

Kawai, A. 1972. Bacterial flora of the bottom sediments in the sea. Mar. Sci. (Kaiyo Kagaku) 4:94–97.

Kawai, A., and I. Sugahara. 1972. Microbiological studies on nitrogen fixation in aquatic environments. VII. Some ecological aspects of nitrogen fixing bacteria. Bull. Jap. Soc. Sci. Fish. 38:291−297.

Kawai, A., M. Sugiyama, R. Shiozaki, and I. Sugahara. 1971. Microbiological studies on the nitrogen cycle in aquatic environments. I. Effect of oxygen tension on the microflora and the balance of nitrogenous compounds in the experimental aquarium. Mem. Res. Inst. Food Sci. Kyoto Univ. 32:7−15.

Kawai, A., Y. Yoshida, and M. Kimata. 1965. Biochemical studies on the bacteria in the aquarium with a circulating system. II. Nitrifying activity of the filter sand. Bull. Jap. Soc. Sci. Fish. 31:65−71.

Kimata, M., A. Kawai, and Y. Ishida. 1960. The method for sampling of marine bottom muds. Bull. Jap. Soc. Sci. Fish. 26:1227−1230.

Manabe, T. 1969. New modification of Lubochinsky's indophenol method for direct microanalysis of ammonia-N in seawater. Bull. Jap. Soc. Sci. Fish. 35:897−906.

Matsuyama, M., and Y. Saijo. 1971. Studies on biological metabolism in a meromictic Lake Suigetsu. J. Oceanogr. Soc. Jap. 27:197−206.

Richards, F. A., and R. A. Kletsch. 1965. Determination of ammonia (and amino acid). Bull. Fish. Res. Bd. Can. 125 (2nd edn.):83−87.

Skerman, V. D., and I. C. MacRae. 1957. The influence of oxygen availability on the degree of nitrate reduction by *Pseudomonas denitrificans.* Can. J. Microbiol. 3:505−530.

Strickland, J. D. H., and T. R. Parsons. 1965. Determination of nitrate. Bull. Fish. Res. Bd. Can. 125 (2nd edn.):73−78.

Tomiyama, T., and K. Kanzaki. 1951. A semi-micro method for determination of sulfide contained in muddy deposits. Bull. Jap. Soc. Sci. Fish. 17:115−121.

NITROGEN FIXATION IN THE MARINE ENVIRONMENT: THE EFFECT OF ORGANIC SUBSTRATES ON ACETYLENE REDUCTION

Y. MARUYAMA, T. SUZUKI, and K. OTOBE

Department of Agricultural Chemistry
University of Tokyo
and
Ocean Research Institute
University of Tokyo

Nitrogen fixation in seawater samples and sediment samples obtained from various sea areas was investigated by the acetylene-reduction method. Properties of acetylene reduction in the sediment samples such as optimum pH, time course of the reaction, effect of NaCl, oxygen and acetylene concentrations, effect of organic compounds, temperature, and storage of the reaction under both aerobic and anaerobic conditions were examined. Ethylene production was greatest at $30°C$ in sediment samples from 30 m $(16°C)$ in Tokyo Bay, whereas sediments from 1370 m $(2.5°C)$ in Sagami Bay showed the greatest ethylene production at $20°C$.

Similar rates of acetylene reduction were found under anaerobic conditions as were found under aerobic conditions. However, some differences were observed in K_m for acetylene reduction, optimum NaCl concentration, substrate effect, and stability.

The addition of organic compounds such as glucose, mannitol, sucrose, and pyruvate stimulated ethylene production in most seawater and sediment samples. This stimulative effect of organic substrates was more notable in open seawater samples than in coastal seawater samples. The effect of each compound varied with the sediments and with aerobic or anaerobic conditions. Glucose promoted ethylene production in all sediment samples tested.

Low rates of acetylene reduction were found: 31 nmole C_2H_4/m^3/day at $20°C$ (290 ng N fixed/m^3/day) in seawater samples from three stations in Tokyo Bay and 0.1–1.1 nmole C_2H_4/g/day at $3°C$ (0.88–10 ng N/g/day) in sediment samples from six stations in Tokyo Bay (the mean value of five stations in another experiment was 1.45 nmole C_2H_4/day at $20°C$–13.5 ng N/g/day). If organic substrates were supplied to the samples, the values were higher.

Taxonomic features of the marine microorganisms grown on non-nitrogenous media containing glucose, mannitol, and sucrose as carbon sources were different in some respects from those of marine heterotrophic bacteria.

Evidence for close correlation between the number of the bacteria and nitrogen fixation was found in samples from various sea areas.

INTRODUCTION

Biological nitrogen fixation has a very important significance for the nitrogen cycle within the biosphere. Although oceans and seas constitute a much greater part of the biosphere than land, only limited information is available with respect to nitrogen fixation in the ocean.

Nitrogen fixation by a blue-green alga, *Trichodesmium* sp., has been reported in *in situ* experiments in the tropical Atlantic Ocean (Goering *et al.*, 1966), and recently Brooks *et al.* (1971) have reported acetylene reduction in an estuarine environment.

In a previous paper (Maruyama *et al.*, 1970), the distributional mode of nitrogen-fixing bacteria in the open seas was extensively investigated by selective cultural methods using non-nitrogenous media. From that data, nitrogen-fixing bacteria appeared to be widely but unevenly distributed at all depths in seawater and sea-bottom sediments.

The nitrogen-fixing activity of seawater samples and sediment samples obtained from various sea areas was examined by the acetylene-reduction method. Rates of acetylene reduction in those samples were generally low; however, the addition of organic substrates remarkably stimulated the activity of many of those samples. The process of nitrogen fixation is known to require energy, and thus heterotrophic nitrogen-fixing bacteria are dependent on an organic carbon substrate for nitrogen fixation.

This paper is concerned with these substrate effects on acetylene reduction by seawater or sediment samples and on the properties of bacteria capable of growing on non-nitrogenous media containing these substrates as carbon sources.

MATERIALS AND METHODS

Sampling Location

Seawater samples and sea-bottom sediment samples were collected during cruises by the research vessels *Tanseimaru* and *Hakuhomaru* of the Ocean Research Institute, University of Tokyo: KT-69-4, 28 May to 1 June 1969; KT-70-14, 4 November to 8 November 1970; KT-71-12, 14 August to 17 August 1971; KT-71-19, 28 November to 1 December 1971; KT-72-7, 12 June to 15 June 1972; KT-72-8, 20 June to 22 June 1972; KH-69-4, 22 August to 14 October 1969; KH-70-4, 1 July to 24 August 1970; and KH-71-2, 10 May to 18 May 1971. The sampling stations of these cruises are shown in Fig. 1.

Enumeration and Isolation of Bacteria

The plate-culture method using the two formulas for non-nitrogenous media was adopted in the investigation for the purpose of counting and isolating nitrogen-fixing bacteria. The composition of the media A and N, and medium P for heterotrophic bacteria has been described in a previous paper (Maru-

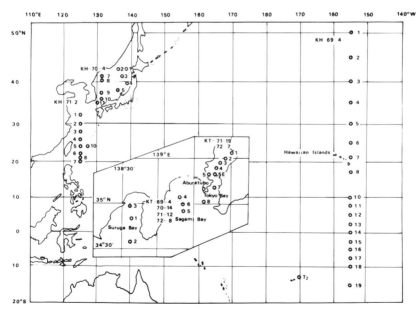

Fig. 1. Location of sampling stations.

yama *et al.*, 1970). The method of sampling and enumeration of bacterial numbers attached to plankton and in seawater has also been described (Maruyama *et al.*, 1970). Taxonomic studies with the purified bacteria were made following the method described by Colwell (1970).

Measurements of Nitrogen Fixation

Nitrogen-fixing activity was measured by the acetylene-reduction method. Seawater samples (10–30 liters), obtained by using the Van-Dorn water sampler, were passed through type HA Millipore filters to concentrate the microorganisms. The filters were broken into small pieces and suspended with the filtered seawater in 20- to 40-ml glass vessels with a stopcock or rubber serum stoppers for gas exchange. Sediment samples (1–10 g of wet weight) obtained by using a gravity core sampler or the Smith-McIntyre bottom sampler were suspended with filtered seawater in the reaction vessels. In the experiment on Aburatsubo Inlet sediments, Tris-HCl buffer (pH 8.0) containing 3% NaCl was used as the suspending solution.

Air was removed by evacuation and replaced by a gas mixture containing acetylene (7 cm Hg) and oxygen (4–10 cm Hg, usually 5 cm Hg) with the balance argon for aerobic conditions (abbreviated as O_2 in the text), or containing acetylene (10 cm Hg) with the balance argon for anaerobic conditions (abbreviated as $-O_2$). The evacuation and refilling procedure was repeated three times.

In most experiments on substrate addition, 0.05 ml of a substrate mixture solution containing 5 μmole each of glucose, mannitol, sucrose, and Na-pyruvate (collectively abbreviated as S or GMSP), or a solution containing 20 μmole of a single substrate was added to the reaction vessel. The vessels were incubated at various temperatures (3–20°C) as close as possible to the *in situ* temperature. Relatively long incubations, usually from several hours to days and occasionally for a month, were necessary to detect significant amounts of ethylene. Controls with the same volume of filtered seawater as the test samples were subjected to the identical procedures in each of the experiments.

Ethylene formation was measured with a gas chromatographic apparatus, Shimadzu GC-4APF (hydrogen flame ionization detector), fitted with a 3 mm by 100 cm Porapak T column and run at a column temperature of 50°C.

RESULTS

General Properties of Acetylene Reduction by Sediment Samples

Some of the properties of acetylene reduction were examined with sediment samples taken at a station (6–7 meters depth) in Aburatsubo Inlet. The rate of acetylene reduction by the samples was usually 2–4 nmole ethylene produced per day at 20°C per gram of sample.

Maximum ethylene production was found at 10 cm Hg oxygen concentration. Approximately the same amount of ethylene was produced under anaerobic conditions. The optimum pH for acetylene reduction was 8.0 in the pH range 5.0–10.0 examined under both aerobic and anaerobic conditions.

Ethylene production increased with increasing NaCl concentration and reached a maximum at 2–3% NaCl under anaerobic conditions, and 3–4% under aerobic conditions. At NaCl concentrations above this maximum ethylene production decreased rapidly.

The relationship between the rate of acetylene reduction and substrate (acetylene) concentration is shown in Fig. 2. Aerobic activity reached a maximum at 5 cm Hg acetylene, whereas the anaerobic activity peaked at about 15 cm Hg. High concentrations of acetylene (20 cm Hg) were found to inhibit the aerobic acetylene reduction.

Acetylene reduction at 20°C in this sediment was constant with time for 72 hr. Another sediment sample (KT-71-19, station 4) displayed a constant rate for 23 days at 3°C incubation temperature. Anaerobic reduction proceeded at a constant rate for 48 hr in Aburatsubo sediment samples and for 10 days in Tokyo Bay samples, and then decreased gradually in both samples.

Acetylene reduction by sediment samples was greatly influenced by incubation temperature; Fig. 3 shows the comparison of the effect of incuba-

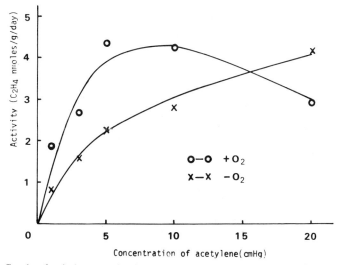

Fig. 2. Graph of ethylene production *versus* acetylene concentration for Aburatsubo Inlet sediment.

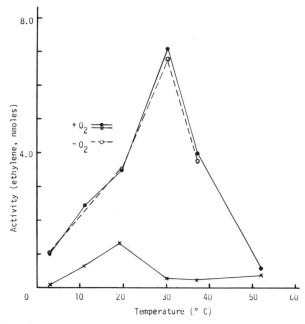

Fig. 3. Effect of temperature on acetylene reduction by sediment samples from Tokyo Bay (KT-71-19, station 4, 32 m depth, ●, ○) and Sagami Bay (KT-71-12, station 6, 1370 m depth, ×).

tion temperature on acetylene reduction in the Tokyo Bay sediment (*in situ* temperature 16°C) with that in the Sagami Bay sediment (*in situ* temperature 2.5°C). Maximum ethylene production in the Tokyo Bay sediment occurred at 30°C compared with 20°C in the Sagami Bay sediment. Similar results were obtained in the Aburatsubo sediment (*in situ* temperature about 15°C) to those in the Tokyo Bay sediment.

The influence of storage at 4°C on acetylene reduction in sediment samples from Aburatsubo was examined. Aerobic acetylene reduction was constant for 3 months storage and then gradually decreased to about one-half after 6 months storage. Anaerobic activity in the samples was stable for a month and 50% of the activity was lost after 4–5 months storage.

Effect of Substrates

Ethylene production by seawater samples and sediment samples obtained from various sea areas was generally low and in many cases undetectable under the present experimental conditions. However, the activity was remarkably stimulated by the addition of substrates (S) in many samples (Table 1). The effect of various compounds on the rate of acetylene reduction was investigated. As shown in Table 2, the reaction was increased about three times by the addition of glucose. On the contrary, $NaNO_3$, known as a potent inhibitor in nitrogen fixation, inhibited ethylene production. Mannitol, sucrose, pyruvate, urea, and ammonium chloride had very little effect. Mannitol, glucose, and yeast extract definitely stimulated anaerobic acetylene reduction, while $NaNO_3$ was again inhibitory.

The activities of sediment samples from Tokyo Bay and Sagami Bay were much more influenced by the addition of substrates than that of Aburatsubo sediments. Glucose and mannitol were highly effective on aerobic acetylene reduction, whereas glucose, sucrose, and mannitol were equally effective on anaerobic reduction in Tokyo Bay samples (Table 3). In the Sagami Bay sediment, pyruvate had the most stimulative effect (Table 4).

Acetylene Reduction by Seawater Samples and Sediment Samples from Tokyo Bay, Suruga Bay, and Sagami Bay

Because the rates of acetylene reduction by these samples were usually very low, long incubation times were necessary to detect ethylene production. The data in Table 5 were obtained after incubation at $20°C$ for 3 days. Although

Table 1. Acetylene reduction by seawater samples and sediment samples

Sample	Station and depth	Ethylene production[a]			
		O_2	$O_2 + S$	$-O_2$	$-O_2 + S$
Sediment	KT-71-12 St. 6				
	0–5 cm (core depth)	0.14	2.9	0.15	—
	10–15 cm (core depth)	0.13	—	0.26	13
Seawater	KH-70-4 St. 5				
	0 m	0.0070	23	—	—
	KH-71-2 St. 7				
	0 m	—	22	—	—
	500 m	—	13	—	—
	1000 m	—	0	—	—

[a] Results given for sediment as nmole/10 g per day at $5°C$ and for seawater as nmole/100 liters per day at $20°C$.

Table 2. Effect of some compounds on acetylene reduction by sediment samples from Aburatsubo Inlet

Compounds	Concentration (mg/g sediment)	Relative activity O_2	$-O_2$
None	—	1.00	1.00
Glucose	40	3.17	2.04
Mannitol	40	0.96	2.62
Sucrose	40	0.78	1.47
Yeast extract	20	1.25	1.95
Urea	20	0.99	0.92
NH_4Cl	20	0.89	1.00
$NaNO_3$	20	0.083	0.15
ATP	10^{-4}[a]	0.57	1.36
Pyruvate	10^{-3}[a]	0.88	1.36

[a] Mole/5g sediment.

Table 3. Effect of organic substrates on acetylene reduction by a sediment sample (KT-71-19, St. 7) from Tokyo Bay

Compounds	Relative activity O_2	$-O_2$
None	1.00	1.00
Glucose	30.7	36.7
Mannitol	52.7	30.6
Sucrose	2.42	35.1
Pyruvate	14.0	1.14
GMSP	36.6	34.2

Table 4. Effect of organic substrates on acetylene reduction by a sediment sample (KT-72-8, St. 6) from Sagami Bay

Compounds	Relative activity
None	1.00
Glucose	17.9
Mannitol	14.2
Sucrose	22.2
Pyruvate	30.5
GMSP	21.9

Table 5. Acetylene reduction by surface-seawater samples from Suruga Bay, Sagami Bay, and Tokyo Bay

Sampling station	Ethylene production (nmole/100 liters/day at 20°C)	
	O_2	$O_2 + S$
KT-72-8		
(Suruga)		
St. 2	3.0	26
St. 3	0.57	7.8
(Sagami)		
St. 6	1.3	16
KT-72-7		
(Tokyo)		
St. 2	3.1	4.7
St. 4	1.7	2.3
St. 5	4.5	5.2

about the same order of ethylene production was found in these samples, acetylene reduction in seawater samples from Tokyo Bay was not as greatly influenced by the addition of organic substrates as those from Suruga Bay or Sagami Bay. The Tokyo Bay samples may have been saturated with organic substrates, as a bloom of phytoplankton was observed when the samples were taken. Assuming a ratio of 1.5 mole of ethylene produced per mole of ammonia fixed, the data in Table 5, calculated as the mean value of three stations, indicate fixation of 290 ng N/m^3 seawater/day at 20°C in Tokyo Bay.

Table 6 shows the ethylene production in sediment samples from Tokyo Bay (incubated for 8 days at 3°C) for various substrate conditions. Using a conversion as for Table 5, a range of values from 0.88 to 10 ng N fixed/g sediment/day at 3°C, with a mean of 3.6 ng N/g/day, was found for aerobic conditions. If appropriate substrates were supplied to the samples, the mean value was raised to 24.7 ng N/g/day. As shown in Fig. 3, acetylene reduction was markedly influenced by incubation temperature. The temperature in Tokyo Bay seems to be over 15°C throughout the year. In recent experiments, the rate of acetylene reduction at 20°C in sediment samples obtained from the same stations in Tokyo Bay (KT-72-7, June, *in situ* temperature 18–21°C) was about four times higher than the results in Table 6, namely, a mean value of 1.45 nmole C_2H_4 produced per gram of sediment per day at 20°C or 13.5 ng N fixed/g/day (186 ng N/g/day when substrates were added).

Table 6. Acetylene reduction
by sediment samples from Tokyo Bay

Sampling station KT-71-19	Ethylene production (nmole/10 g/day at 3°C)			
	O_2	$O_2 + S$	$-O_2$	$-O_2 + S$
St. 1	1.2	6.0	1.4	4.5
St. 2	2.4	4.0	2.7	3.1
St. 3	0.94	1.1	0.53	0.71
St. 4	11	16	5.4	7.4
St. 5E	3.8	53	7.4	23
St. 7	3.6	79	3.5	23

The sediment samples from Suruga Bay and Sagami Bay had less than one-tenth of the acetylene-reduction activity of those from Tokyo Bay.

Properties of Bacteria Growing on Non-nitrogenous Medium N

Bacteria growing on non-nitrogenous media are widely distributed in marine environments. Evidence for close correlation between the number of these bacteria and nitrogen fixation was found in samples from various sea areas. Over 2000 pure cultures of microorganisms growing on medium N were isolated for taxonomic and biochemical studies. Most of the cultured micro-organisms tested (40 strains) showed capability of acetylene reduction, though the activity was very weak when compared with that of *Azotobacter* sp. or *Clostridium* sp. isolated from soils. A great variety of cell morphology was found in these isolated microorganisms and even many yeastlike micro-organisms were isolated. As shown in Table 7, about 60% of the bacteria isolated from seawater samples were gram-positive, in marked contrast to the results that over 90% of heterotrophic bacteria isolated from seawater are gram-negative rods (ZoBell, 1946). About 70% of the bacteria associated with plankton obtained from the Pacific Ocean were gram-negative and, as is the case for marine heterotrophic bacteria, over 90% of the bacteria in the sediments from Suruga Bay, Sagami Bay, and the Japan Sea were gram-negative.

Apparent specific requirements of Ca^{2+}, K^+, or Mg^{2+} for growth were not found in those bacteria tested (199 strains) and none of the gram-positive bacteria (67 strains) required Na^+. Forty-eight of 63 gram-negative bacterial strains tested required Na^+ for their growth.

Glucose utilization was examined by the Leifson test. Most of the bacteria (278) tested were fermentative. About 10% of the bacteria did not oxidize or ferment glucose.

Table 7. Some properties of bacteria
growing on non-nitrogenous medium N^a

Source	Length	Width	Gram stain	
Seawater				
Pacific Ocean	$\leqslant 4\ \mu m$	Slender, short	Negative	64
KH-69-4	188	74	Positive	10
203	$(> 4\ \mu m\ 15)^b$	Oval	Negative	27
		114	Positive	87
Sagami, Suruga	$\leqslant 4\ \mu m$	Slender, short	Negative	46
KT-69-4	157	66	Positive	20
169	$(> 4\ \mu m\ 12)^b$	Oval	Negative	5
		91	Positive	86
Plankton				
KH-69-4	$\leqslant 4\ \mu m$	Slender, short	Negative	57
104	94	74	Positive	17
	$(> 4\ \mu m\ 10)^b$	Oval	Negative	12
		20	Positive	8
Sediment				
KH-70-4	$\leqslant 4\ \mu m$	Slender, short	Negative	32
KT-71-12	36	34	Positive	2
36		Oval	Negative	2
		2	Positive	0

^{a} The figures in each column represent numbers of strains.

^{b} All these microorganisms were gram-positive and many yeast-like organisms
were found.

DISCUSSION

Comparatively low, but detectable, nitrogen-fixing activity was found in
seawater and marine-sediment samples using the acetylene-reduction method.
In some cases, it was necessary to incubate for long periods in order to be
able to measure the ethylene production by gas chromatographic analysis.
The rate of aerobic acetylene reduction with time was constant for 23 days at
$3°C$ and another experiment showed a constant rate for 3 days at $20°C$. The
reaction seemed to be proceeding linearly with further incubation. The
bacteria in the samples were in a steady state, and during the reaction period
their steady state was maintained, since the bacteria could not use acetylene
or ethylene as nutrients.

The fact that the activity of acetylene reduction was stable after 3
months storage at $4°C$ is a great advantage in measuring nitrogen-fixing
activity of the sediment after sampling.

The conditions for acetylene reduction in the sediment samples, such as pH optimum, salt requirement, temperature effect, etc., suggest that the nitrogen-fixing process in sediment samples is adapted to the marine environment.

A similar rate of ethylene production was observed in many sediment samples under both aerobic and anaerobic conditions. However, the reactions were distinct from each other in many respects.

Addition of organic compounds to seawater and sediment samples remarkably stimulated acetylene reduction. A similar experiment was reported in the Waccasassa estuary sediments (Brooks *et al.,* 1971). In their case, acetylene reduction in the sediments was little affected by glucose, but sucrose gave 1.5–2 times enhancement.

The stimulative effect tended to be greater in open-sea samples than in coastal-sea samples. It seems likely that the effect is closely related to the amount of organic carbon substrates in the environment. The effect of each organic compound was different for the sediment samples from different sea areas. The results suggest that the flora of nitrogen-fixing bacteria is somewhat different in these sediments.

Though more extensive experiments are necessary for an accurate evaluation of the amount of nitrogen fixed, values of 290 ng N fixed/m^3 seawater/day at $3°C$, and 3.6 ng N/g sediment/day at $3°C$ or 13.5 ng N/g/day at $20°C$ were found in Tokyo Bay. The values for the sediments were low compared with 3.07 ng N/g/hour found in the Waccasassa estuarine sediments (Brooks *et al.,* 1971). If the appropriate organic substrates are supplied to the Tokyo Bay sediments the values are raised to about the same order as those in the Waccasassa sediments. As almost the same rate of acetylene reduction was found from the top down to 20 cm in the sediment columns in Tokyo Bay, a considerable amount of nitrogen can be expected to be fixed in Tokyo Bay.

Non-nitrogenous media containing glucose, mannitol, and sucrose seem to provide for good growth of bacteria in seawater. These bacteria demonstrated different morphological features from marine heterotrophic bacteria isolated using ZoBell's medium.

ACKNOWLEDGMENTS

The authors wish to express their sincere gratitude to Professor N. Taga, Ocean Research Institute, University of Tokyo, for his encouragement throughout the course of this study.

We are indebted to the officers and crew of R. V. *Hakuhomaru* and R. V. *Tanseimaru,* Ocean Research Institute, University of Tokyo.

This investigation was supported by grants (JIBP) from the Ministry of Education.

LITERATURE CITED

Brooks, R. H., Jr., P. L. Brezonik, H. D. Putnam, and M. A. Keirn. 1971. Nitrogen fixation in an estuarine environment: The Waccasassa on the Florida Gulf Coast. Limnol. Oceanogr. 16:701−710.

Colwell, R. R. 1970. Collecting the data. *In* W. R. Lockhart and J. Liston (eds), Methods for Numerical Taxonomy, pp. 4−21. Amer. Soc. Microbiol. Washington, D.C.

Goering, J. J., R. C. Dugdale, and D. W. Menzel. 1966. Estimates of *in situ* rates of nitrogen uptake by *Trichodesmium* sp. in the tropical Atlantic Ocean. Limnol. Oceanogr. 11:614−620.

Maruyama, Y., N. Taga, and O. Matsuda. 1970. Distribution of nitrogen-fixing bacteria in the central Pacific Ocean. J. Oceanogr. Soc. Japan. 26:360−366.

ZoBell, C. E. 1946. Marine Microbiology, p. 114. Chronica Botanica, Waltham, Massachusetts.

NON-EQUIVALENCE OF PROTEINS FROM MARINE AND CONVENTIONAL SOURCES FOR THE CULTIVATION OF MARINE BACTERIA

CAROL D. LITCHFIELD

Department of Bacteriology
Rutgers University

Six processed fish proteins (skate, sand lance, dogfish, cod, ocean perch, and flounder) have been tested for their use as protein substrates for the determination of microbial proteolysis. Compared with casein and hemoglobin, the fish-protein sources are one-third to one-sixth as effective when tested with the protease system of *Aeromonas proteolytica*.

In addition, two marine bacterial isolates J-217 and J-218 were grown on the dialyzable hydrolysis products of the six fish proteins. When compared with growth on casein, all six fish proteins failed to provide as much of the required amino acids in an available form. A difference in the amount of the available aromatic amino acids required by J-218 was evidenced by the preferred growth of this isolate on casein, followed by sand-lance protein, dogfish, ocean perch, and skate.

These studies indicate that a potential consequence of extracellular proteases in marine sediments may be the survival or limited growth of the obligate heterotrophs frequently encountered there. The importance of these findings to the incorporation of fish proteins into marine bacteriological media is also discussed.

INTRODUCTION

While studying the distribution of proteolytic bacteria in marine sediments, the question arose as to whether protein preparations from marine animals might not be preferable for the demonstration of proteolysis and growth to the more commonly used casein and beef-extract preparations. To date, though, no evaluations of fish proteins or their hydrolysates as nutrient sources for marine bacteria have been published. In order to answer the first question, then, it was necessary to determine (1) whether fish-protein hydrolysates would result in growth comparable with that obtained with a more standard hydrolysate such as casein, (2) whether there were any significant differences among the various fish-protein concentrates, and (3) to what extent these fish proteins were susceptible to enzymic hydrolysis by marine microbial proteases.

Two marine obligate heterotrophs, J-217 and J-218, which require branched-chain amino acids and aromatic amino acids, respectively, were selected to test the growth-promoting potential of six processed fish proteins: skate, sand lance, cod, ocean perch, flounder, and dogfish. These results were compared with growth on casein hydrolysate using a modification (Litchfield, 1973a) of the dialysis culture procedure in which protein plus enzyme were mixed in a dialysis bag suspended in a nitrogen-free medium. Under these conditions growth is dependent on the dialyzable amino acids and peptides which result from the proteolytic digestion of the substrate placed inside the dialysis bag.

In addition to the nutritional studies, the quantitative susceptibility of the six fish proteins to bacterial proteolytic digestion has also been investigated. There are numerous reports of bacterial involvement in fish-protein digestion either in diseases such as fin rot (Oppenheimer, 1958; Hodgkiss and Shewan, 1950; Li and Fleming, 1967) or in processed fish (Marini and Spalla, 1964; Akamatsu, 1959). To quantitate the susceptibility of the various fish proteins, one organism was selected, *Aeromonas proteolytica,* which produces both an extracellular aminopeptidase (Prescott and Wilkes, 1966) and an extracellular endopeptidase (Prescott and Willms, 1960). Culture filtrates from this organism have been used to test the six fish proteins and the data have been compared with the more widely used hemoglobin and casein substrates.

MATERIALS AND METHODS

Bacterial Cultures and Inoculum Preparations

The two obligate heterotrophic isolates J-217 and J-218 have been tentatively identified as species of flavobacteria and will be described more thoroughly elsewhere (Litchfield and Floodgate, unpublished). For inoculum preparation, J-217 and J-218 were inoculated into 100 ml of a medium composed of 0.2% (v/v) pancreatically digested casein (Prescott and Wilkes, 1966), 0.05%

(w/v) sodium glycerophosphate and artificial seawater (Utility Chemical Co., Paterson, New Jersey), pH 7.5 ± 0.1. All cultures were incubated on a New Brunswick Scientific rotary shaker set at 120 rpm at 18°C for 5 days or until the end of the logarithmic growth phase was reached. The cultures were then centrifuged in a Sorvall RC-2B refrigerated centrifuge at 18,000 X g, washed once with sterile artificial seawater, and resuspended to an optical density of 0.52–0.40. Each experimental flask was given an inoculation with a 0.1% (v/v) inoculum.

Dialysis Culture Procedure

The construction of the dialysis culture system has been previously described (Litchfield, 1973a). The nitrogen-free medium consisted of 1% (v/v) glycerol and 0.05% (w/v) sodium glycerophosphate dissolved in the artificial seawater. The pH was adjusted to pH 7.6 ± 0.1, and the media dispensed into naphalometer flasks capped with a tube containing the dialysis bag. Into each dialysis bag was added the appropriate protein at a concentration of 10–20 mg protein. After sterilization for 12 min at 121°C, 32 μg of semipurified *Aeromonas proteolytica* protease was added to the dialysis tubing in each experimental flask, and the enzyme was washed down into contact with the protein with sterile seawater. Control flasks included no additional nitrogen source and protein autoclaved with the protease solution already added. All experiments were done in replicate and reported growth yields are the average of the readings taken on a Spectronic 20 (Bausch and Lomb) spectrophotometer at 560 mμ.

Preparation of Protease System

Cells from a 48-hr culture of *Aeromonas proteolytica* were centrifuged at 17,500 X g for 20 min in a Sorvall RC-2B refrigerated centrifuge. The culture filtrate was made to 80% (w/v) saturation with ammonium sulfate at pH 8.0. The precipitate was filtered, resuspended in 0.1 M phosphate buffer, pH 8.0, and dialyzed against running tap water overnight. This crude culture filtrate was tested for endopeptidase activity and protein concentration and found to have an endopeptidase specific activity of 39.5 toward hemoglobin substrate.

Enzyme Assay and Protein Determinations

Endopeptidase activity was determined using the modified Anson hemoglobin (Anson, 1938) method of Burgum *et al.* (1964). One endopeptidase unit is defined as the change in optical density at 280 mμ per ml enzyme per 10 min assay period at 37°C.

Specific activities were calculated following protein determinations according to the procedure of Lowry *et al.* (1951).

Substrates for Growth and Endopeptidase Activity

Samples of protein concentrates of cod fillet, deboned ocean perch (redfish), deboned flounder, deboned sand lance, deboned skate, and deboned dogfish were supplied by the Fisheries Research Board of Canada, Halifax, Nova Scotia, and were prepared by the Halifax IPA (isopropyl alcohol extraction) procedure. When used in the dialysis culture experiments, there was no further treatment applied to the proteins. When used in the quantitative endopeptidase assays, however, all six fish proteins were subjected to urea denaturation according to the procedure of Burgum *et al.* (1964). Vitamin-free casein and hemoglobin (Nutritional Biochemicals Corp.) were treated similarly for the protease assays.

RESULTS

The marine isolate J-217 has previously been shown to require branched-chain amino acids for growth (Litchfield and Floodgate, unpublished). Figure 1 shows that, of the seven compounds tested, casein provides the best source

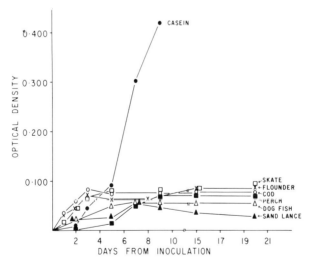

Fig. 1. Growth of J-217 on the dialyzable products of various fish proteins following their digestion by the endopeptidase of *Aeromonas proteolytica*. Into each dialysis bag was added 10–20 mg of protein and 32 μg of semipurified protease. Flasks were incubated at 18°C on a rotary shaker set at 120 rpm. Turbidity resulting from growth was read on a Spectronic 20 spectrophotometer at 560 mμ. There was no increase in turbidity in the two sets of control flasks which consisted of a nitrogen-free medium and the test protein plus autoclaved protease preparation.

of these required nutrients and that there is relatively little difference among the six fish-protein sources. Flounder, though, does supply slightly more of the branched-chain amino acids, either as the free compounds or as peptides, than any of the other fish, while sand lance is by far the least desirable of the fish proteins tested.

Figure 2 depicts the growth patterns for the isolate J-218 which requires aromatic amino acids for growth (Litchfield and Floodgate, unpublished). Again, casein is the best single nutrient source, followed by sand lance, dogfish, and perch. For this organism, though, flounder and cod are both very poor peptide or amino-acid sources.

For both organisms, viable cell counts by standard spread-plate technique indicated an increase of from $3-6 \times 10^3$ cells/ml at inoculation to $3-6 \times 10^5$ cells/ml at an optical density of 0.097.

When cultured in a pancreatically digested casein medium, J-218 had been previously found to produce an extracellular protease (Litchfield,

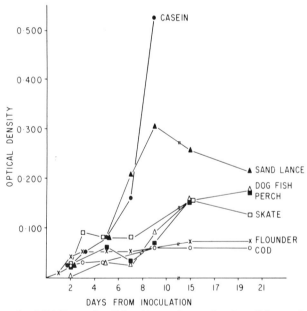

Fig. 2. Growth of J-218 on the dialyzable products of various fish proteins following their digestion by the endopeptidase of *Aeromonas proteolytica*. Into each dialysis bag was added 10–20 mg of protein and 32 μg of semipurified protease. Flasks were incubated at 18°C on a rotary shaker set at 120 rpm. Turbidity resulting from growth was read on a Spectronic 20 spectrophotometer at 560 mμ. There was no increase in turbidity in the two sets of control flasks which consisted of a nitrogen-free medium and the test protein plus autoclaved protease preparation.

Table 1. Comparative susceptibility
of various proteins to hydrolysis by
the endopeptidase of *Aeromonas proteolytica*

Protein[a]	Enzyme units[b]	Percentage hemoglobin activity
Ocean perch	0.85[c]	13.4
Cod fillet	0.92	14.5
Sand lance	0.93	14.7
Dogfish	0.99	15.6
Flounder	1.00	15.8
Skate	1.40	22.1
Casein	2.36	37.3
Hemoglobin	6.33	100.0

[a] All protein sources were urea denatured according to the procedure of Burgum *et al.* (1964).

[b] Enzyme units are defined as the change in optical density at 280 mμ per ml enzyme per 10-min assay period at 37°C.

[c] *Aeromonas proteolytica* endopeptidase was diluted to a concentration of 80.0 μg.

1973*b*). All attempts to demonstrate the production of such an enzyme system when grown in the dialysis culture system and on fish proteins were unsuccessful.

Initially, one might assume that the tremendous differences in the amount of growth obtained on fish proteins is a reflection of the relative susceptibilities of the various proteins to hydrolysis by the endopeptidase of *Aeromonas proteolytica.* That this was not the case is shown in Table 1, for casein is only 2–2½ times more effective as a substrate than the fish proteins. Skate, which is the most efficient protein substrate of the fish tested, does not support growth of either J-217 or J-218 at a level substantially above the others (Figs. 1 and 2). The results in Table 1 indicate that, at least for this endopeptidase, hemoglobin is the superior enzyme substrate. Earlier studies had shown that *Aeromonas proteolytica* endopeptidase can hydrolyze casein and bovine insulin in addition to the hemoglobin (Wilkes *et al.,* 1969), but the latter complex protein is preferred. These studies on fish proteins demonstrate that there are fewer peptide bonds in fish proteins susceptible to hydrolysis by this enzyme.

Another marine sedimentary isolate, J-193, has also been reported to produce an extracellular protease (Litchfield, 1973*b*), and the culture filtrates from this organism were tested on the same substrates. With all six fish proteins, enzyme activity was negligible, while activity was retained toward hemoglobin.

These data, then, seem to indicate that fish proteins are poor substrates for the detection of proteases produced by marine bacteria.

DISCUSSION

Using weanling-rat bioassays, Ousterhout *et al.* (1969) showed that essentially all of the phenylalanine, tyrosine, and tryptophan in the fish proteins they were studying were available for nutritive purposes. However, the branched-chain amino acids isoleucine, leucine, and valine were less so. This perhaps can serve to explain why there is generally better growth of J-218 under conditions reported in this paper than there is of J-217. However, an unspecified fish meal was subjected to pancreatic digestion and found to contain all of the amino acids essential for the growth of bacteria by Scheicher and Sinkovic (1968). In this case, though, they used substantially higher concentrations of the hydrolysates than were used in the present studies.

While there are relatively few studies on the degradation of fish proteins by microbial proteases, the few reports do indicate that fungal proteases and those from streptomyces (pronase) are more effective than bacterial proteases. Olaru *et al.* (1960) noted that a general fish-protein concentrate was more susceptible to proteolysis by *Aspergillus terricola* than was casein. In their study a high titer toxin resulted from growth by *Clostridium perfringens* in the fungal protease digested fish protein. A more extensive study of fish-protein hydrolysis (Hale, 1969) indicated a wide range of activity toward haddock for 20 commercially available proteases. Of those tested, Hale found pronase was the most active with subtilisin, and an undefined preparation (Rhozyme) to be more active than the mammalian and plant proteases trypsin, pancreatin, pepsin, papain, bromelin, and ficin. It is interesting to note, though, that Higashi *et al.* (1965) found that protein prepared from jack mackerel could be degraded by subtilisin but such digestion resulted in a lower nutritive value in rat-feeding studies. Pronase digestion, however, resulted in values comparable with those obtained with casein (Higashi *et al.,* 1965).

It has been suggested (Haynie, 1972) that fish proteins be substituted for the yeast extract-beef extract used in standard nutrient media. General substitution of fish proteins or their hydrolysates for casein or beef extract should be approached with caution because of the considerably more selective nature of the fish proteins. While fish proteins may be useful as an additional nutrient source, the limited growth of the more fastidious organisms, such as tested here, indicates that this group of bacteria could easily be lost if the fish proteins were used in primary isolation media. Evaluation of this possibility is currently underway in our laboratory now.

Furthermore, the extent to which extracellular proteases are produced by microorganisms in the marine environment is not known. Growth of cultures in the laboratory on a rich nutrient source such as casein may result in the induction of extracellular proteolytic enzyme systems which are never expressed under the poor nutrient conditions of the natural environment. However, by the use of the dialysis culture procedure, a mechanism is shown by which such organisms as J-217 and J-218 could obtain nutrients at levels sufficient to maintain their survival, if not growth, in the sediments. If extracellular proteases are produced by bacteria *in situ,* some of the resulting hydrolysis products could become available to the more fastidious members of the microbial community. The possible function in their microenvironment of such obligate heterotrophs is still totally unresolved.

ACKNOWLEDGMENTS

The author would like to thank Dr. R. G. Ackman and Dr. L. Regier of the Fisheries Research Board of Canada, Halifax, Nova Scotia, for kindly supplying the six samples of fish protein used in these studies. Portions of this work were supported by a National Science Foundation Grant GA-32440 and by funds from the Research Council of Rutgers University.

LITERATURE CITED

Akamatsu, M. 1959. Bacteriological studies on the spoilage of fish sausage. I. Number of bacteria present in the meat of fish sausage on the market. Bull. Jap. Soc. Sci. Fish. 25:545−548. (Biol. Abst. 36:49827, 1961).

Anson, M. L. 1938. The estimation of pepsin, trypsin, papain, and cathepsin with haemaglobin. J. Gen. Physiol. 22:79−89.

Burgum, A. A., J. M. Prescott, and R. J. Hervey. 1964. Some characteristics of a proteolytic system from *Phymatotrichum omnivorum.* Proc. Soc. Exp. Biol. Med. 115:39−43.

Hale, M. B. 1969. Relative activities of commercially available enzymes in the hydrolysis of fish protein. Food Technol. 23:107−110.

Haynie, Inc. 1972. Microbiological evaluation of marine proteins. Haynie, Inc., Baltimore, Maryland.

Higashi, H., S. Murayama, T. Onishi, S. Iseki, and T. Watanabe. 1965. Studies on "liquified fish protein". I. Nutritive value of "liquified fish protein". Bull. Tokai Reg. Fish Res. Lab. 43:77−86 (Biol. Abst. 47:96255, 1966).

Hodgkiss, W., and J. M. Shewan. 1950. Pseudomonas infection in a plaice. J. Pathol. Bacteriol. 62:655−657.

Li, M. F., and C. Fleming. 1967. A proteolytic pseudomonad from skin lesions of rainbow trout (*Salmo gairdnerii*). I. Characteristics of the pathogenic effects and the extracellular protease. Can. J. Microbiol. 13:405–416.

Litchfield, C. D. 1973*a*. A laboratory model for the study of the ecological significance of extracellular enzymes. *In* T. Rosswall (ed.), Modern Methods in the Study of Microbial Ecology. Bull. Ecol. Res. Commun. (Stockholm) 17:464–466.

Litchfield, C. D. 1973*b*. Interactions of amino acids and marine bacteria. *In* L. H. Stevenson and R. R. Colwell (eds), Belle W. Baruch Library in Marine Science, Vol. I: Estuarine Microbial Ecology. Univ. South Carolina Press, Columbia, South Carolina, pp. 145–168, 536.

Lowry, O. H., N. J. Rosebrough, A. L. Farr, and R. J. Randall. 1951. Protein measurement with the folin phenol reagent. J. Biol. Chem. 193:265–275.

Marini, F., and C. Spalla. 1964. A new growth factor for a marine bacterium (*Flavobacterium tirrenicum* n. sp.) present in fish meal and a product of microorganisms. G. Microbiol. 12:35–44.

Olaru, A., Zh. Bittner, G. Teodorescu, and S. Fychiu. 1960. Study of toxin formation by *Clostridium perfringens* in a medium produced by hydrolysis of casein and fishbone meal by protease from the fungus *Aspergillus terricola*. Zh. Mikrobiol. Epidemiol. Immunobiol. 31:1880–1885. (Biol. Abst. 37:6552, 1962).

Oppenheimer, C. H. 1958. A bacterium causing tail rot in the Norwegian codfish. Pub. Inst. Mar. Sci. 5:160–162.

Ousterhout, L. E., C. R. Grau, and B. D. Lundholm. 1959. Biological availability of amino acids in fish meals and other protein sources. J. Nutrition 69:65–73.

Prescott, J. M., and S. H. Wilkes. 1966. Aeromonas aminopeptidase: purification and some general properties. Arch. Biochem. Biophys. 117:328–336.

Prescott, J. M., and C. R. Willms. 1960. Some characteristics of the proteolytic system of a marine bacterial species. Proc. Soc. Exp. Biol. Med. 103:410–413.

Scheicher, T., and D. Sinkovic. 1968. Some experiences in the preparations of an enzymic hydrolysate of fish meal protein for microbiological media. Rad. Imunol. Savoda Zagreb 9–10:33–46. (Biol. Abst. 51:137292, 1970).

Wilkes, S. H., B. B. Mukherjee, F. W. Wagner, and J. M. Prescott. 1969. Partial purification and some enzymatic properties of a proteinase from *Aeromonas proteolytica* (33884). Proc. Soc. Exp. Biol. Med. 131:382–387.

ISOLATION AND GROWTH OF A MARINE BACTERIUM
IN LOW CONCENTRATIONS OF SUBSTRATE

A. F. CARLUCCI and SUSAN L. SHIMP

Institute of Marine Resources
University of California, San Diego

A marine bacterium, referred to as 150, has been isolated from water that was taken at a depth of 7 m at the end of the Scripps Institution of Oceanography Pier. This bacterium is a gram-negative, motile rod, capable of growing in extremely low concentrations of dissolved organic matter, such as in unsupplemented seawater. Growth of this bacterium in unsupplemented seawater in the laboratory is faster at 20°C than at 5°C. Bacterium 150 grows more rapidly and to greater cell density in surface than in deep (3000-m) water.

Low-nutrient bacteria may be important in the regeneration of nutrients in the sea, especially in the warmer upper waters and possibly in deep waters.

INTRODUCTION

In the isolation and cultivation of marine bacteria, high concentrations of nutrients are usually included in the various media employed (ZoBell, 1946; Kriss, 1963; Aaronson, 1970; Rodina, 1972). Often as much as 5 g or more of nutrients such as peptone or glucose are added. Although high-nutrient media have been successful for the culturing of many marine bacteria, their use is unrealistic because the amount of nutrients available in the natural environment is so small. The amount of dissolved organic matter in the sea, which is the energy source for most marine bacteria, is generally less than 1 mg C/liter for upper waters and about 0.5 mg C/liter for deeper waters (Menzel and

Ryther, 1970). With the chemoautotrophic bacteria, the concentrations in the sea of the inorganic, energy-yielding ions are also low; for nitrifying bacteria the concentration of ammonia or nitrite is generally less than 0.2 μM (Vaccaro, 1965).

More recently, studies on the heterotrophic uptake of labeled materials have employed low substrate concentrations (Vaccaro and Jannasch, 1966, 1967; Vaccaro et al. 1968; Hamilton and Preslan, 1970; Andrews and Williams, 1971; and references cited therein). In the laboratory Jannasch (1967, 1968, 1970) employed a chemostat for studying the growth of bacteria in seawater supplemented with low levels of nutrients. This report discusses the isolation and growth of a marine bacterium in unsupplemented seawater.

MATERIALS AND METHODS

Isolation of Low-Nutrient Bacterium

Seawater was obtained off the Scripps Institution of Oceanography Pier using sterile Cobet samplers. Samples were taken at the surface and at a depth of 7 m. One-tenth ml of each sample was aseptically plated in triplicate in enriched (CP) or unenriched (SW) solid media. CP medium contained 5 g Bacto-peptone, 0.1 g Bacto-yeast extract, and 0.01 g $FePO_4$ in 1000 ml seawater. CP and SW solid media contained 15 g/liter Difco agar. Plates were incubated at 20°C in the dark and checked periodically for 2 weeks. One of the two types of colonies that grew in the SW plate was picked and carried through several transfers on SW agar before being transferred to a SW agar slant for maintenance. This bacterium was referred to as 150.

Growth of Low-Nutrient Bacterium

All inocula for the various experiments were 4-day-old cultures of bacterium 150 grown in 10 ml sterile unsupplemented seawater. The cells were washed twice with irradiated (1200 watt UV for approximately 4 hr) artificial seawater (Lyman and Fleming, 1940) by centrifugation and resuspension. Ten ml of either aged surface or 3000 m water were placed in acid-cleaned, screw-cap tubes and inoculated with about 0.1 ml of the final washed suspension. The cultures were incubated at both 20 and 5°C. Cell titers were determined periodically by plating an aliquot of the culture on enriched seawater agar medium.

RESULTS

Isolation of Low-Nutrient Bacterium

Two distinct types of colonies grew on the SW agar plates incubated with water from 7 m. One was an agar decomposer and was not studied further.

Table 1. Growth of bacterium 150
in unsupplemented, autoclaved seawaters[a]

Seawater	Temperature ($^{\circ}$C)	Doubling time (hr)[b]	Maximum density (cells per milliliter)[c]
Surface	20	3.0	2.0×10^6
Surface	5	25.0	5.0×10^5
3000 m	20	6.0	1.0×10^5
3000 m	5	34.0	7.0×10^4

[a] Stored in the dark at room temperature for 1 month.

[b] During logarithmic phase of growth.

[c] Initial density was about 10^3 cells per milliliter.

Bacterial cells were isolated from the other colony using standard micro-biological techniques. The bacterium, referred to as 150, is a gram-negative and motile rod. Over 200 colonies grew on CP agar medium, about 10–20 times more than on the SW agar.

Growth of Low-Nutrient Bacterium

Table 1 shows that bacterium 150 grows to higher cell densities in surface than in deep water. As expected, growth is faster at 20°C than at 5°C. The increased growth in surface water over deep water at the same temperature probably reflects the higher amount of available utilizable organic matter in the former. Total dissolved organic matter in the upper water is about twice that of deep water.

DISCUSSION

Bacterium 150 is capable of growing well in extremely dilute concentrations of organic matter. In other experiments, not reported here, we found that 150 grew to various degrees in seawater previously irradiated with a high intensity UV light, in charcoal-treated seawater (Ryther and Guillard, 1962), and in an artificial seawater prepared in double-distilled water. In the latter case, nutrients were probably added as impurities with the salts.

The bacterium seldom grows in seawater to concentrations where tur-bidity can be seen. If peptone is added to seawater in high amounts (0.5–5 g/liter) cell density reaches about 10^9 cells per milliliter. The bacterium does not grow autotrophically, based on results of experiments with NaH^{14}CO$_3$ uptake.

When the widely studied bacterium *Serratia marinorubra* was tested for growth in unsupplemented seawater under the same conditions as 150, it was found to decrease to 50% of the original inoculum titer in 4 days. Bacteria isolated with high medium-nutrient concentrations generally behave similarly.

We have reported previously that nitrate-reducing bacteria were able to grow in unsupplemented or slightly supplemented (0.5 or 1 mg glucose/liter) seawaters (Carlucci and Schubert, 1969). Demonstrations of nitrification by marine bacteria with low substrate concentrations have also been reported (Carlucci and Strickland, 1968; Carlucci and McNally, 1969). Some of the bacteria would only grow in highly or moderately enriched seawater. Jannasch (1967, 1970) reported that a threshold concentration of organic matter was necessary for marine bacterial growth. It seems that some of the high levels of nutrients needed for growth of some bacteria in the sea would only be found in microzones, i.e., decomposing detritus.

In addition to bacterium 150 we have isolated three other bacteria, 151, 152, and 153 (currently under investigation), from various locations in the Pacific Ocean. It seems reasonable to assume that the sea probably contains many bacteria which can readily utilize the low levels of dissolved organic matter present.

Most of the sea is $5°C$ or less, and our laboratory study shows that bacterium 150 still grew well at this low temperature. Jannasch *et al.* (1971) found extremely low microbial activity in deep water. Further work on the effect of hydrostatic pressure on low-nutrient bacteria is needed.

ACKNOWLEDGMENTS

This work was supported, in part, by the Marine Life Research Program, Scripps Institution of Oceanography's component of the California Cooperative Oceanic Fisheries Investigation, a project sponsored by the state of California, and, in part, by the U. S. Atomic Energy Commission, Contract AT(11-1)GEN 10, P.A. 20.

LITERATURE CITED

Aaronson, S. 1970. Experimental Microbial Ecology. Academic Press, New York.

Andrews, P., and P. J. LeB. Williams. 1971. Heterotrophic utilization of dissolved organic compounds in the sea. III. Measurement of the oxidation rates and concentrations of glucose and amino acids in seawater. J. Mar. Biol. Ass. U.K. 51:111−125.

Carlucci, A. F., and P. M. McNally. 1969. Nitrification by marine bacteria in low concentrations of substrate and oxygen. Limnol. Oceanogr. 14:736−739.

Carlucci, A. F., and H. R. Schubert. 1969. Nitrate reduction in seawater of the deep nitrite maximum off Peru. Limnol. Oceanogr. 14:187–193.

Carlucci, A. F., and J. D. H. Strickland. 1968. The isolation, purification, and some kinetic studies of marine nitrifying bacteria. J. Exp. Mar. Biol. Ecol. 2:156–166.

Hamilton, R. D., and J. E. Preslan. 1970. Observations of heterotrophic activity in the eastern tropical Pacific. Limnol. Oceanogr. 15:395–401.

Jannasch, H. W. 1967. Growth of marine bacteria at limiting concentrations of organic carbon in seawater. Limnol. Oceanogr. 12:264–271.

Jannasch, H. W. 1968. Growth characteristics of heterotrophic bacteria in seawater. J. Bacteriol. 95:722–723.

Jannasch, H. W. 1970. Threshold concentrations of carbon sources limiting bacterial growth in seawater. In D. W. Hood (ed.), Organic Matter in Natural Waters, pp. 321–330. Inst. Mar. Sci., College, Alaska.

Jannasch, H. W., K. Eimhjellen, C. O. Wirsen, and A. Farmanfarmaian. 1971. Microbial degradation of organic matter in the deep sea. Science 171: 672–675.

Kriss, A. E. 1963. Marine Microbiology (Deep Sea). (Trans. by J. M. Shewan and Z. Kabata.) Oliver and Boyd, London.

Lyman, J., and R. H. Fleming. 1940. Composition of seawater. J. Mar. Res. 3:134–146.

Menzel, D. W., and J. H. Ryther. 1970. Distribution and cycling of organic matter in the oceans. In D. W. Hood (ed.), Organic Matter in Natural Waters, pp. 31–54. Inst. Mar. Sci., College, Alaska.

Rodina, A. G. 1972. Methods in Aquatic Microbiology. (Trans. by M. Zambruski and R. R. Colwell.) University Park Press, Baltimore.

Ryther, J. H., and R. R. L. Guillard. 1962. Studies of planktonic marine diatoms. II. Use of Cyclotella nana Hustedt for assays of vitamin B_{12} in seawater. Can. J. Microbiol. 8:437–445.

Vaccaro, R. F. 1965. Inorganic nitrogen in seawater. In J. P. Riley and G. Skirrow (eds), Chemical Oceanography, pp. 365–408. Academic Press, New York.

Vaccaro, R. F., and H. W. Jannasch. 1966. Studies on heterotrophic activity in seawater based on glucose assimilation. Limnol. Oceanogr. 11:596–607.

Vaccaro, R. F., and H. W. Jannasch. 1967. Variations in uptake for glucose by natural populations in seawater. Limnol. Oceanogr. 12:540–542.

Vaccaro, R. F., S. E. Hicks, H. W. Jannasch, and F. G. Cary. 1968. The occurrence and role of glucose in seawater. Limnol. Oceanogr. 13:356–360.

ZoBell, C. E. 1946. Marine Microbiology. Chronica Botanica, Waltham, Massachusetts.

OCCURRENCE AND ACTIVITIES OF CELL-FREE ENZYMES IN OCEANIC ENVIRONMENTS

JUHEE KIM and CLAUDE E. ZOBELL

California State University, Long Beach
and
Scripps Institution of Oceanography
University of California, San Diego

Evidence is presented for the widespread occurrence of various cell-free enzymes in aquatic environments, including saline lakes, seawater, and marine sediments. In such environments cell-free enzymes may contribute to the cycle of elements by catalyzing the transformation of various substances. Seasonal variations in the abundance and activities of cell-free enzymes were observed. The properties of cell-free enzymes are probably determined by the nature of the organisms which produced them, although an abiogenic origin of some is mentioned as being within the realm of possibility.

Hydrostatic pressure, temperature, hydrogen-ion concentration, and salinity are considered as environmental factors which affect the stability and activity of certain enzymes. Alkaline phosphatases, amylases, and certain other hydrolases appear to be able to function at all pressures and temperatures occurring in oceanic environments.

INTRODUCTION

The universal occurrence of enzymes in living organisms and the diversity of such organisms at all depths in the sea (Bruun, 1957; ZoBell, 1954) provoke many questions of fundamental significance concerning the stability and activities of enzymes in oceanic environments. The occurrence of cell-free enzymes in seawater and bottom sediments (Hanson and Kim, 1970) is another important aspect of the problem.

Harvey (1925) was one of the first to suspect the presence of naturally occurring cell-free enzymes in seawater. While working on oxidation reactions in seawater, Harvey observed that certain samples of seawater, filtered free of all cellular material, retained the ability to reduce H_2O_2. Heating such seawater to about 100°C or treating it with enzyme inhibitors destroyed its ability to reduce H_2O_2. The continued reduction of nitrate in stored samples of seawater after the death of most of the bacteria suggested to Kreps (1934) that an organic catalyst diffusing out of dead microbial cells was responsible. Indeed, chemical changes were found to occur in seawater filtered free of bacteria. Kreps postulated that organic catalysts may be more concentrated on the sea floor than in seawater. Keys et al. (1935) also observed residual respiration in seawater after the removal of bacteria by filtration through collodion membranes. In studying reducing conditions in marine sediments, ZoBell (1939) described three types of oxygen consumption which are due to (1) the autooxidizable compounds (compounds in the reduced form), (2) an enzymatic adsorption, and (3) a biological adsorption or respiration.

Nearly two decades elapsed before further studies were attempted to demonstrate cell-free enzymes in the marine environment. Goldschmidt (1959) investigated the disappearance of fish bones and crustacean shells in ocean trenches and concluded that the enzyme responsible for the solubilization was an alkaline phosphatase. Some fragmentary evidence of alkaline phosphatase in seawater was recently reported by Wai et al. (1960). Recently, Strickland and Solozano (1966) reported that there was only a slight amount of esterase activity on sugar phosphates, pyrophosphates, and linear polyphosphates in coastal seawater and none in deep-ocean water. Hanson and Kim (1970) investigated activity of cell-free alkaline phosphatase in the marine environment and emphasized the ecological importance of such cell-free enzymes.

Several important papers on free dissolved enzymes in fresh water were reported by Overbeck and Babenzien (1964) and Reichardt et al. (1967). They detected a few enzymes in lake water: alkaline phosphatase, acid phosphatase, and amylase. From their studies on pH, they concluded that only the activities of alkaline phosphatase and amylase could be ecologically significant in lake water. Berman (1970) determined alkaline phosphatase activities in lake waters which were treated with chloroform to kill living cells. He showed seasonal fluctuations on the enzyme activity and indications of induction-repression or activation and inhibition mechanisms. ZoBell and Kim (1972) reported many instances of microbial enzyme activities at deep-sea pressures and summarized the effects of increased hydrostatic pressures on the enzyme activities.

This paper is concerned with the effects of oceanic environment conditions such as pressure, temperature, pH, and salinity on the activities of cell-free enzymes, including some recovered from saline lakes, sea water, and marine sediments, and some prepared from microbial cells.

MATERIALS AND METHODS

Sample Treatment

Employing a modified J-Z bacteriological water sampler and an orange-peel grab, water and sediment samples were collected from several regions off the coast of Southern California, and from two saline lakes in California, the Salton Sea and Mono Lake. Immediately after being collected, the samples were placed in an ice chest and processed within a few hours in the laboratory. The sediment samples were suspended and washed in sterile 3% NaCl solution to elute the cell-free enzymes. The sample size and washing solutions are indicated in Tables 1 and 2. All samples were processed in a cold room by filtering: first, through filter paper (Whatman No. 1) to remove large debris, followed by filtering through a bacteriological filter (Millipore membrane, 0.22 μm). The filtrates were then dialysed against Carbowax (polyethylene glycol, MW 4000) to concentrate the enzyme for analysis.

Alkaline Phosphatase Assay

The activity of alkaline phosphatase (orthophosphoric monoester phosphohydrolase) was tested by the modified method of Andersch and Szczypinski (1947). An equal volume (1 ml) of dialysate was treated with p-nitrophenyl phosphate (p-NPP) (Calbiochem, La Jolla, California) at a concentration of 12.6 μM, 10 μM $MgCl_2$, and 50 mM Tris-HCl buffers (pH 8.0–9.5). The preparations were incubated at 38°C. At time 0 and 5-min intervals the extinction of p-nitrophenol (p-NP) was measured on the Beckman DU-2 at 410 nm, against a substrate buffer water blank. A control was provided for each assay by heating the dialysate in boiling water for 30 sec. Sterility of the preparations was tested by plating them on culturing media for microorganisms such as Marine agar 2216 (Difco). The activity of the naturally occurring cell-free alkaline phosphatase was compared with a commercial preparation (Calbiochem) of the enzyme.

Maximum enzyme activity (μM/min) was recorded as the amount of pNPP released per minute by 1 ml of dialysate. Protein concentration was determined by measuring the absorbance at 280 and 260 nm and applying the correction formula for a mixed solution of proteins and nucleic acids [C = 1.21 (A_{280}) – 0.27 (A_{260}); C, concentration in mg/ml (Warburg and Christian, 1942)].

Table 1. Enzymes detected in samples of coastal water and sediment off Southern California

Location	Source	Sample volume	Dialysis (%)	Protein (mg/ml)	Enzyme(s) detected
Dana Point	Sediment	600 g + 750 ml[a]	87	0.135	Phosphatase, amylase
	Seawater	600 ml	87	0.090	None
Corona del Mar	Sediment	500 g + 500 ml[a]	ND[b]	0.311	Phosphatase, amylase
	Seawater	500 ml	ND	0.156	None
Laguna Beach	Sediment	500 g + 500 ml[a]	92	0.156	None
	Sediment	500 g + 500 ml[c]	84	0.184	Phosphatase, amylase
Santa Ana River Estuary	Sediment	550 g + 770 ml[a]	97	0.290	None
	Sediment	460 g + 800 ml[a]	85	0.147	Phosphatase
	Seawater	405 ml	81	ND	None
Sunset Beach	Sediment	464 g + 250 ml[c]	64	0.096	Phosphatase
	Sediment	425 g + 250 ml[c]	56	0.086	Amylase
	Seawater	500 ml	72	0.170	Amylase
Huntington Harbor	Sediment	550 g + 500 ml[c]	86	0.095	Phosphatase, amylase, succinate dehydrogenase, proteolytic enzyme
	Seawater	450 ml	91	ND	None

[a] 3% NaCl solution.
[b] ND, no data.
[c] Area seawater.

Table 2. Enzymes detected in samples from two inland saline lakes in California

Location	Source	Sample size	Dialysis (%)	Protein (mg/ml)	Enzyme(s) detected
Salton Sea at Salton City	Sediment	500 g + 500 ml[a]	ND[b]	ND	Phosphatase, amylase, urease
	Water	500 ml	ND	ND	None
	Sediment	200 g + 500 ml[a]	90	0.606	None
	Water	500 ml	90	0.350	None
	Sediment	350 g + 600 ml[a]	83	0.612	None
	Sediment	500 g + 750 ml[a]	76	0.296	Phosphatase, amylase
	Water (bottom)	750 ml	76	0.295	Phosphatase, amylase
Mono Lake at Lee Vining	Sediment	700 g + 500 ml[c]	ND	0.824	None
	Lake water	500 ml	ND	0.080	None

[a] 3% NaCl solution.

[b] ND, no data.

[c] 5% NaCl solution.

The Michaelis-Menten constant (K_m) was determined by measuring the effect of substrate concentration on the initial reaction velocity of cell-free alkaline phosphatase. The velocity was examined over a p-NPP range of 2.62×10^{-4} to 2.10×10^{-3} M. The method of least squares was applies to the experimental data to determine the slope and y-intercept. Michaelis-Menten constants (K_m) were then calculated mathematically from Lineweaver and Burke formulas.

Amylase Assay

The activity of amylase (a-1,4-glycan- 4-glycanohydrolase, a-1,4-glycan malto-hydrolase) was assayed by two methods. One was a modified iodometric procedure of Somogyi (1945). A starch-buffer solution was prepared by dissolving 40 mg of starch in 100 ml of citrate buffers (pH 4.0–7.0) and Tris buffers (pH 7.0–8.5). The reaction mixture contained 3 ml starch-buffer solution and 3 ml dialysate and was incubated at 37°C. Two ml of starch-buffer solution were reacted with 0.15 ml N/10 iodine solution and diluted with 1 ml distilled water. The colored reaction mixture was read at 660 nm against a distilled-water blank for maximum absorption at time 0. The decrease in absorption was recorded over 5-min intervals and placed on a relative scale of percentage activity. Sterility and enzymic controls accompanied each reaction. The other method employed Amylopectin-Azure (Calbiochem) as the substrate and measured the amount of released dye at 595 nm (Kim and ZoBell, 1972).

Urease Assay

Urease activity (urea amidohydrolase) was measured by assaying for NH_3 released from urea at 37°C. Equal volumes (1 ml) of 500 μM urea, 50 mM Tris buffers (pH 7.0–8.5), and dialysate were mixed and allowed to react for 5 min. One ml of the reaction mixture was immediately removed and placed in the center well of a Conway dish containing 2 ml 1 N NaOH; the outer well contained 3 ml 1 N H_2SO_4. The dish was sealed with a frosted lid, lined with Vaseline, and incubated for 15 min. At 5 min intervals, 1 ml of the reaction mixture was processed in a similar fashion. At the end of the incubation period, 2 ml of H_2SO_4 were removed and reacted with 1 ml Nessler's reagent and read at 400 nm against an H_2SO_4-Nessler blank. Ammonium sulfate standards were prepared and processed as above. Commercially prepared Jack Bean Meal Urease (Calbiochem) was used for a reaction control.

Succinate Dehydrogenase Assay

Succinate dehydrogenase (succinate: (acceptor) oxidoreductase) activity was assayed by detecting the transfer of hydrogen from sodium succinate. The hydrogen acceptor used in this assay was a 0.1% solution of nitro-blue tetrazolium. Equal volumes (1 ml) of 100 μM sodium succinate, 50 mM Tris buffers (pH 7.0–8.5), hydrogen acceptor, and dialysate were used. A blue to blue-black color change indicated the reduced form of the tetrazolium salt. Sterility and enzymic controls were performed with each assay.

Proteolytic Enzymes Assay

Proteolytic enzyme activity was assayed by measuring the release of the azo dye from a protein-dye complex, Azocoll (Calbiochem). A reaction solution in equal volumes (1 ml) of Azocoll (5 mg/ml), 50 mM Tris buffers (pH 7.0–8.5), and dialysate, plus 5 drops of 1 M cysteine was incubated in sterile test tubes for periods up to 24 hr at 37°C. The reaction solution was filtered through a Whatman No. 1 filter to stop the reaction and the absorbance was measured at 580 nm against a water-Azocoll filtered blank.

Two bacterial proteolytic enzymes were also employed in this investigation. One was protease from *Bacillus subtilis* and the other, known as pronase, from *Streptomyces griseus.* Both enzymes were obtained from Calbiochem and their activities were determined using Azocoll as their substrates.

Compression

The apparatus and techniques used for the study of pressure effects on the enzymes in this investigation are similar to those that have been described by ZoBell and Oppenheimer (1950) and Kim and ZoBell (1972). The reaction mixtures containing enzymes and their specific substrates were placed in small test tubes. The tubes were sealed with Neoprene stoppers which function as pistons. A glass bead was placed in each tube to promote mixing the reaction mixtures. The tubes were placed in pressure vessels (barokams) and incubated at desired temperatures. During the incubation, the barokams were periodically inverted to facilitate mixing of the reaction mixtures. After incubation, the barokams were decompressed and the enzyme activities were measured at 1 atm.

RESULTS

Alkaline Phosphatase Activity

Cell-free alkaline phosphatases were detected in various regions of the marine environment. The one isolated from Huntington Harbor was used to deter-

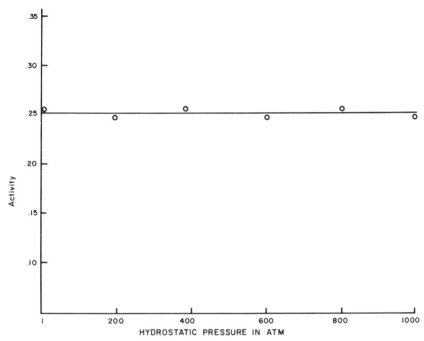

Fig. 1. Activity of cell-free alkaline phosphatase at increased pressure. Activity is expressed as absorbance increase at 410 nm per 15 min at 25°C.

mine the pressure effects on cell-free alkaline phosphatase activity. This enzyme has been characterized. It showed two peaks on the pH curve (pH 8.5 and 10.0), indicating its heterogeneous nature. Its activity was slightly increased in increased salt concentration up to 5%. The enzyme showed maximum activity of 0.81 μM/min and K_m was 3.60 μM. When the enzyme was compressed to 1000 atm at room temperature, its activity was not affected by the increased pressure. The results are shown in Fig. 1.

According to Penniston (1971), the alkaline phosphatase of chicken intestine was not inactivated by pressures up to 800 atm, but at higher pressure it was inactivated, thereby demonstrating a biphasic curve. On the other hand, Morita and Howe (1957) reported that, although the activities of alkaline phosphatase of many species of marine bacteria reacted differently to increased hydrostatic pressure, the enzyme from both of two cultures isolated from true pelagic sediments gave increased phosphatase activity at increased pressure.

AMYLASE ACTIVITY

Cell-Free Amylase

Cell-free amylase activities were detected from water and sediment samples collected from various regions off the coast of Southern California and also from saline lakes as shown in Tables 1 and 2.

In the case of the Sunset Beach samples, amylase was detected in the surface-water samples as well as in the sediment samples. The water samples were collected off the east side of the Sunset Beach jetty and the sediment samples, from an entirely sandy bottom, were taken from approximately 6 m of water in the same region. The amylase activity of the water and sediment samples also showed similar curves with respect to pH (Fig. 2). They both exhibited maximum activity at pH 6.0 and a lesser peak at pH 4.5.

Figure 3 shows the cell-free amylase activity of Huntington Harbor samples against pH and salt concentrations. Sediment and water samples were collected from a back bay and mud flats of Huntington Harbor. The sediment, of a black muddy texture, was obtained from depths of about 1 m,

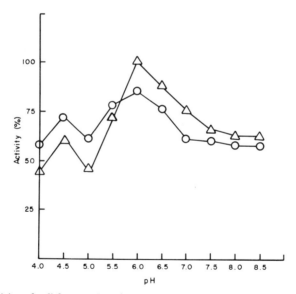

Fig. 2. Activity of cell-free amylase from Sunset Beach *versus* pH. The reaction mixture contained a 1 : 1 mixture of starch-buffer solution, citrate buffers (pH 4.0–6.5), or Tris-HCl buffers (pH 7.0–8.5) and 0.170 mg enzyme solution (protein). Relative activity (%) was recorded in sediment (△———△) and in the surface water (○———○) at 37°C. One hundred per cent activity equals 0.2 mg/ml starch per 5 min.

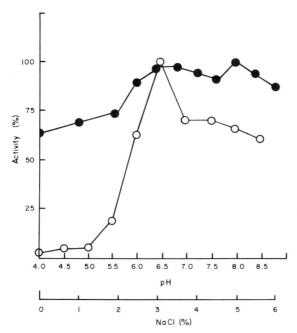

Fig. 3. Activity of cell-free amylase of Huntington Harbor *versus* pH (○────○) and NaCl concentration (●────●). The reaction mixture for pH contained a 1 : 1 mixture of starch-buffer solution, citrate buffers (pH 4.0–6.5) or Tris-HCl buffers (pH 7.0–8.5), and 0.095 mg enzyme solution (protein). The reaction mixture for increased NaCl concentrations contained a 1 : 1 : 1 mixture of starch-buffer solution (pH 6.5), NaCl solution (0–18%), and 0.095 mg enzyme solution (protein). Relative activity (%) was recorded at 37°C. One hundred per cent activity equals 0.2 mg/ml starch per 5 min.

along with surface-water samples from the same area. Amylase from the sediments, over the pH range 4.0–8.5, showed only a single peak activity. Maximum activity was obtained at pH 6.5. There was some activity in the range of pH 7.0–8.5 but little or none at pH 4.0.

The effects of NaCl on the amylase activity were studied over a concentration range from 0 to 6% at its optimum pH 6.5. As the results show in Fig. 3, two peaks were observed, one at 3% and the other at 5% NaCl. This might suggest that the single peak observed in the pH curve was probably a result of two types of amylases, possibly isoamylases. The cell-free amylase is catalytically efficient in seawater.

Mud samples from Laguna Beach showed some activity but most likely the enzyme concentration was very low (Table 1). No amylases were found in the Santa Ana River estuary and only small amounts were found in the Salton Sea (Table 2).

Bacterial α-Amylase

An insolubilized α-amylase of *Bacillus subtilis* (Enzite alpha-amylase, Miles-Seravac, Elkhart, Indiana), was found to be quite stable at pressures up to 1000 atm at 37°C (Table 3). According to Kim and ZoBell (1972), the activity of bacterial amylase at 1000 atm is inhibited somewhat more at pH 8.2 than at pH 6.0. Although deep-sea pressure inhibits the growth of a good many species of marine bacteria, their amylases as well as α-amylase from *Bacillus subtilis* were shown by ZoBell and Hittle (1969) to be fairly stable and active at deep-sea pressures.

Proteolytic Enzyme Activity

Cell-free proteolytic enzymes were detected from sediment samples obtained from the Salton Sea and the Huntington Harbor using Azocoll as its substrate. When two bacterial proteolytic enzymes were tested for their activities at deep-sea pressures in seawater, it was found that both enzymes showed slight inhibition by increased hydrostatic pressure as shown in Fig. 4. Their activities were, though, still significant at 4°C even at 1000 atm.

Other Cell-Free Enzymes Active in the Marine Sediment

Significant activity of urease from the Salton Sea was found by the Conway dish method, and only slight urease activity was detected in Huntington Harbor sediments. Activity of succinate dehydrogenase was observed in Huntington Harbor sediments but the activity was very low.

Table 3. Stability of insolubilized α-amylase at 1 and 1000 atm at 37°C[a]

Pressure (atm)	Absorbance at 595 nm		
	0 hr	24 hr	72 hr
1	0.81	0.83	0.83
1000	0.81	0.77	0.77

[a] The enzymes were removed after decompression at the end of the desired incubation time and tested at 1 atm against Amylopectin-Azure as substrate at 37°C for 30 min. The substrate-enzyme mixtures were periodically mixed and at the end of 30 min they were filtered and the amount of dye released was measured at 595 nm.

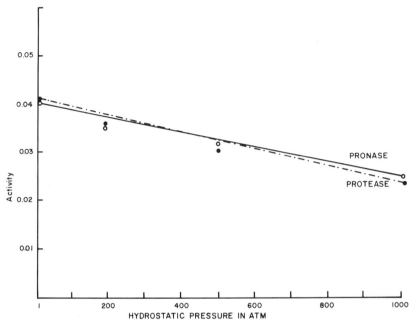

Fig. 4. Activities of protease (*Bacillus subtilis*) and pronase (*Streptomyces griseus*) at increased pressure in seawater. Activity is expressed as absorbance increase at 520 nm per 72 hr at 4°C.

DISCUSSION

Occurrence of Cell-Free Enzymes

Most obvious among the many mechanisms whereby cell-free enzymes may be formed in oceanic environments are (a) the excretion of extracellular enzymes by various organisms and (b) the release of intracellular enzymes from cells undergoing lysis or autolyses. Of course, the possibility of abiogenic synthesis should not be overlooked as a source of cell-free enzymes, since certain organic catalysts have recently been synthesized in the laboratory. Cell-free enzymes would probably remain active in the environment until denatured or inactivated. One might expect to find cell-free enzymes at least as long as any form of life exists in the environment.

In order to demonstrate the occurrence of cell-free enzymes in aquatic environments three criteria must be satisfied: (1) one must preclude the possibility of the enzyme's having been set free from living or dead cells as a result of sampling and/or laboratory procedures; (2) one must show that the catalyst is protein in nature and subject to heat or pH denaturation; (3) the organic catalyst must exhibit reaction specificity such as substrate affinity.

Preparation of the cell-free enzymes from the seawater samples presented a couple of problems. First, the enzymes dissolved in the seawater are highly dilute and require extremely efficient techniques to concentrate them. In our investigations, dialysis was used to concentrate the enzymes but it presented a problem of salt concentration in the enzyme preparations. A method employing a protein-concentration technique through a membrane-filtration procedure, such as the use of an Amicone ultra-filtration concentrator, was found to be satisfactory. The selection of the membranes to be used for filtration should be determined depending upon the molecular weight of the enzymes being studied. One should bear in mind that many of the extracellular enzymes possess relatively low molecular weights.

There appears to be a seasonal variation in the occurrence of certain cell-free enzymes. There was no evidence for alkaline phosphatase or amylase activity in sediment samples collected from the Salton Sea during the hot summer months, whereas such activity in samples collected during the winter months indicated the presence of both enzymes. Such seasonal occurrence was also demonstrated for cell-free enzymes dissolved in freshwater (Overbeck and Babenzien, 1964; Berman, 1970).

In our preliminary studies, little or no correlation was found between the activity of cell-free alkaline phosphatase and microbial biomass in marine sediments when the microbial mass was determined in two ways: by total bacterial plate count of the sediment samples and by determination of adenosine triphosphate (ATP) concentration of the samples using the firefly's luciferin-luciferase system. In working with the free-alkaline phosphatase dissolved in lake waters, Berman (1970) also indicated that the peaks of high phosphatase from 1-m depth did not correspond exactly to observed maxima of *Peridinium*. However, he observed that, with increasing biomass in the photic zone, a concomitant increase of enzymic activity appeared at the 1-m level.

The types and concentrations of the cell-free enzymes are further influenced by the sampling locations, the dominant types of life in that given environment, and density of the life forms. Many of the cell-free enzymes that have been observed during our investigation appear to be isoenzymes which are derived from many different organisms. Alkaline phosphatase, for instance, showed different optimum pH depending upon the location of the sampling, and this may indicate its formation by different species of organisms. Berman (1970) also pointed out that the optimum pH for the free alkaline phosphatase of the lake waters varied seasonally.

Pressure-Temperature Effects

Cell-free enzymes obtained from coastal water and sediments were tested to determine whether they were active at deep-sea pressures. As shown by the data presented in Fig. 1, the free alkaline phosphatase of Huntington Harbor was not affected by increased pressure of up to 1000 atm. Even at 1000 atm, the enzymic hydrolysis of p-nitrophenyl phosphatase was as efficient as that at 1 atm. These observations suggest that cell-free enzymes such as alkaline phosphatase are active in their catalytic role of the biochemical reactions even in the deep-sea environment.

Working with several hydrolytic enzymes of microbial origin, Kim and ZoBell (1972) concluded that enzymic hydrolysis of macromolecules present in the oceanic environment, including the deep sea, would be quite efficient at all pressures and temperatures extant in oceanic environments. For instance, the α-amylase of *Bacillus subtilis* was found to be active in seawater at 1000 atm at 4°C. Although the optimum pH for the activity of cellulase of *Aspergillus niger* is in the neighborhood of 4.5, Kim and ZoBell observed significant activity from the cellulase at 1000 atm at 4°C in seawater. Under the same deep-sea conditions, chitinase from an actinomycete was also found to be active in hydrolyzing chitin. Another hydrolytic enzyme they studied was agarase from a marine bacterium *Pseudomonas atlantica* and they reported that the agarase, unlike the other three enzymes mentioned above, was inhibited as the hydrostatic pressure was increased up to 1000 atm. However, it is important to point out the fact that the agarase was still efficient in catalyzing the hydrolysis of agar in seawater at deep-sea pressure and temperature (1000 atm, 4°C).

Proteolytic enzyme activity was tested in seawater at deep-sea pressure and temperature using two bacterial enzymes, pronase of *Streptomyces griseus* and protease of *Bacillus subtilis* to determine their catalytic efficiency under deep-sea environmental conditions. As shown in Fig. 4, the two enzymes were inhibited by increased pressure up to 1000 atm when incubated at 4°C for 3 days. Since the inhibitory effect of the increased pressure was gradual, not complete at least up to 1000 atm, it can be concluded that proteolysis by cell-free and cellular enzymes would probably be active even in the deep sea. Werbin and McLaren (1951a,b) reported that high pressures of up to 8000 psi, about 544 atm, had an accelerating effect on the rate of protein hydrolysis by crystalline chymotripsin and trypsin.

Table 3 shows the stability of insolubilized α-amylase at 1000 atm in comparison with that at 1 atm. The enzyme, which was suspended in a phosphate buffer solution (pH 7.0), was compressed to 1000 atm and then incubated at 37°C. Another set was incubated at 1 atm. After incubating for 0, 24, and 72 hr, the samples were decompressed and their activity was checked at 1 atm using Amylopectin-Azure as the substrate. The results indicate that the insolubilized amylase was almost as stable at 1000 atm as at 1 atm during the 3 days incubation. Although this observation is rather preliminary, it may indicate that the enzymes adsorbed or bound to the sediment particles may remain active under deep-sea pressure. It also suggests that the enzyme systems in deep-sea organisms may be bound to their membranes or other cellular structures. Morita (1972) hypothesized that the difference in pressure tolerance between organisms may lie in the particulate enzymes such as those attached to membranes (which possess lipoproteins of which the lipid portions are hydrophobic) and other cellular structures. He further speculated that, being associated with cellular structures, the enzymes could then undergo concomitant conformational changes when the associated structures undergo conformational change.

In summarizing the pressure effects on microbial enzyme systems, ZoBell and Kim (1972) pointed out that molecular size, multimerism, and the presence of thiol or other unstable groups in the enzyme are some of the factors which determine the pressure tolerance of enzymes.

Salinity Effects

When the cell-free alkaline phosphatase prepared from the marine sediment samples was tested for its stability in increased salt concentration, it was found that most of the enzymes were not inhibited by salt concentration of up to 6.0% NaCl (Hanson and Kim, 1970), indicating that these cell-free alkaline phosphatases could be active in the sea. In the case of the crude preparation of alkaline phosphatase obtained from *Artemia salina,* a brine shrimp of Mono Lake, Calif., increased salt concentration of NaCl up to 6% showed slight inhibitory effects on the enzyme activity, but it was also noted that the Mono Lake water (salinity is approximately 60 °/oo) inhibited the shrimp alkaline phosphatase activity almost completely. The inhibitory effect of the salt-lake water on the alkaline phosphatase was probably due to various types of salt and other inhibitors dissolved in the Mono Lake water.

In the case of the cell-free amylase prepared from Huntington Harbor sediment samples, the enzyme was tested for its activity in increased NaCl concentrations of up to 6%. As the results in Fig. 3 show, the free-amylase activity was not inhibited. This is probably due to the fact that the chloride ion produces an activation, particularly at alkaline pH, and a shift of the optimum pH toward the alkaline side (Dixon and Webb, 1964). The Huntington Harbor amylase, although it behaves as a homogeneous enzyme in its activity *versus* pH (Fig. 2), may be of a mixture of two different enzymes (with peaks at 3.9 and 5.0% NaCl) in its behavior in increased salt concentration (Fig. 3).

It was also reported by ZoBell and Kim (1972) that crystalline eggwhite lysozyme was partly inhibited by seawater when bacterial cells were used as its substrate, and one of the inhibitory effects of the seawater was due to high concentrations of polycations dissolved in the seawater.

Kim and ZoBell (1972) tested the stability of several extracellular enzymes of microorganisms in seawater and observed that α-amylase, agarase, cellulase, and chitinase were stable in seawater.

pH Effect

Activity of cell-free alkaline phosphatase observed in the sediment samples from Dana Point and Corona del Mar was very similar in response to pH change, both demonstrating one main peak at 8.5 and a lesser peak at pH 10.0, whereas the same enzyme from Huntington Harbor showed only one peak at pH 9.0 and the Salton Sea phosphatase had its maximum activity at pH 10.0. In view of the pH of the oceanic environment, which may range from 7.5 to 8.3, most of the cell-free alkaline phosphatase investigated should be catalytically efficient and play an important role in the mineralization of organic phosphates.

The cell-free amylase obtained from Huntington Harbor and Sunset Beach samples demonstrated high activity at the seawater pH ranges and should be significant in their hydrolytic activity in oceanic environments. When a bacterial α-amylase was tested in seawater (pH 8.2) and also in a phosphate buffer solution (pH 8.0), the purified enzyme showed less activity than at pH 6.0 which is its optimum pH. Nevertheless, its activity at pH 8.0–8.2 was significant (Kim and ZoBell, 1972). However, if the cell-free amylase is either bound or adsorbed to the sediment particles, the optimum pH for the activity of the bound enzyme may be different from the optimum pH observed when the free enzyme is placed in solution, as such examples have been demonstrated by Goldstein *et al.* (1964). They reported that the pH activity profile of insolubilized trypsin at low ionic strength was found to be displaced by approximately 2.5 pH units toward more alkaline pH values, when compared with trypsin under similar conditions.

ACKNOWLEDGMENTS

This investigation was supported by grants from Research Corporation (Grant No. 5242), Long Beach California State University Foundation (Account No.

212.92), and National Science Foundation (Grant No. GB-16605). The authors wish to express their appreciation to Roger B. Hanson for his technical assistance.

LITERATURE CITED

Andersch, M. A., and A. J. Szczypinski. 1947. The use of *p*-nitrophenylphosphate as the substrate in determination of serum acid phosphatase. Amer. J. Clin. Pathol. 17:571–574.

Berman, T. 1970. Alkaline phosphatases and phosphorus availability in Lake Kinneret. Limnol. Oceanogr. 15:663–674.

Bruun, A. F. 1957. Ecology of deep sea and abyssal depths. *In* J. W. Hedgpeth (ed.), Treatise on Marine Ecology and Paleoecology. Geol. Soc. Amer. Mem. 67 1:641–672.

Dixon, M., and E. C. Webb. 1964. Enzymes, 2nd edn. Academic Press, New York.

Goldschmidt, H. 1959. Die Bedeutung der Enzyme in Meerwasser. Chem. Z. 83:442–445.

Goldstein, L., L. Yehuda, and E. Katchalski. 1964. A water-insoluble polyanionic derivative of trypsin. II. Effect of the polyelectrolyte carrier on the linetic behavior of the bound trypsin. Biochemistry 3:1913–1919.

Hanson, R. B., and J. Kim. 1970. Activities of cell-free enzymes dissolved in the marine environment. Bacteriol. Proc. 1970:27.

Harvey, H. W. 1925. Oxidation in seawater. J. Mar. Biol. Ass. U.K. 18:953–969.

Keys, A., E. H. Christensen, and N. Krogh. 1935. The organic metabolism of sea water with special reference to the ultimate food cycle in the sea. J. Mar. Biol. Ass. U.K. 20:181–196.

Kim, J., and C. E. ZoBell, 1972. Agarase, amylase, cellulase, and chitinase activity at deep-sea pressures. J. Oceanogr. Soc. Jap. 28:131–137.

Kreps, E. 1934. Organic catalysts or enzymes in sea water. *In* James Johnson Memorial Volume, pp. 193–292. Univ. Liverpool, England.

Morita, R. Y. 1972. Biochemical aspects. *In* R. Brauer (ed.), Barobiology and the Experimental Biology of the Deep Sea. Univ. North Carolina, Chapel Hill, North Carolina.

Morita, R. Y., and R. A. Howe. 1957. Phosphatase activity by marine bacteria under hydrostatic pressure. Deep-Sea Res. 4:254–258.

Overbeck, J., and H. Babenzien. 1964. Uber den Nachweis von freien Enzymen in Gewasser. Arch. Hydrobiol. 60:107–114.

Penniston, J. T. 1971. High hydrostatic pressure and enzyme activity: inhibition of multimeric enzymes by dissociation. Arch. Biochem. Biophys. 142:322–332.

Reichardt, W., J. Overbeck, and L. Steubing. 1967. Free dissolved enzymes in lake water. Nature (London) 216:1345–1347.

Somogyi, M. 1945. A new reagent for the determination of sugars. J. Biol. Chem. 160:61–68.

Strickland, J. D. H., and L. Solózano. 1966. Determination of monoesterase hydrolysable phosphate and phosphomonoesterase activity in sea water. *In* H. Barnes (ed.), Some Contemporary Studies in Marine Science, pp. 665–674. George Allen and Unwin, London.

Wai, N., T. S. Hung, and Y. H. Lu. 1960. Alkaline phosphatase in sea water. Bull. Inst. Chem. Acad. Sinica 3:1–10.

Warburg, O., and W. Christian. 1942. Isolation and crystallization of enolase. Biochem. Z. 310:384–421.

Werbin, H., and A. D. McLaren. 1951*a*. The effects of high pressure on the rates of proteolytic hydrolysis. I. Chymotrypsin. Arch. Biochem. Biophys. 31:285–293.

Werbin, H., and A. D. McLaren. 1951*b*. The effects of high pressure on the rates of proteolytic hydrolysis. II. Trypsin. Arch. Biochem. Biophys. 32:325–337.

ZoBell, C. E. 1939. Occurrence and activity of bacteria in marine sediments. *In* Recent Marine Sediment (Symposium), pp. 416–427. Amer. Ass. Petrol. Geol., Tulsa, Oklahoma.

ZoBell, C. E. 1954. The occurrence of bacteria in the deep sea and their significance for animal life. Int. Union Biol. Sci. (Copenhagen), Ser. B. (16):20–26

ZoBell, C. E. 1968. Bacterial life in the deep-sea. Bull. Misaki Mar. Biol. Inst. Kyoto Univ. (12):77–96.

ZoBell, C. E. 1970. Pressure effects on morphology and life processes of bacteria. *In* A. M. Zimmerman (ed.), High Pressure Effects on Cellular Processes, pp. 85–130. Academic Press, New York.

ZoBell, C. E., and L. L. Hittle. 1969. Deep-sea pressure effects on starch hydrolysis by marine bacteria. J. Oceanogr. Soc. Jap. 25:36–47.

ZoBell, C. E., and J. Kim. 1972. Effects of deep-sea pressures on microbial enzyme systems. Symp. Soc. Exp. Biol. 26:125–146.

ZoBell, C. E., and C. H. Oppenheimer. 1950. Some effects of hydrostatic pressure on the multiplication and morphology of marine bacteria. J. Bacteriol. 60:771–781.

POTENTIAL MICROBIAL CONTRIBUTION TO THE CARBON DIOXIDE SYSTEM IN THE SEA

RICHARD Y. MORITA, GILL G. GEESEY, and THOMAS D. GOODRICH

Department of Microbiology and School of Oceanography
Oregon State University

INTRODUCTION

Textbooks in oceanography (especially chemical oceanography) present the chemical aspects of the CO_2 system in the ocean (Horne, 1969; Riley and Chester, 1971; Skirrow, 1965) with little or no mention of the microbial involvement. Although chemical oceanographers recognize the importance of biological processes in this system, their discussion on the production of carbon dioxide through respiration (mineralization) is practically nil—nor has the marine microbiologist aided the chemist on this subject.

It has been generally estimated that there is approximately 129×10^{18} g of CO_2 in the ocean (Rubey, 1951). CO_2 fixed by photosynthesis per year is estimated to be approximately 13.6×10^{18} g (Steeman Nielsen, 1952). If there is a stability of the CO_2 system in the ocean, there must be a mechanism to replenish the CO_2 fixed. Approximately 90% of the CO_2 produced from organic matter is due to microbial respiration (mineralization), which is brought about by the large number of microorganisms as well as their rapid rate of metabolism.

Unfortunately, most of the past microbial data obtained in any oceanic water mass does not give us any idea as to the rate at which organic matter is mineralized back to CO_2, NH_3, SO_4, etc. The microbial activity, to a large degree, is responsible for the various oxygen-distribution patterns observed in the ocean (Redfield, 1942; Pytkowicz, 1968) and, probably, the pCO$_2$ values obtained in various parts of the ocean. It becomes important to determine the rate of CO$_2$ production (related directly to the oxygen utilized) in any water mass in order to assess the possible dynamic processes involved. Recently the kinetic technique (Hamilton and Austin, 1967; Hobbie *et al.*, 1968; Hobbie and Wright, 1965; Wright and Hobbie, 1965*a,b*, 1966) has been applied to help determine the rate at which microorganisms can take up a specific ^{14}C compound. This technique is discussed further by Hamilton and Wright in this book. The amount of CO$_2$ liberated from a uniformly labeled compound can· give us an indication of the amount of material mineralized (Harrison *et al.*, 1971). Although the use of these techniques must be interpreted with care, they can give us good clues to the activity of the indigenous microflora under field conditions.

HETEROTROPHIC ACTIVITY IN ANTARCTIC WATERS

For the past 2 years, we have been conducting microbiological studies in antarctic waters. One of the objectives was to determine the heterotrophic potential of the microbes in these waters by use of the kinetic techniques. Data from one station in the Antarctic Convergence are presented in Fig. 1. It should be noted that there is little or no activity below 500 m. The

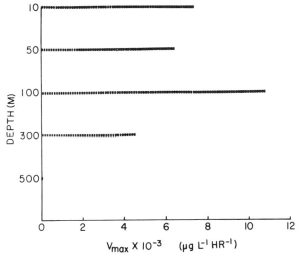

Fig. 1. Depth profile of glutamate uptake in forms at $V_{max} \times 10^{-3}$ (µg/liter/hr) at station 18 (51° 19.00' S, 77° 02.00' E.)

incubation temperature is $-1.0°C$, indicating that high rates of microbial activity do occur in cold environments. Furthermore, many obligately psychrophilic marine bacteria which have been isolated from these waters have maximum growth temperatures of 10 or 15°C.

The V_{max} in Fig. 1 was calculated by the combination of the amount of radioactivity in the cells as well as the radioactivity in the respired CO_2. The rate of $^{14}CO_2$ produced per unit time was experimentally determined in the field. From the experimental data we extrapolated the values to take into consideration larger volumes as well as a longer time period—mainly to illustrate the microbial potential. Although we recognize that the addition of substrate to natural water can change the kinetics of the system (Burnison and Morita, 1973), it should also be realized that amino acids other than glutamate occur in seawater (Park et al., 1962). Therefore, the total CO_2 formed by microbial activity may be even greater than our extrapolated values if all amino acids present in seawater were taken into consideration in the Antarctic convergence. The main objective is to point out the fact that microbial activity should not be neglected as one of the main factors for the pCO_2 or oxygen profiles in the sea.

The lack of heterotrophic activity below 500 m (Fig. 1) should not be taken to indicate that no activity takes place. Li (personal communication)

Table 1. Extrapolated carbon dioxide and ammonia production from heterotrophic uptake studies[a]

CO_2	NH_3[b]
96.8×10^{-3} μg/liter/hr	7.41×10^{-3} μg/liter/hr
96.8 μg/m^3/hr	7.41 mg/m^3/hr
2323.2 μg/m^3/day	177.8 mg/m^3/day
2.3×10^3 g/km^3/day	1.78×10^5 g/km^3/day
1.03×10^6 g/mile^3/hr	8.0×10^7 g/mile^3/hr
2.47×10^7 g/mile^3/day	1.9×10^9 g/mile^3/day

[a] Station 8, near ice shelf; 10 m employing [^{14}C]glutamate as the substrate. Calculated from a V_{max} of 8.7×10^{-3} μg/liter/hr of glutamate. Percentage respiration of this substrate is 73.7. Experiment run at *in situ* temperature of $-1.0°C$. The term "mile" is nautical mile.

[b] All nitrogen released assumed to be NH_3.

Table 2. Carbon dioxide production and pCO_2 values at two stations in upwelled water off the Oregon coast

Depth (m)	pCO_2	CO$_2$ production (μM/hr) From glutamate	From serine
Station DB 1			
5	829.0	3.88×10^{-5}	2.34×10^{-5}
10	844.6	8.26×10^{-5}	3.53×10^{-5}
15	851.0	9.00×10^{-5}	2.80×10^{-5}
25	994.7	1.17×10^{-4}	1.90×10^{-5}
Station DB 25			
5	166.9	1.66×10^{-4}	—
10	—	1.85×10^{-4}	—
15	—	2.98×10^{-4}	—
25	241.4	2.91×10^{-4}	—
50	296.8	1.48×10^{-4}	—
100	476.7	1.52×10^{-3}	—

has calculated that when a water mass sinks in an area of convergence (i.e., the Antarctic water) it does not upwell for approximately 1100 years. This simply indicates that the time scale in which we conduct our experiments is infinitely short compared with 1100 years. Since we are employing an amino acid in the kinetic uptake studies in the Antarctic, the nitrogen liberated should also be taken into consideration. In microbial cells the C:N ratio is approximately 5:1. The C:N ratio of glutamate is also 5:1. If the nitrogen liberated is assumed to be ammonia from glutamate, the amount of ammonia liberated from $^{14}CO_2$ respired from glutamate by the cells can be calculated (Table 1). Again we can see that the extrapolated figures (g/mile^3/day) are large. This nitrogen is then available for the tremendous phytoplankton production in the area of divergence. Furthermore, microbes in these waters will eventually sink and some of the cells will eventually lyse or expire. These dead or lysed cells will also release nutrient material (ectocrine compounds) into the water. In addition, phosphate is probably released. Microbes are known to be rich in phosphate and they possess the ability to hydrolyze both ribonucleic acid and deoxyribonucleic acid (both rich in phosphate) and solubilize inorganic phosphate (Harrison *et al.,* 1972). Furthermore, if these obligately psychrophilic marine bacteria are upwelled into warm waters, they could undergo thermal lysis (Haight and Morita, 1966; Kenis and Morita, 1968) thereby releasing more nutrient material.

The foregoing data help explain the regeneration of nutrients in the Antarctic intermediate water described by Deacon (1933, 1937, 1963).

PRELIMINARY DATA ON UPWELLED WATER

We have used the kinetic method to determine whether or not there was a correction of the high pCO_2 values in upwelled water off the Oregon coast. This upwelled water has high pCO_2 values near the bottom. The chemical oceanographers at Oregon State University believe that the increase in pCO_2 values with depth is due to the presence of organic matter in the sediment. Table 2 gives the pCO_2 values with depth off the Oregon coast and the amount of CO_2 mineralized by the microbial population. Again the CO_2 values were obtained by the kinetic technique. The preliminary data illustrate that microbiologists need to be concerned about the broader problems of microbiology as related to the chemical data obtained by oceanographers.

CONCLUSION

The data presented help illustrate that the stability of the carbon dioxide system in the sea is maintained, to a large degree, by microbial activity. The dynamics of the carbon dixoide system as well as the formation of ammonia must be investigated further.

ACKNOWLEDGMENTS

We would like to acknowledge Drs. Paul Gillespie and Larry Jones for the data obtained in Atlantic waters from which we made our mathematical extrapolations, and Lew Gordon for the pCO_2 values obtained off the coast of Oregon. This investigation was supported by research grants GA-28521 and GV-25057 from the National Science Foundation.

LITERATURE CITED

Burnison, B. K., and R. Y. Morita. 1973. Competitive inhibition for amino acid uptake by the indigenous microflora of Upper Klamath Lake. Appl. Microbiol. 25:103−106.

Deacon, G. E. R. 1933. A general account of the hydrology of the South Atlantic Ocean. Discovery Rep. 7:171−238.

Deacon, G. E. R. 1937. The hydrology of the Southern Ocean. Discovery Rep. 15:1−123.

Deacon, G. E. R. 1963. The Southern Ocean. In N. N. Hill (ed.), The Sea, Vol. 2, pp. 281−296. John Wiley (Interscience), New York.

Haight, R. D., and R. Y. Morita. 1966. Thermally-induced leakage from *Vibrio marinus,* an obligately psychrophilic bacterium. J. Bacteriol. 92:418–423.

Hamilton, R. D., and K. E. Austin. 1967. Assay of relative heterotrophic potential in the sea: the use of specifically labeled glucose. Can. J. Microbiol. 13:1165–1173.

Harrison, M. J., R. E. Pacha, and R. Y. Morita. 1972. Solubilization of inorganic phosphates by bacteria isolated from Upper Klamath Lake sediment. Limnol. Oceanogr. 17:50–57.

Harrison, M. J., R. T. Wright, and R. Y. Morita. 1971. Method for measuring mineralization in lake sediment. Appl. Microbiol. 21:698–702.

Hobbie, J. E., C. C. Crawford, and K. L. Webb. 1968. Amino acid flux in an estuary. Science 159:1463–1464.

Hobbie, J. E., and R. T. Wright. 1965. Bioassay with bacterial uptake kinetics: glucose in freshwater. Limnol. Oceanogr. 10:471–474.

Horne, R. A. 1969. Marine Chemistry. John Wiley (Interscience), New York.

Kenis, P. R., and R. Y. Morita. 1968. Thermally induced leakage of cellular material and viability in *Vibrio marinus,* a psychrophilic marine bacterium. Can. J. Microbiol. 14:1239–1244.

Park, K., W. T. Williams, J. M. Prescott, and D. W. Hood. 1962. Amino acids in deep-sea water. Science 138:531–532.

Pytkowicz, R. M. 1968. Water masses and their properties at 160°W in the Southern Ocean. J. Oceanogr. Soc. Jap. 24:21–31.

Redfield, A. C. 1942. The processes determining the concentration of oxygen, phosphate and other organic derivatives within the depths of the Atlantic Ocean. Pap. Phys. Oceanogr. 9:1–22.

Riley, J. P., and R. Chester. 1971. Introduction to Marine Chemistry. Academic Press, New York.

Rubey, W. W. 1951. Geologic history of sea water. Bull. Geol. Soc. Amer. 62:1111–1147.

Skirrow, G. 1965. The dissolved gases–carbon dioxide. *In* J. P. Riley and G. Skirrow (eds), Chemical Oceanography, Vol. 1, pp. 227–322. Academic Press, New York.

Steeman Nielsen, E. 1952. Production of organic matter in the sea. Nature (London) 169:956–957.

Wright, R. T., and J. E. Hobbie. 1965a. The uptake of organic solutes in lake water. Limnol. Oceanogr. 10:22–28.

Wright, R. T., and J. E. Hobbie. 1965b. The uptake of organic solutes by planktonic bacteria and algae. Trans. Limnol. Oceanogr. Mar. Tech. Soc. 1:116–127.

Wright, R. T., and J. E. Hobbie. 1966. Use of glucose and acetate by bacteria and algae in aquatic ecosystems. Ecology 47:447–464.

V
MICROBIAL INTERACTIONS

SOME THOUGHTS ON MARINE MICROBIAL ECOLOGY

RITA R. COLWELL

Department of Microbiology
University of Maryland

Odum (1972) defines "ecology" as the study of the relation of organisms or groups of organisms to their environment, or the science of the interrelations between living organisms and their environment. In more recent years, the interest in ecology has centered on the dynamics of ecological systems–the interrelations between structure and function–and the examination of principles underlying the patterns observed, i.e., major habitat types, regional community or ecosystem types, etc. (Collier *et al.,* 1973). Ecological systems consist of one or more organisms, together with the various components of the physical and chemical environment with which they are functionally interrelated. These systems vary in degree of complexity. Microbial ecology encompasses the interrelationships between microorganisms and the environment. The microorganisms, in this context, include the viruses, mycoplasma, bacteria, actinomycetes, fungi, algae, and protozoa. The components of the microbial community and the terms of ecological relationships at the microbial level have been defined by Brock (1966) and Alexander (1971). Some of the more practical aspects of microbial ecology are presented by Aaronson (1970), Rodina (1972), and Heukelekian and Dondero (1964).

Since the publication of the now classic volume on marine microbiology by ZoBell (1946), several symposia have been held, mainly to consider the unresolved problems in marine microbiology (Oppenheimer, 1964, 1968), and some attempts to cover the field in the light of more recent developments have been made (Droop and Wood, 1968; Wood, 1965, 1967). Marine microbial ecology remains somewhat diffuse and incompletely understood because the necessary data for understanding the complex ecological relationships occurring in the estuaries and oceans are now being collected. Because of their role in energy cycles, the algae have received the greatest attention of the marine microbial ecologists, with bacteria of the geochemical cycles, i.e., nitrogen, carbon, sulphur, iron, etc., also under scrutiny since the turn of the century. Aside from their involvement in the spoilage of fish and fishery products, the heterotrophic bacteria have not been intensively studied. It is not clearly understood, for example, if all marine animals and plants carry a commensal or symbiotic flora, although the evidence is strong that both types of relationships do indeed exist in the oceans and estuaries.

Of the many possible microbial interactions which deserve consideration, the attachments of bacteria to surfaces, in primary film formation and colonization of marine plant and animal surfaces, nutritional and commensal relationships, and rates of heterotrophic activity *in situ* have been considered in this symposium.

LITERATURE CITED

Aaronson, S. 1970. Experimental Microbial Ecology. Academic Press, New York.

Alexander, M. 1971. Microbial Ecology. John Wiley, New York.

Brock, T. D. 1966. Principles of Microbial Ecology. Prentice-Hall, Englewood Cliffs, New Jersey.

Collier, B. D., G. W. Cox, A. W. Johnson, and P. C. Miller. 1973. Dynamic Ecology. Prentice-Hall, Englewood Cliffs, New Jersey.

Droop, M. R., and E. J. Ferguson Wood (eds). 1968. Advances in Microbiology of the Sea. Academic Press, New York.

Heukelekian, H., and N. C. Dondero (eds). 1964. Principles and Applications in Aquatic Microbiology. John Wiley, New York.

Odum, E. P. 1972. Fundamentals of Ecology. W. B. Saunders, Philadephia, Pennsylvania.

Oppenheimer, C. H. (ed.). 1964. Symposium on Marine Microbiology. Charles C Thomas, Springfield, Illinois.

Oppenheimer, C. H. (ed.). 1968. Unresolved Problems in Marine Microbiology. Marine Biology IV. Proceedings of the Fourth International Interdisciplinary Conference, New York Academy of Sciences, Interdisciplinary Communications Program, New York.

Wood, E. J. F. 1965. Marine Microbial Ecology. Reinhold Publishing Corp., New York. 243 pp.

Wood, E. J. F. 1967. Microbiology of Oceans and Estuaries. Elsevier, New York.

Rodina, A. G. 1972. Methods in Aquatic Microbiology. R. R. Colwell and M. S. Zambruski (eds). University Park Press, Baltimore, Maryland.

ZoBell, C. E. 1946. Marine Microbiology. Chronica Botanica, Waltham, Massachusetts.

PERIPHYTIC MARINE BACTERIA AND THE FORMATION OF MICROBIAL FILMS ON SOLID SURFACES

WILLIAM A. CORPE

Department of Biological Sciences
Barnard College
Columbia University

The first organisms to become attached to clean glass slides submerged in the sea were gram-negative rods isolated and identified as species of *Pseudomonas, Flavobacterium,* and *Achromobacter.* They were largely replaced as the predominant periphytes by bacteria tentatively identified as species of *Caulobacter, Hyphomicrobium,* and *Saprospira.* Diatoms and attached protozoa were observed only after bacterial films had become established. All periphytic marine bacteria isolated as well as pure cultures of *Caulobacter halobacteroides* and *Saprospira grandis* secreted non-diffusible, uronic acid containing carbohydrate material that formed insoluble precipitates with alcian blue and cationic detergents. The polyanions were adsorbed to anion exchange cellulose in the hydroxyl form and were eluted with dilute base. The extracellular carbohydrate materials are believed to be related to the adhesion of bacteria to solid surfaces. Polyanionic material produced by various pseudomonads had a molar ratio of neutral sugar to uronic acid that ranged from 1:3 to 3:1. The products from *Caulobacter halobacteroides* and *Saprospira grandis* had ratios of 5:1 and 20:1, respectively. The importance of periphytic bacteria to other marine life is discussed.

INTRODUCTION

About 40 years ago, various workers studying aquatic environments came to realize that solid surfaces submerged in freshwater or seawater greatly stimulated microbiological activity. Chemically clean glass slides submerged in lake water (Henrici and Johnson, 1935; Smith and ZoBell, 1937) or in seawater (ZoBell and Anderson, 1936) became covered with various kinds of bacteria, other sessile organisms and an assortment of particulate and amorphous organic matter.

Stark *et al.* (1938), Heukelekian and Heller (1940), and Harvey (1941) suggested that solid surfaces served to concentrate nutrients from natural waters, thereby promoting microbial activity. ZoBell (1943) studied the effect of solid surfaces on bacterial activity in some detail. He showed experimentally that inert materials such as glass, plastic, porcelain, and sand, among others, stimulated microbial growth in seawater by adsorption and concentration of nutrients and by providing a solid substratum to which sessile organisms could become attached. He pointed out that the ratio of the numbers of bacteria in water and on solid surfaces seemed to be influenced by the concentration and kind of organic matter present, the proximity of the solid surface to the water mass, the time and temperature of incubation, and the kinds of bacteria present. The firm attachment of bacteria to solid surfaces is followed by growth and formation of microcolonies, which results in the development of a film composed of a variety of bacterial types and their products (Corpe, 1970*a*). The possible means by which bacteria become firmly attached to solid surfaces have been discussed in several papers (Corpe, 1970*b*; Marshall *et al.,* 1971*a,b*), but the exact mechanisms of bacterial-cell adhesion have yet to be discovered.

Microorganisms which occur on any kind of solid or semisolid substratum are commonly referred to as periphyta (ZoBell, 1970). The growth of large populations of periphytic bacteria on the surface results in the formation of a film of cells and their products. Corpe (1970*a*) found that some periphytic bacteria produced a voluminous, extracellular material that surrounded and embedded the cells, apparently cementing them to the glass, and in so doing seemed to create a contiguous film.

Although the observation of film-forming marine bacteria is well documented, their significance has only been surmised. Work in my laboratory has been centered on the primary film-forming bacteria and their function in microfouling of surfaces (Corpe, 1972) and in their enzymic activities (Corpe and Winters, 1972). In the present paper I will try to summarize some of the published and unpublished work from my laboratory concerning the formation of primary bacterial films on glass and other surfaces, and to describe some of the properties of the bacteria concerned which may make them important in the establishment and maintenance of other marine organisms.

MATERIALS AND METHODS

Sampling of Seawater for Periphytic Bacteria

Glass slides were used as the main solid substratum for studying the attachment propensities of marine microorganisms. New slides 50 by 75 mm (Corning Glass Co., Corning, New York) or 1 by 3 inch (Fisher Scientific Co.) were used. At first the slides were soaked in dichromate-H_2SO_4 cleaning solution overnight and then washed exhaustively with distilled water to remove dichromate, but later this was abandoned when it was found that dichromate-treated slides and untreated new slides gave equivalent results. When slides were to be suspended in the sea, pairs of sterile slides were taped together with surgical tape along with a loop of nichrome wire by which the slides were attached to a 6-mm marine steel cable. Sterile slides were handled and taped using aseptic precautions, and were wired to the cable just before the latter was submerged into the sea.

Slides submerged in aquaria or large seawater samples were held in an all-glass staining rack (Wheaton Glass Co., Millville, New Jersey). The racks bearing 10 slides were autoclaved before being submerged.

Sampling Sites and Sources of Water Samples

Groups of slides attached to a cable were submerged in the sea from the end of the pier at Scripps Institution of Oceanography (SIO), La Jolla, California, on two sampling periods during the summers of 1967 and 1968. Slides were also submerged in San Diego Bay at Point Loma and from a Navy Electronics oceanographic research tower in the sea off San Diego. Slides were submerged at various depths, but for the experiments to be reported here depths of between 1 and 5 m below the surface were used.

Slides were also submerged in a 70-gal marine aquarium which was maintained at Barnard College. It was fitted with a combination of devices which regulated light, tide, circulation, filtration, and temperature so as to approximate the natural physical factors which define the neritic biome of *Botryllus* (De Santo, 1967). The seawater, flora, and fauna contained in the aquarium were collected from Long Island Sound. Slides in glass racks were placed in the middle of the aquarium on a glass stand, 6–8 inches from the bottom.

Water samples collected from the New Jersey shore at Sandy Hook State Park and those collected at the Battelle Institute test site at Daytona Beach, Florida, were placed in sterile 3-liter beakers and the rack of sterile slides was submerged and allowed to rest on the bottom of the beaker. The beakers were incubated at room temperature.

When pure cultures of periphytes became available (see below) they were tested for their ability to attach to glass slides in time-course experiments. Slides in glass racks were incubated in sterile Wheaton dishes (Wheaton Glass Co., Millville, New Jersey) to which were added 200 ml of sterile seawater enriched with 0.005% peptone or yeast extract.

Microscopic Examination of Submerged Slides and Determination of the Total Numbers of Cells per Square Centimeter

Slides were removed from seawater aseptically after submergence for various periods of time and rinsed five times with sterile seawater. Preliminary experiments had shown that a minimum of three careful rinses removed most of the unattached organisms. As many rinses as ten with sterile seawater did not produce any further change in the countable population.

The rinsed slides were flooded with 2% (v/v) acetic acid for 2 min after which time they were rinsed in tap water and stained with Huckers crystal violet for 1 min. The excess stain was washed off with tap water and the slides allowed to air dry. Twenty oil immersion fields were counted and the average count per field was multiplied by the total number of oil immersion fields per square centimeter, which for the light microscope (Leitz) employed was 9238.

Microscopic counts of cell suspensions or cultures were done with a Petroff-Hauser counting chamber using the procedure described by Wilson and Knight (1947).

Standard Plate Count

Exposed slides were rinsed five times with sterile seawater and an area of 6–9 cm^2 was vigorously swabbed with a sterile Dacron swab (Scientific Products Co., Raritan, New Jersey). The swab was thoroughly shaken in 5 ml sterile seawater. Dilutions were prepared and inoculated in duplicate into petri plates. Bacto Marine agar (Difco, Detroit, Michigan) was used as the plating medium. Plates were incubated at 25–28°C for a week and colonies were counted and the numbers of viable bacteria per square centimeter were calculated.

Dilution Frequency Test for Estimation of the Numbers of *Caulobacter* Species

Dilution of swabbed slides prepared as described above were inoculated into the seawater broth described by Poindexter (1964), incubated for 2 weeks at 25°C, and examined for stalked caulobacters in the phase microscope. Only a few samples were examined by this means when it was discovered that stalked and filamentous forms seen on submerged slides did not produce colonies on Bacto Marine agar by direct plating.

Isolation of Cultures and Determination of the Predominant Bacterial Flora

Twenty-five per cent of the colonies from plates prepared from the highest dilutions were isolated into nutrient broth prepared with seawater and streaked onto Bacto Marine agar plates. On the basis of colony appearance, character of growth in broth, gram reaction, motility, and microscopic morphology, the predominant type of bacteria were selected for more detailed study. The organisms were maintained on Bacto Marine agar slants. Several samples of seawater were diluted and plated and, after incubation, colonies were isolated and included in some phases of this study.

Characterization of the Isolates

Each of the isolates was tested for its ability to become attached to glass slides submerged in Wheaton dishes containing a sterile glass slide and 200 ml artificial seawater (Instant Ocean Aquarium Systems, Inc., Wickliffe, Ohio) containing 0.005% (w/v) Bacto peptone or yeast extract. Cultures were incubated at room temperature and slides were removed, rinsed, fixed, and stained as described above. The morphology and character of microcolonial growth was noted.

Further characterization of the bacterial isolates was done following procedures described by Shewan et al. (1960) and Shewan (1963). Morphology was observed in broth and agar cultures using the phase contrast microscope and after gram staining by the Hucker modification (Society of American Bacteriologists, 1957). Motility was determined in young agar and broth cultures and flagella were stained after fixation with formaldehyde by the Leifson procedure (Leifson, 1964). Other details of characterization were reported elsewhere (Corpe, 1972).

Cultures of Bacteria Studied

Cultures 4b and T6c were isolated as periphytes from California waters and have been identified as *Pseudomonas atlantica* (Corpe, 1970a). Several other organisms were also isolated as periphytes and identified as *Pseudomonas* species: Ma 8 and Ma 13 from the regulated marine aquarium, Ma 71b from New Jersey shore-water samples and 6a and 6b from water samples collected at Daytona Beach, Florida. Cultures labeled 101 and 103 were isolated from agar plates that had been inoculated with seawater collected on the south shore of Long Island, New York. The latter have several flagella and a vibrio-like appearance. In some phases of the work, *Caulobacter halobac-*

teroides, Poindexter (ATCC 15269), *Hyphomicrobium neptunium,* Liefson (ATCC 19614), and *Saprospira grandis,* Gross (ATCC 23166) were employed. Eleven of the principal isolates described in this work were examined for polysaccharide-producing ability. The marine isolates were grown aerobically in a medium containing Bacto peptone, 0.5% (w/v), and glucose, 1% (w/v), in artificial seawater diluted to provide 8 parts of seawater to 2 parts of distilled water. The organisms were grown at 25–28°C in 500-ml flasks containing 100 ml of medium, mounted on a shaker rotating at 70 rpm. After incubation for 4 days, the cultures were chilled and then centrifuged in a model SS-3 Servall centrifuge with a GSA head (Ivan Sorvall Co., Norwalk, Connecticut) at 9000 rpm. The cells were washed three times with seawater diluted 8 : 2 with distilled water, frozen and lyophilized for determination of dry weight. The culture supernatant fluids were added with stirring to 2 volumes of cold methanol or acetone. The precipitate that formed was allowed to settle in the cold room at 4°C and then the clear supernatant fluid was decanted and the precipitate recovered by centrifugation at 3000 × g. The precipitates were desalted by dialysis against 4 changes (20 liters) of 0.01 N HCl and exhaustively dialyzed against distilled water to remove chloride ion. The small amount of precipitate that persisted or that formed during dialysis was removed by centrifugation. The material did not contain carbohydrate and was discarded. The clear supernatant fluid was neutralized to pH 7 with 0.01 N NaOH, lyophilized, and weighed.

Saprospira grandis was grown in the medium of Lewin (1962) and *Caulobacter halobacteroides* grown in a medium described by Poindexter (1964). Culture supernatant fluids were handled as described above.

Some Analyses of the Polysaccharides

The anthrone method described by Neish (1952) was used as a general test for carbohydrates and as a quantitative method for neutral sugars by reference to a standard curve prepared with glucose. The uronic acid content of the polysaccharides was estimated by the Dische carbazole method according to the procedure given by Kabat and Mayer (1961). Glucuronic acid was used to prepare the standard curve. Samples of polysaccharide were hydrolyzed in sealed tubes with 1 N H_2SO_4 for 0.5, 1, 2, 4, and 6 hr in a boiling-water bath. After the tubes were cooled they were opened, the contents neutralized with 1 N NaOH to pH 7 and the reducing values determined by the Park and Johnson procedure as described by Kabat and Mayer (1961). Hydrolysates were also tested qualitatively for amino sugars by the Elson-Morgan procedure (Kabat and Mayer, 1961).

Samples of the polysaccharides were hydrated and suspended in distilled water and diluted to provide concentrations ranging from 1 mg/ml to 25 ng/ml. One-ml samples were mixed with 0.5 ml of 1% (w/v) alcian blue and after 10 min examined for the presence of a water-insoluble precipitate. The test has been found to be quite specific for acid polyanions (Kang and Corpe, unpublished observations). The polysaccharides were also tested for the formation of water-insoluble complexes with the cationic detergent hexa-decyltrimethylammonium bromide (Eastman Kodak Co., Rochester, New York) using the method of Scott (1965).

The adsorption of polysaccharides by DEAE cellulose in the −OH form was done using the column procedure of Neukom and Kuendig (1965). Columns were eluted with distilled water, 0.05, 0.1, and 0.5 N NaOH, in that order, and successively collected 5-ml fractions analyzed for neutral carbo-hydrates and uronic acids as described above.

RESULTS

Preliminary study of glass slides submerged in seawater confirmed much of the earlier work reported by others; i.e., that firm attachment of bacteria to clean glass slides required a period of incubation of from 6 to 12 hr and that firmly attached cells were not dislodged by rinsing many times in sterile seawater. After submergence of slides for 24 hr or longer, not only single cells but microcolonies of short coccoid rods were seen. With longer periods of submergence (4–5 days) there were many more cells many of which were stalked, budding, or filamentous forms (Corpe, 1970b).

Table 1. Time-course estimation of periphytic bacteria on glass slides submerged in a regulated marine aquarium

Time (hr)	Viable count/cm^2 (X 10^3)	Microscopic count/cm^2 (X 10^5)
16	1.67	2.12
48	13.22	7.27
64	21.18	Nd[a]
91	28.60	45.30
139	25.00	162.00
187	1.92	Nd
211	4.14	Nd
238	2.42	Nd
256	1.12	181.00
281	0.35	Nd

[a] Nd, not determined.

Several time-course experiments were done by submerging sterile slides in a large, regulated, marine aquarium. Pairs of slides were removed daily and fixed and stained for microscopic counts or carefully swabbed and plated on Difco Marine agar. The results of an experiment of this sort are shown in Table 1. The total microscopic count per square centimeter was 2.12 × 10^5/cm^2 after 16 hr, 4.5 × 10^6/cm^2 after 90 hr, and more than 100 × 10^6/cm^2 were estimated at the end of 5 days. It was maintained at about that level through the end of the experiment (10 days). The viable count on the other hand reached a peak of only 28 × 10^3/cm^2 where it was maintained for

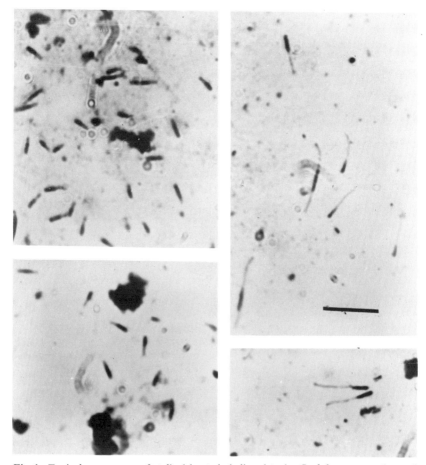

Fig. 1. Typical appearance of stalked bacteria believed to be *Caulobacter* spp. observed on glass slides submerged in seawater for 4–5 days. Stained with crystal violet. The bar represents 5 μm.

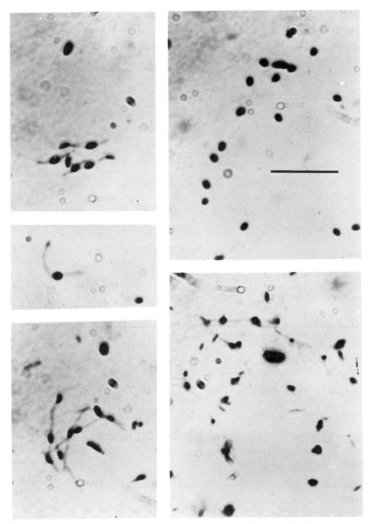

Fig. 2. Typical appearance of budding bacteria identified as *Hyphomicrobium* spp. attached to glass slides submerged in a marine aquarium for 5 days, stained with crystal violet. The bar represents 5 μm.

6 days and then fell to $3 \times 10^2/cm^2$ by the tenth day. The numbers of viable bacteria were always a small fraction of the total count. Vigorous shaking of swabs, used to remove cells from the slides with glass beads, did not produce an increase in viable count by more than a factor of 10.

The periphytic population was initially composed of short rods occurring singly, in small colonies, or short chains. After 48–72 hr, stalked bacteria and budding forms appeared, and by the fourth or fifth day were the predominant type of bacteria. The latter organisms, which are believed to be species of *Caulobacter* and *Hyphomicrobium,* were responsible for the large increase in the total count, but no separate estimate of their numbers was attempted because stalks and hyphal filaments did not always show up with crystal violet staining. The *Caulobacter* species (Fig. 1) were usually attached as

Table 2. Comparison of microorganisms on glass slides submerged at several sites in the sea

Site	Time (hr)	Viable count/cm^2 ($\times 10^2$)	Microscopic count/cm^2 ($\times 10^5$)	Morphological types of attached forms
SIO Pier	24	6.20	5.1	Short, coccoid and slightly curved rods in short chains or clumps
	48	0.79	8.5	Same as above plus few stalked and filamentous forms
	96	44.30	12.0	Mainly stalked and filamentous bacteria, few diatoms
	144	106.40	22.50	Same as above but increased numbers of diatoms
San Diego Bay	24	9.3	28.00	Rods and coccoid rods in clumps
	48	21.00	53.00	Same as above plus stalked bacteria and debris
	96	34.00	Too many to estimate	Stalked bacteria, sessile protozoa, much debris
Sandy Hook, New Jersey	24	11.50	5.21	Short and slightly curved rods
	48	38.00	16.41	Same as above plus stalked and filamentous bacteria
	96	29.00	62.73	Mainly stalked and filamentous bacteria

single cells, and rarely in clusters or colonies. The *Hyphomicrobium* occurred in chains of cells, joined by thin hyphal strands (Fig. 2). Colonies of these organisms were never observed on Difco Marine agar when dilutions of material swabbed from slides was plated directly, so the count of viable bacteria is an estimation of the short, coccoid rods that attached first.

When other sites were studied in a similar way, the events were somewhat similar to those described above. The microscopic count of bacteria showed a steady increase but the viable count at no time even closely approximated the total count (Table 2).

The rate of attachment and growth was different in the various sites sampled, and was probably related to the nutrient levels of the waters and to other environmental factors. Slides submerged in San Diego Bay developed films of bacteria quite rapidly as compared with those submerged from the end of SIO pier or in the water sample from Sandy Hook. In addition, slides submerged in San Diego Bay accumulated a lot of unidentifiable particulate debris after 48 hr which increased in amount and complexity so that after 96 hr a microscopic count was impossible. The viable count was only 34 X $10^2/cm^2$ at 96 hr and did not go higher through 7 days. The organisms to become attached first, again, were short, coccoid or irregularly shaped rods that formed short chains or microcolonies. Stalked and filamentous bacteria appeared next and became predominant.

Filamentous bacteria identified as species of *Saprospira* were the predominant organism on slides submerged from SIO pier for 96 hr. They were less prominent on San Diego Bay slides and not seen at all on slides submerged in Sandy Hook water.

When bacterial films had become established, diatoms often became abundant. Unrinsed slides showed amoebae and an occasional ciliate. Slides submerged in San Diego Bay bore numerous attached ciliates, including species of suctoreans. Slides submerged for 7 days near the bottom often displayed large blooms of a *Zoothamnium* species (Corpe, unpublished observations).

Isolation and Characterization of Primary Film-Forming Bacteria

A representative proportion of the colonies that appeared on pour plates prepared from 1- to 3-day slides exposed at various sites were isolated into seawater nutrient broth and on Difco Marine agar slants. The majority of the isolates were gram-negative, non-spore-forming rods. Using the identification scheme of Shewan *et al.* (1960) and Shewan (1963), they were characterized as species of *Pseudomonas, Flavobacterium,* and *Achromobacter.* Sixty to 90% of the isolates were identified as species of *Pseudomonas,* 10–30% were *Flavobacterium* or *Achromobacter,* and a few organisms from several sites were gram-positive rods or cocci which were not identified further (Table 3).

Table 3. Composition of the periphytic bacterial flora isolated from glass slides submerged in seawater

Site	Slides exposed (days)	Date (mo./yr.)	No. of isolates	Percentage of total				
				Pseudomonas	Flavobacterium	Achromobacter	Others	
Marine aquarium	3	6/67	30	86	14	0	0	
	1	9/67	20	60	10	20	10	
SIO Pier	2	8/67	80	80	16	4	0	
Navy tower	2	8/68	90	90	0	0	10	
Daytona Beach, Fla.	1	4/69	90	90	10	0	0	
	2	4/69	60	60	0	30	10	
Sandy Hook	2	6/70	28	65	17	8	10	

The *Pseudomonas* cultures were characterized with regard to specific physiological differences and have been described elsewhere (Corpe, 1970*a*; 1972). All were gram-negative, non-spore-forming motile rods with a single polar flagellum, were oxidase and catalase positive, and produced no fluorescent pigments. Old colonies on agar are off-white or yellowish in color. All of them grew in seawater nutrient broth at an optimum temperature in the range of 25–28°C. None of the organisms grew in nutrient broth prepared with distilled water.

None of the bacteria isolated from agar plates had the appearance of *Caulobacter, Hyphomicrobium,* or *Saprospira* species in any medium including those used for the enrichment and growth of those organisms. *Caulobacter* and *Hyphomicrobium* species were, however, brought up by enrichment when carefully rinsed slides (4-day) from the marine aquarium were placed in enrichment media (Poindexter, 1964; Hirsch and Conti, 1964); slides submerged for 6 or 24 hr were not.

Attachment and Growth of Pure Cultures on Glass Slides

Attachment and growth of periphytic isolates and other bacteria on glass slides submerged in artificial seawater enriched with 0.005% yeast extract was done and the results of this experiment are summarized in Table 4. The *Pseudomonas* strains became attached and grew at a faster rate than either

Table 4. Attachment and growth of pure cultures of periphytes on glass slides submerged in artificial seawater containing 0.005% yeast extract[a]

	Microscopic count/cm^2 ($\times 10^5$)		
Culture	24 hr	68 hr	120 hr
T6c	<0.09	6.1	2.23
6b	5.40	1.06	0.77
Ma8	3.66	6.48	6.25
Ma71b	0.16	19.73	9.38
103	1.76	2.70	1.06
Caulobacter halobacteroides	<0.09	0.19	0.88
Saprospira grandis	<0.09	0.15	0.93

[a] Two hundred ml of medium inoculated with 3–6000 cells and poured into an all glass dish containing a rack of sterile slides and incubated at 25–28°C.

Caulobacter halobacteroides or *Saprospira grandis*. While the conditions used in this experiment cannot be claimed as more than a general approximation of the natural seawater, the experiment does suggest that the rate of growth may be one factor responsible for the results found in natural conditions. Bacterial isolate 101 is representative of several that were isolated from seawater by direct plating. Since it was not isolated as a periphyte, its attachment and growth was somewhat surprising. After a period of active growth the bacterial count per square centimeter decreased, indicating loss of adhesion.

Carbohydrate Material Secreted by Periphytic Bacteria

When the primary film formers were cultivated in low concentrations (0.005–0.01%) of yeast extract or peptone with glass slides, growth was in large part restricted to the surface of the slide; at least a surface film was apparent though a general turbidity of the medium was not (Corpe, 1970). When slides were fixed and stained with Huckers crystal violet or alcian blue, a lightly stained matrix could often be observed surrounding or embedding the cells. In some cases the stainable material was somewhat restricted to the vicinity of a microcolony, but in other cases the stainable material was fairly extensive and spread about as a continuous film over the surface of the glass (Corpe, 1970a). Extracellular material isolated from cultures of *Pseudomonas atlantica,* represented by T6c and 4b in the present paper, was found to contain a polyanionic carbohydrate composed of several neutral sugars (an-throne positive) and uronic acid among other substances (Corpe, 1970a). It was suggested that this material might be related to the ability of the organisms to attach to solid surfaces and it was reasoned that other periphytic species might produce a similar material, though not necessarily in as large a quantity as found with *Pseudomonas atlantica.*

The various bacterial isolates, including *Caulobacter halobacteroides* and *Saprospira grandis,* were cultivated in broth media, and after various periods of time were centrifuged to remove the cells and the solvent-insoluble carbohydrate material was recovered from the culture supernatant, as de-scribed above. All of the organisms listed in Table 5 secreted extracellular, non-diffusible, acetone-insoluble carbohydrate material, although the quan-tity produced was not large in some instances. The isolated materials invariably contained both neutral sugars and uronic acid but in different molar ratios. The amount of material recovered from culture 101, which was isolated from seawater by direct plating, produced much smaller amounts than any of the periphytic pseudomonads described. *Caulobacter halobacteroides* and *Sapro-spira grandis* material contained less uronic acid in relation to the amount of neutral sugar. The carbohydrate content of the cells was determined as shown

Table 5. Cellular and extracellular
carbohydrate material in cultures of periphytic bacteria

| Culture | Yield per 100 ml of culture[a] | | | | | |
| | Cellular | | | Extracellular | | |
	Cells (mg)	Anthrone carbo-hydrate (μM)	Uronic acid (μM)	Crude polymer (mg)	Anthrone carbo-hydrate (μM)	Uronic acid (μM)
T6c	46	5.8	1.0	130.0	368.0	320.0
6b	46	16.3	3.6	13.4	36.1	19.2
Mal3	66	33.3	3.4	11.3	12.3	12.4
Ma8	65	30.0	4.1	20.5	55.5	22.2
Ma71b	32	1.8	1.2	12.5	15.5	44.4
101	43	2.2	0.3	3.0	0.7	0.8
Caulobacter halobacteroides	Nd[b]	Nd	Nd	21.0	53.0	10.2
Saprospira grandis	Nd	Nd	Nd	34.1	116.0	5.8

[a] μM of anthrone carbohydrate was determined using a standard curve prepared with glucose. Uronic acid was determined by the carbazole method of Dische and galacturonic acid was used as the standard.

[b] Nd, not determined.

in Table 5; again the amount was variable. T6c which produced large amounts of extracellular product contained only a small amount in cells. The ratio of anthrone carbohydrate to uronic acid was different in cellular and extracellular material.

All of the crude extracellular products contained some protein. Aqueous solutions formed insoluble complexes with hexadecyltrimethylammonium bromide and alcian blue. All of the products were hydrolyzed with acid within 2 hr. Longer periods of hydrolysis caused destruction of reducing substances. All hydrolysates gave a positive test for amino sugars with Elson-Morgan reagents. The polyanionic carbohydrate was completely adsorbed to DEAE cellulose in the $-$OH form and was not eluted with water. The carbohydrate was eluted in two to five major peaks with dilute NaOH (Table 6).

Other Properties of Film-Forming Bacteria

Films of bacteria surrounded by a polymer with ion-exchange capacity might be expected to become a center for adhesion of charged particulate materials.

Table 6. Some properties of the extracellular
carbohydrate materials secreted by periphytic marine bacteria[a]

1. Not diffusible through cellophane membranes.
2. Precipitated by alcohols or acetone.
3. Precipitated from aqueous solution with cationic detergents.
4. Form water-insoluble precipitates with aqueous alcian blue.
5. Completely hydrolyzed in 2 hr with 1 N H_2SO_4 at $100°C$ in sealed
   tubes. Reducing substances decomposed with longer treatment.
6. Carbohydrate material adsorbed completely on DEAE cellulose in the −OH
   form. Carbohydrate eluted in fractions with dilute NaOH (0.01−0.5N).

[a] Summary of data from Corpe (1972).

Several marine bacterial cultures were examined for their hydrolytic enzyme
content (Corpe and Winters, 1972). Assays were done for extracellular and
intracellular proteinase, phosphatase, esterase, β-glucosidase, and polysac-
charidase activities. The results of this work are partially summarized in Table
7. While some of the enzyme activities were released into the extracellular
environment of growing cells, the greatest amounts of proteinase, esterase,
and phosphatase were located in the cells and released by cell breakage or by
cell autolysis.

DISCUSSION

Formation of Primary Microbial Films on Glass Slides

Observations of the development of primary microbial films on glass slides
generally confirmed the earlier work (ZoBell, 1943), but the regular sequence
of microbial population changes on slides submerged in all seawater samples,
however, was somewhat unexpected as was the considerable disparity between the microscopic and viable counts (Tables 1 and 2). The latter was
undoubtedly due to the failure of stalked, budding, and filamentous bacteria
to form colonies when dilutions were plated directly on Difco Marine agar.
The method has great limitations with regard to accuracy and reproducibility
even for organisms that grew, which included species of *Pseudomonas, Flavobacterium,* and *Achromobacter.* The question of whether these organisms
were the same as those counted microscopically on 12−24 hr slides could
only be tentatively answered in the affirmative, since the microscopic counts
even at 6 or 12 hr were much higher than the viable count.

The limited growth of these organisms on glass surfaces must be in part
related to the availability and concentration of nutrients at the surface, but

interactions of bacteria involving production of inhibitory substances might also occur (Sieburth, 1967, 1968; Doggett, 1968), together with other micro-environmental factors (Alexander, 1964).

Caulobacter halobacteroides and *Hyphomicrobium neptunium* did not produce microcolonies on glass of the *Pseudomonas* type, and *Caulobacter* and *Saprospira grandis* grew much more slowly than the pseudomonads when the organisms were studied in pure culture (Table 4) in an artificial environment. The conditions used were hardly comparable with natural environments since there was no water movement, opportunity for waste removal, or nutrient replenishment. Because of these factors cell growth per square centimeter by the pseudomonads reached a peak after 3–4 days and then the cells became detached. In another work (Corpe, 1972), it was shown that detachment paralleled decrease in viable count per square centimeter and a decrease in viability of unattached cells in the medium.

Mechanism of Attachment of Bacteria to Surface

When cells settle to a clean surface, a period of incubation is required before firm attachment can be demonstrated. Heat-killed cells or those treated with formaldehyde did not become attached (Corpe, unpublished observations), so it has been considered likely that some sort of metabolic activity is required for them to come into intimate contact with the substratum or to allow them to synthesize a specific adhesive material that would bridge the cell surface and substratum (Corpe, 1970b). Marshal et al. (1971a), using pure cultures of

Table 7. Hydrolytic enzyme mixtures
produced by some film-forming marine bacteria[a]

Culture	Source of enzymes	Type of hydrolase activity found
4b	Extracellular	Proteinase, agarase, alginase
	Intracellular	Proteinase, esterase, phosphatase, β-glucosidase
6b	Intracellular	Proteinase, esterase, phosphatase, carrageenan gum and xanthan gum hydrolase, β-glucosidase
Ma8	Extracellular	Proteinase, esterase, β-glucosidase carrageenan gum hydrolase, alginase, xanthan gum hydrolase
Ma13	Intracellular	Proteinase, esterase, alginase, xanthan gum hydrolase

[a] Summary of some data from Corpe and Winters (1972).

marine bacteria, found that the sorption process has a reversible and irreversible phase. The irreversible attachment of the organisms was believed to be related to their ability to secrete polymeric fibrils which bridge the bacterial surface and the solid substratum, overcoming the repulsion barrier between them (Marshal et al., 1971b).

This writer (Corpe, 1970a) observed that many bacteria that became attached to glass secrete an extracellular slime which surrounds and embeds the cells or their microcolonies. This material stained lightly with crystal violet and alcian blue. In the present work, all of the marine bacteria examined were found to secrete non-diffusible carbohydrate material which was readily precipitated with alcian blue.

Some properties of the secreted carbohydrate material are shown in Table 6. While the amounts secreted by the organisms were variable, the presence of polyanions in the material was clear. Protein was present and several carbohydrate components separable on ion-exchange cellulose were found, so the material cannot be considered homogeneous. The small amount of uronic acid associated with cell carbohydrate may be evidence that residues of polyanionic material are cell bound and not all are produced as loose slime.

Whether material recovered from *Caulobacter* cultures is associated with holdfasts (Poindexter, 1964) has not been determined. It seems probable that the polyanionic material at some time in the growth cycle would be strongly complexed with surface proteins of the cell, so that properties of a holdfast or adhesive fibrils (Marshall et al., 1971b) might be quite different from isolated polyanions.

The fact that bacteria isolated by direct plating of seawater have a periphytic habit and produce small amounts of polyanion may suggest that most marine bacteria are able to exist as periphytes, depending on the nature of the substratum and the nutrients available.

Polyanionic Carbohydrates and Further Development of Microbial Films

The deposition and retention of polyanions to glass or other surfaces must certainly modify the surface charge and other physical properties of the substrata (Baier et al., 1968). Slides which have been coated with the polymers and suspended in seawater seem to be more suitable for entrapment or colonization of bacteria (Corpe, 1970b). Earlier, Sieburth (1967) described the agglutination of an *Arthrobacter* by a gummy material recovered from cultures of a marine pseudomonad.

It is not known if primary film-forming bacteria and their secretions are needed for the colonization of surfaces by other microorganisms, but certainly enrichment of surfaces with cells and adsorbed nutrients will provide a concentrated food supply not otherwise available. Natural surfaces such as sand grains (Meadows and Anderson, 1968), living and dead plants, and animal surfaces (Chan and McManus, 1967; Jones, 1958; ZoBell, 1970) all harbor microbial populations and are sites of intense biochemical activity.

Enzyme Activity of Periphytic Bacteria

The hydrolysis of protein, polysaccharides, esters, and other polymers by enzymes from primary film-forming bacteria was demonstrated (Corpe and Winters, 1972) and is summarized in Table 7. The action of these organisms could have pronounced effects on organotrophs existing on or near the substratum. The importance of bacterial slimes as a part of the bacterial film probably aid in the retention of enzymes and metabolites. The importance of these organisms which are probably representative of a large proportion of marine heterotrophic flora is in the turnover of organic matter in the sea, but because of their periphytic habit they may serve as an important food source for filter-feeding animals (Jannasch, 1954), marine invertebrate animals (Zo-Bell and Feltham, 1938), and bottom-feeding fish (Odum 1968).

ACKNOWLEDGMENTS

This investigation was supported by the Office of Naval Research. The author is grateful to H. Winters, K. Y. Tam, Rita Heller, Louise Slade, Douglas Ritchie, and Charles Goldstein for assistance in carrying out various aspects of this work.

LITERATURE CITED

Alexander, M. 1964. Biochemical ecology of soil microorganisms. Annu. Rev. Microbiol. 18:217–252.

Baier, R. E., E. G. Shafrin, and W. A. Zisman. 1968. Adhesion: mechanisms that assist or impede it. Science 172:1360–1368.

Chan, E. C. S., and E. A. McManus. 1967. Development of a method for the total count of marine bacteria on algae. Can. J. Microbiol. 13:295–301.

Corpe, W. A. 1970a. An acid polysaccharide produced by a primary film forming marine bacterium. Develop. Ind. Microbiol. 11:402–412.

Corpe, W. A. 1970b. Attachment of marine bacteria to solid surfaces. In R. S. Manley (ed.), Adhesions in Biological Systems. Academic Press, New York.

Corpe, W. A. 1972. Microfouling: the role of primary film forming marine bacteria. Proc. Third International Corrosion and Fouling Congress, 2–6 Oct. 1972. National Bureau of Standards, Gaithersburg, Maryland.

Corpe, W. A., and H. Winters. 1972. Hydrolytic enzymes of some periphytic marine bacteria. Can. J. Microbiol. 18:1483–1490.

DeSanto, R. S. 1967. The histochemistry, the fine structure and the ecology of the syntheses of the test in *Botryllus schlosseri* (Pallas) Savigry. Doctoral Dissertation, Columbia Univ. Library, New York.

Doggett, R. G. 1968. New anti-pseudomonas agent isolated from a marine vibrio. J. Bacteriol. 95:1972–1973.

Harvey, H. W. 1941. On changes taking place in sea water during storage. J. Mar. Biol. Ass. U.K. 25:225–233.

Henrici, A. T., and D. E. Johnson. 1935. Studies on fresh-water bacteria. II. Stalked bacteria, a new order of the Schizomycetes. J. Bacteriol. 30:61–93.

Heulelekian, H., and A. Heller. 1940. Relation between food concentration and surface for bacterial growth. J. Bacteriol. 40:547–558.

Hirsch, P., and S. F. Conti. 1964. Biology of budding bacteria. I. Enrichment, isolation and morphology of *Hyphomicrobium* spp. Arch. Mikrobiol. 48:339–357.

Jannasch, H. W. 1954. Okalagische untersuchungen der planktischen bakterien flora in golf von Neapil. Naturwissenschaften 41:42.

Jones, G. E. 1958. Attachment of marine bacteria in zooplankton. U.S. Fish Wildl. Serv. Spec. Sci. Rep. Fish. (279):77–78.

Kabat, E., and M. Mayer. 1961. Experimental Immunochemistry, 2nd edn. Charles C Thomas, Springfield, Illinois.

Leifson, E. 1964. Motile marine bacteria. I. Techniques, ecology and general characteristics. J. Bacteriol. 87:652–666.

Lewin, R. A. 1962. *Saprospira grandis* Gross and suggestions for reclassifying helical, apochlorotic gliding organisms. Can. J. Microbiol. 8:555–563.

Marshall, K. C., R. Stout, and R. Mitchell. 1971a. Selective sorption of bacteria from sea water. Can. J. Microbiol. 17:1413–1416.

Marshall, K. C., R. Stout, and R. Mitchell. 1971b. Mechanism of the initial events in the sorption of marine bacteria to surfaces. J. Gen. Microbiol. 68:337–348.

Meadows, P. S., and J. G. Anderson. 1968. Microorganisms attached to marine sand grains. J. Mar. Biol. Ass. U.K. 48:161–175.

Neish, A. C. 1952. Analytical Methods for Bacterial Fermentations Rep. 46-8-3 Nat. Res. Council Can., Saskatoon, Canada.

Neukom, H., and W. Kuendig. 1965. Fractionation on diethylaminoethyl-cellulose columns. *In* R. L. Whistler (ed.), Methods in Carbohydrate Chemistry, Chap. 5, pp. 14–17. Academic Press, New York.

Odum, W. E. 1968. The ecological significance of fine particle selection by the striped mullet, *Magil cephalus*. Limnol. Oceanogr. 13:92–97.

Poindexter, J. S. 1964. Biological properties and classification of the *Caulobacter* group. Bacteriol. Rev. 28:231–295.

Scott, J. E. 1965. Fractionation by precipitation with quaternary ammonium salts. *In* R. L. Whistler (ed.), Methods in Carbohydrate Chemistry 5, Chap. 11, pp. 38–44. Academic Press, New York.

Shewan, J. M. 1963. The differentiation of certain genera of gram negative bacteria frequently encountered in marine environments. *In* C. H. Oppenheimer (ed.), Marine Microbiology, pp. 499–521. Charles C Thomas, Springfield, Illinois.

Shewan, J. M., G. Hobbs, and W. Hodgkiss. 1960. A determinative scheme for the identification of certain genera of gram negative bacteria, with special reference to the *Pseudomonadaceae,* J. Appl. Bacteriol. 23:379–390.

Sieburth, J. McN. 1967. Inhibition and agglutination of arthrobacters by pseudomonads. J. Bacteriol. 93:1911–1916.

Sieburth, J. McN. 1968. Observations on bacteria planktonic in Narragansett Bay, Rhode Island: a resumé. Bull. Misaki Mar. Biol. Inst. Kyoto Univ. 12:49–64.

Smith, W. W., and C. E. ZoBell. 1937. Direct microscopic evidence of an autochthonous bacterial flora in Great Salt Lake. Ecology 18:453–458.

Society of American Bacteriologists. 1957. Manual of Microbiological Methods. McGraw-Hill, New York.

Stark, W. H., J. Stadler, and E. McCoy. 1938. Some factors affecting the bacterial population of fresh water lakes. J. Bacteriol. 36:653–654.

Wilson, P. W., and S. G. Knight. 1947. Experiments in Bacterial Physiology. Burgess, Minneapolis, Minnesota.

ZoBell, C. E. 1943. The effect of solid surfaces on bacterial activity. J. Bacteriol. 46:38–59.

ZoBell, C. E. 1946. Marine Microbiology. Chronica Botanica, Waltham, Massachusetts.

ZoBell, C. E. 1970. Substratum as an environmental factor for aquatic bacteria, fungi and blue green algae. *In* O. Kinne (ed.), Marine Ecology, Vol. 1, Environmental Factors. John Wiley, New York.

ZoBell, C. E., and D. Q. Anderson. 1936. Observations on the multiplication of bacteria in different volumes of stored sea water and the influence of oxygen tension and solid surfaces. Biol. Bull. Mar. Biol. Lab. Woods Hole 71:324–342.

ZoBell, C. E., and C. B. Feltham. 1938. Bacteria as food for certain marine invertebrates. J. Mar. Res. 1:312–327.

MICROBIAL COLONIZATION OF MARINE PLANT SURFACES AS OBSERVED BY SCANNING ELECTRON MICROSCOPY

JOHN McN. SIEBURTH, RICHARD D. BROOKS, ROBERT V. GESSNER, CYNTHIA D. THOMAS, and J. LAWTON TOOTLE

Narragansett Marine Laboratory
Graduate School of Oceanography
University of Rhode Island

Scanning electron microscopy was used to examine the microbiota which develop on the submerged surfaces of macroscopic plants. This paper summarizes the first year's observations with selected micrographs which show the kinds of microbial assemblages that occur and the apparent differences between substrates and seasons. Driftwood and polypropylene strips were chosen as non-living surfaces to act as controls. The oak driftwood from an exposed rocky point was free of bacteria and algae and showed a definite sequence of fungal development. In contrast, polypropylene strips tied to seaweed fronds were colonized by algal and bacteria-like filaments and pennate diatoms after only 2 weeks immersion.

The grasses also showed markedly different patterns of colonization. The shaded and periodically immersed internodal area of cord grass, *Spartina alterniflora,* showed an initial colonization by the mycelium and hyphopodial appendages of the fungus *Sphaerulina pedicellata.* Sexual stages of this and other fungi developed as this grass matured and senesced during late summer and fall. In contrast, the emerging and submerged surfaces of eelgrass, *Zostera marina,* were colonized mainly by the pennate diatom *Cocconeis scutellum* which formed a unialgal mat. As broken frustules and detritus adhered, a crust was formed which then permitted non-selective colonization.

The seaweeds, unlike the grasses and driftwood, appeared to support lesser population densities but greater diversity. The brown alga *Ascophyllum nodosum,* which is rich in inhibitory polyphenols, was relatively clean during its active growth periods in spring and early summer. During the winter microcolonies of diatoms, yeasts, and the filamentous bacterium *Leucothrix mucor* increased in density. The fine red alga *Polysiphonia lanosa,* which is an epiphyte on *Ascophyllum nodosum,* supports a readily detectable summer epiflora which becomes dense at protected bifurcations. Pennate diatoms, yeasts, and filamentous bacteria increased in density during the winter.

During periods of active growth, marine plants appear to limit the diversity and density of the populations that can develop on their surfaces. Colonization approaches that of inanimate surfaces during periods of dormancy or senescence and especially when the fouling crust isolates the fouling surface from the host.

The major limiting factor in the application of scanning electron microscopy to marine materials is the production of slime layers which obscure the cell surface. In the case of the more gummy seaweeds, like *Chrondrus crispus,* the surfaces are always amorphous and so littered with debris and microorganisms that the micrographs are extremely difficult to interpret. In the case of bacteria, the slime layer sometimes makes their resolution difficult and identification all but impossible.

INTRODUCTION

Heterotrophic and facultatively heterotrophic microorganisms colonize any surface where organic matter is produced or adsorbed. The primary sources of organic matter in the sea are the marine plants. The most obvious habitat for marine microbiologists to study would appear to be any surface where organic matter and microorganisms come in contact. Despite the accepted importance of surfaces (ZoBell, 1946) we microbiologists continue to be mainly preoccupied with studying microorganisms "free" in the water column, which occur in low numbers. This "unattached" or loosely dissociable microflora may be nothing more than transients liberated from a richer substrate which are starving to death while waiting for another surface to colonize.

The microbiota of plant surfaces have been mainly studied by phycologists and, to a lesser extent, by zoologists including protozoologists. The few studies by microbiologists have been mainly concerned with the role of bacteria in fouling. Three main approaches have been taken to study the

Aufwuchs or periphyton of surfaces. One has been to scrape the surface under study and to count and identify the microorganisms in the resulting suspension by light microscopy. This destroys microcolonies and the microzonal relationships between microorganisms. In an attempt to avoid this problem, the second method has been the examination of glass or plastic slides which have been immersed for varying periods of time. This is a variation of the buried-slide technique (Cholodny, 1930) which has been applied to water by a number of investigators including Henrici (1936), ZoBell and Allen (1935), Skerman (1956), and Kriss (1963). Although the glass slide gives an idea of what happens on an inanimate surface, the results can only be applied to discarded glass surfaces such as bottles. The third has been to limit oneself to fine filamentous forms and structures which can be examined in wet mounts (Brock, 1966; Johnson *et al.*, 1971).

Until the development of the scanning electron microscope some 5 years ago, there was no practical way to examine thick, opaque, and three-dimensional preparations. Anything that can be freeze dried, attached to a 12.5-mm diameter aluminum stub, and plated with a palladium-gold alloy can now be examined. Hard solid surfaces are best, while those that are soft and porous, and especially those that have thin, fine structures protruding, can give problems. The instrument permits the resolution of microorganisms as small as 0.2 μm (the limit of the light microscope) and the differentiation of anything with a characteristic form. A major advantage for the microbiologist who is really a bacteriologist is that it forces him to put his microorganisms in perspective with those of his fellow microbiologists who call themselves phycologists, protozoologists, and mycologists.

Much of the material presented in this paper has arisen from student research problems in Sieburth's course on marine microbiology at the University of Rhode Island. Individual papers on the colonization of driftwood (Brooks *et al.*, 1972), *Spartina alterniflora* (Gessner *et al.*, 1972) and *Zostera marina* (Sieburth and Thomas, 1973) are in press. A preliminary form of this paper was presented by Sieburth *et al.* (1972). The purpose of this presentation is to collate and synthesize our observations on the colonization of marine plant surfaces and to show the applications and limitations of scanning electron microscopy for the study of the microflora of natural marine surfaces.

MATERIALS AND METHODS

The driftwood specimens were remnants of oak lobster pots lodged at low tide on the exposed shore of Point Judith, Rhode Island, and were collected in October 1971. *Spartina alterniflora* was collected from southern Rhode

Island salt marshes in the fall of 1971 while *Zostera marina* was collected from the Pattaquamscutt River, Rhode Island, during the fall of 1971 and spring of 1972. Seaweeds were obtained from August 1971 to August 1972 from a densely populated cove adjacent to Camp Varnum some 7 km south of the laboratory. Strips of polypropylene were attached to *Ascophyllum* fronds at the above location in July 1972.

The freshly collected specimens were either fixed in the field or rushed to the laboratory. Fixation in 4% gluteraldehyde ranged from a 10-sec dip to 30-min immersion. After rinsing well with distilled water, representative areas were quick-frozen in isopentane with liquid nitrogen or directly with liquid nitrogen. The frozen sections were freeze dried below 60°C with an Edwards-Pearse tissue dryer (Edwards High Vacuum Ltd., Sussex, England), mounted on SEM stubs with Duco cement containing conductive silver paint, either single or double plated with palladium-gold depending on the specimen, and examined with a Cambridge S-4 scanning electron microscope. The positive micrographs arising from the Polaroid PN Type 55 packs were kept as data records while the 4 by 5 negatives were used with Agfa Brovira I-3 paper to produce figures. (See note added in proof.)

RESULTS

Driftwood which has become lodged in a fixed position is an ideal natural substrate for observing the colonization and decomposition of cellulosic materials in the marine environment. A simplified record of the microscopic appearance of hard, semisoft, and softened wood is shown in Figs. 1 *A, B,* and *C,* respectively. While the wood is in a hard condition, conidia from a number of species can occur in the same microzonal area. By the time the wood is in an intermediate stage of decomposition, conidia of a single species can dominate. When the wood has become soft, the mycelia and conidia are absent and only ascocarps remain. In the rocky and turbulent waters from which these specimens were obtained, diatoms and bacteria were not observed. Another non-living substrate examined has been polypropylene strips tied to fronds of seaweeds. After only 2 weeks immersion, bacteria-like filaments and pennate diatoms (Fig. 1*D*) as well as tufts of green algal filaments (Fig. 1*E*) were well established.

The development of microorganisms on the surfaces of living grasses can vary greatly with the substrate and the environmental conditions. In the shaded and moist internodal area of *Spartina,* mycelia with hyphopodial appendages belonging to the fungus *Sphaerulina pedicellata* develop on the growing plant (Fig. 2*A*). Dense mycelial mats (Fig. 2*C*) and the sexual spores of *Sphaerulina pedicellata* (Fig. 2*B*) and other fungi develop as the plants

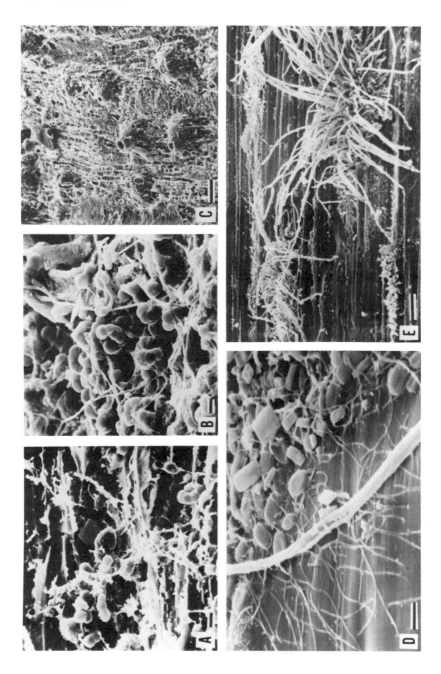

mature and senesce. Bacteria and diatoms do not seem to play a major role until the grass falls, becomes broken, and is submerged. *Zostera,* on the other hand, is initially colonized by a few bacteria and pennate diatoms dominated by *Cocconeis scutellum* (Fig. 2D). An essentially unialgal mat develops (Fig. 2E) upon which broken frustules and detritus adhere to form a crust (Fig. 2F). This crust now supports nonselective colonization by pennate diatoms, filamentous algae, bacteria, and fungi, in addition to nematodes and protozoa. The crust can equal or exceed the biomass of the supporting grass.

The seaweeds do not support such a density of colonization. The diversity of colonization is also restricted and quite repititious. Fig. 3A shows the clean surface of *Ascophyllum nodosum* during its active growing period in early spring. Microzonal areas at this time can also have surface slime with adhering microbial cells (Fig. 3B) and short filaments of *Leucothrix mucor* (Fig. 3G). During winter, when the plant is presumably dormant, pennate diatoms (Figs. 3C,D), yeasts (Figs. 3 E,F) and *Leucothrix mucor* (Fig. 3H) can form dense microzonal areas.

The red alga *Polysiphonia lanosa* is a common epiphyte on *Ascophyllum nodosum*. Since it shares the same habitat it was chosen to compare its seasonal colonization. Although summer specimens of its fine filaments appear barren at first glance (Fig. 4A), closer examination of the bifurcations (Fig. 4B) and the area between the ridges shows a variety of bacteria-sized cells, filaments of *Leucothrix mucor,* yeasts, and pennate diatoms. In winter there is a marked increase in the density of yeast colonies, diatoms, and webs of *Leucothrix mucor* filaments.

Magnifications sufficient to see or identify some microorganisms limit the view to such a small area that one loses perspective. In order to visualize what a larger piece of algal filament would look like in winter, a panorama of a 0.8 mm length of *Polysiphonia lanosa* is shown in Fig. 5. The 2-μm thick filaments of *Leucothrix mucor* (barely visible) have largely coalesced to form thick bundles. The few recurring species of pennate diatoms are quite visible.

Fig. 1. Colonization of non-living surfaces showing a sequence of fungal deterioration of lodged driftwood, and the heterogeneous colonization of polypropylene. *A,* The surface of freshly collected oak driftwood showing mycelium and conidia of multiple species (*Cirrenalia macrocephala* and *Dictosporium pelagica*) in a hardened condition; *B,* domination by a single species (*Cirrenalia macrocephala*) in an intermediate condition; *C,* the absence of mycelia and conidia with the presence of ascocarps of *Leptosphaeria oraemaris* in the softened condition; *D,* a polypropylene strip which had been tied to a frond of *Ascophyllum nodosum* for 2 weeks showing filamentous microorganisms and the pennate diatoms *Cocconeis scutellum* and possibly an *Acanthes* spp.; *E,* tufts of green algal filaments with a scattering of pennate diatoms. Marker bars on *C* and *E* equal 100 μm; all others equal 10 μm.

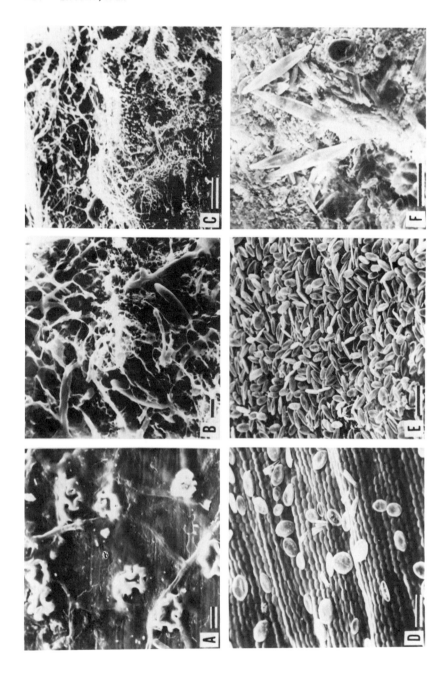

Individual yeast and bacterial cells are difficult or impossible to detect at this magnification.

DISCUSSION

The cellulosic substrates of driftwood and *Spartina alterniflora* showed a definite sequence of fungal development to the exclusion of other micro-organisms. Since a presumably non-nutrient surface like polypropylene is readily colonized by a variety of microorganisms, the refractory nature of the cellulosic substrates does not explain the absence of other microorganisms. It is possible that the dominant fungal microflora is inhibitory and excludes other microorganisms. Another grass, *Zostera marina,* showed a completely different picture of colonization. A unialgal covering of a pennate diatom served as a base for a crust which became heavily colonized by a great variety of microorganisms. It is amazing that this plant can thrive with its central portion completely sheathed. Photosynthesis must be mainly limited to the emerged portion of the blade which is not colonized. It is hard to imagine an adaptive ad-vantage to the host except protection from herbivores. The seaweeds, on the other hand, may be thought of as a photosynthetic root which both adsorbs nutrients and produces "photosynthesate". Heavy colonization like that on *Zostera* would surely interfere with both nutrient adsorption and pho-tosynthesis and severely interfere with the metabolism of the plant. Some seaweeds, like *Ascophyllum* which belongs to the Fucales, are rich in polyphenols. They seem to function as plant-protective substances in two ways. The polyphenols are stored in physodes or fucosan bodies in the outer layers of cells. The ability of these polyphenols to tan polysaccharides as well as proteins may protect tissue damage by water turbulence and grazing animals such as *Littorina littorea* by forming a protective scab. These poly-phenols, which are readily released, are also inhibitory to microorganisms (Sieburth, 1968) and may be responsible for the relatively clean surfaces

---

Fig. 2. Comparison of the microbial colonization of submerged portions of cord grass (*Spartina alterniflora* Loisel) with eelgrass (*Zostera marina* Loisel). The green plants of *Spartina* are initially colonized by the mycelia of *Sphaerulina pedicellata* with their "parasitic" hyphopodia (*A*), later by their ascospores (*B*), and mycelial mats (*C*). The cobblestone surface of *Zostera* after 2 weeks emergence (*D*) becomes colonized by *Cocconeis scutellum* and occasionally by other pennates such as *Licmophora* spp.; later it becomes dominated by *Cocconeis scutellum* (*E*) to form a crust (*F*) which is non-selectively colonized by a number of diatom taxa including *Cocconeis, Navicula, Pleurosigma, Amphora* and possibly *Nitzschia.* Marked bars on *A, B,* and *C* equal 10 μm and on *D, E,* and *F* equal 50 μm.

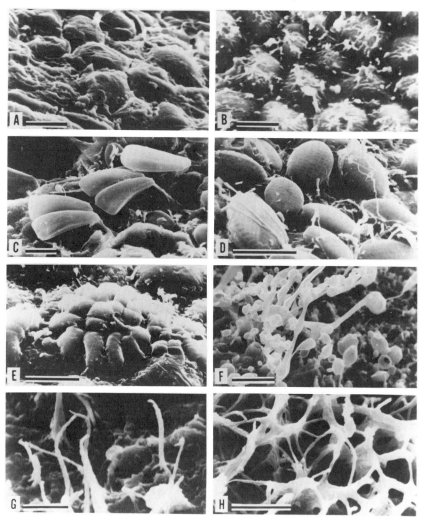

Fig. 3. The microcolonization of the surface of the brown alga *Ascophyllum nodosum* (Loisel) LeJol during the winter. *A,* The clean algal surface (November); *B,* mucoid surface with small microbial cells (November); *C,* the pennate diatom *Rhoicosphenia curvata* (February); *D,* the pennate diatom *Cocconeis scutellum* (February); *E,* a yeast microcolony (January); *F,* pseudomycelium and budding cells of the genus *Candida* (February)—note the dark bud scars on two cells, lower right; *G,* filaments of *Leucothrix mucor* showing individual cells and terminal gonidia (November); *H,* a web of *Leucothrix* filaments (February). All marker bars equal 10 μm.

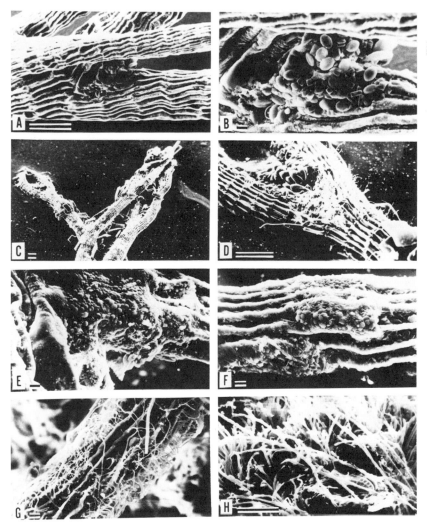

Fig. 4. The filamentous red alga *Polysiphonia lanosa* appears relatively free of an epiflora in summer (*A*) but on close examination shows a mixture of diatoms, yeast and bacteria-like cells (*B*). This plant supports colonization by diatoms in winter (*C* and *D*), yeasts in summer (*E*) and winter (*F*), and dense *Leucothrix mucor* infestation in winter (*G* and *H*). Marker bars on *A, C,* and *D* equal 100 μm; all others equal 10 μm.

Fig. 5. A panorama of an 0.8 mm length of a filament of the red alga *Polysiphonia lanosa* (January specimen) showing coarser bundles and fine individual filaments of the bacterium *Leucothrix mucor* (upper right) and an assortment of pennate diatoms (in the lower left there is the avoid *Cocconeis scutellum*, wedge-shaped *Rhoicosphenia curvata* and possibly *Fragillaria* pennata, while the large central diatom is *Rhabdonema adriaticum*). Marker bar equals 100 μm.

observed on *Ascophyllum nodosum* during its active growing period. The exudation rate for *Ascophyllum nodosum* in spring was some 40 mg C/100 g tissue dry wt/hr. The comparable figure for *Polysiphonia harveyi* was some 0.4 mg C/100 g dry wt/hr (Sieburth, 1969). This negligible value may be due to its dense filamentous structure and its greater degree of spring-summer colonization.

Living-plant surfaces appear to have a markedly reduced diversity in comparison with the epibiota developing on inanimate surfaces such as glass slides. The density of colonization also appears much reduced during periods of active growth. During dormancy and senescence the density and diversity increase but still appear to be limited. An exception is the immersed portion of *Zostera* which supports an epibiota which equals its own biomass. Of what consequence is such a knowledge of the microbial colonization of marine plants?

Macroscopic marine plants play a vital role in the production of organic matter in inshore waters. The microorganisms which produce an Aufwuchs or periphyton may be an important part of this production. In the marshes, the cord grass *Spartina alterniflora* outproduces hybrid corn in productivity. In shallow protected waters, eelgrass *Zostera marina* produces dense stands of growth. The epiflora growing on it can equal or exceed the phytoplankton biomass of the water (Smayda, 1962). In the rocky intertidal and subtidal areas the seaweeds produce biomasses of several kilograms per square meter of area (Blinks, 1955). Both the flowering aquatic plants (Wetzel, 1969) and the seaweeds (Sieburth, 1969) have been shown to release some 30% of their "photosynthesate". Allen (1971) has observed that in ponds this material is immediately taken up by microorganisms in the Aufwuchs or periphyton. Sieburth has examined flowering aquatics *in situ* by snorkeling and scuba. In the proper light a translucent sheath which surrounds these plants can be seen. This is marked in small ponds and protected areas of large ponds where wind-induced turbulence is minimal. Certain areas of marshes and estuaries may also have quiet enough waters for such loose and large assemblages to form. In contrast, most seaweeds require a certain velocity of water movement to supply their nutrient needs (Conover, 1968). It seems highly likely that most of the organic substances released by seaweeds are carried away by the surging waters. The uptake of such materials by suspended particles and their microbiota, and their role in the food chain is another story.

Spartina is harvested by fall and winter winds and falls into the estuaries where it becomes broken and heavily colonized and enters the food chain (Welsh and Carney, 1971). Eelgrass forms a staple of migrating water fowl in the fall. Seaweeds are directly grazed by snails such as *Littorina littorea* and gammarid amphipods. Microbial enrichment of these plants and their remnants may be an important factor in their nutritional value.

This paper attempts to draw together observations made on the microbial epibiota of plants by scanning electron microscopy over a 1-year period. Its applications to describing the microbial habitats of surface films, suspended particles, animal and plant surfaces, and sediments must be obvious. Its limitations may not be. Fungi, yeast, diatoms, and the larger bacterial forms are readily recognized, some even to species. Bacteria can be detected but they are not the neat little bags we expect from light and transmission electron microscopy. A very large microflora consisting of single and paired cocci some 0.2 μm in diameter is missed by scanning surfaces of 500 X and 1000 X magnifications. They can only be detected at magnifications of 2000 X and higher where only a very small area can be seen at a time and where each field has to be refocused. Larger bacteria often appear as loosely stuffed pillows, since we are looking at their outer slime layer and not their more distinct cell wall. Bacteria suspended in water and lying between sand grains are often in microcolonies which are extremely difficult to recognize as such. The application of scanning electron microscopy to natural populations of non-filamentous coccobacillary forms at present is difficult.

ACKNOWLEDGMENTS

The National Science Foundation which not only supported this conference but the acquisition of our scanning electron microscope (GA28903) and the research itself (GB18000) is gratefully acknowledged for making this study and paper possible. We should like also to think Judith Murphy who helped unfold the mysteries of "SEMantha", our new love, and Don Scales who feeds and cares for her.

LITERATURE CITED

Allen, H. L. 1971. Primary productivity, chemo-organotrophy and nutritional interactions of epiphytic algae and bacteria on macrophytes in the littoral of a lake. Ecol. Monogr. 41:97–127.

Blinks, L. R. 1955. Photosynthesis and productivity of littoral marine algae. J. Mar. Res. 14:363–373.

Brock, T. D. 1966. The habitat of *Leucothrix mucor*, a widespread marine microorganism. Limnol. Oceanogr. 11:303–307.

Brooks, R. D., R. D. Goos, and J. McN. Sieburth. 1972. Fungal infestation of the surface and interior vessels of freshly collected driftwood. Mar. Biol. 16:274–278.

Cholodny, N. 1930. Ueber eine neue Methode zur Untersuchung der Boden-mikroflora. Arch. Mikrobiol. 1:620–652.

Cole, G. T., and H. C. Aldrich. 1971. Scanning and transmission electron microscopy and freeze-etching techniques used in ultrastructural studies of hyphomycetes. *In* B. Kendrick (ed.), Taxonomy of *Fungi Imperfecti,* pp. 292–300. Univ. Toronto Press, Toronto.

Conover, J. T. 1968. The importance of natural diffusion gradients and transport of substances related to benthic marine plant metabolism. Bot. Mar. 11:1–9.

Erlandsen, S. L., A. Thomas, and G. Wendelschafer. 1973. A simple technique for correlating SEM with TEM on biological tissue originally embedded in epoxy resin for TEM. *In* Scanning Electron Microscopy 1973, Part 3, pp. 349–356. ITT Research Institute, Chicago.

Gessner, R. V., R. D. Goss, and J. McN. Sieburth. 1972. The fungal microcosm of the internodes of *Spartina alterniflora* Loisel. Mar. Biol. 16:269–273.

Hanic, L. A., and J. S. Craigie. 1969. Studies on algal cuticle. J. Phycol. 5:89–102.

Henrici, A. T. 1936. Studies on fresh water bacteria. III. Quantitative aspects of the direct microscopic method. J. Bacteriol. 32:265–280.

Johnson, P. W., J. McN. Sieburth, A. Sastry, C. R. Arnold, and M. S. Doty. 1971. *Leucothrix mucor* infestation of benthic crustacea, fish eggs and tropical algae. Limnol. Oceanogr. 16:962–696.

Kriss, A. E. 1963. Marine Microbiology (Deep Sea). (Trans. by J. M. Shewan and Z. Kabata.) Oliver and Boyd, London.

Sieburth, J. McN. 1968. The influence of algal antibiosis on the ecology of marine microorganisms. *In* M. R. Droop and E. J. Ferguson Wood (eds), Advances in Microbiology of the Sea, pp. 63–74. Academic Press, New York.

Sieburth, J. McN. 1969. Studies on algal substances in the sea. III. The production of extracellular organic matter by littoral marine algae. J. Exp. Mar. Biol. Ecol. 3:290–309.

Sieburth, J. McN. and C. D. Thomas. 1973. Fouling on eelgrass (*Zostera marina* L.). J. Phycol. 9:46–50.

Sieburth, J. McN., C. D. Thomas, and J. L. Tootle. 1972. Microbial fouling of marine plants. *In* Abst. Annu. Meet. Amer. Soc. Microbiol., p. 76.

Skerman, T. M. 1956. The nature and development of primary films on surfaces submerged in the sea. N. Z. J. Sci. Technol. Sect. B. 38:44–57.

Smayda, T. J. 1962. Some quantitative aspects of primary production in a R.I. coastal salt pond. Proc. 1st Nat. Coastal Shallow Water Res. Conf. Tallahassee Fla., pp. 123–125.

Welsh, B. L., and E. J. Carney. 1971. Integration of computer modeling techniques with laboratory experiments. Univ. Rhode Island Sea Grant Reprint (1)1–5.

Wetzel, R. G. 1969. Excretion of dissolved organic compounds by aquatic macrophytes. BioScience 19:539–540.

ZoBell, C. E. 1946. Marine Microbiology. Chronica Botanica, Waltham, Massachusetts.

ZoBell, C. E., and E. C. Allen. 1935. The significance of marine bacteria in the fouling of submerged surfaces. J. Bacteriol. 29:239.

Note added in proof. Better preparative techniques, including the use of osmium tetroxide, glutaraldehyde fixation, dehydration in a graded ethanol series, and transfer to iso-amyl acetate before CO_2 critical point drying, have vastly improved preparations and the resolution and magnifications obtained. The problems of detecting imbedded microorganisms and confirming external morphological structure by internal ultrastructure has been overcome by combining the techniques of scanning electron microscopy with examination of thick and thin sections by light and transmission electron microscopy, respectively (Cole and Aldrich, 1971; Erlandsen et al., 1973). Another mechanism for limiting the degree of fouling on seaweed surfaces appears to be the sloughing off of the surface layer of the proteinaceous cuticle (Hanic and Craigie, 1969) and the accompanying epibiota to expose a clean surface (J. L. Toottle, M.Sc. Thesis, Univ. Rhode Island, in preparation).

BACTERIAL POPULATIONS ATTACHED TO PLANKTON AND DETRITUS IN SEAWATER

NOBUO TAGA and OSAMU MATSUDA*

Ocean Research Institute
University of Tokyo

Two different cultural and direct microscopic counting methods were employed to estimate the approximate populations of bacteria attached to plankton and detritus and occurring free in seawater, in both neritic and oceanic areas. The average populations of viable bacteria attached to plankton were generally larger (mode 10^5 per liter) than those occurring free in seawater (mode 10^3 per liter) collected from euphotic layers at 14 stations in the Pacific Ocean.

The populations of free-living bacteria in seawater, which were counted by a direct microscopic method, were larger as a rule than those of bacteria attached to detritus particles in both neritic and oceanic waters. The former population was about 6.5 times the mode of the latter observed at 10 stations in the Pacific Ocean. The biochemical properties of both bacterial isolates from plankton and seawater were somewhat different. This might suggest a substantial difference between the bacterial flora of plankton and seawater.

Bacteria in seawater are as a rule distributed in a free-living state or attached to and/or associated with plankton, other organisms, and detritus. A considerable amount of information has been obtained by various authors

* Present address: Department of Fisheries, Faculty of Fisheries and Animal Husbandry, Hiroshima University.

(ZoBell, 1946; Jannasch and Jones, 1959; Jones and Jannasch, 1959; Kriss, 1963; Wood, 1965; Sieburth, 1968; Seki, 1970a,b) with respect to the distribution and population density of bacteria in neritic and oceanic waters, but our knowledge is still scanty with regard to the bacterial flora attached to and/or associated with plankton and detritus in seawater.

In the present paper, we report some results of estimations of the populations of bacterial types attached to plankton and detritus, as compared with free-living bacterial populations in seawater of both neritic and oceanic areas.

MATERIALS AND METHODS

Sampling Locations

The present investigation was carried out at four neritic stations in Suruga and Sagami Bays (Fig. 1) during the cruise KT-68-20 in October 1968, by the R/V *Tansei-maru* of the Ocean Research Institute, University of Tokyo. The 19 oceanic stations along the longitude 155°W in the North and Equatorial Pacific Ocean (Fig. 2) were sampled during the cruise KH-69-4 in August to November 1969 (Marumo, 1970) by the R/V *Hakuho-maru* of the same institute.

Collection of Plankton and Water Samples

Plankton samples were collected by using the Norpac double net, made of two kinds of bolting cloth, XX 13 (0.09-mm mesh) and GG 54 (0.33-mm mesh). Two different sizes of plankton samples were obtained fractionally at the same time by vertical hauls of this net in depths of 150–100 m and 50–0 m. Plankton samples of large size were mainly composed of zooplankton, while those of small size were composed of phytoplankton, including smaller

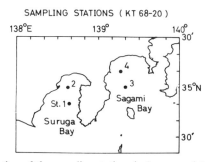

Fig. 1. Location of the sampling stations in Suruga and Sagami Bays.

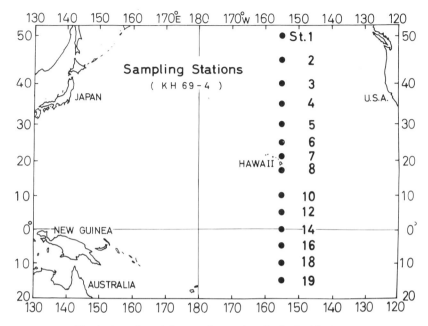

Fig. 2. Location of the sampling stations in the Pacific Ocean.

zooplankton in some cases. Vertical seawater samples were also collected using a Nansen bottle for direct microscopic counting of total and attached bacteria on detritus, and a sterile ORIT-type sampler (Taga, 1968) fixed to the Nansen bottle for viable counts of heterotrophic bacteria in seawater.

Counting of Heterotrophic Bacteria

For viable counts of bacteria attached to plankton, each plankton sample collected by the Norpac net was carefully transferred to a sterilized glass bottle, and a fixed volume was homogenized with a fixed volume of sterilized seawater mixed with 3 ppm of Tween 80 in a sterilized blender. The homogenized sample was diluted in sterile seawater in 10-fold increments from 10^{-3} to 10^{-7} ml, and the dilutions were inoculated in duplicate on agar plates of Medium PPES-II (Taga, 1968) by the spread-plate method of Buck and Cleverdon (1960). For viable counts of bacteria suspended in seawater, duplicate 2- to 50-ml portions of seawater sample were filtered through sterilized Millipore filters (HA-45mm, porosity 0.45 μm). The inoculated filters were placed on agar plates of Medium PPES-II; all plates were incubated at 18°C for 2 weeks before colonies on filters and plates were counted.

Direct Counting of Bacteria and Detritus

A 50- to 100-ml portion of seawater sample was filtered through a sterile Millipore filter (GS-25 mm, porosity 0.22 μm) and salt was rinsed off the filter which was then dried at 80°C. The filter was mounted in a small amount of Cargille immersion oil (R. P. Cargille Laboratories, Inc.) on a slide glass under a cover slip; then the total numbers of free-living bacteria, attached bacteria on detritus, and detritus smaller than 50 μm on the transparent filter were counted with observations of 50 fields by the phase-contrast microscope (Nikon, Japan).

Biochemical Tests on Bacterial Isolates

Bacterial strains isolated from plankton and seawater plate cultures were examined separately for several properties including the hydrolysis of gelatin (Ge), starch (St), chitin (Ch), and tributylin (Li), aerobic production of acid from glucose (Gl), and production of ammonia from peptone (Am). For the first five tests, the same media and procedures were used as described in a previous paper (Taga, 1968). The test for ammonium production by cultures inoculated in the liquid medium 2216E (Oppenheimer and ZoBell, 1952) was detected by Nessler's reagent.

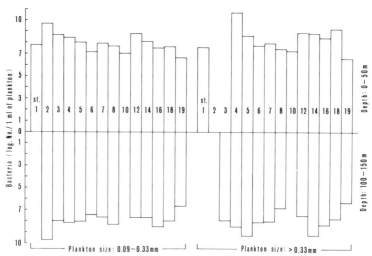

Fig. 3. Viable counts of heterotrophic bacteria attached to plankton in 1 ml of the settling volume, at 14 stations in the Pacific Ocean.

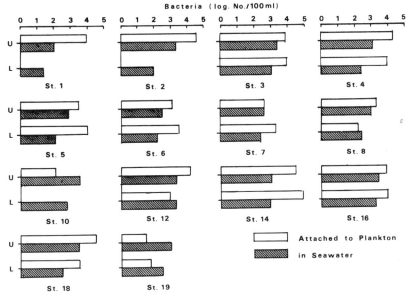

Fig. 4. Viable counts of heterotrophic bacteria attached to plankton and suspended in seawater of upper (*U*, 0–50 m) and lower (*L*, 100–150 m) layers at 14 stations of the Pacific Ocean. Bacterial counts in seawater of upper and lower layers were expressed as averages of those in depths of 0, 20, and 50 m, and 100 and 150 m, respectively. Number of attached bacteria was expressed as the sum of populations attached to two different sizes of plankton.

RESULTS

Viable Count of Bacteria Attached to Plankton

Viable counts of heterotrophic bacteria attached to plankton samples collected from two different depths at 14 stations in the Pacific Ocean are schematically presented in Fig. 3. The counts of viable bacteria attached to plankton in 1 ml of the settling volume were quite high and somewhat variable being in the range 10^7-10^9, with a few exceptions. There was no apparent relationship between the abundance of bacteria and the sampling depths, station sampled, or plankton composition.

From the data shown in Fig. 3, the viable counts of heterotrophic bacteria attached to plankton in 100 ml of seawater were calculated and compared with those of bacteria suspended in seawater of upper (0–50 m) and lower (100–150 m) layers at stations in the Pacific Ocean. The results are shown schematically in Fig. 4. The populations of viable bacteria attached to plankton were larger in most layers of oceanic seawater than those of suspended bacteria. However, exceptions were observed in the upper layers at Stations 10 and 19 and the lower layers at Stations 8, 12, and 19. The frequency distribution of average populations of free and plankton-attached bacteria in 1 liter of seawater were plotted from the data in Fig. 4. From the results given in Fig. 5 it will be seen that the average bacterial population attached to plankton was of the order of 10^5, with a population variation of 10^2-10^5, while that of bacteria occurring free in seawater was of the order of 10^3, with a variation of 10^2-10^4.

Biochemical Properties Found in Bacteria
Isolated from Floras of Plankton and Seawater

Some biochemical differences between the bacterial flora attached to plankton and that occurring free in seawater were investigated with isolates from the neritic areas at both Suruga Bay and Sagami Bay. Figures 6 and 7 give the results schematically showing the occurrence rates of different biochemical properties. Bacteria isolated from plankton were much more active in starch hydrolysis (St) and acid production from glucose (Gl), and less frequently in the hydrolysis of chitin (Ch) than were isolates from seawater. Acid production from glucose (Gl) occurred more frequently in the bacterial flora isolated from phytoplankton (sample size 0.09–0.33 mm) than in the flora isolated from zooplankton (sample size larger than 0.33 mm) at Stations 3 and 4 (see results in Fig. 7).

Appearance of chromogenic bacteria in the bacterial flora of seawater and plankton was also investigated in the neritic and oceanic areas, and their

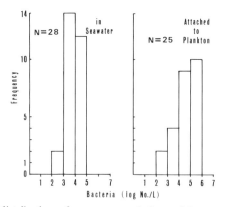

Fig. 5. Frequency distribution of average populations of heterotrophic bacteria, suspended in seawater and attached to plankton, in two layers at 14 stations of the Pacific Ocean.

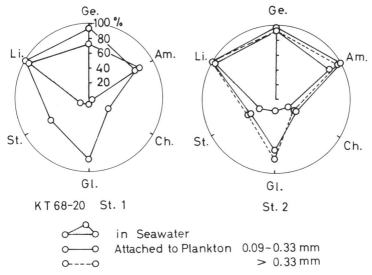

Fig. 6. Schematic comparison of the occurrence rates of several biochemical properties of the heterotrophic bacterial isolates from seawater and those from plankton. Results are expressed as a ratio of the total plate count, at two stations in Suruga Bay. Abbreviations of biochemical properties are given in the text.

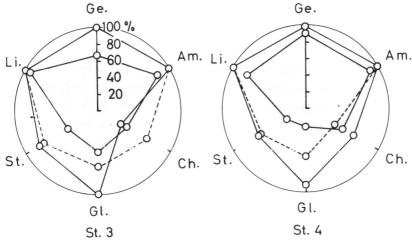

Fig. 7. Schematic comparison of the occurrence rates of biochemical properties of the heterotrophic bacterial isolates from seawater and those from plankton. Results are expressed as a ratio of the total plate count at two stations in Sagami Bay (cf. Fig. 6.). Abbreviations of biochemical properties are given in the text.

differences were apparent (Figs. 8 and 9). As indicated in Fig. 8, the chromogenic bacteria were more abundant in the seawater flora than in the plankton flora in the neritic areas such as Suruga Bay and Sagami Bay. In contrast with Fig. 8, as indicated in Fig. 9, chromogens were a predominant part of the plankton flora rather than of the seawater floras in the oceanic areas of the Pacific Ocean. However, these results, as well as the former ones on the biochemical functions, seem to reveal significant differences between the bacterial flora attached to plankton and that in seawater.

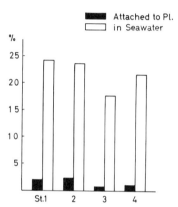

Fig. 8. Appearance of chromogenic bacteria in the bacterial flora from seawater and those from plankton, expressed as a ratio of the total plate count. Samples were collected at depths of 0–200 m in Suruga and Sagami Bays.

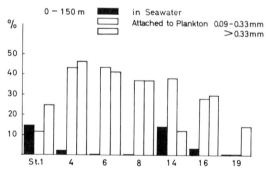

Fig. 9. Appearance of chromogenic bacteria in the bacterial flora from seawater and those from plankton, expressed as a ratio of the total plate count. Samples were collected at depths of 0 to 150 m in the Pacific Ocean.

Table 1. Numbers of bacteria and detritus (abioseston) particles per 100 ml of seawater in Suruga and Sagami Bays

Station	Depth (m)	Direct counts				Viable counts of heterotrophic bacteria (H)	Ratio		
		Total bacteria (T)	Free bacteria (F)	Attached bacteria (A)	Detritus (D)		F/A	A/D	T/H
1	0	3.2×10^5	2.9×10^5	3.0×10^4	2.0×10^3	3.7×10^3	9.7	12.7	85
	20	1.1×10^5	1.0×10^5	1.0×10^4	8.0×10^2	4.3×10^3	10.0	11.6	26
	40	2.0×10^5	1.8×10^5	2.0×10^4	1.4×10^3	1.4×10^3	9.0	13.0	140
	80	2.7×10^5	1.1×10^5	1.6×10^5	5.0×10^3	1.2×10^3	0.69	30.4	220
	125	4.4×10^5	9.5×10^4	3.5×10^5	8.5×10^3	1.9×10^3	0.27	41.3	230
	259	1.3×10^5	1.1×10^5	2.0×10^4	1.7×10^3	8.0×10^2	5.0	11.8	160
	340	4.4×10^5	3.3×10^5	1.1×10^5	1.7×10^4	8.0×10^2	3.0	5.9	550
	681	1.5×10^6	1.5×10^6	0	6.0×10^3				
	867	7.5×10^5	5.7×10^5	1.8×10^5	1.7×10^4		3.2	10.7	
2	0	5.5×10^5	2.1×10^5	3.4×10^5	1.9×10^4	4.1×10^3	0.62	17.7	140
	19	3.0×10^5	2.6×10^5	4.0×10^4	3.0×10^3	2.2×10^3	6.5	3.8	140
	92	6.0×10^5	4.6×10^5	1.4×10^5	2.6×10^4	2.5×10^3	3.3	6.1	240
3	0	3.9×10^5	3.2×10^5	7.0×10^4	3.0×10^3	5.5×10^4	4.6	17.0	7
	48	6.5×10^4	5.1×10^4	1.4×10^4	2.4×10^3	1.9×10^3	3.6	5.2	34
	99	3.0×10^5	2.4×10^5	6.0×10^4	1.5×10^4	2.4×10^3	4.0	5.0	130
	137	6.5×10^5	5.9×10^5	6.0×10^4	9.0×10^3	1.2×10^3	9.8	5.2	550
	908	1.4×10^5	1.1×10^5	3.0×10^4	2.9×10^3	5.0×10^2	3.7	8.5	280

Table 2. Numbers of bacteria and detritus (abioseston) particles per 100 ml of seawater at Station 7 (21°N, 155°W) in the Pacific Ocean

Depth (m)	Direct counts				Viable counts of heterotrophic bacteria (H)	Ratio		
	Total bacteria (T)	Free bacteria (F)	Attached bacteria (A)	Detritus (D)		F/A	A/D	T/H
0	6.8×10^5	4.6×10^5	2.2×10^5	1.0×10^4	6.0×10^2	2.1	22	1100
19	8.0×10^5	7.0×10^5	1.0×10^5	1.0×10^4	1.0×10^2	7.0	10	8000
48	7.8×10^5	7.2×10^5	6.0×10^4	3.0×10^3	5.3×10^2	12.0	20	1500
96	4.2×10^5	3.4×10^5	8.0×10^4	5.0×10^3	1.6×10^2	4.3	16	2600
144	5.5×10^5	5.4×10^5	1.0×10^4	2.0×10^3	3.6×10^2	54.0	5	1500
192	4.7×10^5	3.9×10^5	8.0×10^4	6.0×10^3	1.3×10^2	4.9	13.3	3600
383	3.7×10^5	2.6×10^5	1.1×10^5	5.0×10^3	4.8×10^1	2.4	22	7700
575	5.0×10^5	4.0×10^5	1.0×10^5	6.0×10^3	3.0×10^0	4.0	16.7	167,000
768	5.8×10^5	5.2×10^5	6.0×10^4	5.0×10^3	1.0×10^1	8.7	12	58,000
963	3.1×10^5	2.8×10^5	3.0×10^4	4.0×10^3	2.2×10^1	9.3	7.5	14,000
1462	3.6×10^5	3.2×10^5	4.0×10^4	4.0×10^3	4.2×10^2	8.0	10	860
1947	2.4×10^5	2.0×10^5	4.0×10^4	2.0×10^3	1.6×10^2	5.0	20	1500
2917	5.7×10^5	5.5×10^5	2.0×10^4	3.0×10^3	8.7×10^1	27.5	6.7	6600
3888	4.3×10^5	3.5×10^5	8.0×10^4	6.0×10^3	9.0×10^0	4.4	13.3	48,000
5341	2.3×10^5	1.7×10^5	6.0×10^4	8.0×10^3	3.3×10^1	2.8	7.5	7000

Table 3. Overall range of variation in the numbers of bacteria and detritus (abioseston) particles per 100 ml of seawater at Stations 1–4, 6–8, 10, 12, and 14 in the Pacific Ocean

Depth (m)	Direct counts				Viable counts of heterotrophic bacteria (H)	Ratio		
	Total bacteria (T)	Free bacteria (F)	Attached bacteria (A)	Detritus (D)		F/A	A/D	T/H
0– 160	2.1×10^5 – 2.7×10^6	1.7×10^5 – 2.6×10^6	9.0×10^3 – 2.6×10^5	6.0×10^2 – 2.1×10^4	1.0×10^1 – 8.3×10^3	1.9– 99.0	5.0– 83.3	25– $82{,}000$
180– 5400	9.4×10 – 1.5×10^6	7.0×10^4 – 1.4×10^6	5.0×10^3 – 4.0×10^5	6.0×10^2 – 1.8×10^4	1.0×10^0 – 8.0×10^3	0.25– 95.3	2.5– 50.0	16– $400{,}000$

Bacterial Population Attached to Detritus in Seawater

By using direct microscopic counting, numbers of total bacteria, free-living bacteria, bacteria attached to detritus, and the amount of detritus in seawater were investigated in samples collected from the neritic and oceanic areas. The results are summarized in Tables 1, 2, and 3, respectively. As shown in Tables 1 and 3, the overall range of variation in the numbers of total bacteria (T), free-living bacteria (F), and detritus (D) was roughly the same in the neritic and oceanic areas, whereas the numbers of attached bacteria on detritus (A) and the viable counts of heterotrophs (H) in their overall ranges were a little higher in the neritic areas (Table 1) than in the Pacific Ocean (Table 3). In addition, as summarized in Table 3, the populations of free-living bacteria and bacteria attached to detritus and the numbers of detrital particles in 100 ml of seawater were estimated to be in the ranges $7.0 \times 10^4 - 2.6 \times 10^6$, $5.0 \times 10^3 - 4.0 \times 10^5$, and $6.0 \times 10^2 - 2.1 \times 10^4$, respectively, in depths of 0–5400 m at 10 stations in the Pacific Ocean.

Figures 10 and 11 show schematically the estimated ranges of the numbers of bacteria and detritus at different stations of the Pacific Ocean. From Fig. 10 it is clear that the population of free-living bacteria (F) was obviously larger than the population of bacteria attached to detritus (A) with exceptions at Stations 4 and 10 in the euphotic layers. By contrast with Fig. 10, as indicated in Fig. 11, the distinctions between the populations of free-living bacteria and attached bacteria were not apparent in the aphotic layers at Stations 1, 2, 3 and 4, located at the northern area of the Western North Pacific central water mass. Such distinctions between them were clearly observed in samples collected at the other six stations.

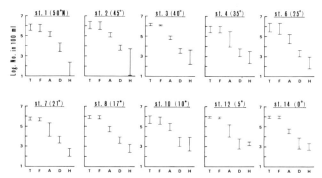

Fig. 10. Numbers of total bacteria (T), free-living bacteria (F), attached bacteria on detritus (A), detritus (D), and heterotrophic bacteria (H) in seawater, expressed as their ranges, at a depth of 0–160 m in the Pacific Ocean.

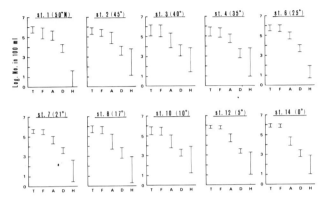

Fig. 11. Numbers of total bacteria (*T*), free-living bacteria (*F*), attached bacteria on detritus (*A*), detritus (*D*), and heterotrophic bacteria (*H*) in seawater, expressed as their ranges, at a depth of 180–5400 m in the Pacific Ocean.

In order to understand more clearly the overall abundance of free-living bacteria (*F*) and attached bacteria on detritus (*A*), the frequency distribution of the ratios of F to A in all the layers observed in the Pacific Ocean was plotted as shown in Fig. 12. The result indicates that the average of the ratios of F to A was 6.5, with a range of values of 1.9–99 (Table 3) in both euphotic and aphotic depths. Although the free-living bacteria in seawater appear to be more numerous as a rule than the bacteria attached to detritus, it should be noted in Fig. 12 that the attached bacteria were more numerous in several aphotic depths at Stations 1, 2, 3 and 4 in the Pacific Ocean as well as in a few euphotic depths of Suruga Bay (Table 1).

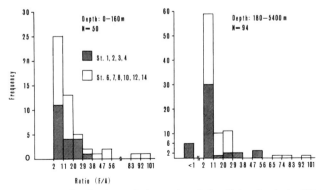

Fig. 12. Frequency distribution of the ratio of free-living bacteria (*F*) to attached bacteria on detritus (*A*) in seawater from the Pacific Ocean.

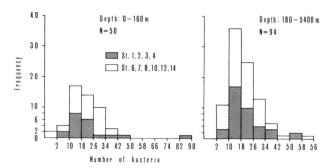

Fig. 13. Frequency distribution of the average number of bacteria (*A*) attached to one particle of detritus (*D*), estimated from the ratio of A/D, in seawater from the Pacific Ocean.

Figure 13 shows the frequency distribution of the ratios of A to D, which was plotted to provide an estimate of the average number of bacteria (*A*) attached to one particle of detritus (*D*) in seawater in the Pacific Ocean. It will be seen from the figure that the mode of the average bacterial numbers attached to detritus was 14 per particle in all depths from 0 to 5400 m, with a range of values from 5 to 83 (Table 3).

DISCUSSION

The importance of adsorption to microbial processes in marine environment has been stressed by ZoBell (1946) and Wood (1965). ZoBell (1943) demonstrated that many sessile bacteria grew preferentially or exclusively attached to solid surfaces in seawater. It is considered likely that most bacteria rarely occur free in seawater but are attached to plankton and other particles (Wood, 1965). Jones and Jannasch (1959), in fact, showed that particles in neritic waters were covered or closely surrounded by microbes; bacteria were apparently attached to dead diatoms in the same environment. Although our knowledge of bacterial populations attached to plankton or detritus particles is scanty, Wood (1965) gave figures of $8.9 \times 10^2 - 3.75 \times 10^4$ bacteria per milliliter of plankton. Recently, Simidu *et al.* (1971) presented viable counts of bacteria in the range $2.8 \times 10^5 - 1.1 \times 10^7$ per milliliter of plankton samples collected from neritic waters. In the present investigation, however, the bacterial populations attached to plankton in oceanic areas were relatively higher, with a range $10^7 - 10^9$ bacteria per milliliter of plankton. The discrepancy between the populations obtained by us and by former workers is considered to have resulted mainly from differences in the sampling seasons or areas, and the methods of bacterial enumeration.

The bacterial flora of the intestines of various marine fish caught on the Japanese coast was almost exclusively species of *Vibrio* (Okuzumi and Horie, 1969). Berland *et al.* (1969) reported the genera *Pseudomonas, Flavobacterium,* and *Achromobacter* to be the most important bacteria associated with marine algae in culture. Simidu *et al.* (1971) found that the main constituent of the bacterial flora from plankton was the *Vibrio-Aeromonas* group, and suggested the existence of a different bacterial flora in plankton than in seawater. These findings are in agreement with the results obtained in the present study, which indicate some differences in biochemical properties between the bacterial flora of plankton and those of seawater.

In conclusion, as far as our results in oceanic areas are concerned, the viable counts of bacteria attached to plankton were quite large, with an average of 10^5 per liter of seawater, as compared with those that occur free in seawater from the euphotic layer, which average about 10^3 per liter. On the other hand, the free-living bacteria in seawater as measured by direct counts were more abundant in most cases than the bacteria attached to detritus particles in neritic and oceanic areas. In fact, the ratio of the former over the latter was 6.5.

ACKNOWLEDGMENTS

This research was supported in part by a grant from the Japanese Ministry of Education and was carried out as part of JIBP-PM. The authors wish to express their sincere gratitude to Dr. K. Ohwada, Ocean Research Institute, University of Tokyo, for his collaboration in the collection of samples and counting of bacteria during the cruise of the R/V *Tansei-maru,* and to Miss Kazuko Arimura, of the same institute, for her able assistance in direct counting of bacteria during the course of this work. Thanks are due to the officers and crew of the R/V *Tansei-maru* and R/V *Hakuho-maru,* Ocean Research Institute, University of Tokyo, for their cooperation during the cruises.

LITERATURE CITED

Berland, B. R., M. G. Bianchi, and S. Y. Maestrini. 1969. Study of bacteria associated with marine algae in culture. I. Preliminary determination of species. Mar. Biol. 2:350–355.

Buck, J. D., and R. C. Cleverdon. 1960. The spread plate as a method for the enumeration of marine bacteria. Limnol. Oceanogr. 5:78–80.

Jannasch, H. W., and G. E. Jones. 1959. Bacterial populations in sea water as determined by different methods of enumeration. Limnol. Oceanogr. 4:128–139.

Jones, G. E., and H. W. Jannasch. 1959. Aggregates of bacteria in sea water as determined by treatment with surface active agents. Limnol. Oceanogr. 4:269–276.

Kriss, A. E. 1963. Marine Microbiology (Deep Sea). (Trans. by J. M. Shewan and Z. Kabata.) Oliver and Boyd, London.

Marumo, R. (ed.) 1970. Preliminary report of the Hakuho Maru Cruise KH-69-4 (IBP Cruise). Ocean Res. Inst. Univ. Tokyo.

Okuzumi, M., and S. Horie. 1969. Studies on the bacterial flora in the intestines of various marine fish. Bull. Jap. Soc. Sci. Fish. 35:93–100.

Oppenheimer, C. H., and C. E. ZoBell. 1952. The growth and viability of sixty-three species of marine bacteria as influenced by hydrostatic pressure. J. Mar. Res. 11:10–18.

Seki, H. 1970a. Microbial biomass on particulate organic matter in seawater of the euphotic zone. Appl. Microbiol. 19:960–962.

Seki, H. 1970b. Microbial biomass in the euphotic zone of the North Pacific subarctic water. Pacific Sci. 24:269–274.

Sieburth, J. McN. 1968. Observations on planktonic bacteria in Narragansett Bay, Rhode Island; a résumé. Bull. Misaki Mar. Biol. Inst. Kyoto Univ. (12):49–64.

Simidu, U., Ashino, K., and E. Kaneko. 1971. Bacterial flora of phyto- and zoo-plankton in the inshore water of Japan. Can. J. Microbiol. 17:1157–1160.

Taga, N. 1968. Some ecological aspects of marine bacteria in the Kuroshio current. Bull. Misaki Mar. Biol. Inst. Kyoto Univ. (12):65–76.

Wood, E. J. F. 1965. Marine Microbial Ecology. Chapman and Hall, London.

ZoBell, C. E. 1943. The effect of solid surfaces upon bacterial activity. J. Bacteriol. 46:39–56.

ZoBell, C. E. 1946. Marine Microbiology. Chronica Botanica, Waltham, Massachusetts.

PRODUCTION AND UTILIZATION OF DISSOLVED VITAMINS BY MARINE PHYTOPLANKTON

A. F. CARLUCCI

Institute of Marine Resources
University of California, San Diego

Marine phytoplankters released vitamins into the culture medium during their growth in laboratory experiments. Of the six phytoplankters tested, only the non-vitamin requirer, *Dunaliella tertiolecta,* excreted vitamin B_{12} , thiamine, and biotin. The other non-vitamin requirer, *Phaeodactylum tricornutum,* produced thiamine and biotin. The B_{12} requirers, *Stephanopyxis turris* and *Skeletonema costatum,* released thiamine and biotin. *Gonyaulax polyedra* released thiamine and possibly biotin. *Coccolithus huxleyi,* a thiamine requirer, produced biotin with low (10 ng/liter) and high (120 ng/liter) concentrations of thiamine in the medium, and B_{12} and biotin with the high thiamine concentration. If starved cells of vitamin-requiring phytoplankters were added to the same culture vessel as the producer they often grew well when the required vitamin was excreted.

In the marine environment, a major portion of dissolved vitamin concentrations may arise from phytoplankton production, especially where the biomass of these organisms is relatively high.

INTRODUCTION

Most of the marine phytoplankters that are cultured in the laboratory require one or more vitamins for their growth (Lewin, 1961; Provasoli, 1963, 1971;

Strickland, 1965). The required vitamins are, in order of importance, B_{12}, thiamine, and biotin. Well over 50% of the diatoms that have been studied in the laboratory require vitamin B_{12}. Although there have been a number of attempts to correlate vitamin distribution in the sea with the presence (or absence) of specific phytoplankters (Menzel and Spaeth, 1962; Spurnov and Murauskaya, 1964; Carlucci, 1970; Propp, 1970), the role of vitamins in marine ecology has not been fully evaluated.

Carlucci (1970) reported on vitamin distribution-phytoplankter composition in a number of coastal water samples. It was concluded that the phytoplankton population as a whole was not influenced by dissolved vitamin concentrations, but that individual species may have been affected. The major problem in these, and all related studies, was that the vitamin requirements of most of the phytoplankters were not known. Some samples contained as many as 50 or more different species. The data were treated statistically, and even when certain known vitamin-requiring phytoplankters, such as *Skeletonema costatum* (B_{12}) and *Coccolithus huxleyi* (thiamine), were abundant their presence could not be correlated with decreases of the required vitamin in the water. The one exception was a small bloom of the B_{12}-requiring dinoflagellate *Gonyaulax polyedra*. At this time *Gonyaulax polyedra* represented over 95% of the plant biomass and its growth could be correlated with a decrease of B_{12} in the water.

An important observation made in the coastal plankton study was that, in many samples, when phytoplankton biomass was high, the concentration of dissolved vitamins was also high (Carlucci, 1970). It was concluded that perhaps the phytoplankters were producing (excreting) vitamins. This conclusion was later verified in the laboratory where specific axenic phytoplankters were shown to produce vitamins under carefully controlled conditions (Carlucci and Bowes, 1970*a,b*). Furthermore, it was also shown that the excreted vitamins could be utilized by vitamin-requiring phytoplankters (Carlucci and Bowes, 1970*b*).

This communication describes the results of experiments dealing with vitamin production and utilization by phytoplankton; some of the data have been discussed previously (Carlucci and Bowes, 1970*a,b*).

MATERIALS AND METHODS

The marine phytoplankters used in this study included the following: *Amphidinium carterae* and *Gonyaulax polyedra* (dinoflagellates); *Cyclotella nana, Phaeodactylum tricornutum, Skeletonema costatum,* and *Stephanopyxis turris* (diatoms); *Dunaliella tertiolecta* (green alga); *Coccolithus huxleyi* (chrysomonad). The vitamin requirements (if any) of these phytoplankters, the starvation procedures used, the media employed, and experimental procedures have been given in detail (Carlucci and Bowes, 1970a,b).

RESULTS

Vitamin Production

Table 1 summarizes the data from all experiments on vitamin production. Each of the six phytoplankters tested produced at least one vitamin in the culture medium. *Dunaliella tertiolecta* was the only phytoplankter which produced all three vitamins. *Dunaliella tertiolecta* is a non-vitamin requirer; the other non-vitamin requirer tested, *Phaeodactylum tricornutum,* did not produce B_{12}. With the other phytoplankters, production of the non-required vitamins was observed, except in the case of *Gonyaulax polyedra* where biotin excretion was inconclusive.

Four of six phytoplankters tested require vitamins for growth: *Skeletonema costatum, Stephanopyxis turris,* and *Gonyaulax polyedra* require B_{12}, and *Coccolithus huxleyi* requires thiamine. All of these phytoplankters except *Gonyaulax polyedra* were tested for vitamin production with limiting and non-limiting concentrations of the required vitamin in the media. *Gonyaulax polyedra* was difficult to culture in the laboratory and was tested only at the non-limiting concentration of B_{12}. Table 2 shows that *Skeletonema costatum* and *Stephanopyxis turris* produced biotin and thiamine with both 2 and 12 ng B_{12} per liter. *Gonyaulax polyedra* excreted thiamine with 12 ng B_{12} per liter. *Coccolithus huxleyi* released both B_{12} and biotin with 120 ng thiamine per liter and only biotin with 10 ng thiamine per liter.

Table 3 gives the amount of vitamin produced by a phytoplankter cell during the logarithmic phase of growth with a high concentration of the required vitamin. On a per cell basis, the data indicated that the larger

Table 1. Vitamin production by marine phytoplankters

Phytoplankter	Vitamin produced
Coccolithus huxleyi	B_{12}, biotin
Dunaliella tertiolecta	B_{12}, thiamine, biotin
Gonyaulax polyedra	Thiamine, biotin
Phaeodactylum tricornutum	Thiamine, biotin
Skeletonema costatum	Thiamine, biotin
Stephanopyxis turris	Thiamine, biotin

Table 2. Production of vitamins by phytoplankters
growing with low and high concentrations of the required vitamin[a]

Phytoplankter	Vitamin production			
	B_{12} requirers			
	2 ng B_{12} /liter		12 ng B_{12} /liter	
	Thiamine	Biotin	Thiamine	Biotin
Skeletonema costatum	+	+	+	+
Stephanopyxis turris	+	+	+	+
Gonyaulax polyedra[b]			+	?
	Thiamine requirer			
	10 ng Thiamine/liter		120 ng Thiamine/liter	
	B_{12}	Biotin	B_{12}	Biotin
Coccolithus huxleyi	−	+	+	+

[a] From Provasoli and Carlucci, 1974.
[b] Not tested with 2 ng B_{12} /liter.

phytoplankters, *Stephanopyxis turris* and *Gonyaulax polyedra,* excreted greater amounts of the vitamin than the smaller phytoplankters such as *Skeletonema costatum.*

Vitamin Utilization in Mixed Culture

When starved cells of a vitamin-requiring phytoplankter were added to the same culture vessel as the vitamin producer, it was observed in many cases that both the producer and utilizer grew independently of each other, the utilizer taking up the released vitamin. Table 4 summarizes the data where vitamin production and utilization occurred in two-phytoplankter systems. In cases where as many as three different starved vitamin-requirers were added to the same culture vessel as the producer there were often indications of vitamin utilization, but these effects did not last long (Carlucci and Bowes, 1970b).

DISCUSSION

The production (excretion) of dissolved vitamins into the culture medium by phytoplankton observed in all these experiments supports the conclusion made from the coastal plankton study (Carlucci, 1970). This conclusion was that the phytoplankton were producing vitamins *in situ.* Figure 1 shows

Table 3. Amount of vitamin produced per
cell per day during logarithmic phase of
growth of various phytoplankters with non-
limiting concentrations of the required vitamins[a]

Phytoplankter	Vitamin	pg/cell/day
Skeletonema costatum	Thiamine	1.1×10^{-5}
Stephanopyxis turris	Thiamine	4.5×10^{-3}
Gonyaulax polyedra	Thiamine	4.8×10^{-3}
Skeletonema costatum	Biotin	1.9×10^{-6}
Stephanopyxis turris	Biotin	4.2×10^{-4}
Coccolithus huxleyi	Biotin	6.0×10^{-6}
Coccolithus huxleyi	B_{12}	4.0×10^{-6}

[a] From Carlucci and Bowes, 1970*a*.

typical curves for vitamin production and phytoplankton growth; the data used to plot the curves summarize the results of a number of experiments, including those reported previously (Carlucci and Bowes, 1970*a*). More recently, Ohwada and Taga (1972) have attributed some of the vitamin concentrations found in freshwater to the activities of phytoplankton.

In the sea the concentrations of dissolved vitamins have previously been reported to arise mainly from the activities of bacteria (Burkholder and Burkholder, 1956; Burkholder, 1958, 1963; see Provasoli, 1963). The results of recent studies (Carlucci, 1970; Carlucci and Bowes, 1970*a,b*) indicate that, in addition to the contribution of bacteria to the dissolved vitamin concentrations, phytoplankter release may also be important. In some situations, such as in coastal waters where phytoplankton biomass may be several orders of

Table 4. Vitamin production and utilization in two-phytoplankter system[a]

Producer	Utilizer	Vitamin produced
Dunaliella tertiolecta	*Coccolithus huxleyi*	Thiamine
	Amphidinium carterae	Biotin
Phaeodactylum tricornutum	*Amphidinium carterae*	Biotin
Skeletonema costatum	*Coccolithus huxleyi*	Thiamine
Coccolithus huxleyi	*Cyclotella nana*	B_{12}

[a] From Carlucci and Bowes, 1970*b*.

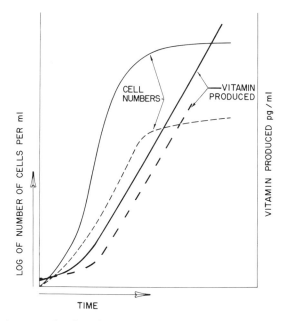

Fig. 1. Typical curves showing the relationship between phytoplankter growth and vitamin production with limiting (- - -) and non-limiting (———) concentrations of the required vitamin.

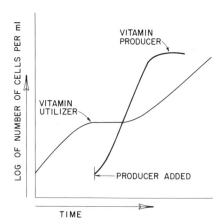

Fig. 2. Typical curves showing growth of vitamin-starved utilizer before and after the addition of the vitamin producer to the same culture vessel.

magnitude higher than bacterial biomass, vitamin release by phytoplankton is probably the more important source of dissolved vitamin.

In laboratory cultures phytoplankter release of dissolved vitamins into the medium may be attributed to two processes (Carlucci and Bowes, 1970*a*). During early and logarithmic phases of growth the release is due to excretion, and in old cultures it is due to excretion and liberation by lysed cells after death. Young cells were always healthy when observed with the phase microscope.

Figure 2 shows typical curves which depict the relationship between vitamin producer and utilizer in a two-phytoplankter system. The data summarize the results of a number of experiments (Carlucci and Bowes, 1970*b*). In this case the producer phytoplankter is generally added after the utilizer is sufficiently starved of the required vitamin.

Attempts to add two or three starved utilizers to the same culture vessel as the producer usually failed to show beneficial effects to any of the utilizers. If there was a beneficial effect it was of short duration. Often inhibition of the utilizer was observed when growth of the utilizer was compared with a control culture (with the required vitamin added). The inhibition was probably due to release of toxic substances into the medium (Carlucci and Bowes, 1970*b*). It is also possible that phytoplankter release of toxic substances may be important in the sea in restricting the growth of other phytoplankters.

Vitamin-phytoplankter interaction in the sea is a complex problem to study. The amount of dissolved vitamin found at any one time represents a balance between that which is being produced and that which is being utilized. These processes are being influenced by physical, chemical, and biological factors. In order to know more about the influence of vitamins on the species composition of phytoplankton blooms, more information on the vitamin requirements of the phytoplankters and their abilities to produce or excrete vitamins is needed.

ACKNOWLEDGMENTS

This work was supported, in part, by the Marine Life Research Program, Scripps Institution of Oceanography's component of the California Cooperative Oceanic Fisheries Investigation, a project sponsored by the State of California, and, in part, by the U. S. Atomic Energy Commission, Contract AT(11-1)GEN 10, P.A. 20.

The author thanks Mrs. Peggy Bowes for excellent experimental assistance.

LITERATURE CITED

Burkholder, P. R. 1958. Studies on B vitamins in relation to productivity of Bahia Fosforescente, Puerto Rico. Bull. Mar. Sci. Gulf Carib. 8:201–223.

Burkholder, P. R. 1963. Some nutritional relationships among microbes of sea sediments and waters. *In* C. H. Oppenheimer (ed.), Symposium on Marine Microbiology, pp. 133–150. Charles C Thomas, Springfield, Ill.

Burkholder, P. R., and L. M. Burkholder. 1956. Vitamin B_{12} in suspended solids and marsh muds collected along the coast of Georgia. Limnol. Oceanogr. 1:202–208.

Carlucci, A. F. 1970. Vitamin B_{12}, thiamine, and biotin. *In* J. D. H. Strickland (ed.), The Ecology of the Plankton off La Jolla, California, in the Period April through September, 1967. Bull. Scripps Inst. Oceanogr. 17:23–31.

Carlucci, A. F., and P. M. Bowes. 1970a. Production of vitamin B_{12}, thiamine, and biotin by phytoplankton. J. Phycol., 6:351–357.

Carlucci, A. F., and P. M. Bowes. 1970b. Vitamin production and utilization by phytoplankton in mixed cultures. J. Phycol. 6:393–400.

Lewin, R. A. 1961. Phytoflagellates and algae. *In* W. Ruhland (ed.), Handbuch Pflanzenphysiologie, Vol. 14, pp. 401–417. Springer-Verlag, Berlin.

Menzel, D. W., and J. P. Spaeth. 1962. Occurrence of vitamin B_{12} in the Sargasso Sea. Limnol. Oceanogr. 7:151–154.

Ohwada, K., and N. Taga. 1972. Vitamin B_{12}, thiamine, and biotin in Lake Sagami. Limnol. Oceanogr. 17:315–320.

Propp, L. N. 1970. Seasonal dynamics of vitamin B_{12} and phytoplankton variability in the Dalnezelenetskaya Guba (Inlet) of the Barents Sea. Okeanologiya 1:851–857.

Provasoli, L. 1963. Organic regulation of phytoplankton fertility. *In* M. N. Hill (ed.), The Sea, Vol. 2, pp. 169–219. John Wiley (Interscience), New York.

Provasoli, L. 1971. Nutritional relationships of marine organisms. *In* J. D. Costlow (ed.), Fertility of the Sea, pp. 369–382. Gordon and Breach, New York.

Provasoli, L., and A. F. Carlucci. 1974. Vitamins and growth regulators. *In* W. D. P. Stewart (ed.), Algal Physiology and Biochemistry. Blackwell Scientific Publications, Oxford. (In press.)

Spurnov, A. T., and Z. A. Murauskaya. 1964. The content of vitamin B_{12} in water of Sevastopol Bay and its possible ecological importance. Tr. Sevastopol Biol. Sta. Akad. Nauk SSSR 17:342–345.

Strickland, J. D. H. 1965. Production of organic matter in the primary stages of the marine food chain, *In* J. P. Riley and G. Skirrow (eds), Chemical Oceanography, Vol. 1, pp. 447–610. Academic Press, New York.

ADSORPTION OF THE MARINE BACTERIOPHAGE PM2
TO ITS HOST BACTERIUM

EUGENE H. COTA-ROBLES*, JUDITH P. GREGORY,
and THOMAS E. RUCINSKY

Department of Microbiology
The Pennsylvania State University

The lipid-containing marine bacteriophage PM2, which was originally isolated and characterized by Espejo and Canelo (1968a), has been described consistently as exhibiting a slow rate of adsorption to its host bacterium, *Pseudomonas* BAL-31 (Espejo and Canelo, 1968b; Lara, 1970; Dahlberg and Franklin, 1970). Espejo and Canelo (1968b) suggested that the high ionic strength of the adsorption medium could conceivably play an important role in the slow adsorption rate of PM2. As Dahlberg and Franklin (1970) have emphasized, kinetic studies of PM2 infection have been complicated by the non-synchronous pattern of infection which results from slow adsorption of phage to host cell.

Recently striking advances have been made in elucidating the structural organization of a marine pseudomonad by MacLeod and his co-workers (Forsberg *et al.*, 1970a,b). Such studies have shown clearly that it is possible to obtain authentic protoplasts of these gram-negative bacteria. Since one of the long-range objectives of this laboratory is to make a detailed comparison of the lipoproteins in the membrane of PM2 with the lipoproteins of the cytoplasmic membrane of the host cell, we have decided to apply the

* Present address: University of California, Santa Cruz, California.

techniques of the MacLeod Group to BAL-31 since this bacterium is itself a marine pseudomonad. Our initial findings have shown that a rather simple manipulation of the cells markedly increases the rate of adsorption of PM2 to BAL-31.

MATERIALS AND METHODS

The cultivation of BAL-31 and of PM2 has been performed as previously described (Cota-Robles *et al.,* 1968). High-titer preparations of PM2 were prepared utilizing the polyethylene glycol method of Yamamoto *et al.* (1970). Phage stocks were stored in 0.02 M Tris buffer pH 8.1 containing 1.0 M NaCl.

Adsorption was routinely measured after mixing washed cells (see below) and phage at a given multiplicity of infection (*moi*) at 28°C in a given medium by scoring for the disappearance of free phage from the supernatant fluid following centrifugation at 1500 g for 5 min (Adams, 1959). The infection medium routinely used was AMS medium which is composed of 26.0 g NaCl, 12 g $MgSO_4 \cdot 7H_2O$, 0.7 g KCl, and 1.5 g $CaCl_2 \cdot 2H_2O$ per liter of distilled water. Occasionally adsorption was monitored by assaying for the loss of viability of host cells following infection with PM2. Turbidometric analysis of cell lysis as well as phage production were also used to verify that a given infection was productive.

Exponentially growing cells of BAL-31 at a concentration of approximately 1×10^8 cells per milliliter were washed aseptically by centrifugation and resuspension in an equivalent volume of the appropriate solution, followed by recentrifugation and resuspension in a given medium. Electron microscopy was performed as described earlier (Cota-Robles *et al.,* 1968) with the substitution of low-viscosity epoxy resin for the vestopal embedding medium.

RESULTS

Table 1 depicts the adsorption of PM2 to normal and washed cells of BAL-31 when phage is added to cells at a *moi* of 0.1 at 28°C. It is clear that cells washed with 0.1 M NaCl adsorb PM2 much more rapidly than do unwashed cells. The rate of adsorption of PM2 to BAL-31 can be seen to increase by one order of magnitude if the cells are merely washed once with 0.1 M NaCl. Washing cells in either 0.5 M NaCl, 0.25 M NaCl, or AMS medium alone also led to rates of adsorption significantly greater than those rates exhibited by unwashed cells. However, we have chosen to utilize 0.1 M NaCl as the standard washing medium.

Table 1. Adsorption of PM2
to washed and unwashed cells of BAL-31

Time (min)	Percentage unadsorbed phage	
	Normal cells	Washed cells
0	100	100
5	77.6	26.8
10	54.2	8.2
15	49.6	2.8
20	41.8	< 1
25	18.9	< 1

Espejo and Canelo (1968*b*) originally studied the replication of PM2 not only in cells grown in a complex medium such as that used in this investiga-tion, but also in cells grown in a minimal medium consisting of mineral salts, including ammonium salts supplemented with 0.2% glucose. We have found that cells grown in such a minimal medium also adsorb PM2 slowly and that following a single washing such cells adsorb PM2 rapidly.

Phase microscopic examination of cells removed from the growth medi-um and resuspended in 0.1 M NaCl demonstrated extensive retraction of the protoplast from the cell wall, i.e., "plasmolysis". This is in confirmation of the report of Thompson *et al.* (1970) that a marine pseudomonad undergoes plasmolysis upon suspension in a potassium-free medium. However, washing once in 0.1 M NaCl does not affect the viability of BAL-31.

The increased adsorption of PM2 by cells removed from the growth medium and washed once with 0.1 M NaCl could be the result of the removal of an agent or substance which interferes with adsorption. Such a substance

Table 2. Effect of medium upon the
adsorption of PM2 to BAL-31 cells

Medium	Percentage unadsorbed phage after 10 min
Unwashed cells, in AMS-salts	49.6
Washed cells	
In AMS-salts	6.2
In fresh AMS-NB	6.8
In partially spent AMS-NB	6.0

could be present in the fresh medium or it could be produced by the cells. The possibility that an inhibitory agent could be present in either fresh or partially spent growth medium was eliminated since it was found that, after washing cells in 0.1 M NaCl and resuspending them in either fresh or partially spent medium, one observes the same degree of adsorption of PM2 which is characteristic of washed cells infected in AMS medium (Table 2).

The possibility that a capsular material could be coating unwashed cells and thus interfering with adsorption was considered. Careful and critical light microscopic examination of cells both washed and unwashed stained by Anthony's method was repeatedly performed in the laboratory of Dr. L. E. Casida, Jr., of this department. Such examinations proved to be negative, i.e., capsular material was not observed. Electron microscopic examination of cells revealed that the L layer of the cell wall of unwashed cells demonstrated numerous invaginations or blebs, while washed cells had a uniformly taut L-layer. Moreover, capsular material could not be observed by electron microscopy, although we have detected a delicate structure which is comparable with the loosely bound layer of the envelope of the marine pseudomonad B16 (Fig. 1).

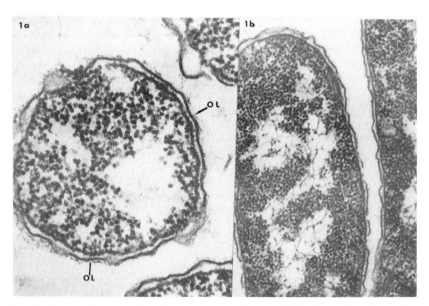

Fig. 1. Thin sections of cells of the marine pseudomonad BAL-31. *a*: Unwashed cells; note loose outer layer (*OL*) at the periphery of the cell. *b*: Washed cells; note breaks in the L-layer of the cell wall. × 48,240.

Table 3. Effect of temperature upon
the adsorption of PM2 to BAL-31 cells

	Rate of adsorption (ml/min)		
	4°C	18°C	28°C
Unwashed cells	4.58×10^{-10}	–	4.19×10^{-10}
Washed cells	4.12×10^{-10}	2.16×10^{-9}	3.0×10^{-9}

The removal of a loosely bound outer layer from the surface of the bacterial cell by washing could explain the increased rate of adsorption of PM2 to washed cells. However, we have found that, if washed cells are resuspended in their own washings, adsorption of PM2 occurs at the same rapid rate demonstrated by washed cells suspended in fresh media. Thus, adsorption is not inhibited by the readdition of any material removed by washing.

Influence of Temperature on Adsorption

We have found that both washed and unwashed cells adsorb PM2 slowly at 4°C. PM2 adsorbs to unwashed cells at 4°C at a rate comparable with that rate demonstrated at 28°C. However, the adsorption of PM2 to washed BAL-31 cells is markedly influenced by temperature. The data in Table 3 are a comparison of the rate of adsorption of PM2 to BAL-31 when the infection is performed at various temperatures. It can be seen that PM2 adsorbs to washed BAL-31 cells 10 times more rapidly at 28°C than at 4°C.

Influences of Ionic Strength and Ionic Species on Adsorption

The rapid adsorption of PM2 to washed cells in a medium of an ionic strength which is comparable with that of seawater has prompted us to investigate the influence of ionic strength upon phage adsorption. We have found that at 28°C washed cells adsorb PM2 at a relatively rapid rate in media of an ionic strength varying from 0.17 to 0.68 μm. These data are presented in Table 4. It should be noted that we have as yet been unable to study adsorption at truly low ionic strengths since the virion is inactivated very rapidly at concentrations of NaCl below 0.1. The data in Table 4 do demonstrate that even in double-strength AMS the rate of adsorption of PM2 to washed BAL-31 is somewhat more rapid than it is to unwashed cells.

Our studies in this area have also been concerned with investigating the role the various cations in AMS-salts play in the adsorption of PM2 to

Table 4. Effect of ionic strength of medium
upon the adsorption of PM2 to washed BAL-31 cells

Ionic strength (μm)	Percentage free phage after 10 min
0.17	13.0
0.34 (normal)	5.04
0.68	37.4
0.34 (unwashed control)	54.0

BAL-31. Table 5 clearly shows that, if the composition of AMS is altered prior to adsorption, one does not appreciably affect the rate of adsorption unless one omits K^+ ions from the medium. Thus adsorption appears to be independent of the concentration of Na^+, Ca^{2+}, or Mg^{2+} ions. Table 6 shows that the concentration of K^+ ions in the adsorption medium markedly influences adsorption of PM2. It should be noted that cells of BAL-31 undergo plasmolysis when suspended in AMS containing low levels of K^+ ions (Thompson *et al.*, 1970).

The fact that a system is now available which insures synchrony of infection has permitted us to reinvestigate the gross parameters of the latent

Table 5. Effect of the deletion of
various components of AMS-salts medium upon
the adsorption of PM2 to washed BAL-31 cells

Component deleted	Percentage free phage after 10 min	Rate of adsorption (ml/min)
KCl	51.7	6.6×10^{-10}
$CaCl_2 \cdot 2H_2O$	8.2	2.4×10^{-9}
$MgSO_4 \cdot 7H_2O$	1.6	3.4×10^{-9}
NaCl	2.1	3.2×10^{-9}
Control: complete AMS-salts	2.4	3.4×10^{-9}

Table 6. Adsorption of PM2 to washed cells of
BAL-31 in media containing varying molarities of KCl

Concentration of KCl (M)	Percentage unadsorbed phage	Rate of adsorption (ml/min)
0.1	9.7	1.68×10^{-9}
0.01 (normal)	7.05	1.55×10^{-9}
0.001	22.5	9.91×10^{-10}
0.0001	71.9	2.6×10^{-10}

period of PM2 in BAL-31. Washed cells infected with PM2 at a *moi* of 4 undergo an abrupt cessation of growth (measured by light scattering) followed by a sharp lysis 45 min after infection. Unwashed cells continue to increase in turbidity until lysis is initiated, at which time a less steep decline in turbidity indicates the lysis of the cells. Careful measurements of the one-step growth curve of PM2 in washed BAL-31 show a latent period of 43–44 min which yields a burst size of approximately 200 virions.

Electron Microscopy

We have encountered some technical difficulties in resolving the adsorption of purified PM2 to BAL-31 by electron microscopy. This may be related to the clumping of PM2 in high-titer preparations obtained by concentration in polyethylene glycol. However, we have been successful in overcoming these difficulties by infecting washed cells in an unpurified PM2-induced lysate of BAL-31. Figure 2 reveals that PM2 appears to adsorb by fusion of the virion to the L-membrane of the cell wall.

DISCUSSION

It must be noted in all candor that this work was initiated on the premise that washing of cells would remove phage receptor sites to permit the intimate characterization of the interaction of the lipoidal virions of PM2 with their receptor sites. This premise has not been substantiated, although in unpublished work we have been able to remove the L-layer of the cell wall by the method of Forsberg *et al.* (1970*a,b*). Cells with their L-layer removed (mureinoplasts) are unable to be infected with PM2. However, the purified L-layer does not inactivate PM2. Although we have not yet been able to

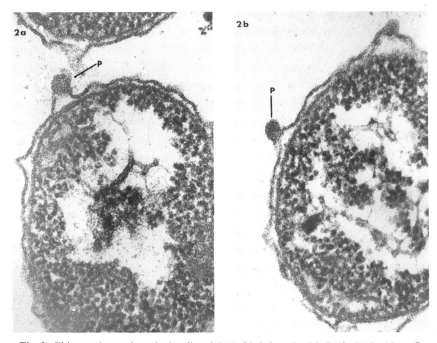

Fig. 2. Thin sections of washed cells of BAL-31 infected with PM2. PM2 virion (*P*) appears to be fused with the L-layer of the cell wall. × 84,240.

achieve our aim of describing the mechanism of adsorption of PM2 to BAL-31, we have demonstrated that a single wash in NaCl does convert cells into a state which permits rapid and effective adsorption.

Forsberg *et al.* (1970*a,b*) have described a procedure which permits them to remove and separate the cell-wall layer of a marine pseudomonad, B16. They have demonstrated that the various layers of the envelope of B16 can be removed sequentially following washing in NaCl, shaking in sucrose, and treatment with lysozyme. In addition, Forsberg *et al.* (1970*a*) have identified a loosely bound outer layer composed of carbohydrate, protein, and lipid which is removed by the initial wash in NaCl. They have substantiated the presence of this loose outer layer by chemical analysis. The findings presented herein regarding the increased adsorption rate of PM2 to washed cells of another marine pseudomonad, BAL-31, do offer further support to the observations of Forsberg *et al.* (1970*a,b*) that a marine pseudomonad has a loosely bound outer cell-wall layer. Although our findings show clearly that a single wash permits PM2 to adsorb rapidly to BAL-31, we have demonstrated that the material removed by washing does not inactivate the virions or interfere with adsorption. Thus, it appears as if the loosely bound outer layer must somehow physically mask the normal phage receptor sites only when the layer actually is in place on the cell surface. Moreover, it appears that the loose outer layer must not form a uniform covering over the entire cell envelope since unwashed cells of BAL-31 are able to adsorb PM2 with significant efficacy. Thus, it seems that the loose outer layer functions only as a partial barrier to bacteriophage infection. The electron microscopic evidence we present does support the view that the loose outer layer is not a uniform covering which coats the entire cell wall.

The most interesting finding presented herein is the electron microscopic evidence which suggests that the entire virion of PM2 appears to fuse with the L-layer of the cell wall. Thus, the infectious process of PM2 does indeed appear to be unique among bacterial systems and may involve hydrophobic bonding between lipoidal components in its envelope and that of the L-layer of the bacterial cell wall. We have shown that the high ionic strength of the medium does not grossly interfere with adsorption, a matter that would pertain if the adsorption process involved classical electrostatic bonding rather than hydrophobic interaction. We have not been able to ascertain whether extremely low ionic strength influences adsorption of PM2 since the virion is itself inactivated at low salt concentration. However, we have been able to obtain rapid adsorption in media devoid of magnesium ions and sodium ions, respectively. Such ions contribute significantly to the ionic strength of our synthetic seawater adsorption medium.

Thus the fact that adsorption occurs rapidly at a relatively high ionic strength offers further evidence that the union between virion and host cell may indeed involve hydrophobic interactions. The putative role that potassium ions play in adsorption of PM2 is not clear. However, at this time we are not yet convinced that potassium ions play a direct role in the adsorption of PM2 to BAL-31 since potassium ions influence the association between cell wall and cytoplasmic membrane of marine pseudomonads. The fact that cells of BAL-31 and B16 demonstrate retraction of the cytoplasm from the wall in the absence of potassium ions has led us to consider the possibility that both the cell wall and cytoplasmic membrane must be juxtaposed for effective adsorption of PM2 to occur. However, we must not neglect to consider that potassium ions could indeed influence a hydrophobic interaction between the virion envelope and the L-layer of the cell wall.

LITERATURE CITED

Adams, M. H. 1959. Bacteriophages. John Wiley (Interscience), New York.
Cota-Robles, E. H., R. T. Espejo, and P. W. Haywood. 1968. Ultrastructure of bacterial cells infected with bacteriophage PM2, a lipid-containing bacterial virus. J. Virol. 2:56—68.

Dahlberg, J. E., and R. M. Franklin. 1970. Structure and synthesis of a lipid-containing bacteriophage. IV. Electron microscopic studies of PM2-infected *Pseudomonas* BAL-31. Virology 42:1073–1086.

Espejo, R. T., and E. S. Canelo. 1968*a*. Properties of bacteriophage PM2: a lipid-containing bacterial virus. Virology 34:738–747.

Espejo, R. T., and E. S. Canelo. 1968*b*. Origin of phospholipid in bacterio-phage PM2. J. Virol. 2:1235–1240.

Forsberg, C. W., J. W. Costerton, and R. A. MacLeod. 1970*a*. Separation and localization of cell wall layers of a gram-negative bacterium. J. Bacteriol. 104:1338–1353.

Forsberg, C. W., J. W. Costerton, and R. A. MacLeod. 1970*b*. Quantitation, chemical characteristics, and ultrastructure of three outer cell wall layers of a gram-negative bacterium. J. Bacteriol. 104:1354–1368.

Lara, J. C. 1970. The infectious process of a unique lipid-containing bacterial virus: bacteriophage PM2. Ph.D. Thesis. Univ. California, Riverside.

Thompson, J., J. W. Costerton, and R. A. MacLeod. 1970. K$^+$-dependent deplasmolysis of a marine pseudomonad plasmolyzed in a hypertonic solution. J. Bacteriol. 102:843–854.

Yamamoto, K. R., B. M. Alberts, R. Benzinger, L. Lawhorne, and G. Treiber. 1970. Rapid bacteriophage sedimentation in the presence of polethylene glycol and its application to large-scale virus purification. Virology 40:734–744.

BACTERIAL FLORA IN THE DIGESTIVE TRACTS OF MARINE FISH

HIROSHI SERA, YUZABURO ISHIDA, and HAJIME KADOTA

Laboratory of Microbiology
Department of Fisheries
Kyoto University

Observations of the bacterial flora in the digestive tracts of red sea bream (*Pagrus major*), a fish with a developed stomach, and file-fish (*Stephanolepsis cirrhifer*), a fish with an undeveloped stomach, fed with diets of known composition revealed the following facts.

1. The population of heterotrophic bacteria in the intestines of red sea bream increased exponentially with time after diet ingestion. In the stomachs, however, the bacterial population was practically constant throughout the period of cultivation. The total number of bacteria in both the intestines and the stomachs was not influenced by the composition of the diet.

2. The bacterial strains (567 strains) isolated from the diets, environmental seawater, and the gastric and intestinal contents of red sea bream and file-fish were taxonomically divided into eight groups. Bacteria of groups I (genus *Vibrio*) and II (genus *Pseudomonas*) were motile with a single polar flagellum. Group I fermented sugars; group II did not. Growth of bacteria of group I was inhibited by vibriostatic compound. Group I was further divided into subgroups, I-a_1, I-a_2, I-b, I-c, I-d, I-e and I-f, based on biochemical characteristics. Subgroup I-a_2 showed the same properties as I-a_1 except for gas formation from sugars. Bacteria in groups III

and IV were peritrichous rods, of which the former utilized sugars and the latter did not. Bacteria in group V (*Pseudomonas*) were lophotrichous and were able to utilize sugar oxidatively. Bacteria in group VI (*Achromobacter*) were non-motile rods. Those in the other groups (VII and VIII) were gram-positive cocci.

3. Of the above mentioned eight groups, bacteria of subgroups I-a_1 and I-a_2 were always dominant in both the stomach and intestine of red sea bream more than 6 hr after ingestion of diet, irrespective of the diet composition. In the case of file-fish, the intestinal flora was mainly composed of group VI and partly of subgroups I-d and I-3. The bacterial flora in the digestive tracts of file-fish was similar in composition to those of the diets.

4. In comparison with the other groups, representative strains of subgroups I-a_1 and I-a_2 were highly resistant to the environmental conditions encountered in the stomach, i.e., the strong acidity and the high concentration of gastric juices, and in the intestine, i.e., the high concentration of bile acids.

The above results suggest that some specific species of *Vibrio* (I-a_1 and I-a_2) are indigenous to the digestive tracts of marine fish which have a highly developed digestive system, but fish which have an undeveloped stomach system have no indigenous bacterial flora in their digestive tracts.

INTRODUCTION

In order to clarify the role of gastro-intestinal bacteria in the nutrition of fish, it is important to have knowledge of the indigenous bacterial flora of the digestive tracts of fish.

Several investigators (Liston, 1955, 1956, 1957; Colwell, 1962; Simidu and Hasuo, 1968*a,b;* Aiso *et al.,* 1968; Okuzumi and Horie, 1969) have reported that the dominant bacteria in the digestive tracts of marine fish are limited to some specific species of *Vibrio*. Most of these studies, however, were carried out using commercial fish which were caught from the sea; they did not deal with the formation of the indigenous bacterial flora in the digestive tracts.

In the present study we attempted to elucidate the mechanism involved in the formation of indigenous bacterial flora in the digestive tracts of fish during cultivation under regulated conditions.

MATERIALS AND METHODS

Organisms

One-year-old red sea bream snapper (*Pagrus major*) was employed as a representative of fish having a developed stomach system. As an example of fish with an undeveloped stomach system, file-fish (*Stephanolepsis cirrhifer*) was used. In both cases, 16 fish were kept in a culture tank (65 by 65 by 105 cm^3) in which running seawater (ca. 26°C) was circulated.

After starvation for 5 days, red sea bream were fed with one of the following diets (Table 1) once or twice a day. The F-diet contained, in 100 g (wet weight), 97 g fish meat and 3 g carboxymethyl cellulose (CM-cellulose). The S-diet contained 72 g fish meat, 25 g potato starch, and 3 g CM-cellulose. The C-diet contained 82 g fish meat, 15 g powdered chitin, and 3 g CM-cellulose.

Sampling of Gastric and Intestinal Contents

Fish which had been fed for a definite period of days were killed 4 hr after their last feed, unless otherwise mentioned. Gastric and intestinal contents of these fish bodies were aseptically collected by use of sterilized glass rods.

Enumeration of Bacteria

The total number of heterotrophic bacteria in the gastric and intestinal contents was estimated by the pour-plate method. A 2.5% NaCl solution was used for diluting the sample. Incubation was carried out at 25°C under aerobic conditions. A medium, containing 3 g precipitated chitin, 15 g peptone, 5 g beef extract, 25 g NaCl, and 15 g agar in 1000 ml (pH 7.0) was employed as the standard basal medium for the enumeration of bacteria.

Characterization of the Bacteria Isolated

Shape of cells, gram stain, motility, and flagellation of cells after 24 and 48 hr incubation at 22°C were employed as the principal characteristics for taxonomic grouping (Harrigan and McCance, 1966). Ability to utilize glucose and *N*-acetylglucosamine was also tested for characterization, according to

Table 1. Composition of the diet[a]

Component	F-diet	S-diet	C-diet
Fish meat	97 (89)	72 (39)	82 (54)
Potato starch		25 (53)	
Powdered chitin			15 (38)
CM-cellulose	3 (11)	3 (8)	3 (8)

[a] Given in g/100 g wet weight. Figures in parentheses indicate g/100 g dry weight.

the method of Hugh and Leifson (1953). The cytochrome oxidase test was done according to the method of Kovacs (1956). Ability to hydrolyze starch, casein and chitin, and to produce ammonia and indol was determined by ordinary procedures.

Sensitivities of isolated bacteria to low pH and high concentration of bile were examined to distinguish indigenous gastrointestinal bacteria from the others; the isolated strains were inoculated in the basal medium with or without addition of bile powder at a concentration of 2.0% at pH 5.5 or 7.5, and incubated at 25°C. The growth was assessed by measuring turbidity of cultures after 1-, 3-, and 8-day incubation using a Leitz colorimeter, model M with A filter. The sensitivities to low pH and to high concentration of bile were expressed as the S.h. value and the S.b. value, respectively.

$$\text{S.h. value} = \frac{\text{Turbidity of the pH 5.5 culture}}{\text{Turbidity of the pH 7.5 culture}}$$

$$\text{S.b. value} = \frac{\text{Turbidity of the culture containing 2\% bile powder}}{\text{Turbidity of the bile-free culture}}$$

Classification of the Bacteria

Classification of the bacteria isolated was made according to the scheme proposed by Shewan et al. (1960).

Tests for Viability of the Bacteria at
Low pH and in the Presence of Gastric Juice

Bacteria grown at 25°C were harvested by centrifugation in the logarithmic phase and resuspended in 2.5% NaCl solution. A 1-ml portion of the cell suspension was added to 4 ml of 2.5% NaCl-McIlvaine buffer (pH 4.0 or 7.0), containing 2 ml of 2.5% NaCl solution or 2 ml of gastric juice obtained from red sea bream. After 5 hr incubation at 25°C, the number of viable cells was determined by the MPN method.

Examination of Effects of Bile Acids on Growth

Bacteria to be tested were inoculated into the above-mentioned basal medium (pH 7.0) after addition of bile powder, glycocholate, deoxycholate, cholate, or Tween 80 in concentrations of 0.1 and 0.5%, and incubated at 25°C on a rotary shaker. All of the media used were sterilized by filtration with a Seitz filter. The growth was expressed as the increase in turbidity at 600 nm measured with a Hitachi spectrophotometer, model 101.

Detection of Ability of Bacteria to Decompose Bile Acids

A 60-ml culture incubated at 25°C for 18–24 hr on a rotary shaker was centrifuged at 7000 × g to separate cells from supernatant media. The sedimented cells were resuspended in a solution composed of 2 ml of the supernatant and 2 ml of 2.5% NaCl-phosphate buffer (pH 7.0) containing 0.5% cholate or glycocholate. After being kept for 48 hr at 25°C, the suspension was centrifuged, and the supernatant fluid was directly subjected to thin-layer chromatography, according to the method of Aries and Hill (1970a). Benzene-dioxane-acetic acid (70 : 20 : 2) and chloroform-acetic acid-acetone (7 : 1 : 2) were used as solvent systems. Spots of various bile acids were detected by spraying concentrated H_2SO_4 or 10% alcoholic phosphomolybdic acid.

RESULTS AND DISCUSSION

Effect of Diet Composition on Bacterial Flora in Digestive Tracts of Red Sea Bream (Sera and Kimata, 1972)

Red sea bream were fed with F-, S-, and C-diets for 5, 15, and 25 days. Viable counts of total heterotrophic bacteria and chitin-decomposing bacteria in the gastric and intestinal contents were examined. As will be seen in Table 2, the numbers of heterotrophs and chitin decomposers in both the gastric and intestinal contents fluctuated randomly. The majority of the heterotrophs were able to decompose chitin regardless of the diet composition. Such a fluctuation in bacterial number cannot be ascribed to a difference in the diet composition, but to a difference in the extent of digestion of diet.

Morphological and biochemical properties of bacteria isolated from the gastric and intestinal contents of the fish fed with F-, S-, and C-diets for 25 days were examined in comparison with those of bacteria isolated from the environmental seawater and from the C-diet. As shown in Table 3, most of the strains isolated from the gastric and intestinal contents were gram-negative, asporogenous rods with a single polar flagellum, and they were able to produce ammonia from peptone and to hydrolyze chitin. Almost none of the strains isolated from seawater in which the fish were cultivated resembled those from the digestive tracts with respect to their ability to produce indol from peptone and to hydrolyze starch and casein. The strains isolated from the C-diet were different from the others in many respects.

It can be said from these results that many characteristics of bacteria isolated from the digestive tracts were similar to each other and were independent of the diet composition.

Table 2. Total number of heterotrophic bacteria (TB) and number of chitin-decomposing bacteria (CB) in the contents of the digestive tracts of red sea bream, the diet, and environmental seawater (August 1968)[a]

Duration of feeding (days)	Dietary group	Experiment	Stomach		Intestine	
			TB	CB	TB	CB
5	F	I	1.3×10^5	2.2×10^4	1.5×10^5	2.5×10^4
	S	I	—	—	2.0×10^6	2.4×10^5
		II	—	—	1.1×10^7	2.8×10^6
	C	I	1.7×10^6	6.0×10^5	5.6×10^8	2.9×10^8
		II	1.5×10^6	6.1×10^5	1.1×10^9	6.7×10^8
15	F	I	2.0×10^4	6.3×10^3	1.7×10^6	9.0×10^5
		II	1.6×10^4	3.2×10^3	1.5×10^6	8.2×10^5
	S	I	1.5×10^6	1.1×10^5	6.7×10^6	5.0×10^6
		II	1.4×10^6	8.3×10^5	1.2×10^7	1.0×10^7
	C	I	8.0×10^6	5.2×10^6	6.4×10^7	4.8×10^7
		II	1.6×10^6	1.1×10^6	4.0×10^7	2.5×10^7
25	F	I	8.7×10^5	6.7×10^5	4.7×10^7	3.8×10^7
		II	1.2×10^4 b	7.2×10^2	6.7×10^4 b	2.4×10^4
	S	I	1.1×10^5	9.9×10^4	1.3×10^9	8.4×10^8
		II	8.0×10^5	6.7×10^5	7.5×10^8	6.0×10^8
	C	I	1.0×10^6	4.0×10^5	1.4×10^6	1.1×10^6
		II	1.2×10^4 b	9.0×10^3	3.7×10^6	2.3×10^6
Seawater			7.6×10^3	6.4×10^3		
C-diet			1.2×10^5	2.2×10^4		

[a] Results are given as number of cells per gram of contents.
[b] Yeasts were found in these samples.

Table 3. Morphological and biochemical properties of the strains isolated from the contents in the digestive tracts of red sea bream, environmental seawater, and the diet (C-diet) (August 1968)

| | No. of positive strains[a] | | | | | | | |
| | Stomach | | | Intestine | | | | |
Characteristic	F-diet (48)	S-diet (48)	C-diet (39)	F-diet (43)	S-diet (54)	C-diet (67)	Seawater (30)	Diet (21)
Gram-negative	40	48	36	37	54	66	30	17
Shape: Rod	40	48	36	37	54	66	30	16
Coccus	8	0	3	6	0	1	0	5
Motility	39	48	35	35	54	66	30	13
Flagellation:								
Single polar	37	48	33	31	54	64	30	11
Peritrichous	2	0	1	1	0	2	0	1
Cytochrome oxidase	40	48	37	21	54	64	30	11
NH_3 from peptone	48	48	39	43	54	66	30	21
Indol production	1	0	1	3	0	2	30	2
Hugh and Leifson test (glucose)								
Fermentation	39	48	32	32	54	66	30	11
Oxidation	0	0	0	3	0	0	0	1
Gas production	1	0	1	2	0	1	0	2
Hydrolysis of chitin	35	48	34	31	54	64	30	9
casein	4	1	7	5	0	3	27	2
starch	1	0	4	2	0	3	30	1
Yeast detected	9	0	1	2	0	0	0	1

[a] Numbers in parentheses indicate total number of bacteria from source.

In order to make clear whether or not the isolated bacteria were spe-cifically resistant to the environmental conditions (low pH and high concen-tration of bile acids) encountered in fish stomachs and intestines, the effects of pH and a high concentration of bile acids on the growth of the isolates were examined (Table 4). The degree of resistance of the bacteria to high concentrations of bile acids and low pH after 3 days cultivation was expressed as S.b. and S.h. values. A large portion (82–95%) of the strains isolated from the digestive tracts exhibited S.b. and S.h. values of 0.5 and above, although 73% and 50% of the strains isolated from environmental seawater and 86% and 37% of the strains isolated from the diet exhibited S.b. and S.h. values of below 0.5. These results indicate that bacteria in the digestive tracts of marine fish are, in general, more resistant to high concentrations of bile and low pH (5.5) than those found in seawater and diet.

Changes in Bacterial Flora of the Digestive Tracts of Red Sea Bream with Time Lapse after Ingestion (Sera and Ishida, 1972a)

The number of bacteria and the pH of the gastric and intestinal contents were examined at 0.5, 3, 6, 10, and 16 hr after ingestion (Fig. 1). Although the total number of heterotrophs in the stomach was not markedly changed during cultivation, the ratio of chitin-decomposing bacteria to total hetero-trophs gradually increased with time following feeding. The pH of the stomach fluid decreased from 5.6 to 3.7 immediately after ingestion of the diet, and remained constant (between 3.5–3.7) after 3 hr following ingestion. In the intestine, the total number of heterotrophs increased exponentially

Table 4. Effects of addition of bile powder and low pH on the growth of bacteria isolated from the contents of the digestive tracts of red sea bream, the diet, and environmental seawater[a]

Isolated from	No. of strains examined	Bile		Low pH value	
		S.b.<0.5	S.b.≥0.5	S.h.<0.5	S.h.≥0.5
Stomach	145	26	119	10	135
Intestine	166	25	141	9	157
C-diet	22	19	3	8	14
Seawater	30	22	7	15	15

[a]

$$\text{Sh. value} = \frac{\text{Turbidity of the culture of pH 5.5}}{\text{Turbidity of the culture of pH 7.5}}$$

$$\text{S.b. value} = \frac{\text{Turbidity of the culture containing bile powder (2\%)}}{\text{Turbidity of the bile-free culture}}$$

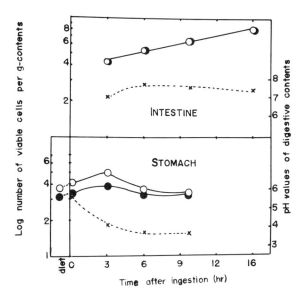

Fig. 1. Changes in number of total heterotrophs (○), chitin-decomposing bacteria (●), and pH (×) in the digestive tracts of red sea bream following ingestion of the diet (fish meat).

after ingestion. Most of these bacteria were able to decompose chitin. Sixteen hr after ingestion, the population of the bacteria had exceeded 10^8 cells per gram intestine content. The pH of the intestine remained constant between 7.0 and 7.8.

Table 5 shows characteristics of the bacteria which were isolated from gastric and intestinal contents of fish sampled at different intervals after ingestion, and from the diet (F-diet). Most of these bacteria were gram-negative, asporogenous rods which were able to produce ammonia from peptone. Most of the strains isolated from the gastric contents 6 and 10 hr after ingestion, and from the intestinal contents 3, 10, and 16 hr after ingestion, were very similar to one another with respect to the properties examined. They were also similar to those of the strains found to be originally dominant in the digestive tract of the fish, as described in the previous section, except for the ability to produce gas from glucose.

The effects of high concentration of bile and low pH on the growth of these strains are shown in Table 6. In most strains isolated from the gastric contents 6 and 10 hr after ingestion, and from the intestinal contents 3, 10, and 16 hr after ingestion, both S.b. and S.h. values were 0.5 and greater.

From these data, it is suggested, in the case of fish with highly developed stomachs, that growth of most of the bacteria derived from the diet and seawater was inhibited by the strong acidity of the stomach and the high concentration of bile in the intestine, and only a small proportion of the bacteria survived the conditions in the stomach and grew exponentially in the intestine.

Table 5. Morphological and biochemical properties of strains isolated from the contents in the digestive tracts of red sea bream and the diet (August 1970)

Characteristics	Diet	No. of positive strains							
		Time of sampling after ingestion (hr)							
		Gastric contents				Intestinal contents			
		0.5	3	6	10	3	6	10	16
Total number of strains	15	23	21	17	16	20	22	22	14
Gram-negative	13	15	21	14	15	20	22	22	14
Shape: Rod	12	15	21	14	15	20	22	22	14
Coccus	3	8	0	3	0	0	0	0	0
Motility	5	5	21	14	15	20	22	18	14
Flagellation: Single polar	2	3	1	12	15	20	22	18	14
Peritrichous	3	2	20	2	0	0	0	0	0
Cytochrome oxidase	7	13	14	13	15	20	22	22	14
NH_3 from peptone	14	23	21	17	16	20	22	22	14
H_2S from cystine	4	13	21	15	15	20	22	22	14
NO_3 reduction	6	11	21	16	15	20	22	22	14
Indol production	0	3	0	0	0	1	19	1	0
Hugh and Leifson test									
Glucose: Fermentation	4	3	1	12	15	20	22	22	14
Oxidation	1	7	0	0	1	0	0	0	0
Gas production	0	0	1	12	15	19	2	10	10
NGA:[a] Fermentation	3	3	1	12	15	20	22	22	14
Oxidation	1	2	0	0	0	0	0	0	0
Gas production	0	0	1	12	15	19	2	10	10
Hydrolysis of chitin	3	4	21	12	15	20	22	22	14
casein	6	9	20	1	0	1	19	0	0
starch	5	7	0	1	0	1	19	0	0

[a] NGA: N-acetylglucosamine.

Table 6. Effect of high concentration of bile powder and low pH on the growth of bacteria isolated from the gastric and intestinal contents of red sea bream and from the diet

Source of isolates	Time after feeding (hr)	No. of strains examined	Bile		Low pH	
			S.b.<0.5	S.b.≥0.5	S.h.<0.5	S.h.≥0.5
Gastric contents	0.5	23	19	4	13	10
	3	21	20	1	0	21
	6	17	3	14	0	17
	10	16	1	15	0	16
Intestinal contents	3	20	2	18	2	18
	6	22	19	3	2	20
	10	22	3	19	4	18
	16	14	0	14	0	14
Diet		15	8	7	7	8

Bacterial Flora of the Digestive Tracts of File-Fish, a Fish with Undeveloped Stomach (Sera and Ishida, 1972*b*)

The total number of heterotrophic bacteria in the contents of the fore-half (from throat to the middle part of the intestine) and those of the post-half (from the middle to the anal part of the intestine) of the digestive tracts, and in the diet (shellfish meat) were 1.7×10^6, 2.1×10^6, and 3.0×10^6 cells per gram, respectively.

Table 7. Morphological and biochemical properties of bacteria isolated from the contents in the digestive tracts of file-fish and the diet

Characteristics	No. of positive strains		
	Fore-intestine	Post-intestine	Diet
Number of bacteria tested	25	13	11
Gram-negative	25	13	10
Shape: Rod	25	13	10
Coccus	0	0	1
Motility	16	5	3
Single polar flagellum	13	5	3
Peritrichous flagella	3	0	0
Cytochrome oxidase	25	12	10
NH_3 production from peptone	25	13	11
H_2S production from cystine	25	13	7
NO_3 reduction	21	13	4
Indol production	10	6	2
Hugh and Leifson test			
Fermentation of glucose	20	7	5
Oxidation of glucose	1	0	1
Gas production	1	0	0
Fermentation of NGA[a]	20	7	4
Oxidation of NGA	0	0	0
Gas production	1	0	0
Hydrolysis of chitin	11	4	2
starch	12	4	2
casein	6	4	5

[a] NGA: N-acetylglucosamine.

Table 8. Effect of low pH (5.5) and high concentration of bile powder on the growth of bacteria isolated from the digestive contents of file-fish and from the diet

Source of isolates	Number of strains examined	Bile		Low pH	
		S.b.$<$0.5	S.b.$\geqslant$0.5	S.h.$<$0.5	S.h.$\geqslant$0.5
Fore-intestine	25	21	4	11	14
Post-intestine	13	7	6	7	6
Diet (shellfish)	11	10	1	6	5

Taxonomic characteristics of these bacteria are shown in Table 7. These data indicate that most of the isolates were gram-negative, asporogenous rods, and were able to produce ammonia from peptone, to produce hydrogen sulfide from cysteine, and to oxidize cytochrome c. With respect to the other properties, these isolates were not similar to one another.

Sensitivity of these bacteria to the high concentration of bile and the low pH value was remarkably different from that of the bacteria isolated from the intestine of red sea bream; more than 50% of the former were inhibited by the high concentration of bile and the low pH, as shown in Table 8, but most of the latter were resistant to these factors.

Considering the above results and the fact that the pH of intestinal contents was 6.5–6.7 in the fore-half and 7.3–7.5 in the post-half of the digestive tracts of file-fish, the pH values in the digestive tracts of file-fish probably is not involved in the selection of exotic bacteria, unlike the case of red sea bream. Therefore, it is hypothesized that the microflora in the intestines of file-fish is composed of various kinds of bacteria which are derived from the diet, in contrast to the case of a fish with a highly developed stomach.

Classification of Isolated Bacteria (Sera and Ishida, 1972c)

As shown in Table 9, all the strains (567 strains) isolated from the digestive tracts of red sea bream and file-fish, the diet, and the environmental seawater were taxonomically divided into eight major groups, each of which was further divided into several subgroups, based on the differences in several properties. Bacteria assigned to the groups I (*Vibrio*) and II (*Pseudomonas*) were motile with a single polar flagellum. Group I fermented sugars, and group II did not. Growth of bacteria belonging to group I was inhibited by

Table 9. Grouping of strains isolated from the contents in the digestive tracts of fish, diet and environmental seawater[a]

Shape	Gram stain	Motility	Flagellation	NH₃ from peptone	Indole production	Cytochrome oxidase	Hugh & Leifson test (glucose)	Hydrolysis of Chitin	Hydrolysis of Casein	Hydrolysis of Starch	Group	No. of strains
Rod	−	+	Single polar	+	−	+	F	+	−	− or ±	I-a₁	280
					−	+	F(g)	+	−	− or ±	I-a₂	72
					−	+	F	−	+	−	I-b	3
					−	+	F	−	−	(+)	I-c	7
					+	+	F	+	+	(+)	I-d	63
					+	+	F	+	−	(+)	I-e	6
					+	+	F	−	−	−	I-f	1
					−	(+)	−	(+)	+	−	II-a	7
					−	(+)	−	−	(+)	−	II-b	2
			Peri.	+	−	−	F(g)	−	−	(+)	III-a	6
					−	+	0	−	−	−	III-b	1

Morphology	Group							No. of strains
	IV-a	−	+	+	−	(+)	−	22
	IV-b	(+)	+	−	−	±	−	2
	IV-c	−	−	−	−	+	−	2
Lopho.	V	−	−	−	0	(+)	−	4
Rod	VI-a	+	+	−	F	+	+	1
	VI-b	−	−	+	F	+	+	1
	VI-c	(+)	−	−	F(g)	+	(+)	2
	VI-d	−	−	+	F	+	−	6
	VI-e	(+)	−	−	F	+	−	6
	VI-f	+	+	−	0	+	−	1
	VI-g	−	−	−	0	(+)	−	5
	VI-h	(+)	(+)	−	−	+	−	26
Diplo.	VII-a	−	−	−	F(g)	−	−	1
	VII-b	−	(+)	−	−	−	−	10
Micro.	VIII-a	−	−	−	0	(+)	−	4
	VIII-b	+	+	−	0	−	−	1
	VIII-c	+	+	−	−	−	−	1
	VIII-d	−	(+)	−	−	−	−	14
	VIII-e	−	−	−	−	−	−	3

[a] Abbreviations: F, fermentation; (g), gas production; O, oxidation; (+), some strains were negative.

vibriostatic compound. Group I was further divided into subgroups I-a$_1$, I-a$_2$, I-b, I-c, I-d, I-e, and I-f, based on the biochemical characteristics. Subgroup I-a$_2$ showed similar properties to I-a$_1$ except for an ability to produce gas from sugars. Bacteria of groups III and IV were peritrichous rods, of which the former utilized sugars and the latter did not. Bacteria of group V (*Pseudomonas*) were lophotrichous and were able to utilize sugars oxidatively. Bacteria of groups VI (*Achromobacter*) were non-motile, gram-negative rods. The other groups (VII and VIII) were gram-positive cocci.

On the basis of the grouping described above, a relationship between the bacterial groups and their sources can be deduced. As shown in Fig. 2, the bacterial flora in the stomach and intestine of red sea bream consisted of bacteria of subgroup I-a$_1$, irrespective of the composition of diets given to the fish. In contrast to this, the bacterial flora in the diet was composed of several groups (I-a$_1$, II, III, V, VI and VII), and that of the environmental seawater was constituted of bacteria belonging to subgroup I-d.

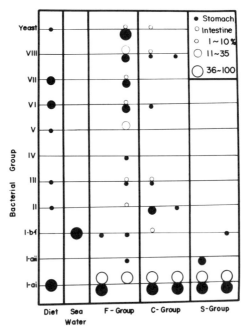

Fig. 2. Frequencies of isolation of each of the bacterial groups from different sources. F-, C-, and S-groups mean the contents of digestive tracts of red sea bream fed with F-, C-, and S-diets, respectively. The size of circles indicates the frequency of isolation from each sample.

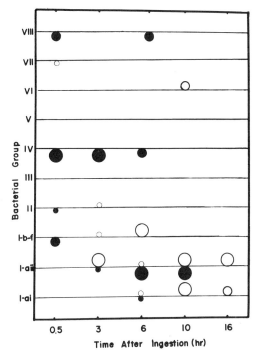

Fig. 3. Changes in isolation frequencies of each bacterial group in the contents of red sea bream with time following ingestion of diets. Symbols are the same as in Fig. 2.

Changes in bacterial flora in the digestive tracts of red sea bream with time following ingestion are shown in Fig. 3. In the stomach, only group I-a$_2$ became dominant 6 hr following ingestion. In the intestine, groups I-a$_1$ and I-a$_2$ almost exclusively predominated 10 hr following ingestion.

Bacterial flora in the fore- and post-parts of the digestive tracts of file-fish were mainly composed of subgroups I-d and I-e, and group VI. The organisms were very similar to those in the diet (shellfish meat), as shown in Fig. 4.

Based on their sensitivity to the high concentration of bile and to the low pH, all the strains isolated could also be divided into two groups (Table 10). Subgroups I-a$_1$ and I-a$_2$ were more or less resistant to both the high concentration of bile and the low pH value. The growth of bacteria belonging to the other groups, however, was inhibited in the presence of 2% of bile powder and/or at pH 5.5.

It became obvious from the taxonomic examination of the isolates that the indigenous bacterial flora in the stomach and intestine of red sea bream is

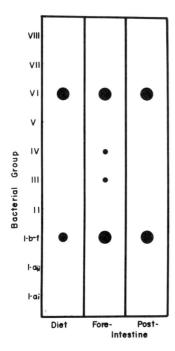

Fig. 4. Frequencies of isolation of each bacterial group from the contents of intestines of file-fish and from diets. Symbols are the same as in Fig. 2.

Table 10. Effect of high concentration of bile powder and low pH on the growth of each bacterial group

Bacterial group	No. of strains			
	S.b. value		S.h. value	
	<0.5	≥0.5	<0.5	≥0.5
I-a₁	25	255	3	277
I-a₂	2	70	1	71
I-b—I-f	69	11	41	39
II	8	1	1	8
III	5	2	1	6
IV	24	2	4	22
V	3	1	1	3
VI	31	17	29	19
VII	9	2	7	4
VIII	15	8	14	9

Table 11. Comparison of a specific group of *Vibrio* isolated from intestines of marine fish, as dominant flora, by several investigators[a]

Characteristics	Sera and Ishida (1972c) *Vibrio* sp. I-a	Liston (1955) Gut group vibrio	Okuzumi and Horie (1969) *Vibrio* sp. I	Aiso *et al.* (1968) *Vibrio* sp.
Shape	Rod	Pleomorphic rod	Rod	Rod
Gram stain	−	−	−	−
Motility	+	+	+	+
Flagellation	Single polar	Single polar	Single polar	Single polar
NH$_3$ from peptone	+	?	+	?
Indole production	−	?	−	?
Cytochrome oxidase	+	?	+	+
Hugh and Leifson test	F	F	F	F
Gas production	(+)	(+)	−	(+)
Nitrate reduction	+	+	+	?
H$_2$S production	+	?	−	?
Hydrolysis of chitin	+	?	+	?
starch	− or ±	?	−	?
Sensitivity to bile acid	R	?	?	R

[a] Abbreviations: F, fermentative; R, resistant; (+) indicates that some strains are negative.

constituted of only one or, at most, two species of *Vibrio* which were assigned to subgroups I-a_1 and I-a_2. These results seem to be consistent with those on the bacterial flora of fish intestines reported by Liston (1955, 1956, 1957), Colwell (1962), Okuzumi and Horie (1968), Simidu and Hasuo (1968*a*) and Aiso *et al.* (1968), as summarized in Table 11.

Effects of pH Value and Gastric Juice on Survival of the Bacteria Indigenous to Digestive Tracts (Sera *et al.*, 1972)

Nine strains of the isolates were used as the representative of the flora of digestive tracts of fish (Table 12) to examine the effects of pH and of gastric juice on the selection of bacteria in digestive tracts.

With respect to the effects of low pH on viability, there was a remarkable difference between the bacteria which were regarded as indigenous to the digestive tracts (strains C_{14} and E_2) and the other strains (Table 13). The former two strains were almost unaffected by pH 4.0 with or without added gastric juice, while the others, except strain B_1, showed rapid loss of viability. In the presence of gastric juice, a marked decrease in the viability was observed with strains F_1 and L_7. These results indicate that the selection of bacteria taking place in the stomach can be attributed to the low pH value and to the presence of gastric juice; the bacteria which had survived these extreme conditions became dominant in the intestinal flora.

Table 12. Sources of the strains used

Strain	Bacterial group		Source[a]	Time (hr)[b]
C_{14}	I-a_1	(*Vibrio* sp.)	GC of red sea bream	6
E_2	I-a_2	(*Vibrio* sp.)	IC of red sea bream	3
F_1	I-d	(*Vibrio* sp.)	IC of red sea bream	6
L_{12}	III	(Peritrichous rods)	IC of file-fish	4
B_1	IV	(Peritrichous rods)	GC of red sea bream	3
L_7	VI-b	(*Achromobacter* sp.)	IC of file-fish	4
M_4	VI-h	(*Achromobacter* sp.)	IC of file-fish	4
A_{28}	VII	(Cocci)	GC of red sea bream	0.5
A_4	VIII	(Cocci)	GC of red sea bream	0.5

[a] Abbreviations: GC, gastric contents; IC, intestinal contents.

[b] Figures indicate the time in hours following ingestion of the diets (fish meat paste for red sea bream, shellfish meat for file-fish).

Table 13. Loss of viability of each strain after
5 hr incubation at pH 4.0 and 7.0 in the presence or
absence of gastric juice prepared from red sea bream[a]

	No. of viable cells (cells per milliliter)			
		After 5 hr incubation		
		pH 4.0		
Strain	Before incubation	(−)	(+)	pH 7.0
C_{14}	2.4×10^8	9.6×10^6	4.8×10^6	2.4×10^8
E_2	2.4×10^8	2.4×10^8	4.8×10^7	4.8×10^8
F_1	1.2×10^9	2.4×10^6	2.4×10^5	1.2×10^9
L_{12}	4.4×10^7	$< 10^2$	3.7×10^2	2.3×10^6
B_1	2.4×10^8	7.7×10^7	4.4×10^7	2.3×10^6
L_7	9.6×10^7	9.6×10^4	$< 10^2$	1.2×10^8
M_4	7.7×10^7	3.7×10^2	4.4×10^3	7.7×10^7
A_4	4.4×10^7	2.3×10^3	9.6×10^3	4.4×10^7
A_{28}	9.6×10^7	9.6×10^3	2.4×10^4	4.4×10^7

[a] (−) in the absence of gastric juice; (+) in the presence of gastric juice.

Effect of Free and Conjugated Bile Acids
on the Growth of Bacteria Isolated from Stomach and Intestine

The growth of the bacteria (strains C_{14} and E_2) which were regarded as indigenous to the digestive tracts in the presence of free and conjugated bile acids was compared with that of the other strains (M_4, F_1, B_1 and L_7).

The relative cell yields (the ratio of the cell yield in the presence of bile acid to that without added bile acid) in cases of 0.1% or 0.5% free bile acids (cholate and deoxycholate), conjugated bile acids (bile powder and glyco-cholate), and Tween 80 (as a detergent), are shown in Fig. 5. With the addition of conjugated bile acids, the relative cell yields of strains C_{14}, E_2, and M_4 increased, while those of the other strains decreased. The free bile acids, especially deoxycholate, however, inhibited the growth of all the strains except M_4.* In all cases the relative cell yields increased with the

* Degradation products of bile acids of these strains were examined using thin-layer chromatography. None of these bacteria degraded bile acids, in contrast to the bacteria indigenous to the intestines of mammals (Norman and Widström, 1964; Midtvedt and Norman, 1967; Hill and Drasar, 1969; Aries et al., 1969; Aries and Hill, 1970a,b).

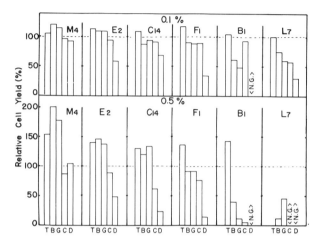

Fig. 5. Effects of conjugated and free bile acids and Tween 80 on the cell yields of each strain. *T*, Tween 80; *B*, bile powder; *G*, glycocholate; *C*, cholate; *D*, deoxycholate; *N.G.*, no growth.

addition of Tween 80. Thus, the inhibitory effect of bile acids for strains F_1, B_1 and L_7 cannot be ascribed to surface-active effects.

Of the strains C_{14}, E_2 and B_1, which were able to survive in the stomach, strain B_1 was remarkably inhibited by bile acids. Of the strains M_4, E_2 and C_{14}, which were resistant to bile acids, only strain M_4 lost viability by being left at low pH.

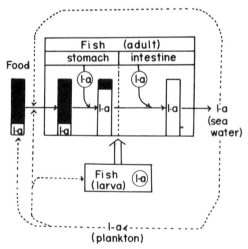

Fig. 6. Scheme for the possible origins of the indigenous bacteria (*Vibrio*, I-a_1 and I-a_2) in the digestive tracts of red sea bream.

Based on the above-mentioned data, the process involved in the formation of the bacterial flora indigenous to the digestive tracts of red sea bream can be speculated upon as summarized in the scheme illustrated in Fig. 6. It is supposed, as stated above, that the bacteria assigned to group I-a (subgroups I-a$_1$ and I-a$_2$ are included) were introduced from foods or environmental seawater and became the dominant organism in the digestive tract by their ability to survive its selective environment. Another possibility is that these bacteria may have been originally indigenous to the digestive tract of the fish having been introduced via food plankton (Simidu *et al.*, 1971) when the fish were in their larval stages.

LITERATURE CITED

Aiso, K., U. Simidu, and K. Hasuo. 1968. Microflora in the digestive tract of inshore fish in Japan. J. Gen. Microbiol. 52:361—364.

Aries, V., J. S. Crowther, B. S. Drasar, and M. J. Hill. 1969. Degradation of bile salts by human intestinal bacteria. Gut 10:575—576.

Aries, V., and M. J. Hill 1970*a*. Degradation of steroids by intestinal bacteria. I. Deconjugation of bile acids. Biochim. Biophys. Acta 202:526—534.

Aries, V., and M. J. Hill. 1970*b*. Degradation of steroids by intestinal bacteria. II. Enzymes catalysing the oxidoreduction of the 3-, 7- and 12-hydroxyl groups in cholic acid, and the dehydroxylation of the 7-hydroxyl group. Biochim. Biophys. Acta 202:535—543.

Colwell, R. R. 1962. The bacterial flora of Puget Sound fish. J. Appl. Bacteriol. 25:147—158.

Harrigan, W. F., and M. E. McCance. 1966. Laboratory Methods in Microbiology. Academic Press, New York.

Hill, M. J., and D. S. Drasar. 1968. Degradation of bile salts by human intestinal bacteria. Gut 9:22—27.

Hugh, R., and E. Leifson. 1953. The taxonomic significance of fermentative *versus* oxidative metabolism of carbohydrates by various gram negative bacteria. J. Bacteriol. 66:24—26.

Kovacs, N. 1956. Identification of *Pseudomonas pyochanea* by the oxidase reaction. Nature (London) 178:703.

Liston, J. 1955. A group of luminous and non-luminous bacteria from the intestine of flatfish. J. Gen. Microbiol. 12:i.

Liston, J. 1956. Quantitative variations in the bacterial flora of flatfish. J. Gen. Microbiol. 15:305—314.

Liston, J. 1957. The occurrence and distribution of bacterial types on flatfish. J. Gen. Microbiol. 16:205—216.

Midtvedt, T., and A. Norman. 1967. Bile acid transformations by microbial strains belonging to genera found in intestinal contents. Acta Pathol. Microbiol. Scand. 71:629—638.

Norman, A., and A. Widström. 1964. Hydrolysis of conjugated bile acids by extracellular enzymes present in rat intestinal contents. Proc. Soc. Exp. Biol. Med. 117:442–444.

Okuzumi, M., and S. Horie. 1969. Studies on the bacterial flora in the intestines of various marine fish. Bull. Jap. Soc. Sci. Fish. 35:93–100.

Sera, H., and Y. Ishida. 1972a. Bacterial flora in the digestive tracts of marine fish. II. Changes of bacterial flora with time lapse after ingestion of diet. Bull. Jap. Soc. Sci. Fish. 38:633–637.

Sera, H., and Y. Ishida. 1972b. Bacterial flora in the digestive tracts of file-fish. Bull. Jap. Soc. Sci. Fish. 38:651–652.

Sera, H., and Y. Ishida. 1972c. Bacterial flora in the digestive tracts of marine fish. III. Classification of isolated bacteria. Bull. Jap. Soc. Sci. Fish. 38:853–858.

Sera, H., Y. Ishida, and H. Kadota. 1972. Bacterial flora in the digestive tracts of marine fish. IV. Effect of H[+] concentration and gastric juices on survival of the indigenous bacteria. Bull. Soc. Sci. Fish. 38:859–863.

Sera, H., and M. Kimata. 1972. Bacterial flora in the digestive tracts of marine fish. I. Bacterial flora of fish, red sea bream snapper and crimson sea bream, fed three kinds of diets. Bull. Jap. Soc. Sci. Fish. 38:50–55.

Shewan, J. M., G. Hobbs, and W. Hodgkiss. 1960. A determinative scheme for the identification of certain genera of gram-negative bacteria, with special reference to the *Pseudomonadaceae*. J. Appl. Bacteriol. 23:379–390.

Simidu, U., K. Ashino, and E. Kaneko. 1971. Bacterial flora of phyto- and zoo-plankton in the inshore water of Japan. Can. J. Microbiol. 17:1157–1160.

Simidu, U., and K. Hasuo. 1968a. Salt dependency of the bacterial flora of marine fish. J. Gen. Microbiol. 52:347–354.

Simidu, U., and K. Hasuo. 1968b. An improved medium for the isolation of bacteria from marine fish. J. Gen. Microbiol. 52:355–360.

A CLASSIFICATION OF *VIBRIO* AND *AEROMONAS* STRAINS FROM NORMAL AND INFECTED FISH

USIO SIMIDU* and EMIKO KANEKO

Institute of Food Microbiology
Chiba University

About ten years ago fish farming in seawater ponds became a booming industry in Japan, and many cases of fish epizootics have been reported in the fish farming ponds. Many vibrio strains were isolated from cases of ulcer disease in shellfish and in cultured and wild fish such as yellowtails, mackerels, puffers, ayu-fish, rainbow trout, and abalones.

We carried out a taxonomic study on these vibrios along with other *Vibrio* and *Aeromonas* strains that were isolated from normal marine fish. The experiment was done following the general procedures used in numerical taxonomy. Morphological and physiological characteristics were examined for 115 strains with some emphasis on morphology and on the utilization of organic acids as a sole carbon source. The characteristics were coded as 191 features of two states. Similarity between two strains was calculated according to the simple matching coefficient of Sokal and Michener (1958). The clustering of strains according to the similarity values was done with both a single-linkage and average-linkage method (Sokal and Sneath, 1963), and the results were compared with the estimates of the inhomogeneity of population following the method of Rogers and Tanimoto (1960).

* Present address: Ocean Research Institute, University of Tokyo.

Inasmuch as the results obtained are rather voluminous, we show here only a few outstanding features. Figure 1 illustrates the result of the clustering with the average-linkage method. This figure clearly shows that most of the fish pathogens or presumptive pathogens examined in this study constitute a single species identical with *Vibrio parahaemolyticus.* The so-called *Vibrio alginolyticus,* which is VP test positive and ferments suctose, also falls into this same species cluster. We referred to this species as phenon 1. Another strain of *Vibrio* from infected rainbow trout seemed to belong to *Vibrio anguillarum-Vibrio ichthyodermis,* while the other three strains from ayu-fish are situated between this latter species and phenon 1.

Apart from the fish pathogens, there was a distinct group of aerogenic, oxidase-negative strains which we denoted as phenon 2. These strains corresponded to *Photobacterium phosphoreum* as recently described by Hendrie *et al.* (1970). This phenon was shown to be quite a homogeneous group as is seen in Fig. 2. Phenon 1, on the contrary, did not show such a homogeneity (Fig. 2).

It is interesting to note that most members of phenon 2 were originally isolated from the gut or gills of normal fish. About one-fifth of these isolates

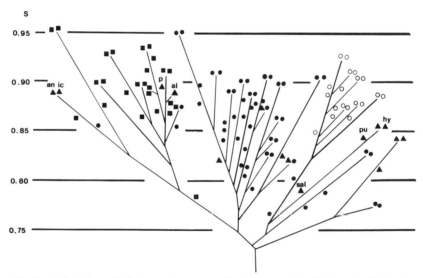

Fig. 1. A clustering of vibrios and aeromonads from normal and infected fish by the average linkage method. Each symbol shows that the particular strain enters into the cluster at the similarity value given in the scale on the left. (■), Strains from infected fish; (●,○), strains from normal fish (including a few strains from seawater; (○), phenon 2; (▲), standard strains; an, *Vibrio anguillarum*; ic, *Vibrio ichthyodermis*; p, *Vibrio parahaemolyticus*; al, *Vibrio alginolyticus*; sal, *Aeromonas salmonicida*; pu, *Aeromonas punctata*; hy, *Aeromonas hydrophila.*

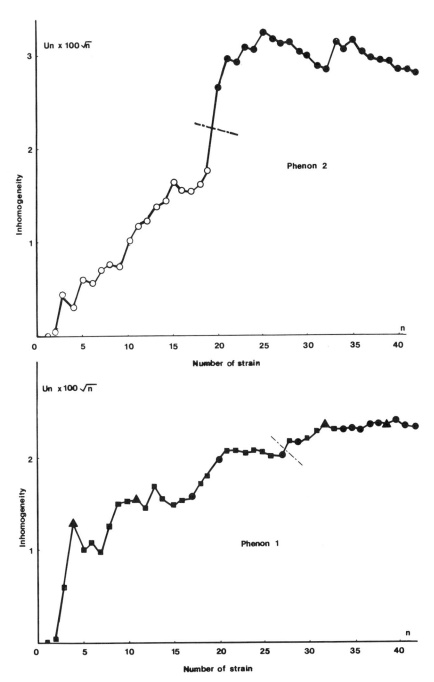

Fig. 2. Increase in the inhomogeneity of two populations, phenon 1 and 2, with the addition of strains. Symbols are the same as in Fig. 1.

were luminescent, although this frequency of occurrence is far from that described as "usually luminescent" (Hendrie *et al.,* 1970).

Not all of the strains from fish intestines, of course, belonged to this species. Ishida, in the preceding report, showed the heterogeneity of the intestinal flora. However, the existence of oxidase-negative strains and of strains with peritrichous flagellation among the *Vibrio-Aeromonas* group will be a further interesting element to consider when comparing the aerobic intestinal flora of marine animals and land animals. While McElroy and Seliger (1962) present a different explanation, bacterial luminescence may be important for the attraction of fish to decaying materials in seawater.

LITERATURE CITED

Hendrie, M., W. Hodgkiss, and J. M. Shewan. 1970. The identification, taxonomy and classification of luminous bacteria. J. Gen. Microbiol. 64:151–169.

McElroy, W. D., and H. H. Seliger. 1962. Biological luminescence. Sci. Amer. 207:76–89.

Rogers, D. J., and T. T. Tanimoto. 1960. A computer program for classifying plants. Science 132:1115–1118.

Sokal, R. R., and C. D. Michener. 1958. A statistical method for evaluating systematic relationships. Univ. Kansas Sci. Bull. 38:1409–1438.

Sokal, R. R., and P. H. A. Sneath. 1963. Principles of Numerical Taxonomy. Freeman, San Francisco, California.

NUTRITIONAL INTERRELATION BETWEEN BACTERIA AND PHYTOPLANKTON IN A PELAGIC ECOSYSTEM

NOBUHIKO TANAKA, MASAMI NAKANISHI, and HAJIME KADOTA

Department of Fisheries
Faculty of Agriculture
Kyoto University
and
Otsu Hydrobiological Station
Faculty of Science
Kyoto University

The production of phytoplankton and bacteria in the water column of Lake Biwa was investigated in an attempt to understand the mechanisms involved in the nutritional interrelation between phytoplankton and bacteria. The distribution of bacteria and phytoplankton in the water column and the responses of these microorganisms to environmental conditions indicated that the excretion of organic matter from phytoplankton plays an important role in the nutrition of bacteria. The excretory products, of which glycollic acid was the main component, apparently served as the principal source of carbon and energy for bacteria in the water.

INTRODUCTION

Phytoplankton often contribute to the production of bacteria in natural waters by supplying organic substrates for growth. A study of the production of phytoplankton and bacteria in the water column of Lake Biwa was undertaken in an attempt to understand the mechanisms involved in the nutritional interrelation between phytoplankton and bacteria.

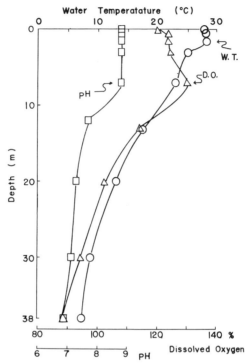

Fig. 1. Vertical distribution of temperature, dissolved oxygen, and pH value in the water column at station Bj-1, Lake Biwa, 24 July 1970.

MATERIALS AND METHODS

Sampling Location

The study was conducted at Shiozu Bay. The bay is located in the northern part of Lake Biwa, which is the largest lake in Japan. The bay receives a relatively small amount of allochthonous organic matter from inflowing waters, the greater part of organic matter present in this bay being supplied by the autochthonous production. In the pelagic water of the bay, the thermocline was usually found at a depth of 10–15 m in the summer (Fig. 1). However, the water was practically homogenous with respect to temperature, concentration of dissolved oxygen, and pH (Fig. 2) in the winter season.

Direct Microscopic Counting of Bacteria

Four ml of the water sample was filtered through a 0.45 μm Millipore filter and the membrane was treated with 10% nitric acid for 2 min. The filter was then stained with 0.3% acid fuchsin for 4 min; this was followed by 0.2% methylene blue for 1 min. After drying, the filter was then subjected to microscopic examination.

Counting of Glycollic-Acid-Utilizing Bacteria

The glycollic-acid-utilizing bacteria were counted by the extinction dilution method using a liquid medium prepared with 1.5 g $NaNH_4HPO_4 \cdot 4H_2O$, 1.0 g KH_2PO_4, 0.2 g $MgSO_4 \cdot 7H_2O$, 2.0 g sodium glycollate, and 1 ml vitamin solution per liter (pH 7.0). The total number of heterotrophic bacteria was counted with a medium containing 5.0 g peptone and 1.0 g yeast extract per liter (pH 7.6).

Determination of Organic Carbon, Organic Nitrogen, and Chlorophyll a

Suspended particulate material was collected by filtering 3 liters of water through a heat-treated glass fiber filter (porosity, 1 μm). Particulate carbon and nitrogen collected with the filter were determined with a Hitachi 026 CHN analyzer. Chlorophyll a was determined by UNESCO's standard method (UNESCO, 1969).

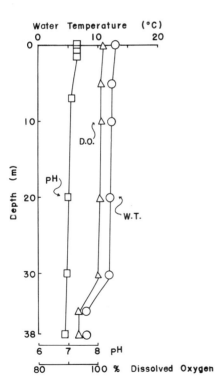

Fig. 2. Vertical distribution of temperature, dissolved oxygen, and pH value in the water column at station Bj-1, Lake Biwa, 27 November 1970.

In Situ Estimation of Photosynthetic Activity

Photosynthetic assimilation of CO_2 was estimated by comparing uptake in "clear" and "dark" 100-ml BOD bottles. Approximately 5 μCi NaH $^{14}CO_3$ was added to each bottle. The bottles were returned to the depths from which the water samples were collected and incubated for 3–4 hr. Following incubation, 10-ml aliquots from the bottles were filtered through Millipore filters (0.45 μm). The damp filters were exposed to HCl vapors for 1 min and then put into vials containing 10 ml dioxane-based fluor (Bray, 1960). The radioactivity was determined using a Packard Tri-Carb Model 2002 Scintillation Spectrometer. The specific activity of samples was determined by the channel-ratio method (Wang and Willis, 1965; Ward and Nakanishi, 1971).

Excretion of Organic Matter from Phytoplankton

The excretion of organic matter by phytoplankton was measured by first incubating 100-ml samples in BOD bottles with 5 μCi NaH $^{14}CO_3$ for 3–4 hr under *in situ* conditions. The water was then filtered with the Millipore filter (0.45 μm) and the cell-free filtrate assayed for radioactive organic compounds. The filtrate was acidified with 2 N HCl and purged with CO_2-free air for 30 min. The CO_2-free filtrate was then neutralized with NaOH and an aliquot removed for liquid-scintillation counting in Bray's fluor (Bray, 1960). The remainder was subjected to ion-exchange resins (Amberlite IR 120 H$^+$ type and Amberlite IR 45 $CO_3{}^{2-}$ type). The products adsorbed to Amberlite IR 120 resin and to Amberlite IR 45 resin were eluted with 2 N NH$_4$OH and 2 N (NH$_4$)$_2$CO$_3$, respectively. One-ml aliquots of the eluates were employed to determine the radioactivity. The method of Bulen *et al.* (1952) was employed to identify the organic acids eluted from the Amberlite IR 45 resin.

Uptake of Glycollic Acid by Bacteria

The heterotrophic uptake of glycollic acid was measured in 50-ml Winkler bottles containing 45-ml water sample together with 0.5-ml sodium glycollate solution and 0.5-ml of 1-[^{14}C] sodium glycollate solution (0.5 μCi/ml). The samples were incubated in the dark at *in situ* temperatures for 2 hr. One-half ml of neutralized formaldehyde solution (37%) was added to stop the activity and the samples were filtered. The filters were washed twice with approximately 20 ml distilled water and placed in vials containing 10 ml Bray's solution and counted with the Packard liquid-scintillation counter. Control tests were made for each concentration of glycollate by fixing a sample with neutralized formaldehyde solution immediately after introduction of the labeled glycollate.

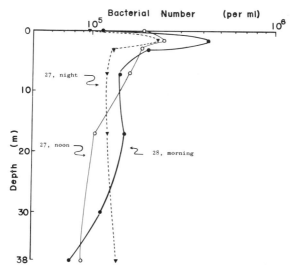

Fig. 3. Vertical distribution of bacteria, counted by the direct microscopic method, in the water column at station Bj-1, Lake Biwa, 27–28 November 1969.

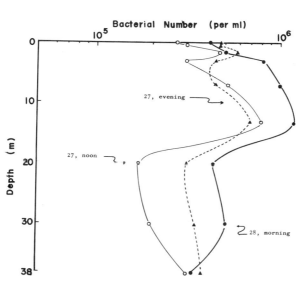

Fig. 4. Vertical distribution of bacteria, counted by the direct microscopic method, in the water column at station Bj-1, Lake Biwa, 27–28 July 1971.

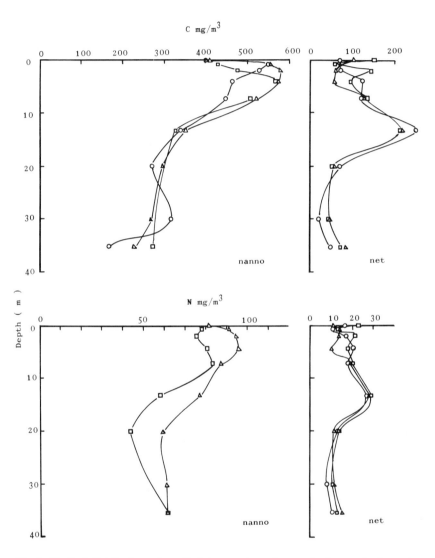

Fig. 5. Vertical distribution of particulate organic carbon and nitrogen in the water column at station Bj-1, Lake Biwa, 27–28 July 1971. Left-hand figures show the distribution in small-sized seston that can be filtered through a XX17 plankton net, and right-hand figures show the distribution in larger-sized seston which cannot pass through the net.

RESULTS AND DISCUSSION

Vertical Distribution of Bacteria

Highest bacterial densities were usually found at depths of 1.5–3 m in winter (Fig. 3) and 1.5–3 m and 10–15 m in summer (Fig. 4). The peak bacterial count found at 10–15 m in summer probably reflects the accumulation of particulate organic matter near the thermocline. The particulate organic matter could be used for attachment of the bacteria as well as an oxidizable substrate for their growth. The peak found at a depth of 1.5–3 m throughout the year may be related not only to the distribution of particulate organic matter but also to the supply of soluble organic substrates excreted by phytoplankton. The distribution of organic carbon and nitrogen (Figs. 5 and 6) suggest that the populations of bacteria are closely related to the distribution of organic substrates.

Vertical Distribution of Phytoplankton and their Activity

The maximum phytoplankton biomass, estimated on the basis of the concentration of chlorophyll a, was found to coincide with the thermocline in

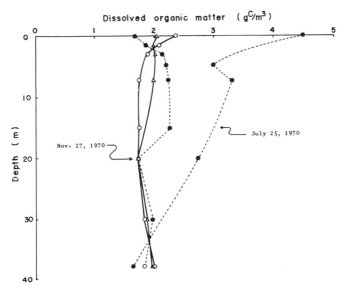

Fig. 6. Vertical distribution of dissolved organic matter in the water column at station Bj-1, Lake Biwa. (o———o), Morning; (△———△), noon; (● - - - ●), evening.

summer (Fig. 7). In winter, however, a distinctive peak in phytoplankton biomass was not observed (Fig. 8). The biomass of nannoplankton was always larger than that of net plankton at any depth.

The photosynthetic activity exhibited a different distribution pattern than that observed for the phytoplankton biomass. Maximum activity was found at a depth of about 3 m in summer and approximately 1.5 m in winter. Diurnal fluctuations in the vertical distribution patterns of chlorophyll a and photosynthetic activity are illustrated in Figs. 9 and 10. These data indicate that the distribution of photosynthetic activity was influenced by light intensity rather than phytoplankton biomass. The distribution of organic matter-excreting activity of phytoplankton in the water column also differed from that of the phytoplankton biomass (Fig. 11).

The effects of light intensity on the photosynthetic and excretion activity of phytoplankton were examined using concentrated phytoplankton pop-

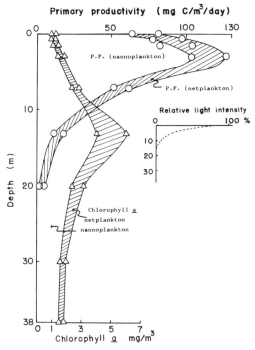

Fig. 7. Vertical distribution of primary productivity and chlorophyll a in the water column at station Bj-1, Lake Biwa, 24–25 July 1970. P.P. indicates the primary productivity.

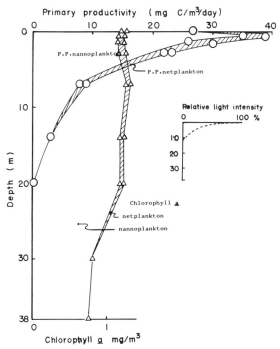

Fig. 8. Vertical distribution of primary productivity and chlorophyll a in the water column at station Bj-1, Lake Biwa, 27 November 1970. P.P. indicates the primary productivity.

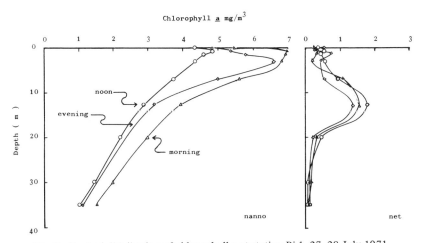

Fig. 9. Vertical distribution of chlorophyll a at station Bj-1, 27–28 July 1971.

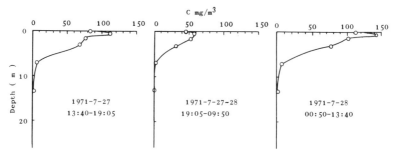

Fig. 10. Vertical distribution of primary productivity at station Bj-1, 27–28 July 1971.

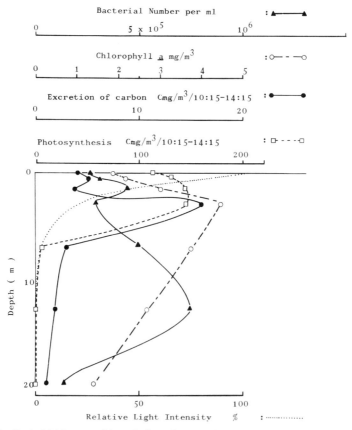

Fig. 11. Bacterial biomass, chlorophyll a, photosynthesis, and excretion of organic substances by plankton in water column at station Bj-1, Lake Biwa, 30 July 1971.

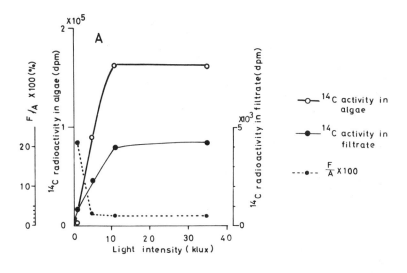

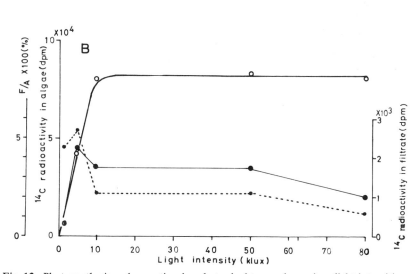

Fig. 12. Photosynthesis and excretion by phytoplankton under various light intensities. Plankton samples employed were obtained by filtering surface water through XX17 net. Incubation time (A,B), 3.5 hr; chlorophyll a (A) 213 mg/m³, (B) 326 mg/m³; water temperature (A) 21°C, (B) 12.5°C; date and station (A) 3 October 1971, Bj-1, (B) 2 December 1971, Bj-1; dominant species (A,B), *Staurastrum dorsidentiferum.*

ulations collected from the central part of Shiozu Bay. Concentrated phyto-plankton populations were prepared by filtering surface water samples through a XX17 plankton net and resuspending the plankton in a small amount of sample water supplemented with $NaH^{14}CO_3$. The dominant species in the phytoplankton populations employed was *Staurastrum dorsidentiferum*. After 3–4 hr of incubation under various light intensities, the samples were filtered through 0.45 μm Millipore filters. The filters and the filtrates were analyzed with respect to the photosynthetic activity and the organic-matter-excreting activity. The results of this experiment (Fig. 12) revealed that both the excretion activity and the photosynthetic activity increased with an increase of light intensity within certain limits. The ratio of

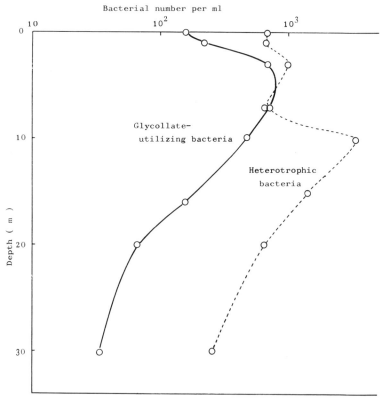

Fig. 13. Vertical distribution of patterns in the water column of glycollic-acid-utilizing bacteria and total heterotrophic bacteria. Counting, extinction dilution method; incubation, 20°C for 17 days (glycollate-utilizing bacteria) and 20°C for 7 days (total hetero-trophs).

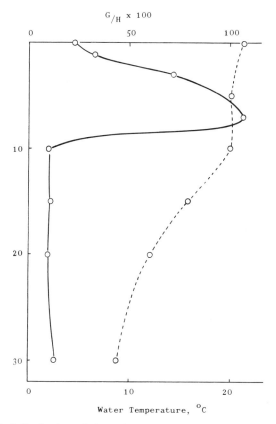

Fig. 14. Vertical distribution of the ratio of number in 1 ml of glycollic-acid-utilizing bacteria (*G*) to total number of heterotrophic bacteria (*H*) in 1 ml. The dotted line shows the vertical changes of water temperature.

excretion to photosynthesis under the illumination of relatively low light intensities was generally larger than that under the high light intensities. Glycollic acid was the major component excreted by the concentrated phytoplankton samples.

Nutritional Interrelation Between Bacteria and Phytoplankton

Since glycollic acid can be used for the growth of bacteria, we examined the vertical distribution patterns of heterotrophic and glycollate-utilizing bacteria (Figs. 13 and 14). Two peaks in the distribution of heterotrophic bacteria were observed, one at about 3 m and another at 12 m. One broad peak at a depth of 3–7 m in the pattern of glycollate-utilizing bacteria was observed

(Fig. 13). The peak in glycollate-utilizing bacteria corresponded closely to the maximum excretion of organic carbon seen in Fig. 11. This observation suggests that the population maximum of bacteria found at a depth of approximately 3–7 m resulted from the abundant supply of soluble organic matter, especially glycollic acid excreted by phytoplankton.

It is generally thought that in euphotic zones of oligotrophic waters the excretion of organic matter from phytoplankton plays an important trophic role in the production of bacteria. The excretory products probably serve as principal sources of carbon and energy for bacterial production. Studies of the kinetics of bacterial uptake of glycollate in the surface waters revealed that the magnitude of uptake was similar to that of the production of this compound (Fig. 15). These data agree with those reported by Wright (1970) on Gravel Pond, Massachusetts.

The peak in heterotrophic bacterial population observed at 10 m depth (Fig. 13) probably reflected the attachment of bacteria to particulate organic

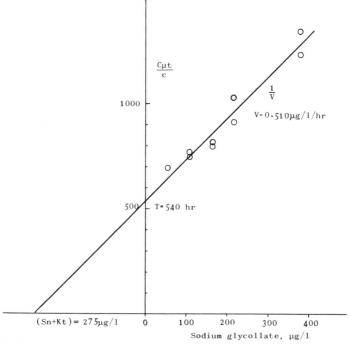

Fig. 15. Uptake of glycollate by surface water collected from Lake Biwa, 19 July 1972. Incubation was made at *in situ* temperature (24.5°C); C, CPM of 1 μCi; c, CPM of sample filter; μ, fraction of a μCi added to sample; t, incubation time.

matter. The microflora at a depth of approximately 3 m may be different from that to be found at the depth of the thermocline.

LITERATURE CITED

Bray, G. A. 1960. A simple efficient liquid scintillator for counting aqueous solutions in a liquid scintillation counter. Anal. Biochem. 1:279–285.

Bulen, W. A., J. E. Varner, and R. C. Burrell. 1952. Separation of organic acids from plant tissues. Anal. Chem. 24:187–190.

UNESCO. 1969. Determination of photosynthetic pigments in sea-water, 2nd edn. Monogr. Oceanogr. Methodol. UNESCO Paris 1:69.

Wang, C. H., and D. L. Willis. 1965. Radio-Tracer Methodology in Biological Science. Prentice-Hall, Englewood Cliffs, New Jersey.

Ward, F. J., and M. Nakanishi. 1971. A comparison of Geiger-Müller and liquid scintillation counting methods in estimating primary productivity. Limnol. Oceanogr. 16:560–563.

Wright, R. T. 1970. Glycollic acid uptake by bacteria. In D. W. Hood (ed.), Organic Matter in Natural Waters, pp. 521–536. Inst. Mar. Sci., College, Alaska.

DISTRIBUTIONS OF YEASTS IN THE WATER MASSES OF THE SOUTHERN OCEANS

JACK W. FELL*

Rosenstiel School of Marine and Atmospheric Science
University of Miami

The marine-occurring yeasts of the southern Indian, southern Pacific and Antarctic Oceans were studied to determine the species present and to examine the relationship of species distributions to environmental conditions. Thirty-three species of three classes were identified: 2 *ascomycetes*, 7 *basidiomycetes*, and 24 *deuteromycetes*. Of these, only eight species (*Debaryomyces hansenii, Candida natalensis, Leucosporidium scottii, Leucosporidium antarcticum, Torulopsis austromarina, Torulopsis norvegica, Torulopsis* sp. and *Sympodiomyces parvus*) were considered to be prevalent. One species, *Debaryomyces hansenii*, was widely dispersed in the study region while the distributions of the other species were limited to specific water masses or to certain geographic regions. Several species, *Leucosporidium antarcticum, Rhodosporidium bisporidiis, Rhodosporidium diobovatum, Rhodosporidium malvinellum, Sympodiomyces parvus, Torulopsis austromarina* and *Torulopsis* sp., may be endemic to these southern high-latitude regions. Viable yeast cell counts varied among the study regions, ranging from 1 per liter to 200 per liter with averages of 1−14 per liter.

INTRODUCTION

The occurrence of phytoplankton and zooplankton species in specific water masses of the open ocean has been the subject of considerable study, and the

---

* Contribution No. 1718 from the University of Miami, Rosenstiel School of Marine and Atmospheric Science.

specificity in distribution of many of these organisms is such that they are used as water-mass indicators (Johnson and Brinton, 1963). In contrast, very little is known about the distributions of microorganisms, particularly the fungi. In the present study we examined the relationship of yeast distributions to the specific water masses in the Antarctic, southern Pacific and southern Indian Oceans. This region is particularly suited for this type of study owing to the fairly small changes in temperature and salinity from surface to bottom waters and to the relative freedom from terrestrial influences.

The southern oceans have the water-mass structure depicted in Fig. 1. For a detailed discussion of this region, the papers by Gordon (1967, 1971) and Gordon and Goldber (1970) are recommended. The objectives of our program were to examine yeast populations in each of these water masses during longitudinal transects. The basic questions being asked were as follows: what species inhabit antarctic oceanic regions and what, if any, water mass- and geographic-species relationships exist? The data from these collections have not been completely analyzed; however, we do have sufficient information to present a discussion of these concepts.

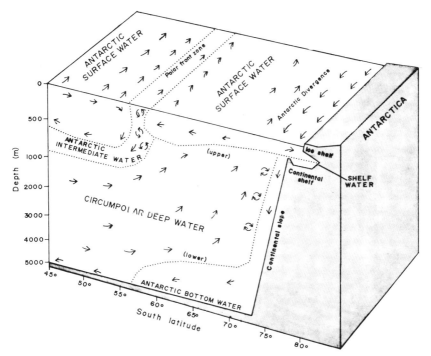

Fig. 1. Schematic representation of water masses and core layers in the antarctic and subantarctic. (From Gordon and Goldberg, 1970).

METHODS

Collections were made during six cruises of the U. S. Naval Ship *Eltanin* (Fig. 2): cruise 23 from approximately 56°S to 64°S and 94°W to 131°W, April–May 1966; cruise 24 from 35°S to 45°S and 125°W to 150°W, July–Aug. 1966; cruise 26 off New Zealand, Nov.–Dec. 1966; cruise 27 from 45°S to 78°S and 147°E to 180°E, Jan.–Feb. 1967; cruise 35 from 38°S to 60°S and 117°E to 133°E, Aug.–Sept. 1968; cruise 38 along 150°E from 40°S to 64°S, March–May 1969. Specific details of the individual station sites of *Eltanin* cruises 23, 24, 26, and 27 were presented by Jacobs and Amos

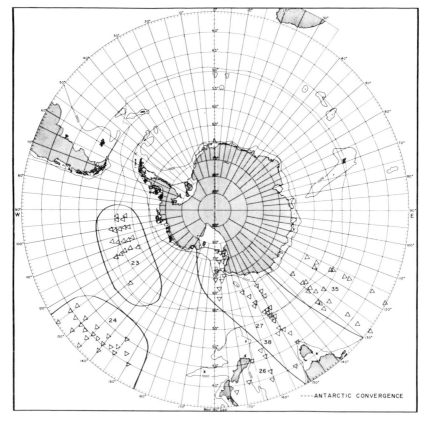

Fig. 2. Station location for yeast collections during *Eltanin* cruises 23, 24, 26, 27, 35, and 38.

(1967), cruise 35 by Jacobs *et al.* (1970), and cruise 38 by Jacobs *et al.* (1972).

Water samples were obtained with the aseptic 2-liter sampling device designed by Niskin (1962). In general, the following water depths were sampled: 2, 10, 25, 50, 100, 150, 200, 400, 600, 800, and 1000 m, and every additional 250 m to the bottom. Subsamples of 500 ml and 1000 ml from each water sample were filtered through 0.45 µm porosity cellulose-ester filters. The filters with the entrapped organisms were placed on agar media containing 2.3% Difco nutrient agar, 0.2% agar, 2% glucose, and 0.1% yeast extract with the pH adjusted to 4.5 with HCl. The agar plates were incubated at 12°C. The resulting yeast colonies were enumerated, subcultured, and returned to Miami under refrigeration. Yeast identifications followed the techniques reviewed by van der Walt (1970).

Coordinated biological-hydrographical data were collected by placing the Niskin samplers 2 m below the Nansen bottles that were used for collecting physical and chemical information. The hydrology was evaluated using the data published by Jacobs and his co-workers. Standard T-S diagrams were prepared and the water-mass envelopes estimated from the information presented by Gordon (1971), although for presentation purposes we did not include his transitional zones. The species distributions were plotted on the T-S diagrams and the water-mass envelopes overlaid.

RESULTS AND DISCUSSION

All three classes of yeasts, the ascomycetes, basidiomycetes and deuteromycetes (Table 1), were obtained in the collections. The majority of these species were rarely isolated, i.e., they were obtained in less than a dozen of the 2100 water samples. Several species (*Debaryomyces hansenii, Candida natalensis, Leucosporidium scottii, Leucosporidium antarcticum, Torulopsis austromarina, Torulopsis norvegica, Torulopsis* sp., and *Sympodiomyces parvus*) were quite prevalent and an examination of their occurrences and distributions suggests several factors that influence the biogeography of marine yeasts.

Few Species of Ascomycetes Inhabit Oceanic Regions

Two species of asomycetes were obtained in the collections, *Debaryomyces hansenii* and *Debaryomyces vanriji,* although the latter species was rarely isolated. This apparent rarity of ascomycetous yeasts is typical of most offshore regions. Meyers *et al.* (1967a) reported *Debaryomyces hansenii* and *Hansenula jadinii* from the Black Sea, Meyers *et al.* (1967b) isolated *Debaryomyces hansenii* and *Hanseniaspora uvarum* from the North Sea, while Fell

Table 1. Yeasts present in southern oceanic regions

	Fungi imperfecti
	Candida intermedia
	Candida natalensis
Ascomycetes	*Candida parapsilosis*
	Candida polymorpha
Debaryomyces hansenii	*Candida rugosa* like
Debaryomyces vanriji	*Candida scottii*
	Candida utilis
	Candida vanriji
	Cryptococcus albidus
	Cryptococcus albidus var. *diffluens*
	Cryptococcus infirmo-miniata
	Cryptococcus laurentii
Basidiomycetes	*Cryptococcus laurentii* var. *flavescens*
Leucosporidium antarcticum	*Cryptococcus laurentii* var. *magnus*
Leucosporidium scottii	*Cryptococcus macerans*
Rhodosporidium bisporidiis	*Rhodotorula glutinis*
Rhodosporidium capitatum	*Rhodotorula minuta*
Rhodosporidium dacryoidum	*Rhodotorula pilimanae*
Rhodosporidium	*Rhodotorula rubra*
infirmo-miniatum	*Sporobolomyces salmonicolor*
Rhodosporidium malvinellum	*Sympodiomyces parvus*
	Taphrina ? sp.
	Torulopsis austromarina
	Torulopsis inconspicua
	Torulopsis norvegica
	Taphrina? sp.
	Trichosporon cutaneum

(1967) obtained *Debaryomyces hansenii, Hanseniospora uvarum, Pichia fermentans,* and *Saccharomyces cerevisiae* from the Indian Ocean. In all of these studies *Debaryomyces hansenii* was the most prevalent species.

Basidiomycetes Occur in Marine Environments

It has been assumed that basidiomycetes are virtually absent from the sea (Wilson, 1960), although there have been a few reports of their presence on plant substrates. Feldmann (1959) observed the basidiomycete *Melanotaenium ruppiae* on the sea grass *Ruppia maritima.* Another basidiomycete, *Nia vibrissa,* may have a world-wide distribution, occurring on submerged wood panels, mangrove (*Rhizophora*) roots, and *Spartina* stems and rhizomes (Kohlmeyer and Kohlmeyer, 1971).

The present observations (Table 1) demonstrate that basidiomycetes are intrinsic members of the oceanic marine mycoflora. These basidiomycetes are of interest owing to their life cycles, which resemble the smut-fungi (Fell, 1970). The latter are terrestrial organisms that cause severe destruction to cereals and other commercially important crops. It is not known whether or not the marine basidiomycetous yeasts have a role in the ocean that is analogous to that of the terrestrial smut-fungi.

The two genera that were found in this study, *Leucosporidium* and *Rhodosporidium,* are the perfect stages of certain species of the yeast genera *Candida* and *Rhodotorula,* respectively. Taxonomic separation of these basidiomycete genera employs the traditional criteria for classification of the imperfect yeasts (Lodder, 1970).

Deuteromycetes Have the Greatest Diversity of Species

Twenty-one species have been identified and several more have yet to be determined. Some of the identified species, particularly *Candida scottii* and members of the genera *Rhodotorula* and *Cryptococcus* are probably the imperfect stages of basidiomycetes. This affinity is suggested by the previously mentioned life-history studies and by the base composition of DNA (Meyer and Phaff, 1970). While some of the other species may be imperfect stages of ascomycetes, extensive testing for sexual responses in all of the imperfect yeasts found in this study proved to be negative.

Certain Species are Ubiquitous in Oceanic Waters

Debaryomyces hansenii is possibly the most widely dispersed species and does not have any apparent geographic or hydrographic limitations. As depicted in Fig. 3A, *Debaryomyces hansenii* was found in all the water masses and was recovered during all of the cruises. The cell counts were 1–40 cells per liter with an average of 5 per liter. The occurrences were sporadic through the water column. As previously mentioned, this species has been isolated from a wide variety of oceanic regions, and it is also a ubiquitous terrestrial inhabitant.

Species with Distributions that Correspond to
Specific Geographic and Hydrographic Conditions

In the absence of a circumpolar study of yeast distributions, it is difficult to present any definitive conclusions on this statement. However, the present data indicate distribution patterns dependent on hydrographic and geographic conditions, although it is not always possible to separate the two factors. Several species are discussed in order to show the types of distribution patterns that were obtained.

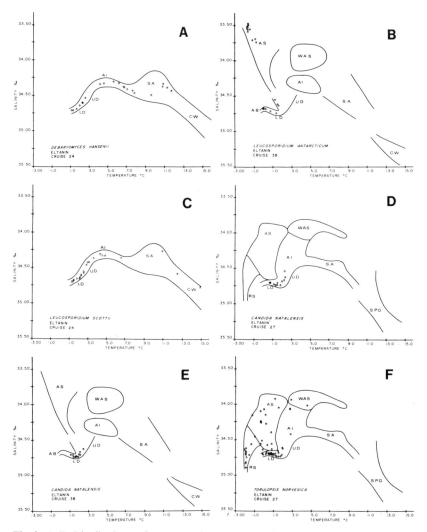

Fig. 3, *A–F.* Distributions of yeast species as compared with the temperature-salinity envelopes of the principal water masses. Abbreviations: *AB*, antarctic bottom water; *AI*, antarctic intermediate water; *AS*, antarctic surface water; *CW*, central water; *LD*, lower deep circumpolar water; *RS*, Ross seawater; *SA*, subantarctic water; *SPG*, South Pacific gyre; *UD*, upper deep circumpolar water; *WAS*, warmed antarctic surface water.

Leucosporidium antarcticum (Fig. 3*B*) was distributed through several water masses but was restricted to one location (64°S, 150°E) adjacent to the pack ice. Concentrations were 1–22 cells per liter, averaging 9 per liter. The majority of the isolations were from the upper 100 m, although the species was recovered from depths of 3000 m. The species was not found farther northward (62°S) in areas free from ice. The only other report of this species is from the waters adjoining the Antarctic Peninsula (Fell *et al.*, 1969), which also have abundant sea ice.

Leucosporidium scottii was collected in all of the water masses from 74°S to 35°S in quantities from 1 to 12 cells per liter (average of 2 per liter). It was isolated in the Pacific sector during cruises 23, 24 (Fig. 3*C*), 26, and 27 but was not obtained at any of the stations east of 158°E, nor was it at any of the sites inhabited by *Leucosporidium antarcticum.*

Candida natalensis is an example of a species concentrated in a narrow longitudinal zone. It was obtained from 149°E to 177°E, 50–64°S, at depths from 245 to 3755 m in the lower and upper deep waters (Fig. 3, *D* and *E*). The cell concentrations were 1–6 cells per liter averaging 2 per liter. There was only one isolate outside this region (131°W, 57°S). *Candida natalensis* was recorded from cruises 27 and 38 which were both in the vicinity of 150°E but were taken 2 years apart. In both instances, *Candida natalensis* was in lower and upper deep waters which indicates that prevalent species can be consistent members of the mycoflora.

Similarly, *Torulopsis norvegica* was in a narrow longitudinal zone, 151°E to 175°W, 53°–77°S, where it was quite abundant, 1–200 cells per liter, averaging 30 per liter. In contrast to *Candida natalensis,* which was confined to deep waters, *Torulopsis norvegica* was found in all of the water masses at temperatures below 5°C (Fig. 3*F*). Populations of *Torulopsis norvegica* were of greater density than any other species in this study and they were only observed during *Eltanin* cruise 27. *Torulopsis norvegica* was not found on *Eltanin* cruise 38, which contradicts the statement made of *Candida natalensis* concerning the stability of the mycoflora. The reason for the discrepancy is not apparent, although it may reflect seasonal differences. *Eltanin* cruise 27 was in January–February while *Eltanin* cruise 38 was in March–May.

Torulopsis sp. (Fig. 4*A*) was also more widely distributed than *Candida natalensis,* extending from antarctic surface waters to lower and upper deep circumpolar water in the region of 45°–64°S, 147°E–94°W. Population densities were 1–3 cells per liter, at an average of 1 per liter. Compared with the distribution of *Torulopsis norvegica, Torulopsis* sp. was not found in Ross Sea derived water.

Torulopsis austromarina ranges from the Ross Sea on cruise 27 (Fig. 4*B*) to lower deep, upper deep, and antarctic surface waters on cruise 38 (Fig.

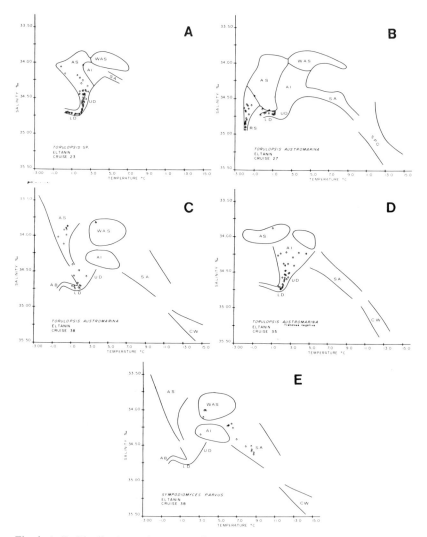

Fig. 4, *A–E.* Distributions of yeast species as compared with the temperature-salinity envelopes of the principal water masses. Abbreviations: *AB,* antarctic bottom water; *AI,* antarctic intermediate water; *AS,* antarctic surface water; *CW,* central water; *LD,* lower deep circumpolar water; *RS,* Ross seawater; *SA,* subantarctic water; *SPG,* South Pacific gyre; *UD,* upper deep circumpolar water; *WAS,* warmed antarctic surface water.

4*C*). Viable cell counts were 1–4 per liter with an average of 1 per liter. The possibility of geographically separated physiological races is suggested by the distribution patterns of *Torulopsis austromarina*. One of the characteristics of *Torulopsis austromarina* from the cruise 27 and 38 collections is the ability to utilize trehalose. In contrast the organisms isolated on cruise 35 (Fig. 4*D*) were unable to assimilate that carbon compound. In all other respects the organisms appeared to be identical. Utilization of trehalose is considered to be an important characteristic in the diagnosis of species of *Torulopsis* (van Uden and Buckley, 1970), suggesting an ecological separation of two physiologically similar species or races. We noted a similar situation in the Indian Ocean (Fell, 1967) where *Candida atmosphaerica* was ubiquitous in all of the water masses sampled, whereas a closely related *Candida* sp. which differed in the inability to utilize trehalose was found only in the southern Indian Ocean.

All of the above-mentioned antarctic species occurred in waters with temperatures less than 5°C. In contrast, *Sympodiomyces parvus* (Fig. 4*E*) is a more northerly species, inhabiting warmed antarctic surface waters in the vicinity of the polar front and northward to the antarctic intermediate and subantarctic waters. Apparently the species is not capable of inhabiting the underlying deep waters. *Sympodiomyces parvus* was isolated during cruises 23, 27, 35, and 38 from 44° to 61°S in numbers of 1–3 per liter, averaging 1 per liter.

Quantitative Distributions

The yeast cell counts ranged from 1 to 200 viable cells per liter and the occurrences varied with individual cruises. Cruise 23 was a 20-station study of the Antarctic Convergence. A total of 417 water samples was examined. Thirty-one per cent contained yeasts with an average for the positive samples of 4 per liter.

Cruise 27, a transect from the Ross Sea to 45°S, had a total of 30 stations and 344 water samples; 37% were positive with an average of 14 per liter.

Cruise 38, a transect at 150°E, had 18 stations and 366 samples; 32% were positive with an average of 3 per liter. In contrast, during cruise 35, which was in the Indian Ocean region, only 15% of the 393 water samples (22 stations) were positive, with an average cell count of 1 per liter.

Cruise 24, a study of the subtropic convergence, had 22 stations and 455 water samples; 18% were positive with an average of 2 per liter. Cruise 26, in the Tasman Sea, had only 7 stations; 22% of the 130 samples were positive with an average concentration of 1 per liter.

The percentage frequency of occurrence of yeasts in these high-latitude southern waters is considerably less than those recorded for other areas. In the North Sea (Meyers *et al.*, 1967*a*) yeasts were found in 99% of the samples in numbers ranging from < 10–> 3000 per liter (averages were not given). In the Black Sea (Meyers *et al.*, 1967*b*), 48% of the samples were positive and the average population density was 5–10 per liter. Indian Ocean water samples (Fell, 1967) were 65% positive for yeasts in average concentrations 36 per liter.

Populations of *Torulopsis norvegica* were observed in the greatest densities, averaging 30 per liter, which accounts for the high average for *Eltanin* cruise 27. *Leucosporidium antarcticum* (9 per liter) and *Debaryomyces hansenii* (5 per liter) were next most abundant species, although *Debaryomyces hansenii* was the only one of the three species that was relatively ubiquitous in the study area. The remaining species were present in concentrations of 1–2 per liter.

Psychrophilism in Antarctic Waters

The majority of the species isolated were able to reproduce in laboratory culture at 24°C. Exceptions to this were *Torulopsis* sp., with a maximum growth temperature of 12–15°C, *Leucosporidium antarcticum*, 17°C, and *Torulopsis austromarina*, 18–21°C. *Torulopsis* sp., in addition to the low growth temperature, has an unidentified nutritional requirement. It was not possible to grow the organism on the standard synthetic testing media (Wickerham, 1951) and it has been quite difficult to maintain the cultures on the various organic isolation and maintenance media. As this is one of the more abundant species, it raises a question as to whether some species are not isolated owing to their inability to reproduce on our isolation medium.

Species Endemic to Southern Marine Waters

Of the 26 species that have been identified in this study, seven have not been reported from other marine or terrestrial sources. *Torulopsis austromarina*, *Torulopsis* sp., and *Leucosporidium antarcticum*, because of their low temperature requirements, may be restricted to antarctic waters, although it would be interesting to examine arctic regions. The apparent endemism of *Rhodosporidium bisporidiis*, *Rhodosporidium dacryoidum* and *Rhodosporidium malvinellum* may be fortuitous, as studies on basidiomycetous yeasts are relatively recent. However, we have been searching for basidiomycetous yeasts in tropical waters and have not found these species, although other species (*Rhodosporidium diobovatum*, *Rhodosporidium sphaerocarpum* and *Rhodosporidium toruloides*) inhabit the tropics.

Sympodiomyces parvus, which we recorded from the polar front and northward, appears to be endemic to that particular region as there have not as yet been any reports of the genus in other marine or terrestrial environments.

CONCLUSIONS

The results indicate that some species have distinct distribution patterns. In particular, *Leucosporidium antarcticum* inhabits the area of pack ice, *Leucosporidium scottii* the Pacific sector of the study region, *Candida natalensis* the lower and upper deep waters in a narrow geographic range, *Sympodiomyces parvus* the polar front and northward, and the two physiological varieties of *Torulopsis austromarina* were geographically separated. *Torulopsis norvegica* was also found in a narrow geographic zone, but was widespread in all of the water masses below 5°C. In some cases the distributions could be temperature dependent, as a few species have a tendency to be psychrophilic. However, temperature and salinity are probably not the primary factors in determining distributions since most yeasts have wide temperature and salinity tolerances. With the exception of the low-temperature yeasts, most of these antarctic organisms grew at 24°C in the laboratory. The studies of the genera *Torulopsis* (van Uden and Vidal-Leiria, 1970) and *Candida* (van Uden and Buckley, 1970) reported maximum salinity tolerances varying from 1 to 22% NaCl, which is far above the concentrations found in the ocean. Sodium chloride requirements are rare among fungi and there is no evidence, to our knowledge, to indicate that the 2 °/oo salinity difference in this oceanic region is sufficient to inhibit or stimulate yeast growth. Possibly the limitations are due to nutrition; the yeasts could be saprophytes or parasites on a specific group or groups of marine plants or animals.

The yeast population densities were lower than reported from other regions. This may be due to the extreme distance from inshore areas, where yeasts are abundant and carried offshore, and to a general low productivity of antarctic regions. Although studies of antarctic productivity are far from complete, Bunt (1971) in his literature review has indicated that certain regions of the antarctic have quite low productivity levels. If this is the case, it could be reflected in the quantities of heterotrophic yeasts.

ACKNOWLEDGMENTS

This research was supported by the National Science Foundation through the Offices of Biological Oceanography and Antarctic Research. Technical assistance was provided by Ingrid L. Hunter and Adele S. Tallman. Dr. Arnold L. Gordon, Lamont-Doherty Geological Observation, is gratefully acknowledged for his discussions on antarctic water masses.

LITERATURE CITED

Bunt, J. 1971. Microbial productivity in polar regions. Symp. Soc. Gen. Microbiol. 21:333–354.

Feldmann, G. 1959. Une ustilaginale marine parasite due *Ruppia maritima* L. Rev. Gen. Bot. 66:35–40.

Fell, J. W. 1967. Distribution of yeasts in the Indian Ocean. Bull. Mar. Sci. 17(2):454–470.

Fell, J. W. 1970. Yeasts with heterobasidiomycetous life cycles. *In* D. G. Ahearn (ed.), Recent Trends in Yeast Research, pp. 49–66. Georgia State University, Atlanta.

Fell, J. W., A. Statzell, I. Hunter, and H. J. Phaff. 1969. *Leucosporidium* gen. nov. The heterobasidiomycetous stage of several yeasts of the genus *Candida*. Antonie van Leeuwenhoek 35:433–462.

Gordon, A. L. 1967. Structure of Antarctic waters between 20°W and 170°W. Antarctic Map Folio Series, Folio 6. Amer. Geogr. Soc., New York.

Gordon, A. L. 1971. Oceanography of Antarctic waters. Antarct. Res. Ser. 15:169–203.

Gordon, A. L., and R. D. Goldberg. 1970. Circumpolar characteristics of Antarctic waters. Antarctic Map Folio Series, Folio 13. Amer. Geogr. Soc., New York.

Jacobs, S. S., and A. F. Amos. 1967. Physical and chemical oceanographic observations in the Southern Oceans. Tech. Rep. No. 1-Cu-1-67, Lamont Doherty Geological Observatory of Columbia University, Palisades, New York.

Jacobs, S. S., P. M. Bruchhausen, and E. B. Bauer. 1970. *Eltanin* Reports. Cruises 32–36, 1968. Hydrographic stations, bottom photographs, current measurements. Lamont-Doherty Geological Observatory of Columbia University, Palisades, New York.

Jacobs, S. S., P. M. Bruchhausen, F. L. Rosselot, A. L. Gordon, A. F. Amos, and M. Belliard. 1972. *Eltanin* Reports. Cruises 37–39, 1969; 42–46, 1970. Hydrographic stations, bottom photographs, current measurements, nephelometer profiles. Lamont-Doherty Geological Observatory of Columbia University, Palisades, New York.

Johnson, M. W., and E. Brinton. 1963. Biological species, water-masses and currents. *In* M. N. Hill, The Sea, Vol. 2, pp. 381–414. John Wiley (Interscience), New York.

Kohlmeyer, J., and E. Kohlmeyer. 1971. Marine fungi from tropical America and Africa. Mycologia 63(4):831–861.

Lodder, J. (ed.) 1970. The Yeasts. North-Holland, Amsterdam.

Meyer, S. A., and H. J. Phaff. 1970. Taxonomic significance of the DNA base composition in yeasts. *In* D. G. Ahearn (ed.), Recent Trends in Yeast Research. Georgia State University, Atlanta.

Meyers, S. P., D. G. Ahearn, and F. J. Roth, Jr. 1967a. Mycological investigations of the Black Sea. Bull. Mar. Sci. 17:576–596.

Meyers, S. P., D. G. Ahearn, W. Gunkel, and F. J. Roth, Jr. 1967b. Yeasts from the North Sea. Mar. Biol. 1(2):118–123.

Niskin, S. J. 1962. A water sampler for microbiological studies. Deep-Sea Res. 9:501–503.

van Uden, N., and H. Buckley. 1970. The genus *Candida. In* J. Lodder (ed.), The Yeasts. North-Holland, Amsterdam.

van Uden, N., and M. Vidal-Leiria. 1970. The genus *Torulopsis. In* J. Lodder (ed.), The Yeasts. North-Holland, Amsterdam.

van der Walt, J. P. 1970. Criteria and methods used in classification. *In* J. Lodder, The Yeasts. North-Holland, Amsterdam.

Wickerham, L. J. 1951. Taxonomy of yeasts. U.S. Dep. Agr. Tech. Bull. No. 1029.

Wilson, I. M. 1960. Marine fungi: a review of the present position. Proc. Linn. Soc. (London) session 171, 1958–1959:53–70.

DISTRIBUTION OF HETEROTROPHIC BACTERIA IN A TRANSECT OF THE ANTARCTIC OCEAN

W. J. WIEBE and C. W. HENDRICKS

University of Georgia
Department of Microbiology

Bacterial distribution in water columns at five localities in a transect from 64°S lat. to 40°S lat. at 150°E long. was examined. The results are summarized as follows. There were reproducible differences with depth in bacterial distribution in the upper 1000 m and these differences appeared associated with water masses. Below about 1000 m, colony-forming units showed little variation from locality to locality. We hypothesize that the water masses regulate the number of colony-forming units present, but that the types of cells reflect surface conditions.

INTRODUCTION

Biogeographical investigations on algal, zooplankton, and fish species in the sea have been conducted since the very beginning of modern oceanography. These studies have provided insights into the distribution of organisms in the sea and, in addition, also provided information about their functions (see for example, Walsh, 1971; Alverson et al., 1964; Johnson and Brinton, 1963). Similar studies with bacteria have been attempted only by Kriss and co-workers (Kriss, 1963; Kriss et al., 1967, Kriss et al., 1969). Because of the sampling and incubation procedures used, the data of Kriss and co-workers have been the subject of some controversy concerning their reliability (see

"Discussion"). Thus, biogeographical studies of bacteria in the sea comparable with those for other organisms are not available.

During cruise 38 of the USNS *Eltanin,* we had the opportunity to examine bacterial distribution in the zone from 64° to 40°S latitude at 150°E longitude covering the regions of the Antarctic Divergence, Antarctic Convergence, and Subtropical Convergence. In this region the water masses and currents are well defined. This provides us with an opportunity to examine bacterial distribution with regard to water masses and currents under near-ideal conditions. Furthermore, there are few data on the bacteria in this region of the world; indeed, Knox (1970) has placed the role of bacteria in this system at the top of his priorities for future investigations.

For this study we employed viable plate count procedures. (In this paper we refer to the count data as colony-forming units, CFU). We recognize that these procedures do not account for all bacteria present in a particular sample. However, for comparative purposes we need to make only two assumptions about the bacterial populations in the water column: first, at any particular depth the bacterial types capable of being cultured are the same from day to day, and second, at any particular depth the distribution of bacteria as singles, pairs, clumped, attached to detritus, etc., will remain the same over our time on station.

In this paper we consider three questions concerning bacterial distribution. First, do water masses contain recognizably different CFUs or types of bacteria? Second, are CFU data reproducible when multiple casts are taken at the same station? Third, do bacteria from surface waters sink in the water column?

MATERIALS AND METHODS

Water was collected with Niskin sterile bacteriological samplers (Niskin, 1962). The filled samplers were stored in racks on deck; the processing of samples was undertaken immediately and finished within 2–3 hr. The deck temperature was 0–8°C at localities 7–11; at locality 12 it was about 15°C. Water temperature at the conclusion of processing never rose above 10°C at the first five localities; at locality 12, water samples from below 200 m never rose above 15°C. Locality, station numbers and locations are listed in Table 1. Hydrographic data were collected by J. Fell (unpublished data). See A. L. Gordon (1967, 1970) for a general hydrographic description of this region.)

Samples were removed from the bags aseptically, filtered through 0.45 μm (HA) 47 mm gridded Millipore filters and placed on agar plates. Incubation was at 3 ± 1°C, except those samples from warmer waters (8–16°C), which were incubated at 12°C. The growth medium was 0.1% Bacto-peptone,

Table 1. Latitude and
longitude of sampled water columns

Locality no.	Station no.	Latitude	Longitude
7	1	64° 18'S	150° 03'E
	2	64° 12'S	150° 01'E
	3	64° 09'S	150° 12'E
8	4	62° 00'S	149° 58'E
	5	61° 41'S	149° 56'E
	7	61° 55'S	150° 04'E
9	8	57° 26'S	149° 34'E
	9	57° 18'S	149° 35'E
	10	57° 32'S	149° 15'E
10	11	54° 12'S	150° 19'E
	12	53° 58'S	151° 15'E
	13	53° 40'S	151° 37'E
11	14	49° 41'S	152° 30'E
	15	49° 43'S	152° 24'E
	17	49° 46'S	152° 32'E
12	18	40° 00'S	152° 00'E
	19	40° 00'S	152° 00'E

0.1% Bacto-yeast extract, 0.1% Pan-Mede, 1.1% Bacto-agar, pH 7.4, in sea-water. ("Bacto-" products from Difco Laboratories, Detroit, Michigan; Pan-Mede from Paines and Byrne Ltd., Greenford, England.) Colony counts were made after 1–3 weeks and also after returning to our laboratory at the University of Georgia. Plates were transported to the United States in cold chests with ice always present and temperatures were always below 10°C at the time of re-icing.

A cold tray was used to maintain agar-plate temperatures below 10°C while working in a room temperature laboratory (Wiebe and Hendricks, 1971).

For pure-culture studies isolates were chosen from four depths in the water column. Isolates were randomly selected and purified by three sequential streakings. The pure-culture results reported here concern temperature growth response using procedures outlined by Colwell and Wiebe (1970). A detailed study of the taxonomic distribution patterns will be reported in a subsequent publication.

RESULTS

The distribution patterns of bacteria are shown in Fig. 1*A* and *B* for the complete water column with an expanded scale for the upper 200 m. Below about 1000–1500 m, counts were uniform at all localities, and there were no discernible relationships to water masses. In the upper waters, however, bacterial distribution in water columns showed reproducible differences at all localities except station 7. While the CFU per milliliter was very small (from < 1 to a few hundred) there was good agreement between separate stations at a single locality, including surface waters. Locality 7 samples displayed the greatest variability in cell counts from top to bottom of any locality. Nutrients (El Sayed, Weisberg and Fryxell, unpublished data) also were more variable at locality 7 than at any other locality. Currents and water masses in this region are somewhat confusing; it is the zone of the Antarctic Divergence and of the region where cold surface water sinks to form the antarctic bottom water. The variability in CFU and nutrients from station to station appears to reflect the variability of the water masses in this region. At locality 8

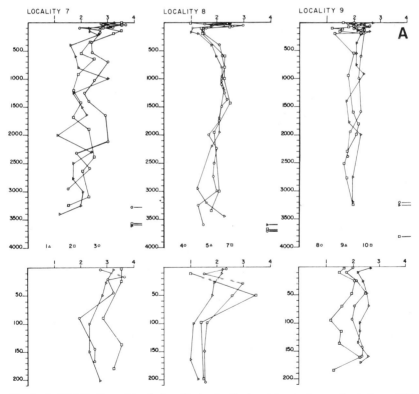

Fig. 1, *A*. Distribution of bacteria with depth. Horizontal axis, log no. bacteria per 100 ml; vertical axis, depth in meters. Numbers at the bottom (e.g. 1Δ) indicate station number.

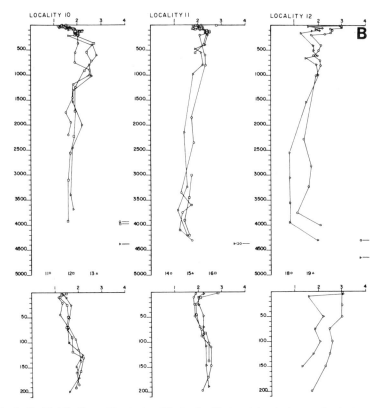

Fig. 1, *B.* Distribution of bacteria with depth. Horizontal axis, log no. bacteria per 100 ml; vertical axis, depth in meters. Numbers at the bottom (e.g. 11○) indicate station number.

three distinct water masses were seen, antarctic surface water to about 100 m rising Pacific deep water, 100–400 m, and circumpolar water below 400 m. These regions correspond to high numbers of CFU at the surface, lower numbers of CFU in the 100–400 m region and stable higher numbers of CFU in circumpolar water. Locality 9, which is just south of the Antarctic Convergence, and an area of sinking water, shows better than 10^2 per 100 ml all the way down the water column, except at station 10 (see below). At localities 10n12 the upper 200n400 m of water are within the subantarctic surface-water zone. This water flows northward at the surface to the Subtropical Convergence; it then sinks to about 400 m and returns southward, rising to the surface just north of the Antarctic Convergence. Between localities 9 and 10 in the region of the Antarctic Convergence, sinking surface water forms the northward-moving antarctic intermediate water, defined hydrographically by the salinity minimum. In the upper 200 m the increasing numbers of CFU with depth at locality 10 may reflect the organic-nutrient-poor subantarctic waters rising to the surface and mixing with the organic-nutrient-rich, sinking antarctic intermediate water. At the surface CFU (upper 200–300 m) were higher at localities 11 and 12 than locality 10, perhaps in response to the increasing primary production in the northward-moving surface water (Jitts and Carpenter, unpublished data). In the 400–1000 m region at these three localities, numbers of CFUs were higher than either directly above or below, the highest counts being found at locality 10 where the forming antarctic intermediate water has most recently descended, and the lower counts at the more northernly localities (11, 12) as the current moves north.

While we attempted to remain on exact position for each station in a locality, this was not always possible. At locality 10, station 11 was some 50 nautical miles south of the other two casts; and at locality 9 station 10 was 6–14 nautical miles south of the other two stations. In both cases counts in the upper 1000 m differed from the other stations at each locality. It is not possible to prove that the change in position is the reason for the observed deviations, but the fact remains that in each case where different values were found there was some physical displacement. While in many areas of the Southern Ocean such small distances might be expected to be of trivial importance, in this region of the Antarctic, because of the strong vertical moments of water masses and the narrowness and wandering nature of the polar front, small changes in location can matter very much.

We examined the growth-temperature ranges of isolates, since sampling localities started with water columns of uniformly cold water and surface temperatures increased northward. Because bacteria respond rapidly to environmental changes (Sieburth, 1968) we have used temperature as one convenient physiological marker for possibly tracing the origin of bacteria. Growth-temperature ranges for isolates from four different depths are shown in Table 2. Except at locality 7 at 1413–1502 m and locality 8 at 2 m, which contained a large number of eurythermic (wide temperature range) isolates, all depths in localities 7–11 yielded mostly stenothermic (restricted temperature range) isolates. Samples from localities 9–11 yielded roughly the same types with the majority having a maximum growth temperature below 25°C. However, locality-12 isolates showed a large increase in eurythermic isolates from surface to bottom. Temperature data showed that there was a difference between Antarctic Ocean and Subtropical Pacific convergence isolates. At localities 7–11, within the Antarctic Ocean, 82.5% of all isolates grew at 20°C while only 27% grew at 25°C and 15% at 30°C. At locality 12, 100% of the isolates grew at 20°C, 91% at 25°C, and 84% at 30°C.

Table 2. Isolate temperature responses

Locality	Depth (m)	Percentage organisms growing at various temperatures								No. of organisms tested
		4°C	15°C	20°C	25°C	30°C	32°C	35°C	37°C	
7	21–25	100	99	81	36	33	3	3	3	70
	136–200	100	62	30	0	0	0	0	0	53
	1413–1502	100	88	80	55	53	27	27	16	64
	2810–3256	100	79	74	7	3	3	3	0	57
8	2	100	100	100	73	73	73	73	50	22
	200–600	100	65	44	22	22	17	17	11	46
	1254–1500	100	100	41	11	11	8	3	3	37
	2500–3345	100	86	36	29	29	17	17	17	42
9	23	100	93	90	18	16	7	6	4	68
	140–372	100	97	95	16	3	0	0	0	64
	1241–1572	100	97	94	14	4	4	3	3	70
	2384–2530	100	93	87	33	24	22	22	16	67
10	2–25	100	97	88	50	5	5	5	0	40
	184–200	100	91	84	46	9	7	7	0	54
	1547–1741	100	100	100	38	17	14	14	2	48
	2170–3918	100	92	90	51	15	15	15	2	39
11	23–25	100	97	87	30	16	13	11	2	62
	142–198	100	96	96	38	16	13	13	0	45
	1289–1864	100	98	98	29	10	9	9	0	63
	1985–3966	100	90	81	32	5	5	5	2	41
12	2–25	100	100	100	100	100	98	98	5	50
	149–200	100	100	100	91	85	64	47	0	32
	1267–1567	100	100	96	88	79	75	73	11	48
	3994–4296	100	100	100	75	44	13	13	6	16

The temperature data (Table 2) indicate that, with regard to temperature range, organisms showed as great or greater similarity vertically than horizontally in the water column, regardless of the water masses being sampled. This was particularly obvious for Locality 12 where the upper limit of growth for isolates was much greater throughout the water column than at any other locality except surface water at locality 8 and 1500 m at locality 7. It might be argued that because the surface water at locality 12 was much warmer than at the other localities (15°C), psychrophiles were killed off. However, deep water never warmed above 15°C and 25% of the organisms from the deep water and 12% from mid-depths did not grow at 25°C. Thus psychrophiles probably were not damaged greatly by our treatment.

Sodium requirements were much more uniform, with 89.5% requiring added sodium for growth, and this requirement was not spared by seawater amounts of magnesium, calcium, or potassium. Of the 10.5% that did not require sodium, 48% were from surface waters at locality 7 and 27% were from surface waters from locality 12.

DISCUSSION

There are several notable features of our data. First, they demonstrate that the number of CFU are stable in a given water mass on a day-to-day basis, that is, numbers from any one day are reasonably representative of the bacterial distribution at a given location. Second, there are reproducible differences in CFU in different water masses in the upper 1000 m. Third, they present presumptive evidence that small particles, in this case bacteria, travel from the sea surface to the bottom. Finally, while the number of CFU appears to be affected or even regulated by water masses and currents, the types of cells most often seen in an entire water column reflect surface types. There appears to be a gradual replacement of types, while relative numbers remain constant.

A major objective of this study was to examine the biogeographical distribution of bacteria in relation to water masses and current. We know of no comparable bacterial data by other authors with which to compare our study. Kriss et al. (1969) attempted a similar type of study, but it is not possible to interpret their results because of the non-sterile techniques used for the collection of samples and the high temperature of incubation (23–28°C) employed. The temperature of incubation they used was particularly unfortunate, since we found that only 27% of our isolates grew at 25°C and only 15% at 30°C. These results might explain the large areas they showed from which they could grow no cells. We find that bacteria, as estimated by viable count procedures, appear to be associated with specific water masses in the upper 1000 m in this region of the World Ocean. It should be emphasized that we are not proposing to use bacterial distribution data for identifying water masses but rather we wished to examine the effect of water masses on bacterial distribution. Viable count procedures are useful for following changes in populations, although we recognize that they are not strictly quantitative.

The distribution of bacterial types along this transect appears to differ somewhat from the distribution of other organisms. Walsh (1971) found two distinct phytoplankton communities, one north and one south of the Antarctic Convergence. Some species, however, e.g., *Fragilariopsis antarctica,* were ubiquitously distributed. Fell and Hunter (unpublished data) examined yeast distribution on *Eltanin* cruise 38 and found that many, but not all, species were similarly restricted in distribution. Johnson and Brinton (1963) reviewed the biogeographical distribution of zooplankton species and also found that the Antarctic Convergence presented a major barrier to species distribution, although again some species were ubiquitous. By both physiological and taxonomic criteria the bacteria in this study showed a high degree of similarity among isolates from localities 8–11, which span the Antarctic Convergence, and less similarity with localities 7 and 12, which are at the extremes of the transect. Using our techniques we did not find specific or indicator bacterial types associated exclusively north or south of the Antarctic Convergence. It should be noted that while taxonomic reliability at the genus level appears quite good in a large number of studies of marine bacteria, it is notoriously poor for discriminating between species. It may well be that our powers of discrimination are too coarse to resolve differences.

In this study we chose to examine aerobic (respiring), organic matter utilizing (heterotrophic) bacteria, so that we might relate our study to those on community oxygen consumption and primary production. There are of course many bacterial groups that do not fall into these categories. Pomeroy, Wiebe, and Thomas (unpublished data) examined the relationship between bacterial numbers and respiration and ATP levels and found no recognizable correlations. Respiration varied with temperature more than any other single parameter and ATP values gave variable results. In fact respiration values for multiple samples showed much greater variation than CFU numbers. Christian (1972) calculated that even if one hundred times more cells were present in samples than we counted and if they were maximally metabolizing they would still not account for the level of respiration measured (Pomeroy, Wiebe, and Thomas, in preparation). This is true even though the procedure for concentration of samples for respiration measurement loses about two thirds of the total biomass and more heterotrophic than autotrophic organisms (Holm-Hanson *et al.,* 1970).

Taxonomic and physiological studies of isolates (Wiebe, unpublished data) fron this transect support the idea that bacteria sink in antarctic water columns. Beside the growth-temperature range studies, a numerical taxonomic analysis has been performed on 1260 isolates. This technique permits a direct phenotypic comparison between isolates, based on the computer technique described in Sokal and Sneath (1963). For each species group present a Hypothetical Median Organism (HMO) was designated, using the technique discussed in Liston *et al.* (1963). The HMO is considered a representative of the entire group. Thus by comparing the HMO of the groups, we can reduce the number of ,organisms dealt with to a manageable size, in this case from 1260 to 108. In this manner it was possible to compare groups from one locality and station with all others. At both localities 7 and 12 fewer of the species groups had linkages with the mid-locality groups (8–11), that is, they showed more similarity vertically than horizontally. Other differences between localities also were noted. For example, at locality 7 pigmented colonies on the viable count plates were seen at all depths but were greatly reduced in number at other stations. These results are being prepared for publication.

What are the possible mechanisms involved in replacement and sinking? The role of particulate matter and the sinking of particles in the sea has been reviewed recently by Riley (1970). Particles of the size and specific gravity of bacteria probably do not sink rapidly. The estimated sinking time of $2-6~\mu m$ particles to the depth of 4000 m is 110 years (Riley, 1970). They could sink if attached to larger particles, and, while undoubtedly some sinking takes place this way, few attached bacteria were observed in this region (Wiebe and Pomeroy, 1972). The mixing effects of current movements might explain some downward transfer but cannot explain how bacteria from the surface arrive at the bottom, since water movement in this region, except perhaps at locality 7, does not carry to the bottom. The most likely mechanism of vertical bacterial movement is via zooplankton migrations and the sinking of fecal pellets, although data are lacking for confirmation of this hypothesis.

Bacterial "disappearance" is more difficult to account for. They may sink to the sediment, be grazed, or die. Presently there are too few data to decide how this takes place and the mechanisms mentioned are not of course mutually exclusive. The observation that the types of bacteria change while numbers remain constant in marine environments has been noted before. Wiebe and Liston (1972) found similar results during studies of sediment bacteria off the Oregon-Washington coast. Numbers of bacteria per gram of sediment at a given location were quite constant but there was recognizable taxonomic diversity. They suggested that while numbers were controlled by available carbon, the types present fluctuated in response to other (unknown) features.

It has been suggested that bacteria might serve as an important alternate food source for zooplankton (Riley, 1970). Knox (1970) cited the question of bacteria as food for animals as an important problem to be answered in the Antarctic ecosystem. Our values of 1–1000 CFU per ml are extremely low and preclude bacteria being an important, generally distributed, particulate energy source. Even if the actual number of cells were a thousand times greater they still would not supply the energy calculated to be necessary for filtering organisms (Jorgensen, 1966). Bacteria may be important in localized zones, however. Recently, Sorokin (1971) has found very high bacterial population in zones of less than 1 m thickness. Thus, in these localized regions bacteria might constitute an important energy source, but it is our conclusion that in most open-ocean waters bacteria are of limited direct importance as food for marine animals.

ACKNOWLEDGMENTS

We should like to thank Dr. L. R. Pomeroy for advice during *Eltanin* cruise 38 and during the writing of this paper and Beverly Mannes for invaluable technical assistance. This research was supported by Grant GA-1477 from the U. S. Antarctic Research Program of the National Science Foundation.

LITERATURE CITED

Alverson, D. L., A. T. Pruter, and L. L. Ronholt. 1964. A study of demersal fishes and fisheries of the Northeastern Pacific Ocean. H. R. MacMillan Lectures in Fisheries. Institute of Fisheries, University of British Columbia, Vancouver.

Breed, R. S., E. G. D. Murray, and N. R. Smith (eds.). 1957. Bergey's Manual of Determinative Bacteriology, 7th edn. Williams & Wilkins, Baltimore, Maryland.

Christian, R. R. 1972. The effects of temperature upon the reproduction and respiration of selected marine psychrophiles. M.S. Thesis. Univ. Georgia.

Colwell, R. R., and W. J. Wiebe. 1970. "Core" characteristics for use in classifying aerobic, heterotrophic bacteria by numerical taxonomy. Bull. Ga. Acad. Sci. 28:165–185.

Gordon, A. L. 1967. Structure of Antarctic waters between 20°W and 170°W. Antarctic Map Folio Series, Folio 6. Amer. Geogr. Soc., New York.

Gordon, A. L. 1970. Circumpolar characteristics of Antarctic waters. Antarctic Map Folio Series, Folio 13. Amer. Geogr. Soc., New York.

Hobbie, J. E., O. Holm-Hanson, T. T. Packard, L. R. Pomeroy, R. W. Sheldon, J. P. Thomas, and W. J. Wiebe. 1972. Distribution and activity of microorganisms in ocean water. Limnol. Oceanogr. 17:544–555.

Holm-Hanson, O., T. T. Packard, and L. R. Pomeroy. 1971. Efficiency of the reverse-flow filtration technique for concentrating marine plankton. Limnol. Oceanogr. 15:832–835.

Johnson, M. W., and E. Brinton. 1963. Biological species, water masses and current. In M. N. Hill (ed.), The Seas, Vol. 2, pp. 381–414. John Wiley (Interscience), New York.

Jorgensen, C. B. 1966. Biology of Suspension Feeding. Pergamon Press, New York.

Knox, G. A. 1970. Antarctic marine ecosystems. In M. W. Holdgate (ed.), Antarctic Ecology, Vol. I, pp. 69–96. Academic Press, New York.

Kriss, A. E. 1963. Marine Microbiology (Deep Sea). (Trans. by J. M. Shewan and Z. Kabata.) Oliver and Boyd, London.

Kriss, A. E., I. E. Mishustina, and M. N. Lebedeva. 1969. Bacterial population densities (heterotrophs) in the water column of the Southern and Indian Oceans. Mikrobiologiya 38:511–517.

Kriss, A. E., I. E. Mishustina, I. N. Mitskevich, and E. V. Zemtsova. 1967. Microbial Populations of Oceans and Seas. St. Martin's Press, New York.

Liston, J., W. J. Wiebe, and R. R. Colwell. 1963. Quantitative approach to the study of bacterial species. J. Bacteriol. 84:1061–1070.

Niskin, S. J. 1962. A water sampler for microbiological studies. Deep-Sea Res. 9:501–503.

Pomeroy, L. R., and R. E. Johannes. 1968. Occurrence and respiration of ultraplankton in the upper 500 meters of the ocean. Deep-Sea Res. 15:381–391.

Riley, G. A. 1970. Particulate organic matter in the sea. Adv. Mar. Biol. 8:1–118.

Sieburth, J. McN. 1968. Observations on bacteria planktons in Narragansett Bay, Rhode Island; A resumé. Bull. Misaki Mar. Biol. Inst. Kyoto Univ. 12:49–64.

Sokal, R. R., and P. H. A. Sneath. 1963. Principles of Numerical Taxonomy, W. H. Freeman, San Francisco.

Sorokin, J. 1971. On the role of bacteria in the productivity of tropical oceanic waters. Int. Revue Ges. Hydrobiol. 56:1–48.

Walsh, J. J. 1971. Relative importance of habitat variables in predicting the distribution of phytoplankton at the ecotone of the Antarctic upwelling ecosystem. Ecol. Monogr. 41.291–309.

Wiebe, W. J., and C. W. Hendricks. 1971. Simple reliable cold tray for the recovery and examination of thermosensitive organisms. Appl. Microbiol. 22:734–735.

Wiebe, W. J., and J. Liston. 1972. Studies of the aerobic, non-exacting, heterotrophic bacteria of the benthos. In A. T. Pruter and D. L. Alverson (eds). The Columbia River Estuary and Adjacent Ocean Waters: Bioenvironmental Studies, pp. 281–312. University of Washington Press, Seattle.

Wiebe, W. J., and L. R. Pomeroy. 1972. Microorganisms and their association with aggregates and detritus in the sea: A microscopic study. Mem. Ist. Ital. Idrobiol. 29(suppl.):325–352.

ECOLOGICAL STUDIES OF *VIBRIO PARAHAEMOLYTICUS*

RITA R. COLWELL and TATSUO KANEKO

Department of Microbiology
University of Maryland

Ecological data obtained for Chesapeake Bay and the Atlantic Ocean coastal region showed the distribution of *Vibrio parahaemolyticus* to be restricted to the estuary. *Vibrio parahaemolyticus* was not found in samples collected off the South Carolina and Georgia coasts, even from samples collected 4–5 miles from shore.

INTRODUCTION

A great number of physical, chemical, geological, and biological factors of the marine environment, namely, salinity, temperature, pressure, organic nutrients, turbidity, dissolved gases, light, water movement, ionizing radiation, etc., affect the distribution of organisms in the marine environment. Very little is known about the distribution of individual bacterial species in the oceans and estuaries of the world. Usually, total counts of viable, aerobic, heterotrophic bacteria are done without attempting to identify and classify the species of bacteria involved. It is well known that the number of bacteria decreases from shore to open ocean and from surface water to deep water, increasing at or just above the water-sediment interface (ZoBell, 1946).

There are problems in following the distribution of a given species in the marine environment, not the least of which is the difficulty in identifying and classifying marine bacteria. *Vibrio parahaemolyticus* is a prime example of such difficulties. The confusion in identification of the organism, even after a more concise definition was provided in 1964 (Sakazaki, 1964), makes it difficult to assess the available information properly. *Vibrio parahaemolyticus, Vibrio alginolyticus,* and *Vibrio anguillarum* are discriminated using several characteristics, such as growth at 7% and 10% NaCl, acid production from sucrose and cellobiose, Voges-Proskauer (VP) reaction, and swarming on agar.

A large number of papers have appeared in the literature on the distribution of *Vibrio parahaemolyticus,* mainly from the epidemiological viewpoint, seeking to determine the source of the organism after fish catches had been landed. Extensive studies of the distribution of *Vibrio parahaemolyticus* in water, sediment, and marine animals such as plankton, fish, and shellfish samples, from not only the marine but also freshwater environment, i.e., upper river regions, have been done. Unfortunately, the studies of the distribution and incidence of the organism were not developed into complete analyses of the ecology of *Vibrio parahaemolyticus.*

The work of Takikawa (1956) on the pathogenic halophilic bacteria stimulated survey studies of these halophiles in the marine environment, as well as in fish markets, restaurants, etc., for epidemiological purposes. The first major survey was done by Miyamoto (1960), who sampled seawater and plankton in each season of the year in Tokyo and Sagami Bay, with the main focus on epidemiological search for the source of the food poisoning occurring along the Tokai area of the eastern Japanese coast in 1959. He speculated that an unusual "bloom" of plankton, which often occurred off the eastern coast of Japan as a result of the encounter of a warm current (Kuroshio) with a cold current (Oyashio), permitted concurrent development of an unusual "bloom" of the causative agent, finally resulting in contamination of the fish with the pathogenic halophilic bacteria. From the early 1960's, many workers, including food, sanitary, and fisheries microbiologists, began survey studies of the distribution of the halophilic bacteria. Seawater, marine sediments, and marine animals such as fish, plankton, and shellfish were sampled.

Aoki (1967*b*) reported on the isolation of *Vibrio parahaemolyticus* from plankton, fish, seawater, and sediment collected in the middle Pacific Ocean, as well as in the coastal region off Honolulu, Hawaii. Aoki (1967*a*) also reported the occurrence of *Vibrio parahaemolyticus* in the open sea off Southeast Asia, from seawater, plankton, and fish samples. Yasunaga and Kuroda (1964) reported isolating the organism from tuna caught in the open sea of the Indian Ocean. However, Horie *et al.* (1963*a,b,* 1964) reported that no bacteria belonging to biotype-1 could be detected from plankton collected in the pelagic zone of the Pacific Ocean, whereas biotype-2 strains were found to be widely distributed in plankton samples. Biotype-3 strains were not isolated from plankton, but were detected in young fish and in the gastric contents of fish and squid.

Thus, a commonly accepted observation is that *Vibrio parahaemolyticus* is rarely isolated from the marine environment during the winter months. To summarize the information on the distribution of *Vibrio parahaemolyticus,* this organism is easily isolated from seawater, sediment and plankton samples, or fish collected in coastal areas, with especially high counts found at the mouths of rivers during the summer months.

In the United States, the first isolation of *Vibrio parahaemolyticus*-like organisms was reported by Ward (1968) who used the arabinose medium of Horie *et al.* (1963*b*). Frozen sediment samples collected from south coastal areas of the United States were analyzed. Baross and Liston (1968) were the first to report isolation of *Vibrio parahaemolyticus* in the United States in sediments collected from Puget Sound and the Washington coast during the summer months. Approximately 10% of the *Vibrio* spp. proved to be *Vibrio parahaemolyticus* or related organisms. In Chesapeake Bay, Krantz *et al.* (1969) isolated *Vibrio parahaemolyticus* from lethargic and moribund crabs retained in commercial tanks during the shedding stage. Kaneko and Colwell (1973) reported that *Vibrio parahaemolyticus* could be isolated from several areas in Chesapeake Bay, where it was detected in sediment and plankton samples. Thus, *Vibrio parahaemolyticus* can be isolated from coastal and estuarine regions in several parts of the world.

One question which remains is whether *Vibrio parahaemolyticus* is restricted to the coastal regions or can be isolated from the open sea. The purpose of this study was to examine this question of the distribution of *Vibrio parahaemolyticus* in the open sea.

MATERIALS AND METHODS

The media and procedures followed in sampling have been described elsewhere (Kaneko and Colwell, 1973; Kaneko, 1973).

Samples were taken from 21 August 1971, through 27 August 1971, aboard the R/V *Eastward,* the Duke University Research Vessel. Four transects, Charleston, Wassow Sound, Sapelo Island, and Fernandina Beach, off the South Carolina and Georgia coasts, were studied. The transects were located on the continental shelf; the lengths of the transects were approximately 60–70 miles, with each transect comprising five stations (see Fig. 1 and Table 1).

RESULTS AND DISCUSSION

Results of bacteriological analyses for the Sapelo Island transect, off the Georgia coast, are given in Figs. 2, 3, and 4. Total, viable, aerobic, hetero-

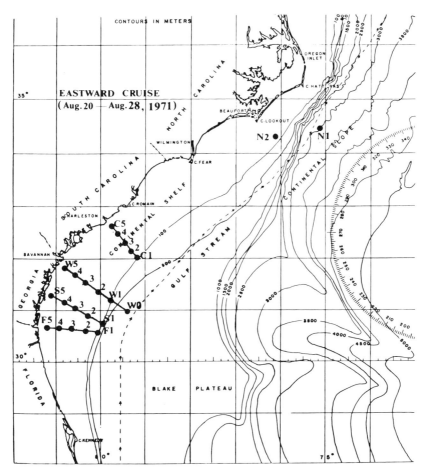

Fig. 1. Stations sampled during the R/V *Eastward* cruise (April and August 1971). Stations C1 to C5 were located on the Charleston, South Carolina, transect; W1 to W5 on the Wassow Sound transect, S1 to S5 on the Sapelo Island transect, and F1 to F5 on the Fernandina transect. N1 and N2 were stations off North Carolina.

trophic bacteria (TVC) of water samples (per 100 ml) at the station closest to shore (station 5), 4–10 miles from shore, were approximately 10^5. In general, the TVC decreased with distance from shore. Some fluctuations were detected at stations between 1 and 5. Presumptive vibrio (PV) and presumptive *Vibrio parahaemolyticus* (PVP) counts were enumerated using inoculated plates incubated at two temperatures, 25 and $37°C$, since marine bacteria, in general, do not grow at $37°C$. Counts at $25°C$ were considered to be total counts for each sample.

Table 1. Location of stations on the R/V *Eastward* cruise in August 1971

Station no.	Sampling date	Latitude	Longitude	Depth (m)	Water temperature (°C)		Samples collected
					Surface	Bottom	
C5[a]	8/21/71	32°39' N	79°40' W	10	28.7	27.5	Plankton
C4	8/21/71	32°37' N	79°28' W	20	28.8	28.2	Surface water and sediment
C3	8/21/71	32°29' N	79°16' W	25	28.7	24.5	
C2	8/21/71	32°22' N	79°02' W	49	28.0	22.8	Plankton
C1	8/21/71	32°18' N	78°50' W	180	27.6	11.2	Surface water and sediment
W5[b]	8/22/71	31°53' N	80°49' W	9	28.7	28.1	Surface water, sediment and plankton
W4	8/22/71	31°41' N	80°34' W	17	29.0	27.0	Surface water, sediment and plankton
W3	8/22/71	31°32' N	80°19' W	25	28.4	24.8	
W2	8/22/71	31°23' N	80°02' W	40	28.4	22.8	Surface water and sediment
W1	8/23/71	31°12' N	79°47' W	200	28.7	<10.0	
W0	8/23/71	31°04' N	79°21' W	700	28.5	<10.0	Surface water and bottom water

S5[c]	8/24/71	31°20' N	81°11' W	7	30.4	28.7	Surface water, sediment and plankton
S4	8/24/71	31°13' N	80°53' W	17	27.5	26.4	Surface water and sediment
S3	8/25/71	31°05' N	80°35' W	34	27.8	<24.0	Surface water, sediment and plankton
S2	8/25/71	30°58' N	80°12' W	30	27.5	23.4	Surface water
S1	8/25/71	30°51' N	79°59' W	180	29.1	<8.0	Surface water and sediment
F5[d]	8/27/71	30°43' N	81°10' W	15	27.2	25.5	Surface water and plankton
F4	8/27/71	30°44' N	81°02' W	20	28.0	25.0	
F3	8/27/71	30°43' N	80°40' W	27	27.7	22.5	Surface water
F2	8/27/71	30°41' N	80°32' W	40	27.4	<22.5	Surface water and plankton
F1	8/27/71	30°41' N	80°03' W	190	29.3	9.0	Surface water and plankton

[a] C5, C1 on Charleston transect.
[b] W5, W0 on Wassow Sound transect.
[c] S5, S1 on Sapelo Island transect.
[d] F5, F1 on Fernandina Beach transect.

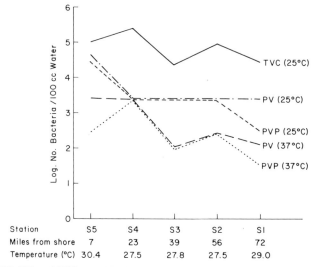

Fig. 2. TVC, PV, and PVP per 100 ml surface water (2–5 m) in samples collected off the Georgia coast. Sapelo Island transect.

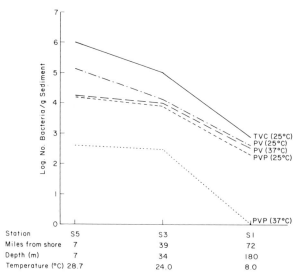

Fig. 3. TVC, PV and PVP in sediment samples (per gram) collected off the Georgia coast. Sapelo Island transect.

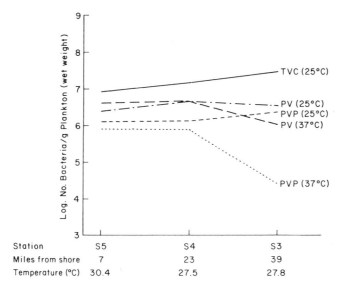

Fig. 4. TVC, PV, and PVP associated with plankton (per gram, wet weight) collected off the Georgia coast. Sapelo Island transect.

The distribution of PVP (total or 37°C) counts decreased rapidly with distance from shore, especially PVP (37°C) counts, which indicated a very limited distribution, approximately 10^4 at station 5 to 10^2 at station 1. However, confirmed *Vibrio parahaemolyticus* was not isolated from water samples collected at the stations examined in this study.

The TVC data for the sediment samples, in general, gave similar results as the water samples, i.e., TVC, PV (25°C), and PVP (25°C), for the Sapelo Island transect showed a continuous decrease with distance from shore, and confirmed *Vibrio parahaemolyticus* was not isolated from any of the sediment samples.

Although the plankton populations sampled were made up of complex communities, the bacteriological analyses were done on the whole plankton samples, without separation of zooplankton from phytoplankton, mainly because of limited time available aboard ship. The number of bacteria associated with plankton did not show any characteristic pattern, as did the sediment and water samples (see Fig. 4). No confirmed *Vibrio parahaemolyticus* strains were isolated from the plankton.

These data, in conjunction with data obtained for Chesapeake Bay samples (Kaneko and Colwell, 1973) and other coastal region stations (Kaneko, 1973), lead us to conclude that *Vibrio parahaemolyticus* is specifically an estuarine organism. Confirming results were reported by Scheffers and Golten (1973) in their search for *Vibrio parahaemolyticus* along the Dutch coast.

As for the origin of *Vibrio parahaemolyticus,* several workers (Yasunaga and Kuroda, 1964; Aoki, 1967*a*) reported isolating the organism from samples collected in the open ocean. However, the difficulties and confusion in the identification of *Vibrio parahaemolyticus* often lead to misinterpretation of the distribution of the organism. In this study, the distribution of *Vibrio parahaemolyticus,* as defined by the methods of numerical taxonomy, DNA base composition, and DNA/DNA reassociation (Colwell *et al.,* 1973), was found to be limited strictly to the estuary.

ACKNOWLEDGMENT

This work was supported by Sea Grant Project No. 04-3-158-7, National Oceanic and Atmospheric Administration, U. S. Department of Commerce.

LITERATURE CITED

Aoki, K. 1967*a*. Survey of the distribution of *Vibrio parahaemolyticus* in the Southeast Asia Sea. *In* T. Fujino and H. Fukumi (eds), *Vibrio parahaemolyticus.* 2nd edn., pp. 329–331. Nayashoten, Tokyo, Japan.

Aoki, K. 1967*b*. A survey of the distribution of *Vibrio parahaemolyticus* in the Middle Pacific Ocean. *In* T. Fujino and H. Fukumi (eds), *Vibrio parahaemolyticus.* 2nd edn., pp. 325–327. Nayashoten, Tokyo, Japan.

Baross, J., and J. Liston. 1968. Isolation of *Vibrio parahaemolyticus* from the Northwest Pacific. Nature (London) 217:1263–1264.

Colwell, R. R., T. E. Lovelace, L. Wan, T. Kaneko, T. Staley, P. K. Chen, and H. Tubiash. 1973. *Vibrio parahaemolyticus*–isolation, identification, classification, and ecology. J. Milk Food Technol. 36:202–213.

Horie, S., K. Saheki, T. Kozima, and Y. Sekine. 1963*a*. Oceanographic survey on the distribution of Takikawa's so-called pathogenic halophilic bacteria. Bull. Jap. Soc. Sci. Fish. 29:37–43.

Horie, S., K. Saheki, M. Nara, T. Kozima, Y. Sekine, and T. Takayanagi. 1963*b*. Distribution of Takikawa's so-called pathogenic halophilic bacteria in the coastal sea area. Bull. Jap. Soc. Sci. Fish. 29:785–793.

Horie, S., K. Saheki, T. Kozima, M. Nara, and Y. Sekine. 1964. Distribution of *Vibrio parahaemolyticus* in plankton and fish in the open sea. Bull. Jap. Soc. Sci. Fish. 30:786–791.

Kaneko, T. 1973. Ecology of *Vibrio parahaemolyticus* and related organisms in Chesapeake Bay. Ph.D. Thesis, Georgetown Univ., Washington, D. C.

Kaneko, T., and R. R. Colwell. 1973. Ecology of *Vibrio parahaemolyticus* and related organisms in Chesapeake Bay. J. Bacteriol. 113:24–32.

Krantz, G. E., R. R. Colwell, and T. E. Lovelace. 1969. *Vibrio parahaemolyticus* from the blue crab *Callinectes sapidus* in Chesapeake Bay. Science 164:1286–1287.

Miyamoto, Y. 1960. Oceanographic survey and halophilic bacteria. Shokuhin Eisei Kenyu 119:23–30.

Sakazaki, R. 1964. Sero-typing of *Vibrio parahaemolyticus*. Modern Media, 10:355–363.

Scheffers, W. A., and C. Golten. 1973. A search for marine vibrios along the Dutch coast. Antonie van Leeuwenhoek 39:366.

Takikawa, I. 1956. Certain halophilic bacteria isolated from food poisoning patients. Jap. J. Infect. Dis. 30:439–440.

Ward, B. Q. 1968. Isolation of organisms related to *Vibrio parahaemolyticus* from American estuarine sediments. Appl. Microbiol. 16:543–546.

Yasunaga, N., and M. Kuroda. 1964. Studies of *Vibrio parahaemolyticus*. II. Distribution of *Vibrio parahaemolyticus* in marine sediment and on fish in Southeast Asia. Endemic Dis. Bull. Nagasaki Univ. 6:201–208.

ZoBell, C. E. 1946. Marine Microbiology. Chronica Botanica, Waltham, Massachusetts.

MINERALIZATION OF ORGANIC SOLUTES BY HETEROTROPHIC BACTERIA

RICHARD T. WRIGHT

Department of Biology
Gordon College

The view that bacterial action is unimportant in nutrient regeneration is examined in the light of recent research employing ^{14}C-labeled organic compounds to study natural mineralization processes. Two methods for measuring mineralization are examined on the basis of their assumptions and their usefulness in different environmental situations. The method of Williams and Askew (1968) is simpler, can be used in a wide variety of environments, but involves some uncertain assumptions. The kinetic approach generates more data but does not work in some aquatic systems. The application of these methods to date is reviewed. It is concluded that where nutrients are released from algae and zooplankton in organic form, bacterial mineralization is the pathway whereby the nutrients become available again for primary production.

INTRODUCTION

The participation of heterotrophic microbes in the flow of energy and the cycling of nutrients in aquatic ecosystems is a subject that has received much speculation but relatively little direct investigation until very recently. In summarizing the state of affairs in 1961, Rittenberg (1963) observed that direct information on *in situ* bacterial processes in aquatic systems was needed. He pointed out that the ordinary methods employed by aquatic microbiologists do not yield this. Citing this observation by Rittenberg, Johannes (1968) later stated that progress had indeed been made in understanding the process he called nutrient regeneration, enough to say that direct bacterial action is a relatively unimportant part of this process. This is a surprising statement, for ecologists have long believed that bacterial involvement in nutrient regeneration is crucial. The controversy is one that can be resolved rather easily, as this paper will show.

In the years since Johannes published the observations cited, a new approach to the study of *in situ* microbial activity has become established. This approach employs ^{14}C-labeled organic solutes to measure the uptake and subsequent mineralization of organic solutes at natural concentrations of both solutes and microbes. Several significant steps have contributed to the development of methods. Most important was the initial work by Parsons and Strickland (1962) showing that the uptake of [^{14}C]glucose and acetate by natural microplankton populations could be analyzed with the kinetic scheme derived for enzyme-substrate systems by Michaelis and Menten. Wright and Hobbie (1965, 1966) extended the kinetic approach by adapting a form of the Lineweaver-Burke equation for analyzing the uptake data, and subsequently a number of workers began to use the method and generate further refinements. Prior to 1969, workers were measuring only substrate taken up and retained in the cells. Substrate taken up and subsequently respired, or mineralized, was not detected and hence the method was called into question by Hamilton and Austin (1967).

In the meantime, another line of attack was developed for directly measuring the mineralization of organic solutes. Kadota *et al.* (1966) suggested a method which employed ^{14}C-labeled organic solutes added to a sample, followed by recovery of the $^{14}CO_2$ formed by microbial action on the organic compound used. Williams and Askew (1968) independently developed a similar approach and dealt much more thoroughly with the principles and application of their method than Kadota *et al.* Williams and Askew also pointed out the problem of not knowing how much substrate was respired when using the kinetic approach of Wright and Hobbie. In the meantime, Hobbie and Crawford (1969) introduced a method from cell physiology which measured respired substrate, and the kinetic approach was given new life. In thus dealing with respired substrate, the technique originally viewed as a correction by Hobbie and Crawford is in reality another approach to measuring mineralization of organic solutes by the natural microplankton. In what follows, attention will be given mainly to the process of mineralization—how it is measured, and what the results are to date. Before proceeding, however, it would be wise to define some terms. The term *mineralization* will be taken to mean the oxidation of the carbon in organic substrates to carbon dioxide; hence what is actually measured is *carbon mineralization.* Another term often given the same meaning is *respiration.* As used by Johannes (1968), *nutrient regeneration* means the release of soluble organic or inorganic forms of nitrogen and phosphorus. Clearly, the process as defined is incomplete, since nitrogen and phosphorus in organic form will usually be involved in mineralization. *Nutrient release* would perhaps more accurately define the processes described by Johannes, and perhaps the discrepancy between his conclusions and the generally accepted role of the aquatic bacteria as agents of nutrient regeneration is most easily explained as a matter of semantics. Nonetheless, there still remains the need to measure mineralization and nutrient regeneration *in situ.* The following section will compare two available methods for measuring microbial mineralization of organic solutes.

MINERALIZATION METHODS

Terms and Symbols Used

n Subscript indicating natural rates, concentrations, or turnover times.

a Subscript indicating rates, concentrations, or turnover times due to substrate added in order to make the measurement.

f In an isotope experiment, the fraction of available isotope taken up or respired during the incubation time.

t Incubation time of a measurement using isotope (hr).

S Natural concentration of a given organic solute (μg/liter).

A Concentration of the compound added to make the measurement (μg/liter).

v_r Rate of mineralization of a given compound by the natural micro-organisms (μg/liter hr).

T_r Turnover time of compound due to mineralization, identical to mineralization turnover time as used by Williams and Askew (1968) (hr or days).

v_t Rate of transport of compound into microbial cells (μg/liter hr).

T_t Turnover time of compound due to transport of a compound into the microbial cells (hr).

v_a Rate of incorporation (assimilation) of compound into new cell material or substrate pool (μg/liter hr).

T_a Turnover time of compound due to its incorporation into microbial cells (hr).

The symbols allow for distinguishing between three microbial processes: mineralization of a compound, the incorporation of the compound into new cell material and substrate pools, and the transport of the compound from the external medium into the cells, presumed to be the sum of the first two processes.

Discussion of the Kadota and Williams Methods

With the method suggested in Kadota *et al.* (1966), uniformly labeled [^{14}C] glucose is added to a water or sediment sample at a single concentration, and the $^{14}CO_2$ produced during a short incubation period is subsequently measured. The results are to be expressed as a rate of glucose mineralization, in units of milligrams carbon per cubic meter per day. In terms of the above symbols, the measurement can be given as

$$v_r = f(A)/t. \tag{1}$$

However, an unknown quantity of naturally occurring substrate is always present, and so the measurement should really be expressed as

$$v_r = f(S + A)/t \tag{2}$$

and since S is not known, v_r cannot be calculated. Thus, the "rate" obtained with this method is a relative rate, resulting from mineralization activity but not giving the natural mineralization rate of an organic solute. Even if the natural substrate concentration were known and used in equation (2) with the measurement of mineralization, the problem of the influence of added substrate on the rate of mineralization would have to be considered, as the next method illustrates.

The method suggested by Williams and Askew (1968) provides, instead of a rate, the turnover time for naturally occurring substrate (T_{rn}).* This is a useful value in itself, and, if it can be combined with an independent measurement of the substrate concentration, the natural rate of mineralization can be obtained. Since the method proposes to yield a natural parameter, it should be examined closely.

As with the Kadota method, uniformly labeled substrate is added at a single concentration and after incubation the $^{14}CO_2$ is collected, measured and then related to the radioactivity of applied substrate and the incubation time. The result is expressed as t/f, with units of time. This value is a measure

---

* In later papers, Williams and co-workers have preferred to use the term "respiration rate" in presenting their results, with units of *% respired per day*. This is interchangeable with turnover time given in hours in the following relationship: % respired/day = 24 × $100/T_{rn}$ (hr).

of the natural turnover time if it can be assumed that the increase in substrate concentration created by the added isotope is either negligible or is accompanied by a proportionate increase in the rate of respiration. The assumption can be expressed as:

$$\frac{S + A}{v_{rn} + v_{ra}} = \frac{S}{v_{rn}} = T_{rn}. \tag{3}$$

The use of isotope to obtain T_{rn}, based on this assumption, is valid for the following reasons.

a. The measurement of respiration rate using labeled substrate is as given in equation (2), where v_r is the respiration rate due to the total available substrate $S + A$, and is identical with the term $v_{rn} + v_{ra}$ appearing in equation (3).

b. Equation (2) can be rearranged to give turnover time:

$$\frac{S + A}{v_r} = T_r = \frac{t}{f}. \tag{4}$$

c. From the assumption of proportionate respiration, as expressed in equation (3),

$$\frac{S + A}{v_r} = \frac{S + A}{v_{rn} + v_{ra}} = \frac{S}{v_{rn}} = T_{rn} = \frac{t}{f}. \tag{5}$$

Thus, without knowing either the natural rate of mineralization or the natural substrate concentration, the natural turnover time can be found. Note, however, that the method turns on the critical assumption of direct proportionality between substrate concentration and respiration rate. There are two situations for which the assumption must be considered.

1. The concentration of added substrate is much smaller than the natural concentration. Here the measurement approaches the ideal for a tracer experiment, where the tracer is a negligible fraction of the concentration in question and therefore will not measurably influence the process.

2. The concentration of added substrate is in the same order of magnitude or greater than that of the natural substrate. In this case, the increase in total available substrate created by the addition of the tracer must be accompanied by a proportionate increase in the rate of mineralization.

To know which of these situations one is working under, one must know the natural substrate concentration. It is possible to obtain concentration data for many specific organic solutes, but the techniques are laborious and often not too precise. For solutes such as glucose, the concentration in natural waters appears to be in the range of 10^{-7}–10^{-9} M, potentially overlapping in range with the levels of radioactive substrate needed to be able to measure mineralization (Vallentyne and Whittaker, 1956; Hobbie and Wright, 1965; Vaccaro et al., 1968; Hobbie et al., 1968; Andrews and Williams, 1971). Subsequent work by Williams and his co-workers (Andrews and Williams, 1971) illustrates the problem well. They routinely added 0.2 μg [^{14}C] glucose to 450 ml of sample water, giving a concentration of 0.44 μg/liter added to the natural concentration. Eight analyses of naturally occurring glucose yielded a range in concentration from 0.4 to 5.7 μg/liter, averaging 2.8 μg/liter. Thus, substrate added to perform the mineralization measurements contributed from 7 to 52% of the total substrate present, averaging 14%. Clearly, the second situation cited above will frequently apply, where the added substrate concentration and the natural substrate concentration are in the same order of magnitude. The reasons for this are apparently immutable—the low concentrations of natural substrate, and the limitations of specific activity for ^{14}C-organic compounds.

This being the case, the relationship between mineralization rate and external substrate concentration becomes important. Looking at it in the simplest way, two additional relationships are involved: (1) between transport (uptake) of the substrate and external concentration, and (2) between transport and respiration rate. Regarding the first, sufficient study has been given to pure cultures of bacteria to establish that the normal response of transport rate to substrate concentration follows a typical saturation curve which can be analyzed with Michaelis-Menten kinetics (Cohen and Monod, 1957; Kepes, 1963; Hobbie and Wright, 1965; Vaccaro and Jannasch, 1967; Hamilton et al., 1966; Pardee, 1968). The recent measurements of organic-solute uptake in marine and fresh waters have demonstrated the presence of microbial populations with uptake which follows Michaelis-Menten kinetics and is very effective at the low substrate concentrations found there (Parsons and Strickland, 1962; Wright and Hobbie, 1965, 1966; Vaccaro, 1969). Thus, the relationship between natural uptake and the substrate concentrations applicable to mineralization measurements may be first order and therefore nonlinear. It is not always so, however (e.g., Vaccaro and Jannasch, 1967), and it should be made clear that our understanding of the processes of solute uptake and mineralization by natural microbial populations is presently limited by the lack of any information on what species are involved, and how they interact with each other and with a variable complex of substrates which itself is not understood. However, the only workable model at the present time is the kinetic one just referred to.

Concerning the relationship between transport and respiration of a solute, the simplest assumption would be that respiration is proportional to transport, above some minimum level for endogenous respiration. This is borne out by work with chemostat cultures (Schulze and Lipe, 1964). Uptake and mineralization studies by the present author have shown a proportional relationship between mineralization (respiration) and transport of glucose, acetate, and glycollate. The relationship may change with different seasons and bodies of water, but aside from the usual scatter obtained from most isotope studies, no consistent effect of external substrate concentration can be observed within a set of samples for these substrates. Work performed with K. Burnison, however, indicated that some amino acids show a consistent increase in percentage mineralization (percentage respired) with increasing external concentrations. Burnison (1971) later analyzed the pattern more thoroughly by calculating a slope relating percentage respired to external substrate concentration for 16 amino acids (Table 1). Five of the 16 indicate, as this table shows, an effect of concentration. Thus, there seem to be some substrates which show respiration to be proportional to transport, and some which show increasing respiration with the increased availability of substrate. In these latter, it is noteworthy that total uptake still follows Michaelis-Menten kinetics, and in no case is the separate process of mineralization actually proportional to applied substrate concentration.

Table 1. Slope of regression lines relating percentage respired to substrate concentration, Klamath Lake water, 5 November 1970[a]

Amino acid	Slope	Amino acid	Slope
Tyrosine	2.11	Lysine	0.41
Ornithine	1.88	Asparagine	0.32
Phenylalanine	1.13	Serine	0.31
Arginine	1.10	Glutamic acid	0.27
Leucine	0.79	Proline	0.20
Threonine	0.62	Glycine	−0.36
Alanine	0.47	Valine	−0.36
		Aspartic acid	−0.37

[a] Those lines having slopes greater than 1 represent significant departure from proportionality between transport and respiration of the amino acid, at substrate concentrations in the 1–100 μg/liter range. (From Burnison, 1971.)

Table 2. Calculated data for transport and respiration when respiration is 40% of transported substrate[a]

Added substrate (A) (µg/liter)	$S + A$ (µg/liter)	v_t (µg/liter hr)	v_r (µg/liter hr)	T_r (hr)
0	3.0	0.187	0.075	40
0.2	3.2	0.195	0.078	41
0.3	3.3	0.199	0.080	41.5
0.5	3.5	0.206	0.083	42.5
0.75	3.75	0.214	0.085	44.1
1.0	4.0	0.222	0.089	45
1.5	4.5	0.237	0.095	47
2.0	5.0	0.250	0.100	50
3.0	6.0	0.273	0.109	55
5.0	8.0	0.307	0.123	65
10.0	13.0	0.360	0.144	90
20.0	23.0	0.410	0.164	140

[a]Natural substrate concentration = 3 µg/liter; transport constant for uptake (K_t) = 5 µg/liter; maximum velocity of transport (V_{max}) - 0.5 µg/liter·hr.

In those situations where respiration is proportional to transport, and transport follows saturation kinetics, respiration rate will relate to external substrate concentration in the non-linear manner described with Michaelis-Menten kinetic equations. Continuing with the assumptions, since the substrate increase necessary in order to be able to measure mineralization rate is not accompanied by a proportionate increase in respiration, the effect of any added substrate will be to increase the apparent turnover time. This can best be shown by taking a hypothetical but realistic construct based on a bacterial transport system (Table 2). The data for velocity of transport (v_t) are generated by applying the constants S, V_{max}, and K_t, and the variable A to the adapted Michaelis-Menten equation:

$$v_t = \frac{V_{max}\ (S + A)}{K_t + (S + A)} .\tag{6}$$

Mineralization turnover times (T_r) are calculated on the assumption that respiration is 40% of the substrate transported into the cells. Figure 1, a plot of the turnover time *versus* the ratio of added to total substrate, shows the most significant point to be brought out—the increase in turnover time due to the effect of substrate added to the system in order to make the measurement. This increase can be more directly estimated if V_{max} is known

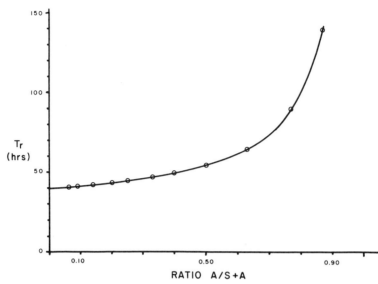

Fig. 1. Plot of mineralization turnover time against the ratio of added substrate to total substrate present. Data taken from Table 2.

from kinetic measurements. By substitution of the formula for turnover time in the Lineweaver-Burke form of the Michaelis-Menten equation, the increase in turnover time due to added substrate (A) is A/V_{max}.

Williams and co-workers are aware of these problems (Williams and Askew, 1968; Williams and Gray, 1970), although they have not dealt with them thoroughly in any of their publications. The approach they suggest is to measure turnover over a range of low concentrations of added substrate, and to work subsequently in the range where turnover time is shown to be independent of the amount of added substrate (and therefore respiration is proportional to substrate concentration). The data they present in support of this approach are not convincing (see Table 3). Of a series of five concentrations of added substrate, only the lowest two, when the added glucose gives concentrations of 0.18 and 0.54 µg/liter, give similar turnover times. The higher concentrations yield increases in turnover time in the manner indicated by Table 2. As this table shows, when added substrate is a small fraction of the total substrate present, the effect on turnover is relatively small. For example, when added substrate is 10% of the natural substrate concentration, turnover time is increased only 4%. The imprecision of the technique is such that a 4% increase would not be detected unless a great many replicates were conducted. There is, therefore, a rationale for the Williams and Askew method that is compatible with microbial systems showing both rate-limiting kinetics and a constant ratio of respiration to total uptake. In effect, it is a tracer experiment with a broader than usual latitude for the tracer concentration. By working in a range where added substrate is 20% or less than naturally occurring substrate, one can obtain results that are probably within 10% of the true value sought. In view of the imprecision of working with isotopes and natural populations (Wright, 1973), a potential 10% error is acceptable. However, when added substrate is great enough to be approximately equal in magnitude to naturally occurring substrate, the apparent turnover time will be considerably higher than the natural turnover time. Of course, it then becomes quite important to know the natural substrate concentration, unless one is willing to perform mineralization measurements frequently over a range of substrate concentrations as Williams and Askew (1968) suggest. This is exactly what is involved in the approach that follows, and it is a method that yields several additional pieces of data of considerable value.

The Kinetic Approach

The reasoning for this approach follows that presented in Wright and Hobbie (1965) for a kinetic approach to measuring the uptake of labeled organic solutes by microorganisms. This paper demonstrated how several natural parameters for uptake could be obtained using isotopically labeled substrate, without knowing either the natural substrate concentrations or the natural rate of uptake. The critical assumption, first made for natural waters by Parsons and Strickland (1962), is that uptake follows a typical saturation

Table 3. The effects of various added amounts of glucose on the mineralization turnover time of glucose in seawater[a]

Added glucose concentration (μg/liter)	Measured turnover time (days)
0.18	10.1
0.54	9.9
1.8	18.6
5.4	41.5
18.0	89.0

[a] From Williams and Askew (1968).

curve which can then be analyzed with Michaelis-Menten kinetic equations. Subsequent research has supported this assumption for many aquatic systems (e.g., Vaccaro and Jannasch, 1967; Allen, 1969; Munro and Brock, 1968; Wetzel, 1968; Vaccaro, 1969; Wright, 1970; Crawford, 1971), but there are many systems where the results do not fit saturation kinetics, and these tend to involve the more oligotrophic systems and winter conditions (e.g., Vaccaro and Jannasch, 1967; Hamilton and Preslan, 1970). Many of the problems encountered by workers using the kinetic approach are discussed in Wright (1973).

As already noted, Hobbie and Crawford (1969) developed a method that corrects for respired substrate by measuring the loss of $^{14}CO_2$ during incubation and gives total transport. Thus, where respiration shows rate-limiting saturation, the modified Michaelis-Menten equation of Wright and Hobbie (1965) can be used for respiration data, and will allow calculation of the natural turnover time for mineralization if measurements are made at several concentrations of added, uniformly labeled substrate. This equation is

$$t/f = (K + S)/V_{max} + A/V_{max} \qquad (7)$$

where K is a constant related to uptake, V_{max} the maximum velocity for respiration (which would apply when transport is at maximum velocity), and the other symbols are as already defined. Each concentration A used will generate a different t/f, and when t/f is plotted against A a straight line should result. Extrapolation back to the y-intercept of the line, where $A = 0$, gives the turnover time for mineralization at the natural substrate concentration (see Wright and Hobbie, 1965, 1966, for details of the derivation). Figure 2 shows such a kinetic plot for the mineralization and assimilation components of uptake by a lake microplankton community.

The kinetic approach yields considerably more data than the previous one discussed. It puts mineralization in the context of total uptake and gives the other kinetic parameters of V_{max} and $(K + S)$. It also provides a rationale for interpreting the data obtained from the activities of a community of microbial plankton, a working model to which much of the mineralization data collected to date can be fitted (Williams and Askew, 1968; Hobbie and Crawford, 1969; Williams and Gray, 1970; Harrison et al., 1971). However, not all work with ^{14}C-labeled organics will yield data consistent with the kinetic model. The approach suggested by Williams and Askew has the advantage of involving considerably fewer samples for collecting the turnover data. It is therefore probably the best approach for oligotrophic waters and many aquatic environments during winter months, when used with the proper precautions.

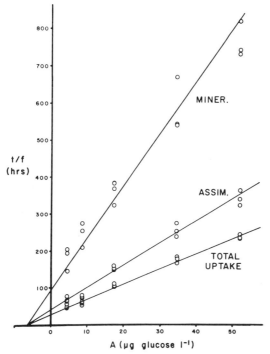

Fig. 2. A kinetic plot of microbial activity on glucose for surface water from Upper Klamath Lake, Oregon, 11 March 1970 (temp. 5°C). Turnover times are 95 hr for mineralization, 42 hr for assimilation, and 28 hr for total activity.

RESULTS OF MINERALIZATION STUDIES

The results discussed here will be directed toward the problems presented early in this paper—mineralization and the regeneration of nutrients.

A major contribution of the recent research using ^{14}C-labeled organic compounds with natural plankton has been to demonstrate the presence of an active population of heterotrophic microbes using low molecular weight organic solutes and, by their activity, keeping the separate solutes at concentrations normally no higher than a few micrograms per liter. Most of this activity can be readily traced to the bacteria (Wright and Hobbie, 1966; Williams, 1970). The work on mineralization using the methods discussed has clearly shown that microbial activity in natural waters is continually producing CO_2 from dissolved organic solutes. Studies to date have involved a number of different compounds.

The compound most frequently studied is glucose. Natural concentrations are normally in the range 0.5–10 μg/liter (Vaccaro *et al.,* 1968; Andrews and Williams, 1971). The only study that has combined mineralization and concentration measurements to give glucose flux is the work of Andrews and Williams (1971) for English Channel water. They found a pronounced seasonal variation in glucose turnover and an annual flux of 6.4 g/n^2, which they estimated to be about 1.5% of the total carbon fixed by phytoplankton. Their study also included measurements of assimilated glucose, based on particulate fixation of labeled glucose and the same assumptions used for mineralization. These results (Williams, 1970) indicated that, of the glucose taken up, approximately 33% is mineralized and the remainder assimilated. Thus, the amount of glucose flux can be increased to 4.5% of the annual phytoplankton production (assumed to be the predominant source of the organic solutes). Mineralization turnover times were as low as 10 hr in summer months.

In a study of the kinetics of heterotrophic activity conducted over an annual cycle in Upper Kalmath Lake, Oregon, the present author found turnover times for mineralization of glucose ranging from 18 to 470 hr, with mineralization values from 11 to 45%, averaging 26%. Interestingly, the higher percentage mineralization figures corresponded in general with warmer temperatures. Of a total of 21 separate kinetic measurements, 19 gave statistically acceptable results (Wright, 1973). Crawford (1971) performed a series of kinetic measurements of glucose uptake and mineralization in Pamlico Estuary, North Carolina, using the kinetic approach, and found 11/12 to be statistically acceptable. He found percentage mineralization ranging from 8 to 17% and averaging at 13%, and mineralization turnover times from 1.7 to 190 hr, lowest in the warmest times of the year. The V_{max} values obtained, a measure of the heterotrophic potential of the population, indicated a capability of utilization of as much as 575 μg/liter/day with an average of 125 μg/liter/day.

Several organic acids have been the subject of uptake and mineralization studies. Andrews and Williams (1971) tried with no success to measure mineralization of ^{14}C-labeled palmitic acid and stearic acid in seawater. The present author found vigorous mineralization of uniformly labeled [^{14}C] acetate in Upper Klamath Lake water. Values for percentage mineralization ranged from 15 to 52%, averaging 29%. Of the 17 occasions when measurements were made with the kinetic method, only 12 gave statistically acceptable results. The major cause of unacceptable data seemed to be sample heterogeneity. All samples were done in triplicate for four concentrations, and it was not unusual to find replicates differing as much as 100% in the context of very active uptake and mineralization. Mineralization turnover times were not particularly low, ranging from 65 to 400 hr, quite comparable with those for glucose in the same lake. V_{max} values for total uptake indicated potential utilization of 2 to 55 µg acetate/liter/day, with an average of 17 µg/liter/day.

Another ^{14}C-labeled organic acid applied to Klamath Lake water was glycollic acid. This substrate gave the least number of acceptable fits to the kinetic equation, only 9/21. However, glycollic acid has consistently yielded the highest percentage mineralization of all substrates reported to date, an average of 72% (and thus a growth yield of only 28%). Except during the spring diatom bloom, glycollate uptake and mineralization was considerably below that for glucose and acetate, as revealed by comparisons of V_{max} for these three substrates. Turnover times were correspondingly higher. Further studies of this substrate in Gravel Pond, Massachusetts, have shown that the epilimnetic microplankton maintain a V_{max} for glycollate uptake and mineralization throughout the summer that ranges from 14 to 65 µg/liter day. A 2-day enrichment experiment confirmed the predictive value of the V_{max} figures by measuring the simultaneous chemical disappearance and microbial uptake of glycollic acid added to lake-water samples. The initial V_{max} of the microbial population in a sample taken in midsummer from the epilimnion was 2.45 µg/liter hr, or 59 µg/liter day. Samples were enriched with 250 or 500 µg/liter of glycollate and then followed for 48 hr for uptake and for chemical disappearance of the glycollic acid. In the 250 µg/liter sample, 125 µg/liter was used, while 200 µg/liter was used in the 500 µg/liter sample after 48 hr of incubation in the dark. The ability of the microbial flora to mineralize glycollic acid seems to be related to the occurrence of this compound as a major algal excretory product, as work by Tanaka, Nakanishi, and Kadota presented elsewhere in this volume indicates.

Another group of substrates for which there is now considerable information on mineralization is the amino acids. This work has direct bearing on nutrient regeneration for it seems quite probable that when an amino-acid molecule is oxidized, nitrogen will be released in the form of ammonia. Two unpublished studies have employed the kinetic approach to uptake and mineralization, working with the amino acids individually (Crawford, 1971; Burnison, 1971). Williams and co-workers, however, used an amino-acid mixture of known composition to investigate mineralization (Williams, 1970; Andrews and Williams, 1971). All three studies have included measurements of ambient amino-acid levels, allowing a direct determination of amino-acid flux. One common finding was that amino-acid concentrations show little seasonal change, whereas amino-acid transport and mineralization by the natural microbial plankton varied as much as two orders of magnitude. Periods of highest activity and shortest turnover time coincided with periods of highest primary production, whereas winter conditions invariably brought on the lowest levels of activity.

Williams and Gray (1970) conducted enrichment experiments and clearly demonstrated a rapid response by the marine heterotrophic bacteria to sudden increases in amino acids. Within a 2-day period, the bacteria were able to accommodate 10- to 100-fold substrate concentration increases, thereby increasing a normal mineralization of 10–50 μg amino acids/liter day to 1000–2000 μg/liter day. Andrews and Williams (1971) estimated that amino-acid mineralization as measured by their methods accounted for some 6% of the total carbon annually fixed by the algae, and at least 12% of the amino acids produced. They found that, of the amount taken up, about 78% of the amino-acid mix was assimilated by the bacteria, with the result that, besides the 6% directly oxidized, an additional 24% of algal annual production went into the food chain via bacterial assimilation of amino acids. Crawford (1971) reported that approximately 10% of algal production during summer months could be accounted for by bacterial assimilation and mineralization of the 12 amino acids he studied.

Both Crawford (1971) and Burnison (1971) studied the amino acids as separate compounds, and so they were able to report a significant difference in the pattern of mineralization of different amino acids. Burnison found that the amino acids with the greatest V_{max} values also had the highest percentage mineralization and hence were preferred as carbon sources. Burnison and

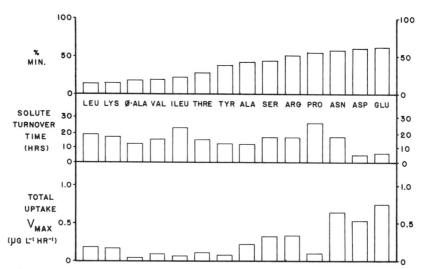

Fig. 3. Percentage mineralization, transport V_{max} and turnover times for amino acids in surface water of Upper Klamath Lake, Oregon, 25 Feb. 1970 (temp. 5°C).

Crawford noted that these are amino acids with entry into the TCA cycle by known metabolic pathways. Figure 3 shows a comparison of percentage mineralization, turnover time, and V_{max} for amino acids in Upper Klamath Lake. The wide range in mineralization for different amino acids indicated here is a finding common to the work of Burnison and Crawford. Glutamic and aspartic acids gave the highest mineralization percentages in both studies, averaging around 50%. With a C : N ratio of 5 : 1 or 4 : 1, these substrates will clearly yield most of their nitrogen as released ammonia when taken up and used for energy by bacteria having a C : N ratio closer to 12 : 1.

CONCLUSIONS

The two basic methods for measuring mineralization of organic solutes have clearly begun to provide information on the importance of the heterotrophic microflora. The choice of which method is better—Williams' technique using one concentration of added ^{14}C-substrate, or the kinetic approach—will depend on the system being studied and the desired data. The kinetic approach requires a greater number of samples and gives more data, yet it does not work in some aquatic systems. Williams' approach is simpler but involves some uncertainties in basic assumptions. These two methods at the present time represent the only options available for obtaining data on the natural mineralization activity of the aquatic heterotrophic microflora. They have not been in use for very long, but the information generated by workers applying them to date has already rendered invalid the idea that the bacteria are unimportant in nutrient regeneration.

The work by Williams (1970) on the size distribution of the organisms responsible for measurable mineralization has shown beyond any reasonable doubt that the bacteria are the major agents of uptake and rapid oxidation of organic solutes in the planktonic environment. Enrichment experiments indicate the rapid response of the naturally occurring heterotrophic bacteria to sudden increases in substrate, such that a few days at most are sufficient for complete removal and substantial oxidation of the artificially added substrate. Obviously, the same fate awaits organic solutes introduced by natural means, whether by excretion from algae or zooplankton, or by autolysis or solution from detritus.

Many of the organic substrates studied to date are simple carbon compounds with no potential algal nutrients other than the carbon itself. Even when the substrate contains phosphorus or nitrogen, the methods discussed deal only with the fate of the carbon. So it is by inference that, for example, amino-acid mineralization can also be considered nutrient regeneration, but the inference is strong and obvious.

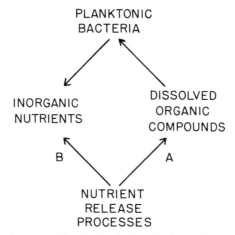

Fig. 4. The role of heterotrophic planktonic bacteria in nutrient regeneration. Pathway *B* indicates direct release of inorganic nutrients by algae and zooplankton. Pathway *A* depicts the mineralizing activity of the bacteria in nutrient regeneration.

Phosphorylated organic solutes have not been subjected to the methods under consideration, but there is little reason to doubt that they will be acted on primarily by the heterotrophic bacteria. The work reported here is only a fragmentary picture of the microbial involvement in mineralization and nutrient regeneration, and certainly much more research is required before a more definitive description of these processes is forthcoming. The information on amino acids, however, clearly indicates bacterial involvement in the return of nitrogen to the inorganic state, whatever its source. In his paper on nutrient regeneration, Johannes (1968) presented a figure depicting pathways of transfer and regeneration of phosphorus and nitrogen in aquatic ecosystems. The work and ideas discussed in the present paper suggest the need to revise this figure to include the concepts shown in Fig. 4. As this figure indicates, it seems only logical to separate the results of nutrient release into organic and inorganic fractions. Nutrient release itself is taken to mean the escape of organic and inorganic compounds from within organisms or organic particles to the water medium in a dissolved state. Much more work needs to be done in quantifying the release processes, both with respect to their sources and to the relative amounts appearing in the organic and inorganic states. There does, however, seem to be sufficient information to confirm the role of the bacteria in planktonic ecosystems as major mineralizers of organic solutes, releasing the nutrients previously locked in organic molecules and hence less available to the algae than to the bacteria until the necessary bacterial action occurs.

ACKNOWLEDGMENTS

I am grateful for support from the National Science Foundation through grant GB 7741. Additional support came from F.W.Q.A. Program No. 16010 to Dr. Richard Y. Morita, whose continued encouragement and help have been vital to my own research interests.

LITERATURE CITED

Allen, H. L. 1969. Chemo-organotrophic utilization of dissolved organic compounds by planktic algae and bacteria in a pond. Int. Rev. Gesamten. Hydrobiol. 54:1−33.

Andrews, P., and P. J. LeB. Williams. 1971. Heterotrophic utilization of dissolved organic compounds in the sea. III. Measurement of the oxidation rates and concentrations of glucose and amino acids in sea water. J. Mar. Biol. Ass. U. K. 51:111−125.

Burnison, B. K. 1971. Amino acid flux in a naturally eutrophic lake. Ph. D. Thesis. Organ State Univ., Corvallis.

Cohen, G. N., and J. Monod. 1957. Bacterial permeases. Bacteriol. Rev. 21:169−194.

Crawford, C. C. 1971. The utilization of dissolved free amino acids by estuarine microorganisms. Ph.D. Thesis. North Carolina State Univ., Raleigh, North Carolina.

Hamilton, R. D., and K. E. Austin. 1967. Assay of relative heterotrophic potential in the sea: the use of specifically labelled glucose. Can. J. Microbiol. 13:1165−1173.

Hamilton, R. D., K. M. Morgan, and J. D. H. Strickland. 1966. The glucose uptake kinetics of some marine bacteria. Can. J. Microbiol. 12:995−1003.

Hamilton, R. D., and J. E. Preslan. 1970. Observations on heterotrophic activity in the Eastern Tropic Pacific. Limnol. Oceanogr. 15:395−401.

Harrison, M. J., R. T. Wright, and R. Y. Morita. 1971. Method for measuring mineralization in lake sediments. Appl. Microbiol. 21:698−702.

Hobbie, J. E., and C. C. Crawford. 1969. Respiration corrections for bacterial uptake of dissolved organic compounds in natural waters. Limnol. Oceanogr. 14:528−532.

Hobbie, J. E., C. C. Crawford, and K. L. Webb. 1968. Amino acid flux in an estuary. Science 159:1463−1464.

Hobbie, J. E., and Wright, R. T. 1965. Bioassay with bacterial uptake kinetics: glucose in freshwater. Limnol. Oceanogr. 10:471−474.

Johannes, R. E. 1968. Nutrient regeneration in lakes and oceans. In M. R. Droop and E. J. Ferguson Wood (eds). Advances in Microbiology of the Sea, Vol. I, pp. 203−213. Academic Press, New York.

Kadota, H., Y. Hata, and H. Miyoshi. 1966. A new method for estimating the mineralization activity of lake water and sediment. Mem. Res. Inst. Food Sci. Kyoto Univ. 27:28–30.

Kepes, A. 1963. Permeases: identification and mechanism. *In* Recent Progress in Microbiology, 8th Int. Congr. Microbiol., Montreal, 1962, pp. 38–48. Univ. Toronto Press, Toronto.

Munro, A. L. S., and T. D. Brock. 1968. Distinction between bacterial and algal utilization of soluble substances in the sea. J. Gen. Microbiol. 51:35–42.

Pardee, A. B. 1968. Membrane transport proteins. Science 162:632–637.

Parsons, T. R., and J. D. H. Strickland. 1962. On the production of particulate organic carbon by heterotrophic processes in sea water. Deep-Sea Res. 8:211–222.

Rittenberg, S. C. 1963. Marine microbiology and the problem of mineralization. *In* C. H. Oppenheimer (ed.), Symposium on Marine Microbiology, pp. 48–60. Charles C Thomas, Springfield, Illinois.

Schulze, K. L., and R. S. Lipe. 1964. Relationship between substrate concentration, growth rate, and respiration rate of *Escherichia coli* in continuous culture. Arch. Mikrobiol. 48:1–20.

Vaccaro, R. F. 1969. The response of natural microbial populations in seawater to organic enrichment. Limnol. Oceanogr. 14:726–735.

Vaccaro, R. F., E. Hicks, H. W. Jannasch, and F. G. Cargy. 1968. The occurrence and role of glucose in seawater. Limnol. Oceanogr. 13:355–360.

Vaccaro, R. F., and H. W. Jannasch. 1967. Variations in uptake kinetics for glucose by natural populations in seawater. Limnol. Oceanogr. 12:540–542.

Vallentyne, J. R., and J. R. Whittaker. 1956. On the presence of free sugars in filtered lake water. Science 124:1026–1027.

Wetzel, R. G. 1968. Dissolved organic matter and phytoplankton productivity in marl lakes. Mitt. Int. Ver. Limnol. 14:261–270.

Williams, P. J. LeB. 1970. Heterotrophic utilization of dissolved organic compounds in the sea. I. Size distribution of population and relationship between respiration and incorporation of growth substrates. J. Mar. Biol. Ass. U. K. 50:859–870.

Williams, P. J. Le B., and C. Askew. 1968. A method of measuring the mineralization by micro-organisms of organic compounds in seawater. Deep-Sea Res. 15:365–375.

Williams, P. J. Le B., and R. W. Gray. 1970. Heterotrophic utilization of dissolved organic compounds in the sea. II. Observations on the responses of heterotrophic marine populations to abrupt increases in amino acid concentration. J. Mar. Biol. Ass. U. K. 50:871–881.

Wright, R. T. 1970. Glycollic acid uptake by planktonic bacteria. *In* D. W. Hood (ed.), Organic matter in natural waters, pp. 521–536. Inst. Mar. Sci., College, Alaska.

Wright, R. T. 1973. Some difficulties in using ^{14}C-organic solutes to measure heterotrophic bacterial activity. *In* H. L. Stevenson and R. R. Colwell (eds), Belle W. Baruch Library in Marine Science, Vol. 1, Estuarine Microbial Ecology. Univ. South Carolina Press, Columbia, South Carolina.

Wright, R. T., and J. E. Hobbie. 1965. The uptake of organic solutes in lake water. Limnol. Oceanogr. 10:22–28.

Wright, R. T., and J. E. Hobbie. 1966. Use of glucose and acetate by bacteria and algae in aquatic ecosystems. Ecology 47:447–464.

SOME PROBLEMS WITH HETEROTROPHIC-UPTAKE METHODOLOGY

B. THOMPSON and ROBERT D. HAMILTON

University of Manitoba
and
Freshwater Institute, Winnipeg

The "relative heterotrophic potential" or "heterotrophic-uptake" technique is an example of a rare class of microbiological methods in that it is devoted to the attempt to resolve the question of *in situ* role and activity. Potentially, it appears to be an extremely powerful tool and, since its initial description by Parsons and Strickland (1962), it has been widely used in various marine and freshwater environments.

In its original form the method was intended to assay the conversion of certain organic compounds into particulate material which would then be available to the higher trophic levels of the marine food chain. However, it was not long until Wright and Hobbie (1966), in a classic paper, improved the method and changed its original intent. These authors suggested that the method would assay the turnover time of the compound under investigation. At this point the method ran into difficulty because it was soon suggested that if one wished to measure the total picture one should, at least, include the loss through the conversion of organic carbon to carbon dioxide (Hamilton and Austin, 1967). That this factor was important was soon confirmed (Hobbie and Crawford, 1969) and its assessment is now included in most applications of the approach.

As more investigators tried the technique, more difficulties were discovered and discussed. It is not our purpose to explore all these problems at this time, for in a recent paper Wright (1973) has dealt with many of them. We highly recommend his discussion to your attention. At this time we propose simply to amplify some questions he has discussed and to present some problems that we have encountered in applying the method to a novel limnological problem. It should be stated at the outset that we do not wish to discourage the use of the approach but simply to provoke investigation of certain critical areas.

In order that the context in which we are using the method is understood we should explain that our current research centers around lakes in the Experimental Lakes Area which is maintained by the Freshwater Institute (Johnson and Vallentyne, 1971). Most of the lakes in this area are oligotrophic but two small lakes (3–5 hectares) have been manipulated and now can be termed eutrophic. In one case nitrogen and phosphorus additions were used, while nitrogen, phosphorus, and carbon (as sucrose, 5.54 g $C/m^2/year$) were added to a second. The additions were made weekly and were applied to the epilimnion of the lakes. Our particular interest was in following the degradation of the added sucrose while using the other treated lake and one untreated lake as reference points.

One of the first problems one should face when proposing to use the heterotrophic-uptake approach is whether the concept of the kinetic model used is valid in this application. Wright (1973) and others have dealt with this problem at length and have explored some of the assumptions that must be accepted and some conditions which must be enforced. It should be emphasized, however, that this is, after all, a tracer method. The amount of labeled compound should be kept small, relative to the unlabeled compound, to prevent radiobiological damage to the organisms. Moreover, additions of metabolites should not approach levels which are completely unnatural. Given the low levels of organics known or presumed to exist in natural waters, one is forced to employ high-specific-activity substrates. The experimenter is always treading a very delicate line and adherence to these constraints is predicated on knowledge or firm estimates of *in situ* concentrations. Usually, however, immediate knowledge of these values is impossible and, in the oceanographic context, is unknown until the ship is many miles down-range or even back at the dock. Therefore, the investigator must be prepared to discard data points which are shown to be untenable or, at least, be prepared to curtail his conclusions. A further problem which may eliminate the use of some data points is that the substrate concentration during the experiment should not change substantially, the accepted figure being 5%.

The conscientious investigator is always gambling between incubating long enough to obtain detectable uptake and incubating too long a time. More-over, if one is not using true tracer technology the correct incubation time will vary with the amount of substrate added. However, Hall (unpublished data) has recently developed a mathematical treatment which may permit the use of data in which the substrate concentration changes significantly during incubation.

Another basic consideration is the possibility of preferential utilization of one carbon isotope over another—the occurrence of an isotope effect. This effect is almost always ignored by most investigators because it would be extremely difficult to determine. However, as the heterotrophic-uptake method is applied to increasingly sophisticated problems it is to be hoped that some effort will be expended in resolving the question of the magnitude of the isotope effect under various field conditions.

A procedural problem that caused us some initial concern was the question of what type of blank to employ. Theoretically a blank should detect all abiological effects. In surveying the literature we found that a number of different fixatives, poisons, and antibiotics have been employed to arrest biological activity in blanks. Some workers employed no treatment save to filter the blank "immediately" after addition of the label, while those using a treatment might filter immediately, or after incubation. Presumably most treatments which would arrest biological activity would also affect the surface phenomena important to blank values. Should the same treatment be applied to the sample one risks cell breakage or leakage, thus also prejudicing the data.

"Immediate" filtration ignores well-documented pulse-labeling experi-ments involving significant biological uptake during very short exposure periods and does not really account for abiological adsorption and absorption effects which may be time dependent. Fortunately, most blank values that we have encountered in our lake experiments have been so low, relative to uptake values, that the variance between different procedures can be con-veniently ignored. However, the question of the proper blank is a problem we commend to the attention of those of you working with situations in which the blank value is a significant fraction of observed uptake values. Indeed, the heterotrophic-uptake procedure encounters a high rate of failure under cer-tain oceanic conditions (Vaccaro and Jannasch, 1967; Vaccaro, 1969; Hamil-ton and Preslan, 1970), especially those conditions where low levels of uptake are encountered. In our experience under such conditions, it is not uncom-mon for blanks to vary unpredictably with the level of added labeled and unlabeled substrate. Perhaps the use of a single blank value is not justified and the use of blanks for each level of added substrate should be considered. Of course, if one is not worried about the distinction between biological and abiological activities, one need not use a blank at all. Presumably abiological effects, which are not artifacts of sampling or of enclosure of the sample in a bottle, would be fully as important a consideration in nature as the biological effects.

Regarding the question of fixation of samples, one can accept some arguments for fixation if one treats the blank in a similar manner and desires to preserve uniformity of practice. There also might be some justification for fixing samples if one must resort to non-scintillation counting equipment, although liquid-scintillation equipment is becoming more readily available and less expensive day by day. If one is using scintillation techniques the usual justification for fixation is that the sample must be dried or it will cloud the counting cocktail and that any biological activity during the drying procedure must be eliminated. In our experience this is unjustified in that dioxane-based cocktails (Schindler and Holmgren, 1971) accept wet membrane filters and will dissolve the filter, thus providing much enhanced counting conditions. Moreover, such cocktails permit long storage of samples, thus eliminating the losses of activity known to be associated with drying samples collected on filters and with dry storage of such samples (Wallen and Geen, 1968).

Earlier we mentioned that, if considerations other than the production of particulate from soluble material is to be subject to investigation, the production of labeled carbon dioxide must be assayed. A number of techniques for the assay of $^{14}CO_2$ have been presented in the literature, most of which are very time consuming and some of which are very difficult to execute.

Hobbie et al. (1968) measured $^{14}CO_2$ with an ion chamber and electrometer, while Williams and Askew (1968) flushed out $^{14}CO_2$ and converted it to $Ba^{14}CO_2$ which was subsequently counted. Hobbie and Crawford (1969) developed a relatively simple analysis for respired $^{14}CO_2$, which has been widely adopted and modified. The experiment was performed in a 25-ml Erlenmeyer flask sealed with a rubber serum stopper. A plastic rod and cup assembly was suspended from the serum stopper and contained a 25 by 50 mm piece of accordion-folded Whatman No. 1 filter paper. At the end of the incubation period, 2 N H_2SO_4 was injected through the serum stopper to halt heterotrophic activity and to lower the pH sufficiently to drive off all the CO_2. $^{14}CO_2$ was then adsorbed in 0.2 ml phenethylamine injected through the serum stopper onto the paper. After shaking the flask for 1 hr at room temperature, the filter paper was removed and counted in a toluene-based scintillation fluor. This method has been extended to the measurement of mineralization in lake sediments (Harrison et al., 1971). Phenethylamine is superior to hyamine, NaOH, or KOH as an absorber of $^{14}CO_2$ which is to be assayed by liquid-scintillation techniques (Woeller, 1961). However, it has been reported (Ragland, 1967) that the counting rate from phenethylamine-[^{14}C] carbonate decreases with time owing to loss of $^{14}CO_2$; losses of up to 50% in 3 hr have been noted. We have not encountered this loss in samples counted at intervals over 6 hr, and again over 24 hr. However, in our experiments phenethylamine concentrations were in great excess over the amount required to trap all the labeled and unlabeled CO_2 liberated.

In an effort to reduce the difficulty of collecting $^{14}CO_2$ following the incubation of a sample as well as in an attempt to reduce the technical difficulty and reproducibility of collection techniques we have investigated a number of procedures. We have been able to reduce the single-sample treatment time to approximately 5 min, while obtaining 97% efficiency, and we can see no reason why a single operator could not control several of our collection devices.

In the technique we employ 100-ml serum bottles which are sealed with conventional serum-bottle stoppers during the incubation of a 50-ml sample. As a collection device we employ a standard scintillation vial thus eliminating any transfer of the normally viscous carbon dioxide trapping solutions. A cut-down silicone rubber stopper is used as a closure for the scintillation vial during collection. Two simple jigs are prepared (cut-down stoppers serve well), each of which holds a single appropriately positioned 4.0-inch sharpened steel cannula (air inlet) and a similarly positioned 1.5-inch disposable hypodermic needle (air outlet). The four needles are linked in a closed loop by *small*-bore silicon rubber tubing (we have encountered CO_2 losses if plastic or rubber tubing is used) which incorporates a sealed, inexpensive, push-pull diaphragm pump similar to the pumps used to aerate aquaria.

Immediately before attempting collection, 5 ml of Woeller's Fluor A (27 ml phenethylamine, 27 ml absolute methanol, 500 mg PPO, and 10 mg POPOP in 100 ml scintillation-grade toluene) is added to the scintillation vial and it is sealed with the silicone stopper. A needle-cannula pair, held in its jig, is inserted through the stopper and the other pair is quickly inserted through the serum-bottle stopper on the incubation vial. Sulfuric acid (0.25 ml 2 N H_2SO_4) is added to the serum bottle to produce pH ~2 and the air pump is activated. Use of HCl is not recommended as it is somewhat volatile and so will degrade the phenethylamine. Air circulation is maintained for 5–10 min, at the end of which time the scintillation vial is removed and 5 ml of Woeller's Fluor B is added (5 g PPO and 100 mg POPOP in 1 liter scintillation-grade toluene). The sample is then ready for counting.

A comparison of this technique with that of Harrison *et al.* (1971) has been attempted and the results are presented in Table 1. Should one wish to employ the former technique our results indicate that, while efficiency of

Table 1. A comparison between collection of $^{14}CO_2$ by diffusion in shaken flasks and by active circulation of the atmosphere in a closed loop which incorporates both the incubation flask and a trap

Type of collection		Percentage label recovered
$^{14}CO_2$ collection on GF/C wick, shaken	10 min	24.55
	35 min	53.06
	1 hr	78.86
	1½ hr	86.54
$^{14}CO_2$ collection by bubbling for	2½ min	80.21
	5 min	96.66
	10 min	97.23
	15 min	94.70
$^{14}CO_2$ collected from 0.1 μCi ^{14}C sucrose (U) after bubbling for 5 min		0.4172
		0.8377

collection is the same, less quenching is encountered if GF/C paper is used as a collection wick rather than Whatman No. 1 filter paper.

Another major problem which we have encountered is the so-called filtration error which appears to be a major source of error in the use of primary productivity techniques. During their assays of primary production, using both ^{14}C and O_2 techniques, Arthur and Rigler (1967) noted that recovery of fixed carbon decreased on a per unit volume basis with increasing amounts of sample filtered. This effect was corrected by filtering and counting aliquots from a sample, plotting the results and extrapolating to 0 ml. Schindler and Holmgren (1971) have applied the technique to a large number of samples and they estimated that, in the Experimental Lakes Area, primary production values vary between 1.2 and 6.1 times greater than uncorrected values.

We were intrigued by these observations and began using the technique in connection with our heterotrophic uptake studies. We have found that filtration correction curves are commonly encountered and that, if one accepts the curve as reality, the corrections to what would be normally accepted uptake values can be quite severe (Fig. 1).

The source of this error is not clear. In their studies of marine systems, Berman and Williams (1971) noted that radioactive solutions could be retained by that portion of the filter which is covered by the flanges of the membrane filtration unit and that this contamination could not be removed

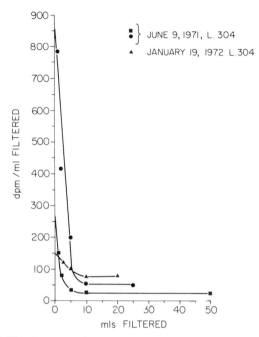

Fig. 1. Typical filtration correction curves obtained from experimental lake 304 using U-[^{14}C] sucrose as a substrate.

by the washing procedure which would normally be employed. We thought that this might be a source of the filtration error in that it would represent a constant error which would become more important the smaller the volume filtered. Consequently we excised the edges of some filters and counted them separately from the inner circles. The results are presented in Table 2 and they appear to agree in part with previous studies. While these data indicate that this phenomenon may be a partial source of the filtration error we do not believe it to be a significant source, for most of our studies involve much higher uptake values than those encountered in these experiments. Moreover, experiments in which normal membrane filters were used together with filters which had hydrophobic edges did not produce materially changed shapes of the observed correction curves. It has also been suggested that $^{14}CO_2$ may somehow be involved in the error, but the use of standard fuming techniques did not affect the patterns we have observed.

Arthur and Rigler (1967) suggested that the filtration error might be due to the breakage of fragile phytoplankton cells during filtration. This seems reasonable in that many people appear to have little regard for the fragility of cells and employ full vacuum and other harsh measures during filtration. Some observations noted during primary productivity studies in the Experimental Lakes Area would tend to support this concept in part, for increased correction curves are generally found when fragile phytoplankton forms are present and decreased corrections are sometimes observed when the phytoplankton is dominated by species which have strong cell walls. However, our experience with heterotrophic uptake is that no mechanical factors, such as variation in degree of vacuum employed during filtration, have any effect upon filtration-correction curves.

Perhaps the most definitive evidence we have against the hypothesis of cell damage is that only rarely have we been able to find a correction curve when doing ATP analyses of lake samples. Even when such curves are obtained they indicate extremely small errors relative to those that ^{14}C techniques produce.

In order to eliminate the problem entirely in phytoplankton primary production studies, Schindler (personal communication) has proposed elimination of filtration by simply acidifying the sample, driving off the $^{14}CO_2$ by bubbling, and counting the solution directly in a cocktail which will accept water. In practice, he has found that the results obtained in this manner agree well with the intercept one generates through the use of a filtration-correction curve.

Schindler (personal communication) has suggested that the correction curve is due to a colloidal organic compound which is excreted by the phytoplankton during incubation. He has found that labeled colloids, which have been separated by gel filtration and fractionating procedures, are taken up by filters and that the uptake displays a saturation phenomenon. Consequently, a "filtration error" is produced. Similar results have been obtained during ^{32}P experiments conducted by Lean (personal communication).

Table 2. Dpm/ml on edge and center of membrane filters as influenced by the amount of sample filtered

Filtered (ml)	Inner circle (dpm/ml)	Outer ring (dpm/ml)	Percentage total dpm on outer ring
50	1665	55	3.189
25	1993	85	4.094
15	2591	50	1.901
5	2452	203	7.640
2	2629	256	8.872
1	2568	319	11.050

Therefore, it would appear that these errors might be associated with some excreted metabolite. In primary productivity studies Schindler's technique would appear to account for them satisfactorily. Unfortunately, it cannot be applied to the heterotrophic-uptake technique, yet any metabolite which was excreted under similar circumstances should be included in any calculations involving the flux of the substrate in question. We cannot suggest any quick solution to the problem save to propose some technique analagous to that of Schindler: one which removes only the added labeled compound such that the radioisotope remaining in particulate material and in solution can be assayed. We commend the problem to the attention of those interested in heterotrophic-uptake work.

LITERATURE CITED

Arthur, C. R., and F. H. Rigler. 1967. A possible source of error in the ^{14}C method of measuring primary productivity. Limnol. Oceanogr. 12:121–124.

Berman, T., and P. J. LeB. Williams. 1971. Notes on the methodology of the radiocarbon techniques for studying algal productivity. Research on the Marine Food Chain. Progr. Rep. Inst. Mar. Resources, Univ. California. (Unpublished manuscript.)

Hamilton, R. D., and K. E. Austin. 1967. Assay of relative heterotrophic potential in the sea: the use of specifically labelled glucose. Can. J. Microbiol. 13:1165–1173.

Hamilton, R. D., and J. E. Preslen. 1970. Observations on heterotrophic activity in the eastern tropical Pacific. Limnol. Oceanogr. 15:395–401.

Harrison, M. J., R. T. Wright, and R. Y. Morita. 1971. Method for measuring mineralization in lake sediments. Appl. Microbiol. 21:698–702.

Hobbie, J. E., and C. C. Crawford. 1969. Respiration corrections for bacterial uptake of dissolved organic compounds in natural waters. Limnol. Oceanogr. 14:528–532.

Hobbie, J. E., Crawford, C. C., and K. L. Webb. 1968. Amino acid flux in an estuary. Science 159:1463–1464.

Johnson, W. E., and J. R. Vallentyne. 1971. Rationale, background, and development of experimental lake studies in northwestern Ontario. J. Fish. Res. Bd. Can. 28:123–128.

Parsons, T. R., and J. D. H. Strickland. 1962. On the production of particulate organic carbon by heterotrophic processes in sea water. Deep-Sea Res. 8:211–222.

Ragland, J. B. 1967. Methods of Combustion. In Preparation of Samples for Liquid Scintillation Counting. Nuclear Chicago Handbook, Nuclear Chicago Corporation, Des Plaines, Illinois.

Schindler, D. W., and S. K. Holmgren. 1971. Primary production and phytoplankton in the Experimental Lakes Area, northwestern Ontario, and other low-carbonate waters, and a liquid scintillation method for determining ^{14}C activity in photosynthesis. J. Fish. Res. Bd. Can. 28:189–201.

Vaccaro, R. F. 1969. The response of natural microbial populations in sea water to organic enrichment. Limnol. Oceanogr. 14:726–735.

Vaccaro, R. F., and H. W. Jannasch. 1967. Variation in uptake kinetics for glucose by natural populations in seawater. Limnol. Oceanogr. 12:540–542.

Wallen, D. G., and G. H. Geen. 1968. Loss of radioactivity during storage of ^{14}C-labeled phytoplankton on membrane filters. J. Fish. Res. Bd. Can. 25:2219–2224.

Williams, P. J. LeB., and C. Askew. 1968. A method of measuring the mineralization by microorganisms of organic compounds in sea-water. Deep-Sea Res. 15:365–375.

Woeller, F. H. 1961. Liquid scintillation counting of ^{14}CO$_2$ with phenethyl-amine. Anal. Biochem. 2:508–511.

Wright, R. T. 1973. Some difficulties in using ^{14}C-organic solutes to measure heterotrophic bacterial activity. In H. L. Stevenson and R. R. Colwell (eds), Belle W. Baruch Library in Marine Science, Vol. 1, Estuarine Microbial Ecology. Univ. South Carolina Press, Columbia, South Carolina.

Wright, R. T., and J. E. Hobbie. 1966. Use of glucose and acetate by bacteria and algae in aquatic eco-systems. Ecology 47:447–464.

INDEX